ENGINEERING FUNDAMENTALS

An Introduction to Engineering

ENGINEERING FUNDAMENTALS

An Introduction to Engineering

Saeed Moaveni

Minnesota State University, Mankato

BROOKS/COLE

THOMSON LEARNING™

Australia • Canada • Mexico • Singapore • Spain • United Kingdom • United States

BROOKS/COLE
THOMSON LEARNING

Publisher: *Bill Stenquist*
Marketing Team: *Christopher Kelly and Mona Weltmer*
Editorial Coordinator: *Valerie Boyajian*
Production Coordinator: *Mary Vezilich*
Production Service: *Martha Emry Production Services*
Manuscript Editor: *Pamela Rockwell*
Permissions Editor: *Sue Ewing*
Cover Design: *Lisa Henry*
Cover Photo: *Corbis*
Interior Illustration: *Atherton Customs*
Interior Design: *Roy Neuhaus Design*
Photo Researcher: *Myrna Engler*
Print Buyer: *Vena Dyer*
Typesetting: *UG / GGS Information Services, Inc.*
Printing and Binding: *Phoenix Color Corporation*

For more information about this or any other Brooks/Cole product, contact:
BROOKS/COLE
511 Forest Lodge Road
Pacific Grove, CA 93950 USA
www.brookscole.com
1-800-423-0563 (Thomson Learning Academic Resource Center)

Printed in the United States of America

10 9 8 7 6 5 4 3 2 1

Library of Congress Cataloging-in-Publication Data
Moaveni, Saeed.
Engineering fundamentals: an introduction to engineering / Saeed Moaveni.
p. cm.
ISBN 0-534-38116-2 (paper)
1. Engineering. I. Title.

TA147.M63 2002
620—dc21

2001043913

To my mother and father

Contents

Preface

This book was written with four basic objectives in mind. The first objective is to give students some ideas about how to make the transition from high school to college smoothly and how to develop good study habits so that they can obtain a good engineering education.

The second objective of this book is to introduce students to the engineering profession and its various branches and to provide some answers to questions that entering college students may have. Does the student really want to study engineering? What is engineering and what do engineers do? What are some of the areas of specialization in engineering? Does she want to become a mechanical engineer, or should she pursue civil engineering? Or would he be happier becoming an electrical engineer? How many different engineering disciplines are there? How will the student know that he has picked the correct field? Will the demand for her area of specialization be high when she graduates and in the years that follow?

The third objective of this book is to emphasize that engineers apply physical and chemical laws and principles and mathematics to design, develop, test, and supervise the production of millions of parts and products as well as services that we use in our everyday lives. This text shows through examples that there are many satisfying and challenging jobs for engineers. However, students should realize that, whereas the activities of engineers are quite varied, there are some personality traits and work habits that typify most of today's successful engineers.

- Engineers are problem solvers.
- Good engineers have a firm grasp of the fundamental principles that they can use to solve many different problems.
- Good engineers are analytical, detailed oriented, and creative.
- Good engineers have a desire to be lifelong learners. For example, they take continuing education classes, seminars, and workshops to stay abreast of new innovations and technologies.
- Good engineers have written and oral communication skills that equip them to work well with their colleagues and to convey their expertise to a wide range of clients.
- Good engineers have time-management skills that enable them to work productively and efficiently.
- Good engineers have good "people skills" that allow them to interact and communicate effectively with various people in their organization.

- Engineers are required to write reports. These reports might be lengthy, detailed technical reports containing graphs, charts, and engineering drawings, or they may take the form of a brief memorandum or an executive summary.
- Engineers are adept at using computers in many different ways to model and analyze various practical problems.
- Good engineers actively participate in local and national discipline-specific organizations by attending seminars, workshops, and meetings. Many even make presentations at professional meetings.
- Engineers generally work in a team environment where they consult each other to solve complex problems. Good interpersonal and communication skills are increasingly important now because of the global market.

Clearly, an interest in building things, or taking things apart, or solving puzzles is not all that is required to become an engineer. In addition to having a dedication to learning and a desire to finding solutions, an engineer needs to foster certain personal attitudes and personality traits.

The fourth objective of this book is to introduce students right away to some of the basic principles and physical laws that they will see over and over in some form or other during the next four years. These are the concepts that every engineer should know regardless of area of specialization. Throughout the book students are reminded that they need to develop a firm grasp of the fundamentals to succeed as students and later as practicing engineers. Many of the problems at the end of each chapter in this book require students to make brief reports so that they become aware that successful engineers need to have good written and oral communication skills. Information collection and proper utilization of that information are also encouraged by asking students to do a number of assignments that require information gathering by using the Internet as well as employing traditional methods. To emphasize the importance of teamwork in engineering and to encourage group participation, many of the assignment problems require group work, and some even require the participation of the entire class.

ORGANIZATION

This book is organized into 16 chapters, which include enough material for a broad selection of topics. The reason for this approach is to give instructors sufficient materials and the flexibility to choose specific topics to meet their needs. Chapter 1 introduces study habits of successful engineering students. Chapter 2 provides an introduction to engineering. Engineering problem solving and presentation is found in Chapter 3. Later chapters deal with common engineering materials, engineering symbols, engineering standards and codes, use of spreadsheets in engineering, engineering drawings, and engineering design—information that is basic and thus an integral part of any introductory course.

In Chapter 1, the transition from high school to college is explained in terms of good study habits that students need to form and suggestions on how they may want to go about budgeting their time effectively. Chapter 2 provides a comprehensive introduction to the engineering profession and its branches. It introduces students to the engineering profes-

Ship returns to:

THOMSON LEARNING
DISTRIBUTION CENTER
7625 EMPIRE DRIVE
FLORENCE, KY 41042

The enclosed materials are sent to you for your review by
JANET HOLLOWAY 800 8762350X7044

SALES SUPPORT

SHIP TO: EDWIN YAZ
MARQUETTE UNIVERSITY
DEPT OF ELECTRICAL & COMP ENGINE
HAGGERTY ENGINEERING 289
PO BOX 1881
MILWAUKEE WI 53201-1881

WAREHOUSE INSTRUCTIONS

SLA: 7 BOX: Staple

LOCATION	QTY	ISBN	AUTHOR/TITLE
K-ASY-035-01	1	0-534-38116-2	MOAVENI ENGINEERING FUNDAMENTALS

INV# 42836882SM
PO# 32099866
DATE: 11/19/01
CARTON: 1 of 1
ID# 4992180

ASSEMBLY
-SLSB

VIA: UP

PAGE 1 OF 1

BATCH: 1303488

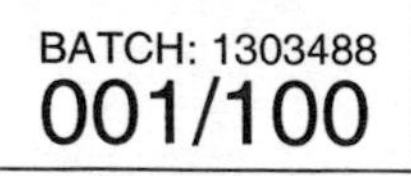
001/100

sion and explains some of the common traits of good engineers. Various engineering disciplines and engineering organizations are discussed. Chapter 3 explains the role and importance of fundamental dimensions and units in the analysis of engineering problems. Basic steps in the analysis of any engineering problem are discussed in detail. The communication skills required for effective presentation of engineering work are also discussed in some detail. Various forms of engineering communication, including homework presentation, brief technical memos, progress reports, detailed technical reports, and research papers are explained.

Chapter 4 introduces length and length-related variables and explains their importance in engineering work. For example, the role of area in heat transfer, aerodynamics, load distribution, and stress analysis is discussed. Measurement of length, area, and volume along with numerical estimation (such as the trapezoidal rule) of these values are presented. Chapter 5 considers time and time-related engineering parameters. Periods, frequencies, linear and angular velocities and accelerations, volumetric flow rates, and flow of traffic are also discussed in Chapter 5. Mass and mass-related parameters such as density, specific weight, mass flow rate, and mass moment of inertia and their role in engineering analysis are presented in Chapter 6. Chapter 7 covers the importance of force and force-related parameters in engineering. What is meant by pressure, modulus of elasticity, impulsive force (force acting over time), work (force acting over a distance), and moment (force acting at a distance) are discussed in detail.

Temperature and temperature-related parameters are presented in Chapter 8. Concepts such as temperature difference and heat transfer, specific heat, and thermal conductivity are covered. Chapter 9 considers topics such as direct and alternating current, electricity, basic circuit components, power sources, and the tremendous role of electric motors in our everyday life. Chapter 10 presents energy and power and explains the distinction between these two topics. The importance of understanding what is meant by work, energy, power, watts, horsepower, and efficiency is emphasized in Chapter 10. Chapter 11 deals with engineering materials and their mechanical, electrical, and thermophysical properties. Solid materials such as light metals, steel, plastics, composites, glass, wood, and concrete, as well as fluid materials such as air and water, are discussed.

Engineering standards and codes and their importance in the engineering profession as well as in our everyday lives are covered in Chapter 12. National and international organizations that develop and maintain these standards and codes are also discussed in Chapter 12. The need for engineering symbols along with examples of actual symbols used in engineering practice, such as electrical, mechanical, logic, and HVAC symbols and abbreviations, are presented in Chapter 13. Chapter 14 discusses the difference between computer programming and using spreadsheets and provides a basic introduction to Microsoft® Excel. A brief introduction to engineering drawing, engineering design, and engineering ethics is given in Chapters 15 and 16.

Relevant everyday examples that students can easily associate with are given in each chapter. Many of the problems at the conclusion of each chapter are hands-on, requiring the student to gather and analyze information. Many of the problems also require the use of the Internet for information collection. Each chapter begins by stating the objectives and concludes by summarizing what the reader should have gained from studying the chapter. A summary of pertinent engineering concepts and formulas is given in the Appendix.

CASE STUDIES—ENGINEERING MARVELS

To emphasize that *engineers are problem solvers* and to emphasize that engineers apply physical and chemical laws and principles, along with mathematics, to *design* products and services that we use in our everyday lives, four case studies are placed throughout the book. These projects are truly engineering marvels. Following Chapter 4, the design of New York City Water Tunnel No. 3 is discussed. The design of the Caterpillar 797 Mining Truck, the largest mining truck in the world, follows Chapter 7. Following Chapter 10, relevant information about the design of the Hoover Dam is discussed. The design of the Boeing 777 is presented in a case study following Chapter 15. There are assigned problems at the end of each case study. The solutions to these problems incorporate the engineering concepts and laws that are discussed in the preceding chapters. Finally, a fifth case study, from the National Society of Professional Engineers, is presented following Chapter 16 to conclude the discussion on engineering ethics.

WEB SITE

There will be a Web site at **http://info.brookscole.com/Moaveni** for the following purposes:

1. There are assignments at the end of each chapter that require information collection and analysis by groups of students or by the entire class. Students' findings for these assignments will be posted on the Web in order to make all the findings available to every engineering student and teacher who uses the book. This will provide additional discussion points for students and faculty. The results may be sent via e-mail directly as attached files, or the groups may elect to send the Web address of their schools where they post their findings; links will then be established to these sites. Please e-mail any pertinent student findings or links to: **saeed.moaveni@mnsu.edu**
2. There are assignments that require the use of large data files and Microsoft® Excel. These data files are available at **http://info.brookscole.com/Moaveni** for students to download.
3. Also posted on the Web site will be additional interesting problems and ideas that may be brought to our attention by students and faculty as they work through the book. Appropriate problems will be posted on the Web site, along with the names of the students, the faculty, and their university or college affiliation.
4. An on-line evaluation sheet relating to problems and topics will be provided so that the materials in this book can be modified continuously. The on-line feedback sheets also allow students and faculty to voice their opinions on other useful topics that they would like to see covered in more detail.
5. Finally, I will post, at the Web site, any necessary corrections that are brought to my attention.

REFERENCES

In writing this book, several engineering books, Web pages, and other materials were consulted. Rather than giving you a list that contains hundreds of resources, I will cite some of the sources that I believe to be useful to you. I think all freshman engineering students should own a handbook in their chosen field. Currently, there are many engineering handbooks available in print or electronic format, including chemical engineering handbooks, civil engineering handbooks, electrical and electronic engineering handbooks, and mechanical engineering handbooks. I also believe all engineering students should own chemistry, physics, and mathematics handbooks. These texts can serve as supplementary resources in all your classes. Many engineers may also find useful the ASHRAE handbook, the *Fundamental Volume,* by the American Society of Heating, Refrigerating, and Air-Conditioning Engineers.

In this book, some data and diagrams were adapted with permission from the following sources:

Baumeister, T., et al., *Mark's Handbook,* 8th ed., McGraw Hill, 1978.
Electrical Wiring, 2nd ed., AAVIM, 1981.
Electric Motors, 5th ed., AAVIM, 1982.
Gere, J. M., *Mechanics of Materials,* 5th ed., Brooks/Cole, 2001.
Hibbler, R. C., *Mechanics of Materials,* Macmillan, 1991.
U.S. Standard Atmosphere, Washington D.C., U.S. Government Printing Office, 1962.
Weston, K. C., *Energy Conversion,* West Publishing, 1992.

ACKNOWLEDGMENTS

I would like to express my sincere gratitude to the editing and production team at Brooks/Cole Publishing, especially the publisher, Bill Stenquist, who made every effort to produce an excellent introductory book in engineering. I am also grateful to Vernon Boes, Mary Vezilich, and Sue Ewing of Brooks/Cole Publishing, Martha Emry of Martha Emry Production Services, copy editor Pam Rockwell, Jim Atherton and the talented team of artists at Atherton Customs, and the composition team at University Graphics.

I am grateful to Minnesota State University for providing me with a sabbatical leave and to Syracuse University for providing me with a comfortable environment during my sabbatical stay at its lovely campus, which allowed me to complete and polish this work. I would also like to thank Dr. Nancy Mackenzie of Minnesota State University, who made valuable editing suggestions during the first draft of the manuscript. I am also thankful to the following reviewers who offered general and specific comments: William Beckwith, Clemson University; William Park, Clemson University; Amir G. Rezaei, West Virginia University Institute of Technology; Ronald L. Thurgood, Utah State University; and Gerald Vinson, Texas A & M University.

I hope you enjoy the book.

Saeed Moaveni

CHAPTER

1 Study Habits of Successful Engineering Students

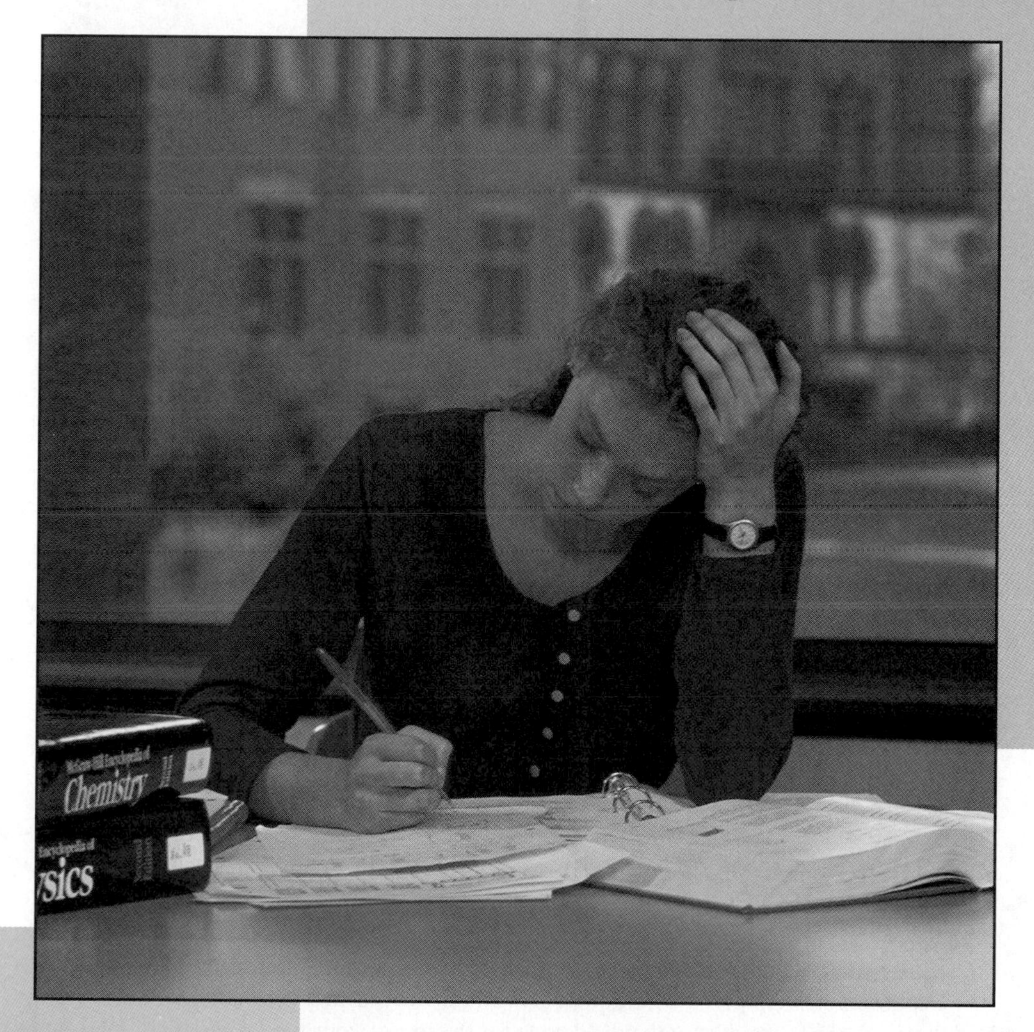

Making the transition from high school to college requires extra effort. In order to have a rewarding education you should realize that you must start studying and preparing from the first day of class, attend class regularly, get help right away, take good notes, select a good study place, and form study groups. You should also consider the time management ideas discussed in this chapter to arrive at a reasonable weekly schedule. Your education is an expensive investment. Invest wisely.

In this chapter, you will be introduced to some very important suggestions that could, if followed, make your engineering education and future career more rewarding. Read this section very carefully, and think about how you can adapt the strategies offered here to get the optimum benefit from your college years. If you encounter difficulty in your studies, reread this chapter for ideas to help you maintain some self-discipline. ■

1.1 MAKING THE TRANSITION FROM HIGH SCHOOL TO COLLEGE

You belong to an elite group of students now because you are studying to be an engineer. According to the *Chronicle of Higher Education*, approximately only 5% of students who graduate with a B.S. degree from universities and colleges across the United States are engineers. You will be taught to look at your surroundings differently than other people do. You will learn how to ask questions to find out how things are made, how things work, how to improve things, how to design something from scratch, and how to take an idea from paper to a machine shop and actually build something.

Some of you may be on your own for the very first time. Making the transition from high school to college may be a big step for you. Keep in mind that what you do for the next four or five years will affect you for the rest of your life. Remember that how successful and happy you are will depend primarily on you. You must take the responsibility for learning; nobody can make you learn. Depending on which high school you attended, you may not have had to study much to get good grades. Therefore, you may need to develop some new habits and get rid of some of your old habits in order to thrive as an engineering student and eventually as a good engineer. The rest of this chapter presents suggestions and ideas that will help you make your college experience successful. Consider these suggestions and try to adapt them to your own unique situation.

1.2 BUDGETING YOUR TIME

Each of us has the same 24 hours in a day, and there is only so much that a person can do on an average daily basis to accomplish certain things. Many of us need approximately 8 hours of sleep every night. In addition, we all need to have some time for work, friends and family, studying, relaxation and recreation, and just goofing around.

Suppose you were given a million dollars when you reached your adulthood and were told that is all the money that you would have for the rest of your life for clothing, food, entertainment, leisure activities, and so on. How would you go about spending the money? Of course, you would make reasonable efforts to spend and invest it wisely. You would carefully budget for various needs, trying to get the most for your money. You would look for good sales and plan to buy only what was necessary, and you would attempt not to waste any money. Think of your education in a similar manner. Don't just pay your tuition and plan to sit in class and daydream. Your education is an expensive investment, one that requires your responsible management. A student at a private university went to her instructor to drop a class because she was not getting the grade she wanted. The instructor asked her how much she had paid for the class. She said that she had spent approximately $2000 for the four-credit class. The instructor happened to have a laptop computer on his desk and asked her the following question: If you bought a laptop computer from a computer store, took it home to install some software on it, and had some difficulty making the computer work, would you throw it in the trash? The student looked at her instructor as if he had asked a stupid question. He explained to her that her dropping a class she had already paid tuition for is similar in many ways to throwing away a computer the first time she has trouble with some software. Try to learn from this example. Generally speaking, for most of us learning is a lot of work at the beginning, and it's not much fun. But often, after even a short period of time, learning will become a joy, something you work at that raises your own self-esteem. Learning and understanding new things can be downright exciting. Let us examine what you can do to enhance your learning during the next few years to make the engineering education you are about to receive a fulfilling and rewarding experience.

Let us begin by performing some simple arithmetic to see how efficient we might be in using our time. With 24 hours in a given day, we have, for a one-week period, 168 hours available. Let's allocate liberal time periods to some activities common to most students. Refer to Table 1.1, when following this example. Notice that the time periods allocated to various activities in this table are very generous and you don't have to deprive yourself of sleep or relaxation or socializing with your friends. These numbers are meant only to give you a reasonable starting point to help you budget your time on a weekly basis. You may prefer to spend an hour a day relaxing during the week and use the additional social hours on weekends. Even with generous relaxation and social time, this sample allows 68 hours a week to devote to your education. A typical engineering student takes 16 semester credits, which simply means about 16 hours a week are spent in the classroom. You still have 52 hours a week to study. A good rule of thumb is to spend at least 2 to 3 hours of studying for each hour of class time, which amounts to at least 32 hours and at most 48 hours a week of studying. Of course, some classes are more demanding than others and will require more time for preparation and homework, projects, and lab work. You still have from 4 to 20 hours a week in your budget to allocate at your own discretion.

TABLE 1.1 An Example of Weekly Activities

Activity	Required Time per Week
Sleeping: (7 days/week) × (8 hours/day)	= 56 (hours/week)
Cooking and eating: (7 days/week) × (3 hours/day)	= 21 (hours/week)
Grocery shopping	= 2 (hours/week)
Personal grooming: (7 days/week) × (1 hour/day)	= 7 (hours/week)
Spending time with family, (girl/boy) friends, relaxing, playing sports, exercising, watching TV: (7 days/week) × (2 hours/day)	= 14 (hours/week)
Total	**= 100 (hours/week)**

TABLE 1.2 An Example of Weekly Schedule for a Freshman Engineering Student

Hour	Monday	Tuesday	Wednesday	Thursday	Friday	Saturday	Sunday
7–8	*Shower/ Dress/ Breakfast*	*Shower/ Dress/ Breakfast*	*Shower/ Dress/ Breakfast*	*Shower/ Dress/ Breakfast*	*Shower/ Dress/ Breakfast*	Extra sleep	Extra sleep
8–9	**CALCULUS CLASS**	**CALCULUS CLASS**	**Study English**	**CALCULUS CLASS**	**CALCULUS CLASS**	*Shower/ Dress/ Breakfast*	*Shower/ Dress/ Breakfast*
9–10	**ENGLISH CLASS**	**Study Intro. to Eng.**	**ENGLISH CLASS**	**Study Intro. to Eng.**	**ENGLISH CLASS**	Grocery shopping	**OPEN HOUR**
10–11	**Study Calculus**	**INTRO. TO ENG. CLASS**	**Study H/SS**	**INTRO. TO ENG. CLASS**	**Study Calculus**	Grocery shopping	**OPEN HOUR**
11–12	**H/SS CLASS**	**Study Chemistry**	**H/SS CLASS**	**Study Chemistry**	**H/SS CLASS**	Exercise	Relax
12–1	Lunch	Lunch	Lunch	Lunch	Lunch	Lunch	Lunch
1–2	**CHEM. CLASS**	**Study Calculus**	**CHEM. CLASS**	**Study Intro. to Eng.**	**CHEM. CLASS**	Relax	Relax
2–3	**Study H/SS**	**CHEM. LAB**	**Study H/SS**	**Study Intro. to Eng.**	**Study H/SS**	**Study**	**Study English**
3–4	**Study Calculus**	**CHEM. LAB**	**Study H/SS**	**Study Calculus**	**Study Calculus**	**Study Chemistry**	**Study H/SS**
4–5	Exercise	**CHEM. LAB**	Exercise	Exercise	Exercise	**Study Chemistry**	**Study Chemistry**
5–6	Dinner	Dinner	Dinner	Dinner	Dinner	Dinner	Relax
6–7	**Study Chemistry**	**Study Calculus**	**Study Calculus**	**Study Calculus**	Relax	Relax	Dinner
7–8	**Study Chemistry**	**Study Calculus**	**Study Calculus**	**Study Chemistry**	**Study Calculus**	**Study Intro. to Eng.**	**Study Calculus**
8–9	**Study Intro. to Eng.**	**Study Chemistry**	**Study Intro. to Eng.**	**Study Chemistry**	Recreation	Recreation	**Study Calculus**
9–10	**Study English**	**Study Chemistry**	**Study English**	**Study English**	Recreation	Recreation	**Study English**
10–11	Relax/Get ready for bed	Relax/Get ready for bed	Relax/Get ready for bed	Relax/Get ready for bed	Recreation	Recreation	Relax/Get ready for bed

You may not be an 18-year-old freshman whose parents are paying most of your tuition. You may be an older student who is changing careers. Or you may be married and have children, so you must have at least a part-time job. In this case, obviously you will have to cut back in some areas. For example, you may want to consider not taking as many credits in a given semester and follow a five-year plan instead of a four-year plan.

Depending on how many hours a week you need to work, you can rebudget your time. The purpose of this time budget example is mainly to emphasize the fact that you need to learn to manage your time wisely if you want to be successful in life. Every individual, just like any good organization, monitors his or her resources. No one wants you to turn into a robot and time yourself to the second. These examples are provided to give you an idea of how much time is available to you and to urge you to consider how efficiently and wisely you are allocating and using your time. The point is that budgeting your time is very important.

With the exception of a few courses, most classes that you will take are scheduled for 50-minute periods, with a 10-minute break between classes to allow students to attend several classes in a row. The other important reason for having a 10-minute break is to allow time to clear your head. Most of us have a limited attention span and cannot concentrate on a certain topic for a long period of time without a break. Taking a break is healthy; it keeps your mind and body working well.

Typically, as a first-term freshman in engineering you may have a course load similar to the one shown here:

Chemistry (3)
Chemistry Lab (1)
Introduction to Engineering (2)
Calculus (4)
English Composition (3)
Humanities/Social Science electives (3)

Table 1.2 is an example of a schedule for a freshman engineering student. You already know your strengths and weaknesses; you may have to make several attempts to arrive at a good schedule that will fit your needs the best. You may also need to modify the example schedule shown to allow for any variability in the number of credits or other engineering program requirements at your particular school.

1.3 DAILY STUDYING AND PREPARATION

You start studying and preparing from the very first day of class! It is always a good idea to read the material that your professor is planning to cover in class ahead of time. This practice will improve both your understanding and retention of the lecture materials. It is also important to go over the material that was discussed in class again later the same day after the lecture was given. When you are reading the material ahead of a lecture, you are familiarizing yourself with the information that the instructor will present to you in class. Don't worry if you don't fully understand everything you are reading at that time. During the lecture you can focus on the material that you did not fully understand and ask questions. When you go over the material after the lecture, everything then should come together. Remember to read before the class and study the material after the class on the same day!

Attend Your Classes Regularly Yes, even if your professor is a bore, you can still learn a great deal from attending the class. Your professor may offer additional

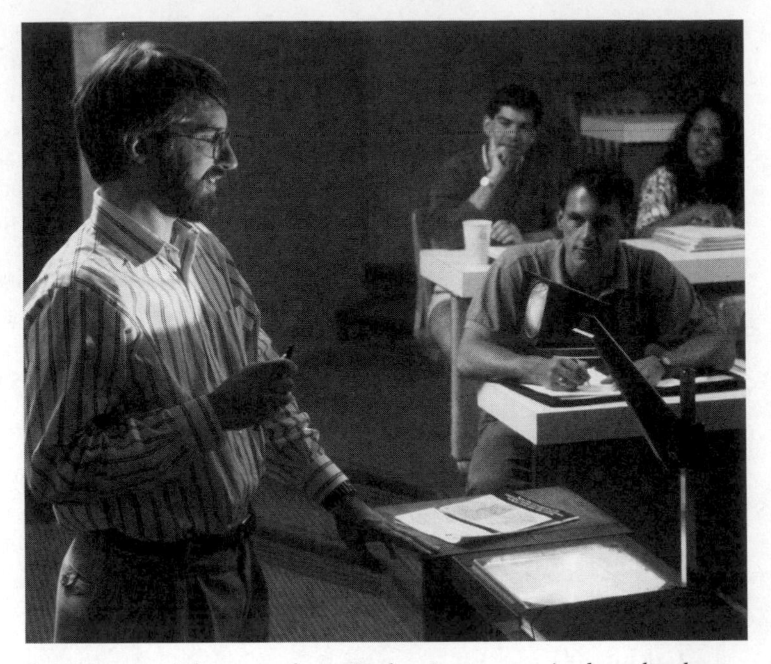

Read the material your professor is planning to cover in class ahead of time.

explanations and discussion of some material that may not be well presented in your textbook. Moreover, you can ask questions in class. If you have read the material before class and have made some notes about the concepts that you do not fully understand, during the lecture you can ask questions to clarify any misunderstanding. If you need more help, then go to your professor's office and ask for additional assistance.

Get Help Right Away When you need some help, don't wait till the last minute to ask! Your professor should have his or her office hours posted on the office door or be available on the Web. The office hours are generally stated in the course syllabus. If for some reason you cannot see your professor during the designated office hours, ask for an appointment. Almost all professors are glad to sit down with you and help you out if you make an appointment with them. After you have made an appointment, be on time and have your questions written down so that you remember what to ask. Once again, remember that most professors do not want you to wait until the last minute to get help!

Any professor can tell you some stories about experiences with students who procrastinate. Recently, I had a student who sent an e-mail to me on Sunday night at 10:05 P.M. asking for an extension on a homework assignment that was due on the following day. I asked the student the next morning, "When did you start to do the assignment?" He replied, "at 10:00 P.M. Sunday night." On another occasion, I had a student who came to my office and introduced himself to me for the first time. He asked me to write him a recommendation letter for a summer job he was applying for. Like most professors, I do not write recommendation letters for students, or anyone, whom I do not actually know. Get to know your professors and visit them often!

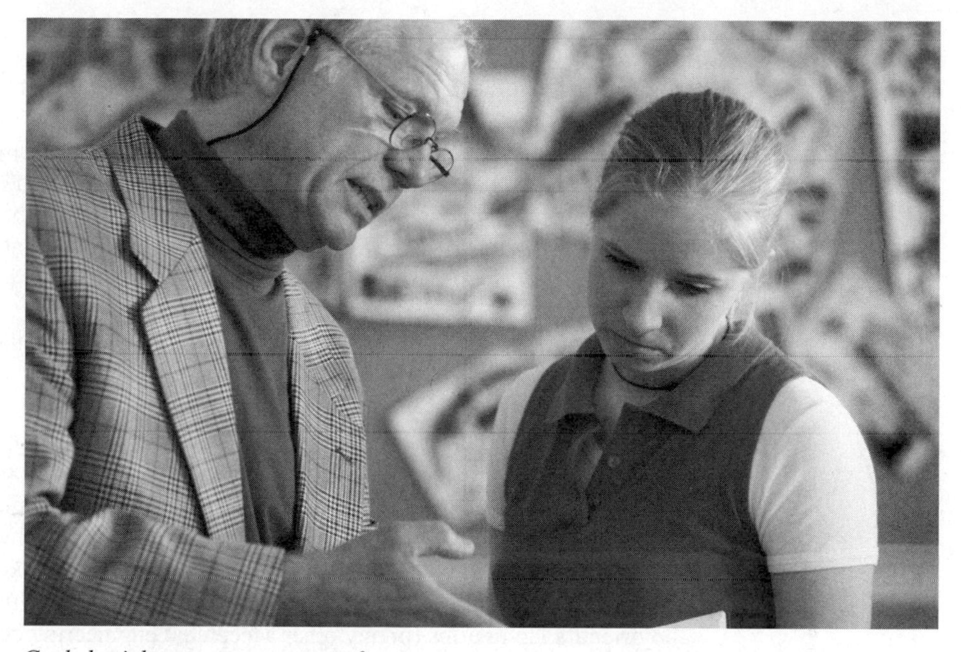

Get help right away—see your professor.

Take Good Notes Everyone knows that it is a good idea to take notes during lecture, but some students may not realize that they should also take notes when reading the textbook. Try to listen carefully during the lectures so you can identify and record the important ideas and concepts. If you have read ahead of time the text materials that your professor is planning to cover in class, then you are prepared to write down notes that complement what is already in your textbook. You don't need to write down everything that your professor says, writes, or projects onto a screen. The point is to listen very carefully and write down only notes regarding the important concepts that you did not understand when you read the book.

Use wirebound notebooks for your notes. Don't use loose papers, because it is too easy to lose some of your notes that way. Keeping a notebook is a good habit to develop now. As an engineer you will need to keep records of meetings, calculations, measurements, etc. with time and date recorded so you can refer back to them if the need arises. Thus it is best that you keep the notes in a wirebound notepad or a notebook with the pages sown into the binding so you won't lose any pages. Study your notes for at least an hour or two the same day you take them. Make sure you understand all the concepts and ideas that were discussed in class before you attempt to do your homework assignment. This approach will save you a great deal of time in the long run! Don't be among those students who spend as little time as possible on understanding the underlying concepts and try to take a shortcut by finding an example problem in the book similar to the assigned homework problem. You may be able to do the homework problem but you won't develop an understanding of the material. Without a firm grasp of the basic concepts, you will not do well on the exams, and you will be at a disadvantage later in life when you practice engineering.

Take good legible notes so that you can go back to them later if you need to refresh your memory before exam time. Most of today's engineering textbooks provide a blank margin on the left and right side of each sheet. Don't be afraid to write in these margins as you study your book. Keep all your engineering books; don't sell them back to the bookstore—some day you may need them. If you have a computer, you may want to type up a summary sheet of all important concepts. Later, you can use the *Find* command to look up selected terms and concepts. You may wish to insert links between related concepts in your notes. Digital notes may take some extra time to type, but they can save you time in the long run when you search for information.

Select a Good Study Place You may already know that you should study in a comfortable place with good lighting. You do not want any distractions while you are studying. For example, you do not want to study in front of a TV while watching your favorite situation comedy. A library is certainly a good place for studying, but you can make your dorm room or your apartment room into a good place for studying. Talk to your roommate(s) about your study habits and study time. Explain to them that you prefer to study in your own room and appreciate not being disturbed while studying. If possible, find a roommate with a declared engineering major, who is likely to be more understanding of your study needs. Remember that a bad place for studying is in your girlfriend's or boyfriend's lap or arms (or any other acceptable engineering configuration). Another useful idea is to keep your desk clean and avoid having a picture of your sweetheart in front of you. You don't want to daydream as you are studying. There is plenty of time for that later.

Form Study Groups Your professor will be the first person to tell you that the best way to learn something is to teach it. In order to teach something, though, you have to first thoroughly understand the basic concepts. You need to study on your own first and

Not a good way to study!

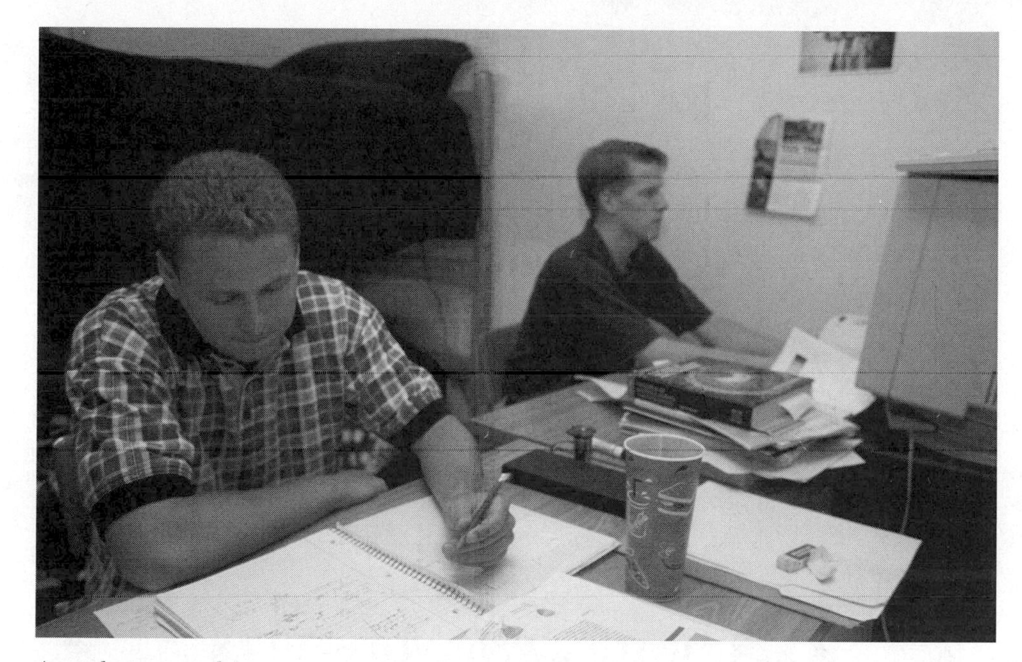

A good way to study!

then get together with your classmates to discuss and explain key ideas and concepts to each other. Everyone in your study group should agree that they need to come prepared to discuss appropriate materials and that they all need to contribute to the discussions. It should be understood that the study groups serve a different purpose than that of a tutoring session. However, if another student in your class asks for assistance, help if you can by explaining new concepts you have learned. If you have difficulty explaining the material to someone else, that could be an indication you don't fully understand the concept yourself and you need to study the material in more detail. So remember that a good way to learn something is to form discussion groups where you explain ideas and concepts in your own words to others in the group. *Be an active learner not a passive learner!*

Prepare for Examinations If you study and prepare from the first day of class, then you should perform admirably on your exams. Keep reminding yourself that there is absolutely no substitution for daily studying. Don't wait until the night before the exam to study! That is not the best time for learning new concepts and ideas. The night before the exam is the time for review only. Just before an exam, spend a few hours reviewing your notes and sample problems. Make sure that you get a good night's rest so you can be fresh and think clearly when you take the exam. It may be a good idea to ask your instructor ahead of time what type of exam it will be, how many questions there will be, or what suggestions she or he has to help you prepare better for the exam. As with any test, be sure you understand what the questions are asking. If there is some ambiguity in the exam questions, ask the instructor for clarification. After you have looked over the exam, you may want to answer the easy questions first and then come back to the more complicated questions.

Join an engineering organization!

1.4 GETTING INVOLVED WITH AN ENGINEERING ORGANIZATION

There are many good reasons to join an engineering organization. Networking, participating in plant tours, listening to technical guest speakers, participating in design competitions, attending social events, taking advantage of learning opportunities through short courses, seminars, and conferences, and obtaining student loans are a few of the common benefits of belonging to an engineering organization. Moreover, good places to learn more about areas of specialization in engineering are the Web sites of various engineering organizations. As you spend a little time reading about these organizations, you will discover that many share common interests and provide some overlapping services that could be used by engineers of various disciplines. You will also note that the primary purpose of these professional engineering organizations is to offer the following benefits:

1. They conduct conferences and meetings to share new ideas and findings in research and development.
2. They publish technical journals, books, reports, and magazines to help engineers in particular specialties keep up-to-date.
3. They offer short courses on current technical developments to keep practicing engineers abreast of the new developments in their respective fields.
4. They advise the federal and state governments on technology-related public policies.
5. They create, maintain, and distribute codes and standards that deal with correct engineering design practices to ensure public safety.
6. They provide a networking mechanism through which you get to know people from different companies and institutions. This is important for two reasons: (1) If there is a problem that you feel requires assistance from outside your organization, you have a pool of colleagues whom you have met at the meetings to help you solve the problem. (2) When

you know people in other companies who are looking for good engineers to hire and you are thinking about something different to do, then you may be able to find a good match.

Find out about the local student chapters of national engineering organizations on your campus. Attend the first meetings. After collecting information, choose an organization, join, and become an active participant. As you will see for yourself, the benefits of being a member of an engineering organization are great!

1.5 OTHER CONSIDERATIONS

Doing Volunteer Work If your study schedule allows, volunteer for a few hours a week to help those in need in your community. The rewards are unbelievable! Not only will you feel good about yourself but you will gain a sense of satisfaction and feel connected to your community. Volunteering could also help develop communication, management, or supervisory skills that you may not develop by just attending school.

Vote in Local and National Elections Most of you are 18 or older. Take your civic duties seriously. Exercise your right to vote, and try to play an active role in your local, state, or federal government. Remember that freedom is not free. Be a good, responsible citizen.

Get to Know Your Classmates There are many good reasons for getting to know one or two other students in your classes. You may want to study with someone from class, or if you are absent from class, you have someone to contact to find out what the assignment is or find out what was covered in class. Record the following information on the course syllabi for all your classes: (1) the name of a student sitting next to you, (2) his or her telephone number, (3) his or her e-mail address.

Get to Know an Upper-Division Engineering Student Becoming acquainted with junior and senior engineering students can provide you with valuable information about their engineering education experience and campus social issues. Ask your instructor to introduce you to a junior or a senior engineering student, and record the following information on your introduction to engineering class syllabus: (1) the name of an upper-division engineering student, his or her telephone number, his or her e-mail address.

SUMMARY

Now that you have reached this point in the text

- You should use the ideas discussed in this chapter to make your transition from high school to college smoothly. You should also consider the time management ideas discussed in this chapter to arrive at a reasonable weekly schedule.
- You should realize that you must start studying and preparing from the first day of class, attend classes regularly, get help right away, take good notes, select a good study place, and form study groups.

- You should know the importance of joining an engineering organization and choose an organization, join, and become and active participant.

PROBLEM

1. Prepare a schedule for the current semester; also prepare two additional alternative schedules. Discuss the pros and cons of each schedule. Select what you think is the best schedule, and discuss it with your instructor or advisor. Consider his or her suggestions and modify the schedule if necessary; then present the final schedule to your instructor. Maintain a daily logbook to keep track of how closely you are following the schedule and where time is being used inefficiently. Write a one-page summary discussing each week's activities that deviated from the planned schedule, and come up with ways to improve or modify its shortcomings. Turn in a biweekly summary report to your instructor or advisor. Think of this exercise as an ongoing test similar to other tests that engineers perform regularly to understand and improve things.

Hour	Monday	Tuesday	Wednesday	Thursday	Friday	Saturday	Sunday
7–8							
8–9							
9–10							
10–11							
11–12							
12–1							
1–2							
2–3							
3–4							
4–5							
5–6							
6–7							
7–8							
8–9							
9–10							
10–11							

CHAPTER

2 Introduction to Engineering

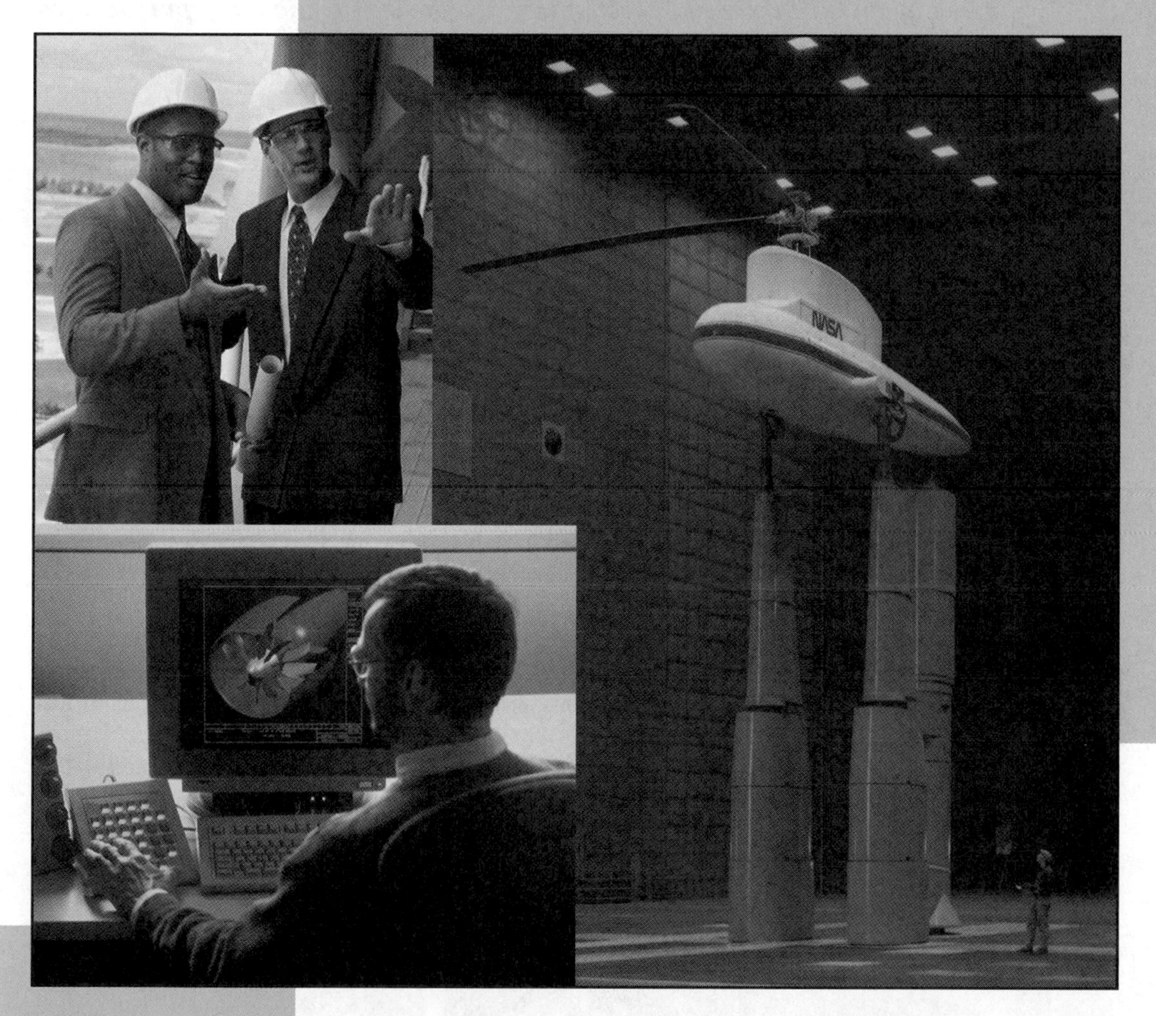

Engineers are problem solvers. Successful engineers possess good communication skills and are team players. They have a good grasp of fundamental physical laws and mathematics. Engineers apply physical and chemical laws and mathematics to design, develop, test, and supervise the manufacture of millions of products and services. They consider important factors such as efficiency, cost, reliability, and safety when designing products. Engineers are dedicated to lifelong learning and service to others.

Possibly some of you are not yet certain you want to study engineering during the next four years in college. You may have the following questions:

Do I really want to study engineering?
What is engineering and what do engineers do?
What are some of the areas of specialization in engineering?
Do I want to become a mechanical engineer, or should I pursue civil engineering?
Or would I be happier becoming an electrical engineer?
How many different engineering disciplines are there?
How will I know that I have picked the best field for me?
Will the demand for my area of specialization be high when I graduate, and beyond that?

The main objective of this chapter is to provide some answers to these and other questions you may have. Additional objectives of this chapter are to introduce you to the engineering profession and its various branches. ■

2.1 ENGINEERING WORK IS ALL AROUND YOU

Engineers make products and provide services that make our lives better (see Figure 2.1). To see how engineers contribute to the comfort and the betterment of our everyday lives, tomorrow morning when you get up, just look around you more carefully. During the night, your bedroom was kept at the right temperature thanks to some mechanical engineers who designed the heating, air-conditioning, and ventilating systems in your home. When you get up in the morning and turn on the lights, be assured that thousands of mechanical and electrical engineers and technicians at power plants and power stations around the country are making certain the flow of electricity remains uninterrupted so that you have enough power to turn the lights on or turn on your TV or radio to watch or listen

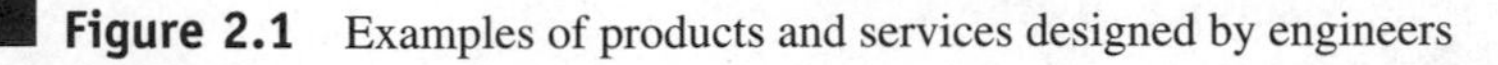

Figure 2.1 Examples of products and services designed by engineers

to the morning news and weather report for the day. The TV or radio you are using to get your morning news was designed by electrical and electronic engineers. There are, of course, engineers from other disciplines involved in creating the final product; for example, manufacturing and industrial engineers. When you are getting ready to take your morning shower, the clean water you are about to use is coming to your home thanks to civil and mechanical engineers. Even if you live out in the country on a farm, the pump you use to bring water from the well to your home was designed by mechanical and civil engineers. The water could be heated by natural gas that is brought to you thanks to the work and effort of chemical, mechanical, civil, and petroleum engineers. After your morning shower, when you get ready to dry yourself with a towel, think about what types of engineer worked behind the scenes to produce the towels. Yes, the cotton towel was made with the help of agricultural, industrial, manufacturing, chemical, petroleum, civil, and mechanical engineers. Think about the machines that were used to pick the cotton, transport the cotton to a factory, clean it, and dye it to a pretty color that is pleasing to your eyes. Then other machines were used to weave the fabric and send it to sewing machines that were designed by mechanical engineers. The same is true of the clothing you are about to wear. Your clothing may contain some polyester, which was made possible with the aid of petroleum and chemical engineers. "Well," you may say, "I can at least sit down and eat my breakfast and not wonder whether some engineers made this possible as well." But the food you are about to eat was made with the help and collaboration of various engineering disciplines, from agricultural to mechanical. Let's say you are about to have some cereal. The milk was kept fresh in your refrigerator thanks to the efforts and work of mechanical engineers who designed the refrigerator components and chemical engineers who investigated alternative refrigerant fluids with appropriate thermal properties and other environmentally friendly properties that can be used in your refrigerator. Furthermore, electrical engineers designed the control and the electrical power units.

Now you are ready to get into your car or take the bus to go to school. The car you are about to drive was made possible with the help and collaboration of automotive, mechanical, electrical, electronic, material, chemical, and petroleum engineers. So, you see there is not much that you do in your daily life that has not involved the work of engineers. Be proud of the decision you have made to become an engineer. Soon you will become one of those whose behind-the-scenes efforts will be taken for granted by billions of people around the world. But you will accept that fact gladly, knowing that what you do will make people's lives better.

Engineers Deal with an Increasing World Population

We as people, regardless of where we live, need the following things: food, clothing, shelter, and water for drinking or cleaning purposes. In addition, we need various modes of transportation to get to different places, because we may live and work in different cities or wish to visit friends and relatives who may live elsewhere. We also like to have some sense of security, to be able to relax and be entertained. We need to be liked and appreciated by our friends and family, as well.

At the turn of the 20th century, there were approximately 6 billion of us inhabiting the earth. As a means of comparison, it is important to note that the world population 100 years ago, at the turn of the 19th century, was 1 billion. Think about it. It took us since the beginning of human existence to reach a population of 1 billion. It only took 100 years to increase the population by fivefold. Some of us have a good standard of living, but some of us living in developing countries do not. You will probably agree that our world would be a better place if every one of us had enough to eat, a comfortable and safe place to live, meaningful work to do, and some time for relaxation.

According to the latest estimates and projections of the U.S. Census Bureau, the world population will reach 9.3 billion people by the year 2050. Not only will the number of people inhabiting the earth continue to rise but the age structure of the world population will also change. The world's elderly population—the people at least 65 years of age—will more than double in the next 25 years (see Figure 2.2).

How is this information relevant? Well, now that you have decided to study to become an engineer, you need to realize that what you do in a few years after your graduation is very important to all of us. You will design products and provide services especially suited

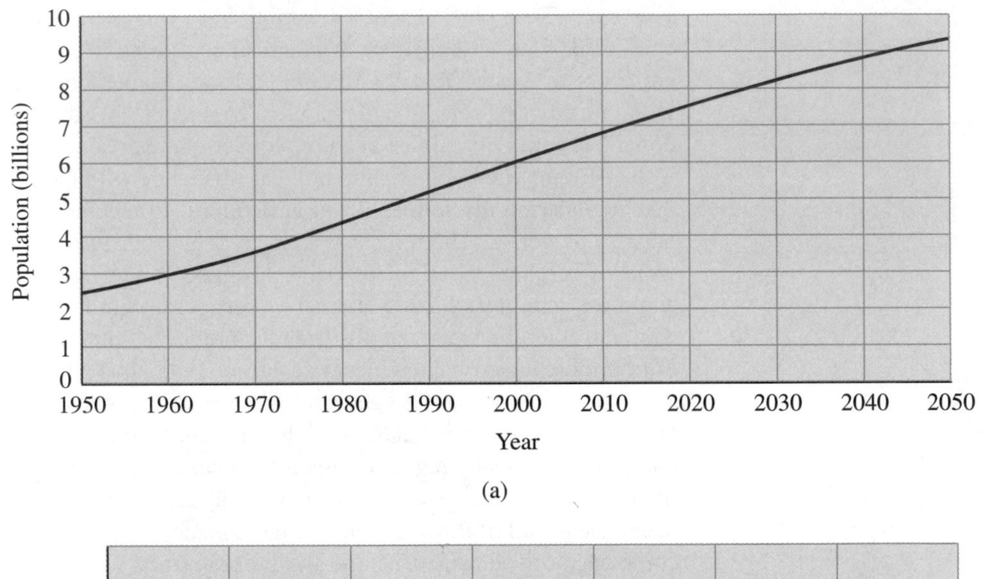

(a)

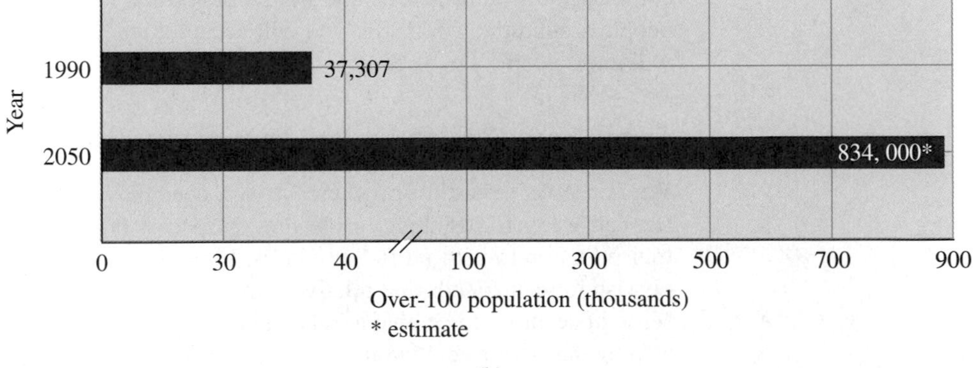

(b)

Figure 2.2 (a) The latest projection of world population growth. (b) The latest estimate of U.S. elderly population growth.

Source: Data from U.S. Census Bureau.

to the needs and demands of an increasing elderly population as well as increased numbers of people of all ages. So prepare well to become a good engineer and be proud that you have chosen the engineering profession in order to contribute to raising the living standard for everyone. Today's world economy is very dynamic. Corporations continually employ new technologies to maximize efficiency and profits. Because of this ongoing change and emerging technologies, new jobs are created and others are eliminated. Computers and smart electronic devices are continuously reshaping our way of life. Such devices influence the way we do things and help us provide the necessities of our lives—clean water, food, and shelter. Moreover, as the age structure of the world population changes, particularly as the number of elderly people increases, new products and services will be required to meet this demand. According to the Bureau of Labor Statistics, U.S. Department of Labor, among the fastest-growing occupations are engineers, computer specialists, and systems analysts. Most computer occupations are projected to grow much faster than they are now.

You need to become a lifelong learner so that you can make informed decisions and anticipate as well as react to the global changes caused by technological innovations as well as population and environmental changes.

2.2 ENGINEERING AS A PROFESSION AND COMMON TRAITS OF GOOD ENGINEERS

In this section, we will first discuss engineering in a broad sense, and then we will focus on selected aspects of engineering. We will also look at the traits and characteristics common to many engineers. Next we will discuss some specific engineering disciplines. As we said earlier in this chapter, perhaps some of you have not yet decided what you want to study during your college years and consequently may have many questions, including: What is engineering and what do engineers do? What are some of the areas of specialization in engineering? Do I really want to study engineering? How will I know that I have picked the best field for me? Will the demand for my area of specialization be high when I graduate, and beyond that?

Don't worry about finding answers to all these questions right now. You have some time to ponder them because most of the coursework during the first year of engineering is similar for all engineering students, regardless of their specific discipline. So you have at least a year to consider various possibilities. This is true at most educational institutions. Even so, you should talk to your advisor early to determine how soon you must choose an area of specialization. Reading the following sections will help you make a decision that you will be happy with. And don't be concerned about your chosen profession changing in a way that makes your education obsolete. Most companies assist their engineers in acquiring further training and education to keep up with changing technologies. A good engineering education will enable you to become a good problem solver throughout your life, regardless of the particular problem or situation. You may wonder during the next few years of school why you need to be learning some of the material you are studying. Sometimes your homework may seem irrelevant, trivial, or out-of-date. Rest assured that you are learning both content information and strategies of thinking and analysis that will equip you to face future challenges, ones that do not even exist yet.

What Is Engineering and What Do Engineers Do?

Engineers apply physical and chemical laws and principles and mathematics to design millions of products and services that we use in our everyday lives. These products include cars, computers, aircraft, clothing, toys, home appliances, surgical equipment, heating and cooling equipment, health care devices, tools and machines that makes various products, and so on (see Figure 2.3). Engineers consider important factors such as cost, efficiency, reliability, and safety when designing these products. Engineers perform tests to make certain that the products they design withstand various loads and conditions. They are continuously searching for ways to improve already existing products as well. They also design and supervise the construction of buildings, dams, highways, and mass transit systems and the construction of power plants that supply power to manufacturing companies, homes, and offices. Engineers play a significant role in the design and maintenance of a nation's infrastructure, including communication systems, public utilities, and transportation. Engineers continuously develop new, advanced materials to make products lighter and stronger for different applications. They are also responsible for finding suitable ways to extract petroleum, natural gas, and raw materials from the earth, and they are involved in coming up with ways of increasing crop, fruit, and vegetable yields along with improving the safety of our food products.

The following represent some common careers for engineers. In addition to design, some engineers work as sales representatives for products, while others provide technical support and troubleshooting for customers of their products. Not all engineers work for private industries; some work for federal, state, and local governments in various capacities. Engineers work in departments of agriculture, defense, energy, and transportation. Some engineers work for the National Aeronautics and Space Administration (NASA). As you can see, there are many satisfying and challenging jobs for engineers.

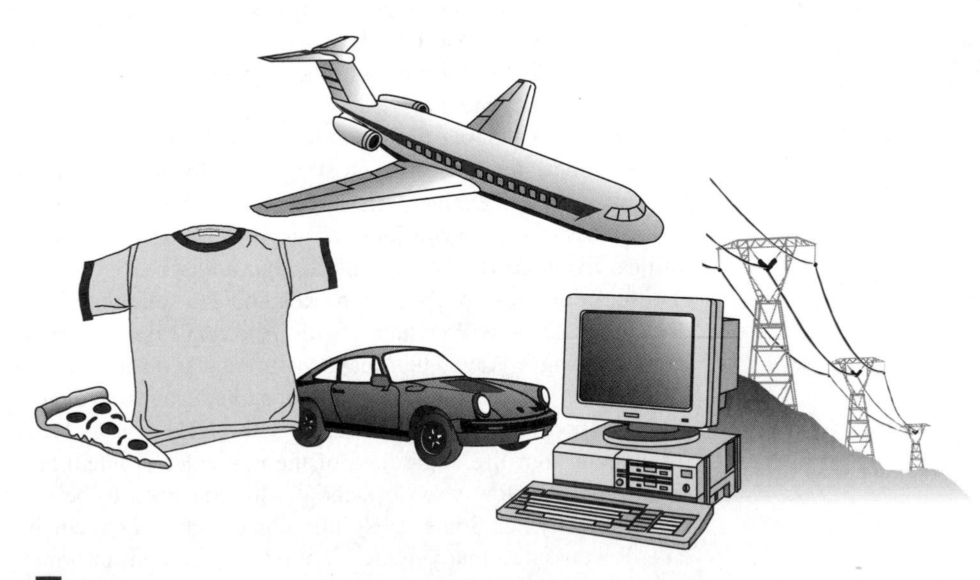

Figure 2.3 As an engineer you will apply physical and chemical laws and principles and mathematics to design various products and services.

Although the activities of engineers are quite varied, there are some personality traits and work habits that typify most of today's successful engineers.

- Engineers are problem solvers.
- Good engineers have a firm grasp of the fundamental principles of engineering, which they can use to solve many different problems.
- Good engineers are analytical, detailed oriented, and creative.
- Good engineers have a desire to be lifelong learners. For example, they take continuing education classes, seminars, and workshops to stay abreast of innovations and new technologies. This is particularly important in today's world because the rapid changes in technology will require you as an engineer to keep pace with new technologies. Moreover, you will risk being laid off or denied promotion if you are not continually improving your engineering education.
- Well-trained engineers are able to work outside their area of specialization in other related fields. For example, a good mechanical engineer with a well-rounded knowledge base can work as an automotive engineer, an aerospace engineer, or as a chemical engineer.
- Good engineers have written and oral communication skills that equip them to work well with their colleagues and to convey their expertise to a wide range of clients.
- Good engineers have time-management skills that enable them to work productively and efficiently.
- Good engineers have good "people skills" that allow them to interact and communicate effectively with various people in their organization. For example, they are able to communicate equally well with the sales and marketing experts and their own colleagues.
- Engineers are required to write reports. These reports might be lengthy, detailed technical reports containing graphs, charts, and engineering drawings, or they may take the form of brief memoranda or executive summaries.
- Engineers are adept at using computers in many different ways to model and analyze various practical problems.
- Good engineers actively participate in local and national discipline-specific organizations by attending seminars, workshops, and meetings. Many even make presentations at professional meetings.
- Engineers generally work in a team environment where they consult each other to solve complex problems. They divide up the task into smaller, manageable problems among themselves; consequently, productive engineers must be good team players. Good interpersonal and communication skills are increasingly important now because of the global market. For example, various parts of a car could be made by different companies located in different countries. In order to ensure that all components fit and work well together, cooperation and coordination are essential, which demands strong communication skills.

Clearly, an interest in building things or taking things apart or solving puzzles is not all that is required to become an engineer. In addition to having a dedication to learning and a desire to find solutions, an engineer needs to foster certain attitudes and personality traits.

These are some other facts about engineering that are worth noting.

- For almost all entry-level engineering jobs, a bachelor's degree in engineering is required.

According to the Bureau of Labor Statistics,

- The starting salaries of engineers are significantly higher than those of bachelor's-degree graduates in other fields. The outlook for engineering is very good. Good employment opportunities are expected for new engineering graduates during 2000–2006.
- Most engineering degrees are granted in electrical, mechanical, and civil engineering, the parents of all other engineering branches.

Visit this book's companion Web site to obtain information about the number of engineers employed by specific area and their mean salaries in recent years.

2.3 ENGINEERING DISCIPLINES AND ENGINEERING ORGANIZATIONS

Now that you have a general sense of what engineers do, you may be wondering about the various branches or specialties in engineering. Good places to learn more about areas of specialization in engineering are the Web sites of various engineering organizations. We explained in Chapter 1 that as you spend a little time reading about these organizations, you will discover many share common interests and provide some overlapping services that could be used by engineers of various disciplines. Following is a list of a few Web sites that you may find useful when searching for information about various engineering disciplines.

American Institute of Aeronautics and Astronautics
www.aiaa.org

American Institute of Chemical Engineers
www.aiche.org

The American Society of Agricultural Engineers
www.asae.org

American Society of Civil Engineers
www.asce.org

American Nuclear Society
www.ans.org

American Society for Engineering Education
www.asee.org

American Society of Heating, Refrigerating, and Air-Conditioning Engineers
www.ashrae.org

American Society of Mechanical Engineers
www.asme.org

Institute of Electrical and Electronics Engineers
www.ieee.org

Institute of Industrial Engineers
www.iienet.org

National Academy of Engineering
www.nae.edu

National Science Foundation
www.nsf.gov

National Society of Black Engineers
www.nsbe.org

National Society of Professional Engineers
www.nspe.org

Society of Automotive Engineers
www.sae.org

Society of Hispanic Professional Engineers
www.shpe.org

Society of Manufacturing Engineering
www.sme.org

Society of Women Engineers
www.swe.org

Tau Beta Pi (All-Engineering Honor Society)
www.tbp.org

NASA Centers
Ames Research Center
www.arc.nasa.gov
Dyrden Flight Research Center
www.dfrc.nasa.gov
Goddard Space Flight Center
www.gsfc.nasa.gov
Jet Propulsion Laboratory
www.jpl.nasa.gov
Johnson Space Center
www.jsc.nasa.gov
Kennedy Space Center
www.ksc.nasa.gov
Langley Research Center
www.larc.nasa.gov
Marshall Space Flight Center
www.msfc.nasa.gov
Glenn Research Center
www.grc.nasa.gov

Inventors Hall of Fame
www.invent.org

U.S. Patent and Trademark Office
www.uspto.gov

For an additional listing of engineering-related Web sites, please see this book's companion Web site: **http://info.brookscole.com/Moaveni**.

What Are Some Areas of Engineering Specialization?

There over 20 major disciplines or specialties that are recognized by professional engineering societies. Moreover, within each discipline there exist a number of branches. For example, the mechanical engineering program can be traditionally divided into two broad

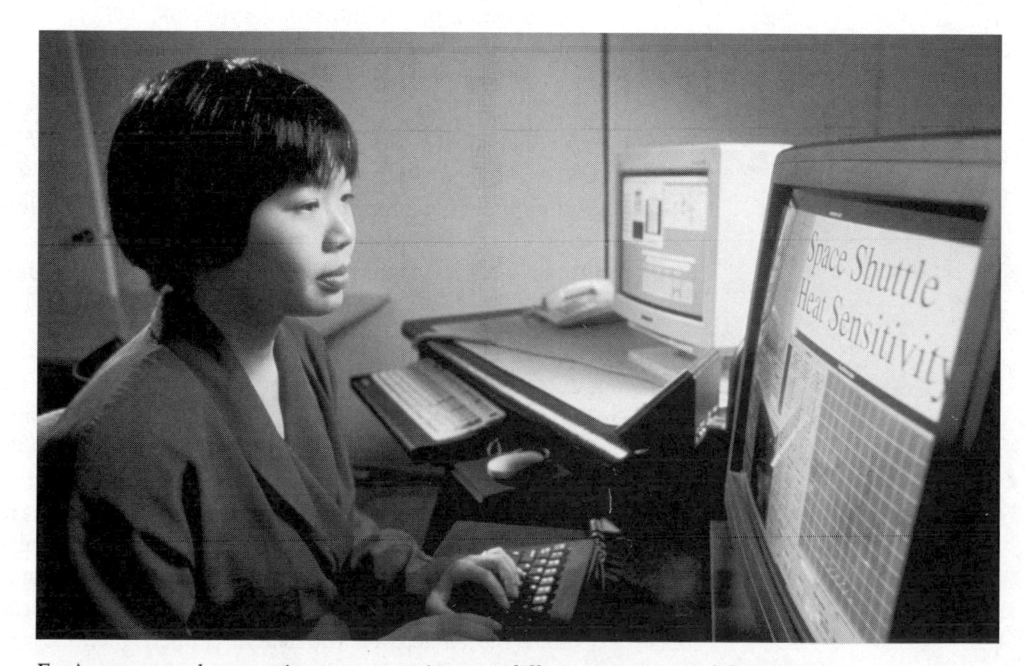

Engineers are adept at using computers in many different ways to model and analyze various practical problems.

areas: (1) thermal/fluid systems and (2) structural/solid systems. In most mechanical engineering programs, during your senior year you can take elective classes that allow you to pursue your interest and broaden your knowledge base in these areas. So, for example, if you are interested in learning more about how buildings are heated during the winter or cooled during the summer, you will take a heating, ventilating, and air-conditioning class. To give you additional ideas about the various branches within specific engineering disciplines, consider civil engineering. The main branches of a civil engineering program normally are environmental, water resources, transportation, and structural. The branches of electrical engineering may include (1) power generation and transmission, (2) communications, (3) control, (4) electronics, and (5) integrated circuits.

Not all engineering disciplines are discussed here, but you are encouraged to visit the Web sites of appropriate engineering societies to learn more about a particular engineering discipline.

Over 300 colleges and universities in the United States offer bachelor's-degree programs in engineering that are accredited by the Accreditation Board for Engineering and Technology (ABET). ABET examines the credentials of the engineering program's faculty, curricular content, facilities, and admissions standards before granting accreditation. It may be wise for you to find out the accreditation status of the engineering program you are planning to attend. ABET maintains a Web site with a list of all accredited programs; visit www.abet.org for more information. According to ABET, accredited engineering programs must demonstrate that their graduates have

- an ability to apply knowledge of mathematics, science, and engineering;
- an ability to design and conduct experiments, as well as to analyze and interpret data;
- an ability to design a system, component, or process to meet desired needs;
- an ability to function on multidisciplinary teams;
- an ability to identify, formulate, and solve engineering problems;
- an understanding of professional and ethical responsibility;
- an ability to communicate effectively;
- the broad education necessary to understand the impact of engineering solutions in a global and societal context;
- a recognition of the need for and an ability to engage in lifelong learning;
- a knowledge of contemporary issues;
- an ability to use the techniques, skills, and modern engineering tools necessary for engineering practice.

Therefore, these are the educational outcomes that are expected of you when you graduate from your engineering program. Bachelor's-degree programs in engineering are typically designed to last four years; however, many students take five years to acquire their engineering degrees. In a typical engineering program, you will spend the first two years studying mathematics, English, physics, chemistry, introductory engineering, computer science, humanities, and social sciences. These first two years are often referred to as *preengineering*. In the last two years, most courses are in engineering, usually with a concentration in one branch. For example, in a typical mechanical engineering program, during the last two years of your studies, you will take courses such as thermodynamics, mechanics of materials, fluid mechanics, heat transfer, applied thermodynamics, and design. Some programs offer a general engineering curriculum; students then specialize in graduate school or on the job.

Many community colleges around the country offer the first two years of engineering programs, which are normally accepted by the engineering schools. Some engineering schools offer five-year master's-degree programs. Some engineering schools, in order to provide hands-on experience, have a cooperative plan whereby students take classes during the first three years and then may take a semester off from studying to work for an engineering company. Of course, after a semester or two, students return to school to finish their education. Schools that offer cooperative programs generally offer full complements of classes every semester so that students can graduate in four years if they desire.

All 50 states and the District of Columbia require registration for engineers whose work may affect the safety of the public. As a first step in becoming a registered professional engineer (PE), you must have a degree from an ABET-accredited engineering program. You also need to take your Fundamentals of Engineering Exam (FE) during your senior year. The exam lasts about eight hours and is divided into a morning and an afternoon section. During the morning session, you will answer multiple-choice questions in chemistry, physics, mathematics, mechanics, thermodynamics, electrical and electronic circuits, and materials science. During the four-hour afternoon session, you will answer multiple-choice questions specific to your discipline, or you may choose to take a general engineering exam. After you pass your FE exam, you need to gain four years of relevant engineering work experience and pass another eight-hour exam (the Principles and Practice of Engineering Exam) given by the state. Candidates choose an exam from one of 16 engineering disciplines. Some engineers are registered in several states. Normally, civil, mechanical, and electrical engineers seek professional registrations.

As a recent engineering graduate, you should expect to work under the supervision of a more experienced engineer. Based on your assigned duties, some companies may have you attend workshops (short courses that could last for a week) or a day-long seminar to obtain additional training in communication skills, time management, or a specific engineering method. As you gain more knowledge and experience, you will be given more freedom to make engineering decisions. Once you have many years of experience, you may then elect to become a manager in charge of a team of engineers and technicians. Some engineers fresh out of college begin their careers not in a specific area of engineering, but in sales or marketing related to engineering products and services.

As already mentioned, there are more than 20 engineering disciplines recognized by the professional societies. However, most engineering degrees are granted in civil, electrical, and mechanical engineering. Therefore, these disciplines are discussed here first.

Civil Engineering Civil engineering is perhaps the oldest engineering discipline. As the name implies, civil engineering is concerned with providing public infrastructure and services. Civil engineers design and supervise the construction of buildings, roads and highways, bridges, dams, tunnels, mass transit systems, and airports. They are also involved in the design and supervision of municipal water supplies and sewage systems. The major branches within the civil engineering discipline include structural, environmental, transportation, water resources, and geotechnical. Civil engineers work as consultants, construction supervisors, city engineers, and public utility and transportation engineers. According to the Bureau of Labor Statistics, the job outlook for graduates of civil engineering is good because as population grows, more civil engineers are needed to design and supervise the construction of new buildings, roads, and water supply and sewage systems. They are also needed to oversee the maintenance and renovation of existing public structures, roads, bridges, and airports.

Electrical and Electronic Engineering Electrical and electronic engineering is the largest engineering discipline. Electrical engineers design, develop, test, and supervise the manufacturing of electrical equipment, including lighting and wiring for buildings, cars, buses, trains, ships, and aircraft; power generation and transmission equipment for utility companies; electric motors found in various products; control devices; and radar equipment. The major branches of electrical engineering include power generation, power transmission and distribution, and controls. Electronic engineers design, develop, test, and supervise the production of electronic equipment, including computer hardware; computer network hardware; communication devices such as cellular phones, television, and audio and video equipment; as well as measuring instruments. Growing branches of electronic engineering include computer and communication electronics. The job outlook for electrical and electronic engineers is good because businesses and government need faster computers and better communication systems. Of course, consumer electronic devices will play a significant role in job growth for electrical and electronic engineers as well.

Mechanical Engineering The mechanical engineering discipline, which has evolved over the years as new technologies have emerged, is one of the broadest engineering disciplines. Mechanical engineers are involved in the design, development, testing, and manufacturing of machines, robots, tools, power generating equipment such as steam and gas turbines, heating, cooling, and refrigerating equipment, and internal combustion engines. The major branches of mechanical engineering include thermal/fluid systems and structural/solid systems. The job outlook for mechanical engineers is also good, as more efficient machines and power generating equipment and alternative energy-producing devices are needed. You will find mechanical engineers working for the federal government, consulting firms, various manufacturing sectors, the automotive industry, and other transportation companies.

The other common disciplines in engineering include aerospace engineering, chemical engineering, petroleum engineering, nuclear engineering, and materials engineering.

Aerospace Engineering Aerospace engineers design, develop, test, and supervise the manufacture of commercial and military aircraft, helicopters, spacecraft, and missiles. They may work on projects dealing with research and development of guidance, navigation, and control systems. Most aerospace engineers work for aircraft and missile manufacturers, the Department of Defense, and NASA. If you decide to pursue an aerospace engineering career, you should expect to live in California, Washington, Texas, or Florida, because these are the states with large aerospace manufacturing companies. According to the Bureau of Labor Statistics, the job outlook for aerospace engineers is expected to grow more slowly than average through the year 2006. One reason for this slower job growth is the decline in Department of Defense expenditures. However, because of population growth and the need to meet the demand for more passenger air traffic, commercial airplane manufacturers are expected to do well.

Chemical Engineering As the name implies, chemical engineers use the principles of chemistry and basic engineering sciences to solve a variety of problems related to the production of chemicals and their use in various industries, including the pharmaceutical, electronic, and photographic industries. Most chemical engineers are employed by chemical, petroleum refining, film, paper, plastic, paint, and other related industries. Chemical engineers also work in metallurgical, food processing, biotechnology and fermentation industries. To meet the needs of the growing population, the job outlook for chemical engineers is also good, according to the Bureau of Labor Statistics.

Manufacturing Engineering Manufacturing engineers develop, coordinate, and supervise the process of manufacturing all types of products. They are concerned with making products efficiently and at minimum cost. Manufacturing engineers are involved in all aspects of production, including scheduling and materials handling and the design, development, supervision, and control of assembly lines.

Manufacturing engineers employ robots and machine-vision technologies for production purposes. To demonstrate concepts for new products, and to save time and money, manufacturing engineers create prototypes of products before proceeding to manufacture actual products. This approach is called *prototyping*. Manufacturing engineers are employed by all types of industries, including automotive, aerospace, and food processing and packaging. The job outlook for manufacturing engineers is expected to be good.

Petroleum Engineering Petroleum engineers specialize in the discovery and production of oil and natural gas. In collaboration with geologists, petroleum engineers search the world for underground oil or natural gas reservoirs. Geologists have a good understanding of the properties of the rocks that make up the earth's crust. After geologists evaluate the properties of the rock formations around oil and gas reservoirs, they work with petroleum engineers to determine the best drilling methods to use. Petroleum engineers are also involved in monitoring and supervising drilling and oil extraction operations. In collaboration with other specialized engineers, petroleum engineers design equipment and processes to achieve the maximum profitable recovery of oil and gas. They use computer models to simulate reservoir performance as they experiment with different recovery techniques. If you decide to pursue petroleum engineering, you are most likely to work for one of the major oil companies or one of the hundreds of smaller, independent companies involved in oil exploration, production, and service. Engineering consulting firms, government agencies, oil field services, and equipment suppliers also employ petroleum engineers. According to the U.S. Department of Labor, large numbers of petroleum engineers are employed in Texas, Oklahoma, Louisiana, Colorado, and California, including offshore sites. Many American petroleum engineers also work overseas in oil-producing regions of the world such as Russia, the Middle East, South America, or Africa.

The job outlook for petroleum engineers is expected to decline through the year 2006, according to Bureau of Labor Statistics, unless oil and gas prices rise enough to encourage increased exploration for oil in the United States. In spite of this fact, if you do decide to study petroleum engineering, employment opportunities for petroleum engineers should be favorable because the number of degrees granted in petroleum engineering has traditionally been low. Also, petroleum engineers work around the globe, and many employers seek U.S.-trained petroleum engineers for jobs in other countries.

Nuclear Engineering Only a few engineering colleges around the country offer a nuclear engineering program. Nuclear engineers design, develop, monitor, and operate nuclear power equipment that derives its power from nuclear energy. Nuclear engineers are involved in the design, development, and operation of nuclear power plants to generate electricity or to power Navy ships and submarines. They may also work in such areas as the production and handling of nuclear fuel and the safe disposal of its waste products. Some nuclear engineers are involved in the design and development of industrial and diagnostic medical equipment. Nuclear engineers work for the U.S. Navy, nuclear power utility companies, and the Nuclear Regulatory Commission of the Department of Energy. Because of the high cost and numerous safety concerns on the part of the public, there are only a few nuclear power plants under construction. Even so, the job outlook for nuclear engineers is not too bad because currently there are not many graduates in this field. Other job opportunities exist for nuclear engineers in the departments of Defense and Energy, nuclear medical technology, and nuclear waste management.

Mining Engineering There are only a few mining engineering schools around the country. Mining engineers, in collaboration with geologists and metallurgical engineers, find, extract, and prepare coal for use by utility companies; they also look for metals and minerals to extract from the earth for use by various manufacturing industries. Mining engineers design and supervise the construction of aboveground and underground mines. Mining engineers could also be involved in the development of new mining equipment for extraction and separation of minerals from other materials mixed in with the desired minerals.

Most mining engineers work in the mining industry, some work for government agencies, and some work for manufacturing industries. The job outlook for mining engineers is not as good as for other disciplines. The mining industry is somewhat similar to the oil industry in that the job opportunities are closely tied to the price of metals and minerals. If the price of these products is low, then the mining companies will not want to invest in new mining equipment and new mines. Similar to petroleum engineers, U.S. mining engineers may find good opportunities outside the United States.

Materials Engineering There are only a few engineering colleges that offer a formal program in materials engineering, ceramic engineering, or metallurgical engineering. Materials engineers research, develop, and test new materials for various products and engineering applications. These new materials could be in the form of metal alloys, ceramics, plastics, or composites. Materials engineers study the nature, atomic structure, and thermophysical properties of materials. They manipulate the atomic and molecular structure of materials in order to create materials that are lighter, stronger, and more durable. They create materials with specific mechanical, electrical, magnetic, chemical, and heat-transfer properties for use in specific applications; for example, graphite tennis racquets that are much lighter and stronger than the old wooden racquets; the composite materials used in stealth military planes with specific electromagnetic properties; and the ceramic tiles on the space shuttle that protect the shuttle during reentry into the atmosphere (ceramics are nonmetallic materials that can withstand high temperatures).

Materials engineering may be further divided into metallurgical, ceramics, plastics, and other specialties. You can find materials engineers working in aircraft manufacturing; various research and testing labs; and electrical, stone, and glass products manufacturers. Because of the low number of current graduates, the job opportunities are good for materials engineers.

SUMMARY

Now that you have reached this point in the text

- You should have a good understanding of the significant role that engineers play in our everyday lives in providing water, food, shelter, and other essential needs.
- You should have a good idea of the common traits and activities of good engineers. Engineers are problem solvers. They possess good oral and written communication skills and have a good grasp of fundamental physical laws and mathematics. They apply physical and chemical laws and mathematics to design, develop, test, and supervise the manufacture of millions of products and services. They are good team players. They consider important factors such as efficiency, cost, reliability, and safety when designing products. Engineers are dedicated to lifelong learning and service to others.
- You should be familiar with the differences among various engineering disciplines. You now know that civil, electrical, and mechanical engineers represent a large percentage of the total number of engineers.

PROBLEMS

1. Observe your own surroundings. What are some of the engineering achievements that you couldn't do without today?
2. Using the Internet, find the appropriate organization for the following list of engineering disciplines. Depending on your personal interests, prepare a brief two-page report about the goals and missions of the organization you have selected.

   Bioengineering
   Ceramic engineering
   Chemical engineering
   Civil engineering.
   Computer engineering
   Electrical engineering
   Electronic engineering
   Environmental engineering
   Industrial engineering
   Manufacturing engineering
   Materials engineering
   Mechanical engineering

3. To increase public awareness about the importance of engineering and to promote engineering education and careers among the younger generation, prepare and give a 15-minute Web-based presentation at a mall in your town. You need to do some planning ahead of time and ask permission from the proper authorities.
4. If your introduction to engineering class has a term project, present your final work at a mall at the date set by your instructor. If the project has a competitive component, hold the design competition at the mall.
5. Prepare a 15-minute oral presentation about engineering and its various disciplines, and the next time you go home present it to the juniors in your high school. Ask your college engineering department and engineering organizations on your campus to provide engineering-related brochures to take along.
6. This is a class project. Prepare a Web site for engineering and its various branches. Elect a group leader, then divide up the tasks among yourselves. As you work on the project, take note of both the pleasures and problems that arise from working in a team environment. Write a brief report about your experiences regarding this project. What are your recommendations for others who may work on a similar project?
7. This is a team project. Prepare a Web-based presentation of the history and future of engineering. Collect pictures, short videos, graphs, and so on. Provide links to major engineering societies as well as to major research and development centers.

8. This is a class project. Each of you is to ask his or her parents to think back to when they graduated from high school or college and to create a list of products and services that are available in their everyday lives now that were not available to them then. Ask them if they ever imagined that these products and services would be available today. To get your parents started, here are few examples: cellular phones, ATM cards, personal computers, airbags in cars, price scanners at the supermarket, E-Z passes for tolls, and so on. Ask your parents to explain how these products have made their lives better (or worse).
9. This is a class project. Each of you is to compile a list of products and services that are not available now that you think will be readily available in the next 50 years. Compile a complete list and present it to the class. You can post your findings on problems 8 and 9 to my Web page or send them to me via e-mail and I will post them so that the rest of the country can look at your results. Or if you send me the URL address, I will establish links to your sites. You may e-mail me at **saeed.moaveni@mnsu.edu**; the URL for this text is **http://info.brookscole.com/Moaveni**.
10. Visit this book's companion Web site to obtain information about the number of engineers employed by specific area and their mean salaries in recent years. Present your findings to your instructor.
11. If you are planning to study chemical engineering, investigate what is meant by each of the following terms: *polymers*, *plastics*, *thermoplastics*, and *thermosetting*. Give at least ten examples of plastic products that are consumed every day. Write a brief report explaining your findings.
12. If you are planning to study electrical engineering, investigate how electricity is generated and distributed. Write a brief report explaining your findings.
13. Electric motors are found in many appliances and devices around your home. Identify at least ten products at home that use electric motors.
14. Identify at least 20 different materials that are used in various products at home.
15. If you are planning to study civil engineering, investigate what is meant by *dead load*, *live load*, *impact load*, *wind load*, and *snow load* in the design of structures. Write a brief memo to your instructor discussing your findings.
16. This is a group project. As you can see from our discussion of the engineering profession in this chapter, people rely quite heavily on engineers to provide them with safe and reliable goods and services. Moreover, you realize that there is no room for mistakes or dishonesty in engineering! Mistakes made by engineers could cost not only money but also more importantly lives. Think about the following: An incompetent and unethical surgeon could cause the death of at most one person at one time, whereas an incompetent and unethical engineer could cause the deaths of hundreds of people at one time. If in order to save money an unethical engineer designs a bridge or a part for an airplane that does not meet the safety requirements, hundreds of peoples' lives are at risk!

    Visit the Web site of the National Society of Professional Engineers and research engineering ethics. Discuss why engineering ethics is so important, and explain why engineers are expected to practice engineering using the highest standards of honesty and integrity. Give examples of engineering codes of ethics. Write a brief report to your instructor explaining your findings.

CHAPTER

3 Engineering Problem Solving and Presentation

A group of engineers preparing for the test flight of the Pathfinder, a solar-powered, remote-piloted aircraft. The Pathfinder was constructed of composite materials, plastics, and foam to create a lightweight aircraft. The aircraft has a wingspan of 30 m and a gross mass of 221 kg. It is capable of moving at a speed of 7 m/s and is powered by a solar array that runs 6 electric motors which turn the propellers. Among other activities, NASA is interested in using the aircraft to study the upper atmosphere without distrubing it.

In this chapter, we will explain fundamental engineering dimensions, such as length and time, and their units, such as meter and second, and their role in engineering analysis and design. Moreover, we will explain what is meant by an engineering system and an engineering component. We will also explain the basic steps involved in the solution of engineering problems: What is the problem? What is to be determined? Can the analysis be simplified by making appropriate assumptions? What are the physical laws and principles that govern and predict the behavior of the given problem? We will also emphasize the fact that you must always find your own ways to verify your solution to a problem. There are no answer books outside the classroom.

We will also explain why it is very important for engineers to know how to communicate well with others both orally and in a written form. We will discuss various ways of presenting your analysis of a problem and the findings by properly presenting your engineering work. ■

3.1 ENGINEERING PROBLEMS AND FUNDAMENTAL DIMENSIONS

The evolution of the human intellect has taken shape over a period of thousands of years. Men and women all over the world observed and learned from their surroundings. They used the knowledge gained from their observations of nature to design, develop, test, and fabricate tools, shelter, weapons, water transportation, and means to cultivate and produce more food. Moreover, they realized that they needed only a few physical quantities to fully describe natural events and their surroundings. For example, the length dimension was needed to describe how tall or how long or how wide something was. They also learned that some things are heavier than other things, so there was a need for another physical quantity to describe that observation: the concept of mass and weight. Early humans did not fully understand the concept of gravity; consequently, the correct distinction between mass and weight was made later.

Time was another physical dimension that humans needed to understand in order to be able to explain their surroundings and to be able to answer questions such as: "How old are you?" "How long does it take to go from here to there?" "How long does it take to cook this food over fire?" The response to these questions in those early days may have been something like this: "I am many many moons old," or "It takes a couple of moons to go from our village to the other village on the other side of mountain." Moreover, to describe how cold or hot something was, humans needed yet another physical quantity, or physical dimension, that we now refer to as temperature.

By now, you understand why we need to formally define physical variables. The other important fact you need to realize is that early humans needed not only physical dimensions to describe their surroundings but also some way to scale or divide these physical dimensions. This realization led to the concept of units. For example, time is considered a physical dimension, but it can be divided into both small and large

portions, such as seconds, minutes, hours, days, years, decades, centuries, millennia, and so on. Today, when someone asks you how old you are, you reply by saying, "I am 19 years old." You don't say that you are approximately 170,000 hours old or 612,000,000 seconds old, even though these statements may very well be true! Or to describe the distance between two cities, we may say that they are 2000 kilometers apart; we don't say the cities are 2,000,000,000 millimeters apart. The point of these examples is that we use appropriate divisions of physical dimensions to keep numbers manageable. We have learned to create an appropriate scale for these fundamental dimensions and divide them properly so that we can describe particular events, the size of an object, the thermal state of an object, or its interaction with the surroundings correctly, and do so without much difficulty.

Today, based on what we know about our physical world, we need seven ***fundamental*** or ***base dimensions*** to correctly express what we know of the natural world. They are ***length, mass, time, temperature, electric current, amount of substance***, and ***luminous intensity***. With the help of these base dimensions we can derive all other necessary physical quantities that describe how nature works.

Let us now turn our attention to the International System of Units that is employed throughout the world. The origin of the present day International System of Units (SI) can be traced back to 1799 with meter and kilogram as the first two base units. By promoting the use of the second as a base unit of time in 1832, Carl Friedrich Gauss (1777–1855), an important figure in mathematics and physics, including magnetism and astronomy, had a great impact in many areas of science and engineering. It was not until 1946 that the proposal for the ampere as a base unit for electric current was approved by the General Conference on Weights and Measures (CGPM). In 1954, CGPM included ampere, kelvin, and candela as base units. The mole was added as a base unit by the 14th CGPM in 1971. A list of SI base (fundamental) units is given in Table 3.1.

TABLE 3.1 A list of SI Base (Fundamental) Units

Physical Quantity		Name of SI Base Unit	SI Symbol
Length	 1.6 m–2.0 m Range of height for most adults	Meter	m
Mass	 50 kg–120 kg Range of mass for most adults	Kilogram	kg
Time	 Fastest person can run 100 meters in approximately 10 seconds	Second	S
Electric current	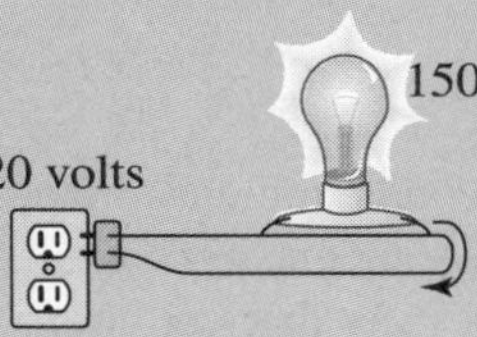 150 watts 120 volts 1.25 amps	Ampere	A
Thermodynamic temperature	Ice water: 273 kelvin  Comfortable room temperature: 295 kelvin	Kelvin	K
Amount of substance	Uranium 238 ← One of the heaviest atoms known Gold 197 Silver 108 Copper 64 Calcium 40 Aluminum 27 Carbon 12 ← Common Carbon is used as a standard Helium 4 Hydrogen 1 ← Lightest atom	Mole	mol
Luminous intensity	 A candle has luminous intensity of approximately 1 candela	Candela	cd

Listed below are formal definitions of base units as provided by the Bureau International des Poids et Mesures.

The ***meter*** is the length of the path traveled by light in a vacuum during a time interval of 1/299,792,458 of a second.

The ***kilogram*** is the unit of mass; it is equal to the mass of the international prototype of the kilogram.

The ***second*** is the duration of 9,192,631,770 periods of the radiation corresponding to the transition between the two hyperfine levels of the ground state of the cesium 133 atom.

The ***ampere*** is that constant current which, if maintained in two straight parallel conductors of infinite length, of negligible circular cross section, and placed 1 meter apart in a vacuum, would produce between these conductors a force equal to 2×10^{-7} newton per meter of length.

The ***kelvin***, a unit of thermodynamic temperature, is the fraction 1/273.16 of the thermodynamic temperature of the triple point of water.

The ***mole*** is the amount of substance of a system that contains as many elementary entities as there are atoms in 0.012 kilogram of carbon 12. When the mole is used, the elementary entities must be specified and may be atoms, molecules, ions, electrons, other particles, or specified groups of such particles.

The ***candela*** is the luminous intensity, in a given direction, of a source that emits monochromatic radiation of frequency 540×10^{12} hertz and that has a radiant intensity in that direction of 1/683 watt per steradian.

You need not memorize the definitions of these units. From your everyday life experiences you have a pretty good idea about some of them. For example, you know how short a time period a second is, or how long a period a year is. However, you may need to develop a "feel" for some of the other base units. For example, How long is a meter? How tall are you? Under 2 meters or perhaps above 5 meters? Most adult people's height is approximately between 1.5 meters and 2 meters. There are exceptions of course. What is your mass in kilograms? Developing a "feel" for units will make you a better engineer. For example, assume you are designing and sizing a new type of hand-held tool, and based on your stress calculation, you arrive at an average thickness of 1 meter. Having a "feel" for these units, you will be alarmed by the value of the thickness and realize that somewhere in your calculations you must have made a mistake. We will discuss in detail the role of the base dimensions and other derived units in the upcoming chapters in this book.

The CGPM in 1960 adapted the first series of prefixes and symbols of decimal multiples of SI units. Over the years, the list has been extended to include those listed in Table 3.2. SI is the most common system of units used in the world. However, in the United States, United States Customary Units (USCS) is the system of units currently in use. Efforts have been under way to convert from USCS to SI usage. Soon, every country in the world, including the United States, will use the International System of Units.

The units for other physical quantities used in engineering can be derived from the base units. For example, the unit for force is the newton. It is derived from Newton's second law of motion. One newton is defined as a magnitude of a force that when applied to 1 kilogram of mass, will accelerate the mass at a rate of 1 meter per second

TABLE 3.2 The List of Decimal Multiples and Prefixes Used with SI Base Units

Multiplication Factors	Prefix	SI Symbol
$1,000,000,000,000,000,000,000,000 = 10^{24}$	yotta	Y
$1,000,000,000,000,000,000,000 = 10^{21}$	zetta	Z
$1,000,000,000,000,000,000 = 10^{18}$	exa	E
$1,000,000,000,000,000 = 10^{15}$	peta	P
$1,000,000,000,000 = 10^{12}$	tera	T
$1,000,000,000 = 10^{9}$	giga	G
$1,000,000 = 10^{6}$	mega	M
$1000 = 10^{3}$	kilo	k
$100 = 10^{2}$	hecto	h
$10 = 10^{1}$	deka	da
$0.1 = 10^{-1}$	deci	d
$0.01 = 10^{-2}$	centi	c
$0.001 = 10^{-3}$	milli	m
$0.000,001 = 10^{-6}$	micro	μ
$0.000,000,001 = 10^{-9}$	nano	n
$0.000,000,000,001 = 10^{-12}$	pico	p
$0.000,000,000,000,001 = 10^{-15}$	femto	f
$0.000,000,000,000,000,001 = 10^{-18}$	atto	a
$0.000,000,000,000,000,000,001 = 10^{-21}$	zepto	z
$0.000,000,000,000,000,000,000,001 = 10^{-24}$	yocto	y

squared (m/s^2). As we mentioned earlier, in the United States, most engineers still use U.S. Customary Units. The unit of length is a foot (ft), which is equal to 0.3048 meter; the unit of mass is pound mass (lb_m), which is equal to 0.453592 kg; and the unit of time is a second (s). In the U.S. Customary system, the unit of force is pound force (lb_f), and 1 lb_f is defined as the weight of an object having a mass of 1 lb_m at sea level and at latitude 45°, where acceleration due to gravity is 32.2 ft/s^2. One pound force is equal to 4.448 newton. Because the pound force is not formally defined using Newton's second law and instead is defined at a specific location, a correction factor must be used in engineering formulas when using U.S. Customary Units.

Examples of common SI units used by engineers are shown in Table 3.3. The physical quantities shown in Table 3.3. will be discussed in detail in the following chapters of this book. Starting in Chapter 4, we will discuss their physical meaning, their significance and relevance in engineering, and their use in engineering analysis. Examples of SI and U.S. customary units used in our everyday lives are shown in Tables 3.4 and 3.5, respectively.

Unit Conversion

Some of you may recall that not too long ago NASA lost a spacecraft called *Mars Climate Orbiter* because two groups of engineers working on the project neglected to communicate correctly their calculations with appropriate units. According to an internal review

TABLE 3.3 Examples of Derived Units in Engineering

Physical Quantity	Name of SI Unit	Symbol for SI Unit	Expression in Terms of Base Units
Acceleration			m/s^2
Angle	Radian	rad	
Angular acceleration			rad/s^2
Angular velocity			rad/s
Area			m^2
Density			kg/m^3
Energy, work, heat	Joule	J	N·m or $kg{\cdot}m^2/s^2$
Force	Newton	N	$kg{\cdot}m/s^2$
Frequency	Hertz	Hz	s^{-1}
Impulse			N·S or kg·m/s
Moment or torque			N·m or $kg{\cdot}m^2/s^2$
Momentum			kg·m/s
Power	Watt	W	J/s or N·m/s or $kg{\cdot}m^2/s^3$
Pressure, stress	Pascal	Pa	N/m^2 or $kg/m{\cdot}s^2$
Velocity			m/s
Volume			m^3
Electric charge	Coulomb	C	A·s
Electric potential	Volt	V	J/C or $m^2{\cdot}kg/(s^3{\cdot}A^2)$
Electric resistance	Ohm	Ω	V/A or $m^2{\cdot}kg/(s^3{\cdot}A^2)$
Electric conductance	Siemens	S	1/Ω or $s^3{\cdot}A^2/(m^2{\cdot}kg)$
Electric capacitance	Farad	F	C/V or $s^4{\cdot}A^2/(m^2{\cdot}kg)$
Magnetic flux density	Tesla	T	$V{\cdot}s/m^2$ or $kg/(s^2{\cdot}A)$
Magnetic flux	Weber	Wb	V·s or $m^2{\cdot}kg/(s^2{\cdot}A)$
Inductance	Henry	H	V·s/A
Absorbed dose of radiation	Gray	Gy	J/kg or m^2/s^2

TABLE 3.4 Examples of SI Units in Everyday Use

Examples of SI Unit Usage	SI Units Used
Camera film	35 mm
Medication dose such as pills	100 mg, 250 mg, or 500 mg
Sports: swimming	100 m breaststroke or butterfly stroke
running	100 m, 200 m, 400 m, 5000 m, and so on
Automobile engine capacity	2.2 L (liter), 3.8 L, and so on
Light bulbs	60 W, 100 W, or 150 W
Electric consumption	kWh
Radio broadcasting signal frequencies	88–108 MHz (FM broadcast band)
	0.54–1.6 MHz (AM broadcast band)
Police, fire	153–159 MHz
Global positioning system signals	1575.42 MHz and 1227.60 MHz

TABLE 3.5 Examples of U.S. Customary Units in Everyday Use

Examples of U.S. Customary Unit Usage	U.S. Customary Units Used
Fuel tank capacity	20 gallons or 2.67 ft^3 (1 ft^3 = 7.48 gallons)
Sports (length of a football field)	100 yd or 300 ft
Power capacity of an automobile	150 hp or 82500 lb·ft/s (1 hp = 550 lb·ft/s)
Distance between two cities	100 miles (1 mile = 5280 ft)

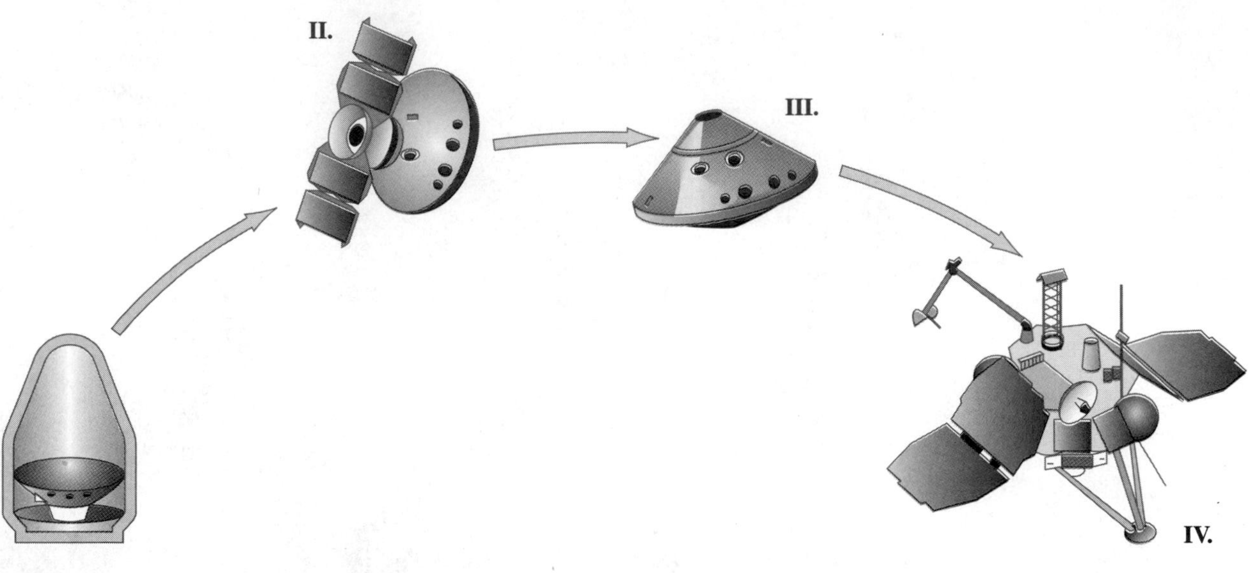

I. Launch
- Delta II 7425
- Launched January 3, 1999
- Launch mass: 574 kilograms

II. Cruise
- Thruster attitude control
- Four trajectory-correction maneuvers (TCM); site-adjustment maneuver September 1, 1999; contingency 5th TCM at entry –24 hours
- Eleven month cruise
- Near-simultaneous tracking with Mars Climate Orbiter or Mars Global Surveyor during approach

III. Entry, Descent, Landing
- Arrival: December 3, 1999
- Jettison cruise stage; microprobes separate from cruise stage
- Hypersonic entry
- Parachute descent; propulsive landing
- Descent imaging of landing site

IV. Landed Operations
- Lands in Martian spring at 76 degrees South latitude, 195 degrees West longitude (76°S, 195°W)
- 90-day landed mission
- Meteorology, imaging, soil analysis, trenching
- Data relay via Mars Climate Orbiter, Mars Global Surveyor, or direct-to-Earth high-gain antenna

Figure 3.1 Mars polar lander mission overview
Source: Courtesy of NASA.

conducted by NASA's Jet propulsion Laboratory, "a failure to recognize and correct an error in a transfer of information between the *Mars Climate Orbiter* spacecraft team in Colorado and the mission navigation team in California led to the loss of the spacecraft." The peer review findings indicated that one team used English units (e.g., foot and pound) while the other used SI units (e.g., meter and kilogram) for a key spacecraft operation. According to NASA, the information exchanged between the teams was critical to the maneuvers required to place the spacecraft in the proper Mars orbit. An overview of the Mars polar lander mission is given in Figure 3.1.

As you can see, as engineering students, and later as practicing engineers, when performing analysis, you will find a need to convert from one system of units to another. It is very important for you at this stage in your education to learn to convert information from one system of units to another correctly. It is also important for you to understand and to remember to show all your calculations with proper units. This point cannot be emphasized enough! **Always show the appropriate units that go with your calculations.** Example 3.1 shows the steps that you need to take to convert from one system of units to another.

EXAMPLE 3.1

A person who is 6 feet and 1 inch tall and weighs 185 pound force (lb_f) is driving a car at speed of 65 miles per hour over a distance of 25 miles. The outside air temperature is 80°F and has a density of 0.0735 pound mass per cubic foot (lb_m/ft^3). Convert all of the values given in this example from U.S. Customary Units to SI units.

Person's height, H

$$H = \left(6 \text{ ft} + (1 \text{ in.})\left(\frac{1 \text{ ft}}{12 \text{ in.}}\right)\right)\left(\frac{0.3048 \text{ m}}{1 \text{ ft}}\right) = 1.854 \text{ m}$$

or

$$H = (1.854 \text{ m})\left(\frac{100 \text{ cm}}{1 \text{ m}}\right) = 185.4 \text{ cm}$$

Person's weight, W

$$W = (185 \text{ lb}_f)\left(\frac{4.448 \text{ N}}{1 \text{ lb}_f}\right) = 822.88 \text{ N}$$

Speed of the car, S

$$S = \left(65 \frac{\text{miles}}{\text{h}}\right)\left(\frac{5280 \text{ ft}}{1 \text{ mile}}\right)\left(\frac{0.3048 \text{ m}}{1 \text{ ft}}\right) = 104{,}607 \text{ m/h} = 104.607 \text{ km/h}$$

or

$$S = \left(104{,}607 \frac{\text{m}}{\text{h}}\right)\left(\frac{1 \text{ h}}{3600 \text{ s}}\right) = 29.057 \text{ m/s}$$

Distance traveled, D

$$D = (25 \text{ miles})\left(\frac{5280 \text{ ft}}{1 \text{ mile}}\right)\left(\frac{0.3048 \text{ m}}{1 \text{ ft}}\right)\left(\frac{1 \text{ km}}{1000 \text{ m}}\right) = 40.233 \text{ km}$$

Temperature of air, T

$$T(°\text{C}) = \frac{5}{9}[T(°\text{F}) - 32]$$

$$T(°\text{C}) = \frac{5}{9}(80 - 32) = 26.7°\text{C}$$

Density of air, ρ

$$\rho = \left(0.0735\,\frac{\text{lb}_\text{m}}{\text{ft}^3}\right)\left(\frac{0.453\text{ kg}}{1\text{ lb}_\text{m}}\right)\left(\frac{1\text{ ft}}{0.3048\text{ m}}\right)^3 = 1.176\text{ kg/m}^3$$

Another important concept that you need to understand completely is that all formulas used in engineering analysis must be dimensionally homogeneous. What do we mean by "dimensionally homogeneous"? Can you, say, add someone's height who is 6 feet tall to his weight of 185 lb_f and his body temperature of 98°F? Of course not! What would be the result of such a calculation? Therefore, if we were to use the formula $L = a + b + c$, in which the variable L on the left-hand side of the equation has a dimension of length, then the variables a, b, and c on the right-hand side of equation should also have dimensions of length. Otherwise, if variables a, b, and c had dimensions such as length, weight, and temperature, respectively, the given formula would be inhomogeneous, which would be like adding someone's height to his weight and body temperature! Example 3.2 shows how to check for homogeneity of dimensions in an engineering formula.

EXAMPLE 3.2

(a) When a constant load is applied to a bar of constant cross section as shown, the amount by which the end of the bar will deflect can be determined from the following relationship:

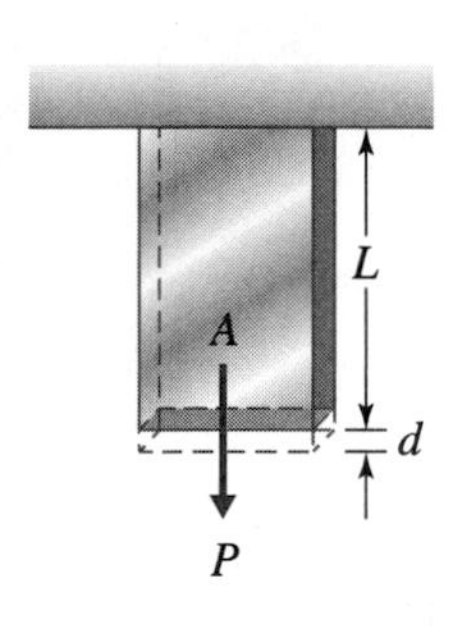

$$d = \frac{PL}{AE} \tag{3.1}$$

where

d = end deflection of the bar in meter (m)

P = applied load in newton (N)

A = cross-sectional area of the bar in m^2

E = modulus of elasticity of the material

What are the units for modulus of elasticity?

For Equation (3.1) to be dimensionally homogeneous, the units on the left-hand side of the equation must equal the units on the right-hand side. This equality requires the modulus of elasticity to have the units of N/m^2, as follows:

$$d = \frac{PL}{AE} \quad \Rightarrow \quad \text{m} = \frac{(\text{N})(\text{m})}{\text{m}^2 E}$$

Solving for the units of E leads to N/m^2, newton per squared meter, or force per unit area.

(b) The heat transfer rate through a solid material is governed by Fourier's law:

$$q = kA\frac{T_1 - T_2}{L} \tag{3.2}$$

where

q = heat transfer rate

k = thermal conductivity of the solid material in watts per meter degree Celsius, W/m·°C

A = area in m^2

$T_1 - T_2$ = temperature difference, °C

L = thickness of the material, m

What is the appropriate unit for the heat transfer rate q?

Substituting for the units of k, A, T_1, T_2, and L, we have

$$q = kA\frac{T_1 - T_2}{L} = \left(\frac{\text{W}}{\text{m}\cdot{}^\circ\text{C}}\right)(\text{m}^2)\left(\frac{{}^\circ\text{C}}{\text{m}}\right) = \text{W}$$

From this you can see that the appropriate SI unit for the heat transfer rate is the watt.

3.2 ENGINEERING COMPONENTS AND SYSTEMS

Every engineered product is made of components. Let us start with a simple example to demonstrate what we mean by an engineering system and its components. Most of us own a winter coat, which can be looked at as a system. First, the coat serves a purpose. Its primary function is to offer additional insulation for our bodies so that our body heat does not escape as quickly and as freely as it would without protective covering. The coat may be divided into smaller components: the fabric comprising the main body of the coat, insulating material, a liner, threads, zipper(s), and buttons. Moreover, each component may be further subdivided into smaller components. For example, the main body of the jacket may be divided into sleeves, collar, pockets, the chest section, and the back section (see Figure 3.2). Each component serves a purpose: The pockets were designed to hold things, the sleeves cover our arms, and so on. The main function of the zipper is to allow us to open and close the front of the jacket freely. It too consists of smaller components. Think once more about the overall purpose of the coat and the function of each component. A well-designed coat not only looks appealing to the eyes but also has functional pockets and keeps us warm during the winter.

Engineering systems are similar to a winter coat. Any given engineered product or engineering system can be divided into smaller, manageable subsystems, and each subsystem can be further divided into smaller and smaller components. The components of a well-designed engineering system should function and fit well together so that the primary purpose of the product is attained. Let us consider another common example. The primary function of a car is to move us from one place to another in a reasonable amount of time. The car must provide a comfortable area for us to sit within. Furthermore, it must shelter us and provide some protection from the outside elements, such as harsh weather and harmful objects outside. The automobile consists of thousands of parts. When viewed

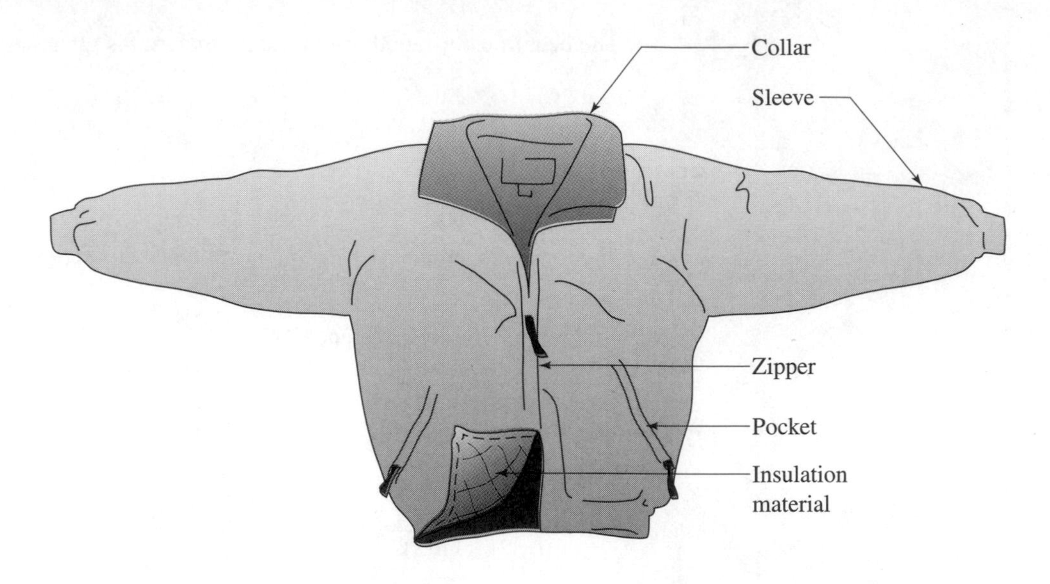

Figure 3.2
A simple system and its components

in its entirety, it is a complicated system. Thousands of engineers have contributed to the design, development, testing, and supervision of the manufacture of an automobile. These include electrical engineers, electronic engineers, combustion engineers, materials engineers, aerodynamics experts, vibration and control experts, air-conditioning specialists, manufacturing engineers, and industrial engineers.

When viewed as a system, the car may be divided into major subsystems or units, such as electrical, body, chassis, power train, and air conditioning (see Figure 3.3). Each major component can be further subdivided into smaller subsystems and their components. For example, the main body of the car consists of doors, hinges, locks, windows, and so on. The windows are controlled by mechanisms that are activated by hand or motors. And the electrical system of a car consists of a battery, a starter, an alternator, wiring, lights, switches, radio, microprocessors, and so on. The car's air-conditioning system consists of components such as a fan, ducts, diffusers, compressor, evaporator, and condenser. Again, each of these components can be further divided into yet smaller components. For example, the fan consists of an impeller, a motor, and a casing. From these examples it should be clear that in order to understand a system, we must first fully understand the role and function of its components.

During the next four or five years you will take a number of engineering classes that will focus on specific topics. You may take a statics class, which deals with the equilibrium of objects at rest. You will learn about the role of external forces, internal forces, and reaction forces and their interactions. Later, you will learn the underlying concepts and equilibrium conditions for designing parts. You will also learn about other physical laws, principles, mathematics, and correlations that will allow you to analyze, design, develop, and test various components that make up a system. It is imperative that during the next four or five years you fully understand these laws and principles so that you can design components that fit well together and work in harmony to fulfill the ultimate goal of a given system. Thus, you can see the importance of learning the fundamentals. If you don't, you are likely to design poor components that, when put together, will result in an even poorer system!

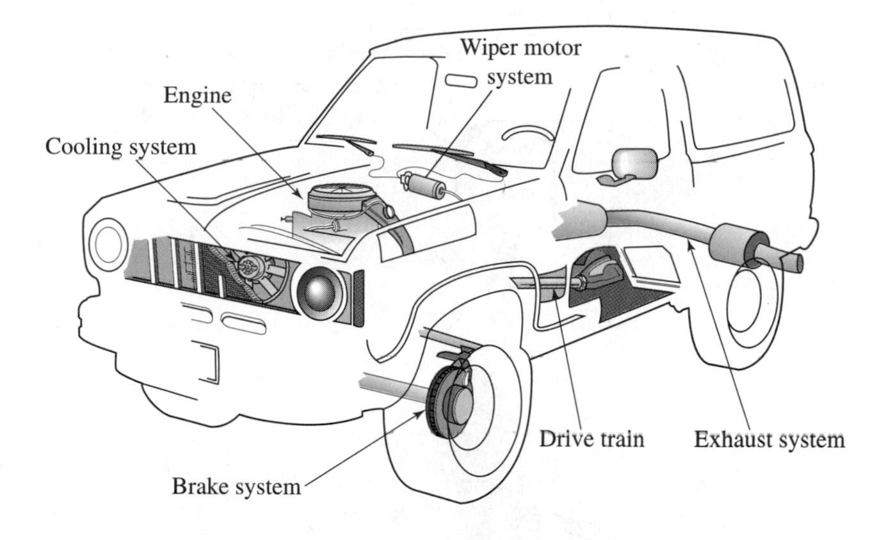

Figure 3.3
An engineering system and its main components

3.3 PHYSICAL LAWS AND OBSERVATIONS IN ENGINEERING

As stated earlier, engineers apply physical and chemical laws and principles along with mathematics to design, develop, test, and mass-produce products and services that we use in our everyday lives. The key concepts that you need to keep in the back of your mind are the physical and chemical laws and principles and mathematics.

Having had a high school education, you have a pretty good idea of what we mean by mathematics. But what do we mean by physical laws? Well, the universe, including the earth that we live on, was created a certain way. There are differing opinions as to the origin of the universe. Was it put together by God, or did it start with a big bang? We won't get into that discussion here. But we have learned through observation and by the collective effort of those before us that things work a certain way in nature. For example, if you let go of something that you are holding in your hand, it will fall to the ground. That is an observation that we all agree upon. We can use words to explain our observations or use another language, such as mathematics, to express our findings. Sir Isaac Newton (1642–1727) formulated that observation into a useful mathematical expression that we know as the universal law of gravitational attraction.

An important point to remember is that the physical laws are based on observations. Moreover, we use mathematics and basic physical quantities to express our observations in the form of a law. Even so, to this day we may not fully understand why nature works the way it does. We just know it works. There are physicists who spend their lives trying to understand on a more fundamental basis why nature behaves the way it does. Some engineers may focus on investigating the fundamentals, but most engineers use fundamental laws to design things.

As another example, when you place some hot object in contact with a cold object, the hot object cools down while the cold object warms up until they both reach an equilibrium temperature somewhere between the two initial temperatures. From your everyday

experience you know that the cold object does not get colder while the hot object gets hotter! Why is that? Well, it is just the way things work in nature! The second law of thermodynamics, which is based on this observation, simply states that heat flows spontaneously from a high-temperature region to a low-temperature region. The object with the higher temperature (more energetic) transfers some of its energy to the low-temperature (less-energetic) object. When you put some ice cubes in a glass of warm soda, the soda cools down while the ice warms up and eventually melts away. You may call this "sharing resources." Unfortunately, we as people do not follow this law closely when it comes to social issues.

To better understand the second law of thermodynamics, consider another example. Some of you may have young children or young brothers and sisters. If you placed the child with some toys in a room that is tidied up and orderly, let the child play with the toys for a while, then came back in a few minutes, you would find toys scattered all over the room in a disorderly way. Why won't you find toys put away nicely? Well, that is because things work spontaneously in a certain direction in nature. These two examples demonstrate the second law of thermodynamics. Things in nature work in a certain direction by themselves.

Engineers are also good bookkeepers. What do we mean by this? Any of us with a checking account knows the importance of accurate record keeping. In order to avoid problems, most of us keep track of the transactions in terms of payments (debits) and deposits (credits). Good bookkeepers can tell you instantly what the balance in their account is. They know they need to add to the recorded balance whenever they deposit some money and subtract from the balance with every withdrawal from the account. Engineers, like everyone else, need to keep track of their accounts. Moreover, similar to bookkeeping a checking account, engineers keep track of (bookkeep) physical quantities when analyzing an engineering problem.

To better understand this concept, consider the air inside a car tire. If there are no leaks, the mass of air inside the tire remains constant. This is a statement expressing

conservation of mass, which is based on our observations. If the tire develops a leak, then you know from your experience that the amount of air within the tire will decrease until you have a flat tire. Furthermore, you know the air that escaped from the tire was not destroyed; it simply became part of the surrounding atmosphere. The conservation of mass statement is similar to a bookkeeping method that allows us to account for what happens to the mass in an engineering problem. What happens if we try to pump some air into the tire that has a hole? Well, it all depends on the size of the hole and the pressure and flow rate of the pressurized air available to us. If the hole is small, we may be able to inflate the tire temporarily. Or the hole may be so large that the same amount of air that we put into the tire comes right back out. To completely describe all possible situations pertaining to this tire problem, we can express the conservation of mass as *the rate at which air enters the tire minus the rate at which the air leaves the tire should be equal to the rate of accumulation or depletion of air inside the tire*. Of course, we will use the physical quantity mass along with mathematics to express this statement. We will discuss the conservation of mass in more detail in Chapter 6.

There are other physical laws based on our observations that we use to analyze engineering problems. ***Conservation of energy*** is another good example. It is again similar to a bookkeeping method that allows us to keep track of various forms of energy and how they may change from one form to another. We will spend more time discussing the conservation of energy in Chapter 10.

Another important law that all of you have heard about is ***Newton's second law of motion***. If you place a book on a smooth table and push it hard enough, it will move. This is simply the way things work. Newton observed this and formulated his observation into what we call Newton's second law of motion. This is not to say that other people had not made this simple observation before, but Newton took it a few steps further. He noticed that as he increased the mass of the object being pushed, while keeping the magnitude

of the force constant (pushing with the same effort), the object did not move as quickly. Moreover, he noticed that there was a direct relationship between the push, the mass of the object being pushed, and the acceleration of the object. He also noticed that there was a direct relationship between the direction of the force and the direction of the acceleration. Newton expressed his observations using mathematics, but simply expressed, this law states that unbalanced force is equal to mass times acceleration. You will have the opportunity to take physics classes that will allow you to study and explore Newton's second law of motion further. Some of you may even take a dynamics class that will focus in greater detail on motion and forces and their relationship. Don't lose sight of the main idea: Physical laws are based on observations.

Another important idea to keep in mind is that a physical law may not fully describe all possible situations. Statements of physical laws have limitations because we may not fully understand how nature works, and thus we may fail to account for all variables that can affect the behavior of things within our natural world. Some natural laws are stated in a particular way to keep the mathematical expressions describing the observations simple. Often, we resort to experimental work dealing with specific engineering applications. For example, to understand better the aerodynamics of a car, we place it inside a wind tunnel to measure the drag force acting on the car. We may represent our experimental findings in the form of a chart or a correlation that can be used for design purposes over a predetermined range. The main difference between laws and other forms of experimental findings is that the laws represent the results of a much broader observation of nature, and almost everything that we know in our physical world obeys these laws. The engineering correlations, on the other hand, apply over a very limited and specific range of variables.

3.4 BASIC STEPS INVOLVED IN THE SOLUTION OF ENGINEERING PROBLEMS

There are four basic steps that must be followed when analyzing an engineering problem: (1) defining the problem, (2) simplifying the problem by assumptions and estimations, (3) performing the solution or analysis, and (4) verifying the results.

Step 1: Defining the Problem

Before you can obtain an appropriate solution to a problem, you must first thoroughly understand the problem itself. There are many questions that you need to ask before proceeding to determining a solution. What is it exactly that you want to analyze? What do you really *know* about the problem, or what are some of the things *known* about the problem? What are you looking for? What exactly are you trying to find a solution to?

Taking time to understand the problem completely at the beginning will save lots of time later and help to avoid a great deal of frustration. Once you understand the problem, you should be able to divide any given problem into two basic questions: What is known? and What is to be found?

Step 2: Simplifying the Problem

Before you can proceed with the analysis of the problem, you may first need to simplify it.

Assumptions and Estimations Once you have a good understanding of the problem, you should then ask yourself this question: Can I simplify the problem by making some reasonable and logical assumptions and yet obtain an appropriate solution? Understanding the physical laws and the fundamental concepts, as well as where and when to apply them and their limitations, will benefit you greatly in solving the problem. It is very important as you take different engineering classes in the next few years that you develop a good grasp of the fundamental concepts in each class that you take.

Step 3: Performing the Solution or Analysis

Once you have carefully studied the problem, you can proceed with obtaining an appropriate solution. You will begin by applying the physical laws and fundamental concepts that govern the behavior of engineering systems to solve the problem. Among the engineering tools in your toolbox you will find mathematical tools. It is always a good practice to set up the problem in *parametric* form, that is, in terms of the variables involved. You should wait until the very end to substitute for the given values. This approach will allow you to change the value of a given variable and see its influence on the final result.

Step 4: Verifying the Results

The final step of any engineering analysis should be the verification of results. Various sources of error can contribute to wrong results. Misunderstanding a given problem, making incorrect assumptions to simplify the problem, applying a physical law that does not truly fit the given problem, and incorporating inappropriate physical properties are common sources of error. Before you present your solution or the results to your instructor or, later in your career, to your manager, you need to learn to think about the calculated results. You need to ask yourself the following question: Do the results make sense? A good engineer must always find ways to check results. Ask yourself this additional question: What if I change one of the given parameters. How would that change the result? Then consider if the outcome seems reasonable. If you formulate the problem such that the final result is left in parametric form, then you can experiment by substituting different values for various parameters and look at the final result. In some engineering work, actual physical experiments must be carried out to verify one's findings. Starting today, get into the habit of asking yourself if your solution to a problem makes sense. Asking your instructor if you have come up with the right answer or checking the back of your textbook to match answers are not good approaches in the long run. You need to develop the means to check your results by asking yourself the appropriate questions. Remember, once you start working for hire, there are no answer books. You will not want to run to your boss to ask if you did the problem right!

3.5 COMMUNICATION SKILLS AND PRESENTATION OF ENGINEERING WORK

As an engineering student, you need to develop good written and oral communication skills. During the next four or five years, you will learn how to express your thoughts,

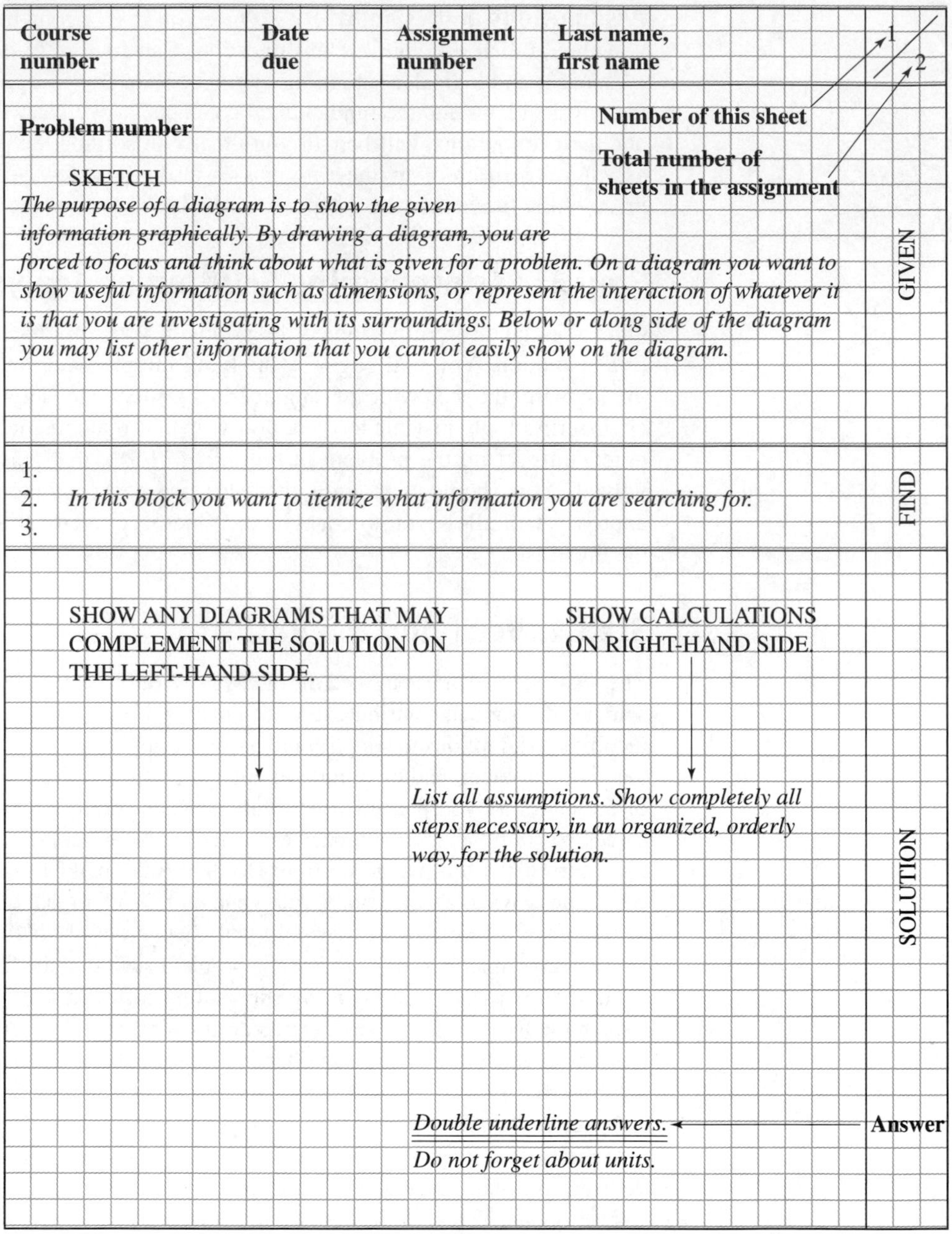

Course number	**Date due**	**Assignment number**	**Last name, first name**	1/2

Problem number

Number of this sheet

Total number of sheets in the assignment

SKETCH

The purpose of a diagram is to show the given information graphically. By drawing a diagram, you are forced to focus and think about what is given for a problem. On a diagram you want to show useful information such as dimensions, or represent the interaction of whatever it is that you are investigating with its surroundings. Below or along side of the diagram you may list other information that you cannot easily show on the diagram.

GIVEN

1.
2. *In this block you want to itemize what information you are searching for.*
3.

FIND

SHOW ANY DIAGRAMS THAT MAY COMPLEMENT THE SOLUTION ON THE LEFT-HAND SIDE.

SHOW CALCULATIONS ON RIGHT-HAND SIDE.

List all assumptions. Show completely all steps necessary, in an organized, orderly way, for the solution.

SOLUTION

Double underline answers. ← **Answer**

Do not forget about units.

Figure 3.4 An example of engineering problem presentation present a concept for a product or a service and an engineering analysis of a problem and its solution, or show your findings from experimental work. Moreover, you will learn how to communicate design ideas by means of engineering drawings or computer-aided modeling techniques. Starting right now, it is important to understand that the ability to communicate your solution to a problem is as important as the solution itself. You may spend months on a project, but if you cannot effectively communicate to others, the results of all your efforts may not be understood and appreciated. Most engineers are required to write reports. These reports might be lengthy, detailed, technical reports containing charts, graphs, and engineering drawings, or they may take the form of a brief memorandum or executive summary. Some of the more common forms of engineering communication are explained briefly next.

Homework Presentation

Engineering paper is specially formatted for use by engineers and engineering students. The paper has three cells on the top that may be used to convey such information as course number, assignment due date, and your name. A given problem may be divided into a "Given" section, a "Find" section, and a "Solution" section. It is a good practice to draw horizontal lines to separate the known information (Given section) from the information that is to be found (Find section) and the analysis (Solution section), as shown in Figure 3.4. Do not write anything on the back of the paper. The grid lines on the back provide scale and an outline for freehand sketches, tables, or plotting data by hand. The grid lines, which can be seen from the front of the paper, are there to assist you in drawing things or presenting tables and graphical information on the front of the page neatly. These grid lines also allow you to present a freehand engineering drawing with its dimensions. Your engineering assignments will usually consist of many problems, thus you will present your work on many sheets, which should be stapled together. Professors do not generally like loose papers, and some may even deduct points from your assignment's total score if the assignment sheets are not stapled together. The steps for presenting an engineering problem are demonstrated in Example 3.3. If you are presenting solutions to simple problems and you think you can show the complete solution to more than one problem on one page, then separate the two problems by a relatively thick line or a double line, whichever is more convenient for you.

EXAMPLE 3.3

Determine the mass of compressed air in a scuba diving tank, given the following information. The internal volume of the tank is 10 L and the absolute air pressure inside the tank is 20.8 MPa. The temperature of the air inside the tank is 20°C. Use the ideal gas law to analyze this problem. The ideal gas law is given by

$$PV = mRT$$

where

P = absolute pressure of the gas, Pa

V = volume of the gas, m^3

m = mass, kg

R = gas constant, $\frac{\text{J}}{\text{kg}\cdot\text{K}}$

T = absolute temperature, kelvin, K

The gas constant R for air is 287 J/kg·K. At this time, do not worry about understanding the ideal gas law. This law will be explained to you in detail in Chapter 8. The purpose of this example is to demonstrate how a solution to an engineering problem is presented. Make sure you understand and follow the steps shown in Figure 3.5.

Progress Report

Progress reports are means of communicating to others in an organization or to the sponsors of a project how much progress has been made and which of the main objectives of the project have been achieved to date. Based on the total time period required for a project, progress reports may be written for a period of a week, a month, several months, or a year. The format of the progress report may be dictated by a manager in an organization or by the project's sponsors.

Executive Summary

Executive summaries are means of communicating to people in top management positions, such as a vice president of a company, the findings of a detailed study or a proposal. The executive summary, as the name implies, must be brief and concise. It is generally no more than a few pages long. In the executive summary, references may be made to more comprehensive reports so that readers can obtain additional information if they so desire.

Short Memos

Short memos are yet another way of conveying information in a brief way to interested individuals. Generally, short memos are under two pages in length. A general format for a short memo follows. The header of the memo contains information such as the date, who the memo is from, to whom it is being sent, and a subject line. This is followed by the main body of the memo.

Date: May 3, 2001
From: Mr. John Doe
To: Members of Project X
Re: Budget Request

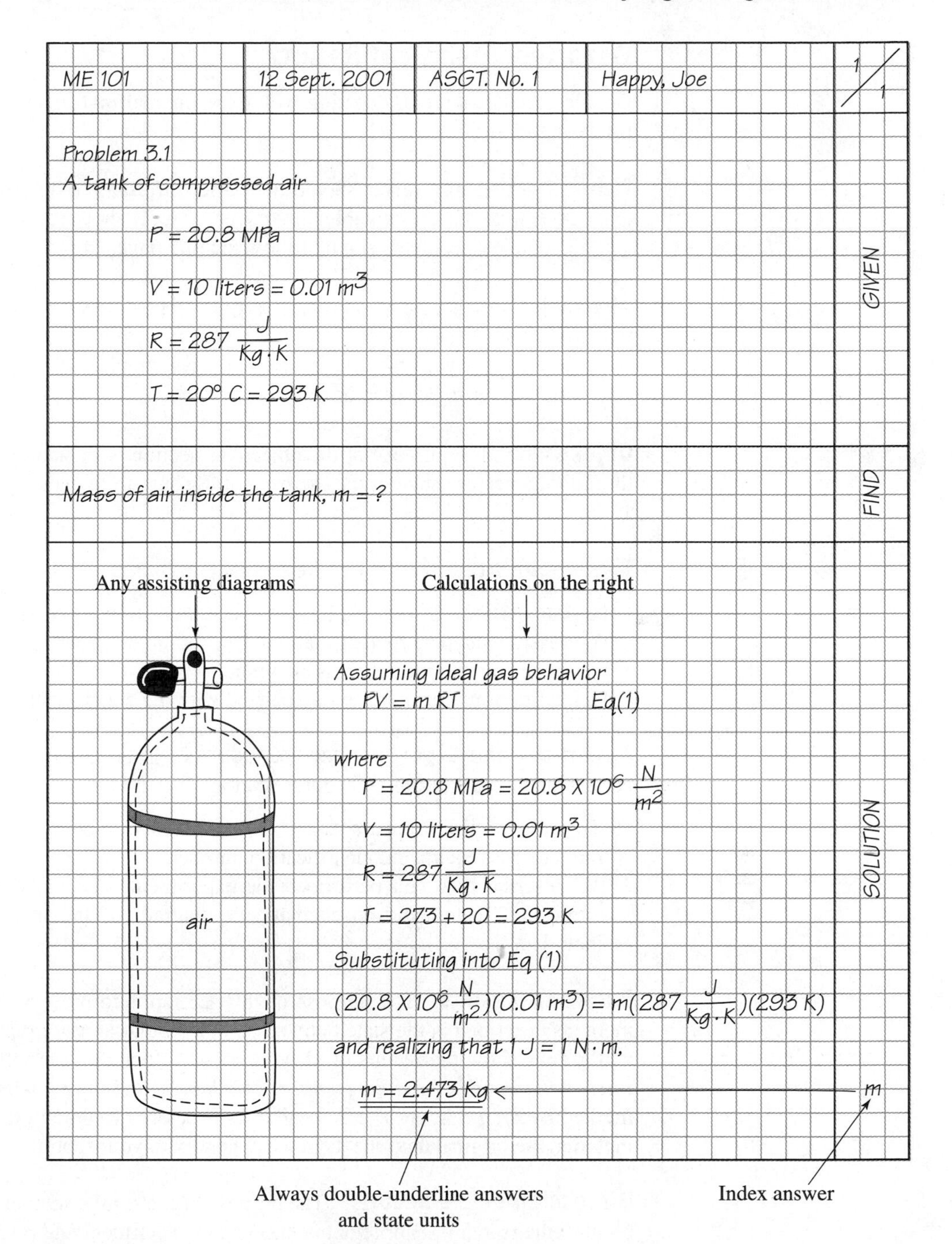

Figure 3.5 An example of engineering homework presentation

Detailed Technical Report

Detailed technical reports dealing with experimental investigations generally contain the following items:

Title The title of a report should be a brief informative description of the report contents. A sample of an acceptable title (cover) sheet is shown on the next page. If the report is long, a table of contents should follow the title page.

Abstract This is a very important part of a report because readers often read this part first to decide if they should read the report in detail. In the abstract, in complete but concise sentences you state the precise objective, emphasize significant findings, and present conclusions and/or recommendations.

Objectives The purpose of the objectives section is to state what is to be investigated through the performance of the experiment. Be sure to list your objectives explicitly (e.g., 1., 2., . . . , etc.)

Theory and Analysis There are several purposes of the theory and analysis section:

- To state pertinent principles, laws, and equations (equations should be numbered);
- To present analytical models that will be used in the experiment;
- To define any unfamiliar terms or symbols;
- To list important assumptions associated with the experimental design.

Apparatus and Experimental Procedures There are two main purposes of the apparatus and experimental procedures section:

1. To present a list of apparatus and instrumentation to be used, including the instrument ranges, least count, and identification numbers.
2. To describe how you performed the experiment. The procedure should be itemized (step 1., 2., etc.) and a schematic or diagram of the instrument setup should be included.

Data and Results The purpose of this section is to present the results of the experiment, as described in the stated objective, in a tabular and/or graphical form. These tables and graphs show the results of all your efforts. Include descriptive information such as *titles, column* or *row headings, units, axis labels*, and *data points* (data points should be marked by ⊙, ⊡, ⚠, etc.). It is sometimes necessary to note in this section that you have included the original data sheets in the appendix to your report.

Discussion of the Results The purpose of the results section is to emphasize and explain to the reader the important results of the experiment and point out their significance. When applicable, be sure to compare experimental results with theoretical calculations.

Conclusions and Recommendations The conclusions and recommendations section compares your objectives with your experimental results. Support your conclusions with appropriate reference materials. Be sure to state recommendations based on the conclusions.

All State University
Department of Mechanical Engineering

Course Title

Experiment No. ____________

Experiment Title ____

____

____

Date Experiment Completed ____________

Students' Names __________________

____

____

Appendix The appendix serves several purposes:

- To provide the reader with copies of all original data sheets, diagrams, and supplementary notes.
- To display sample calculations used in processing the data. The sample calculations should contain the following parts:

A title of the calculation;
A statement of mathematical equation;
Calculation using one sample of data.

References A list of references that have been numbered in the text must be included in the report. Use the following format examples:

For Books: Author, title, publisher, place of publication, date and page(s).

For Journal Articles: Author, title of article (enclosed in quotation marks), name of journal, volume number, issue number, and page(s).

Oral Communication and Presentation

We communicate orally to each other all the time. Informal communication is part of our everyday life. We may talk about sports, weather, what is happening around the world, or a homework assignment. Some people are better at expressing themselves than others are. Sometimes we say things that are misunderstood and the consequences could be unpleasant. When it comes to formal presentations, there are certain rules and strategies that you need to follow. Your oral presentation may show the results of all your efforts regarding a project that you may have spent months or a year to develop. If the listener cannot follow how a product was designed, or how the analysis was performed, then all your efforts become insignificant. It is very important that all information be conveyed in a manner easily understood by the listener.

The oral technical presentation in many ways is similar to a written one. You need to be well organized and have an outline of your presentation ready, similar to the format for a written report. It may be a good idea to write down what you are planning to present. Remember it is harder to erase or correct what you say after you have said it than to write it down on a piece of paper and correct it before you say it. You want to make every effort to ensure that what is said (or sent) is what is understood (or received) by the listener.

Rehearse your presentation before you deliver it to a live audience. You may want to ask a friend to listen and provide helpful suggestions about your style of presentation, delivery, content of the talk, and so on.

Present the information in a way that is easily understood by your audience. Avoid using terminology or phrases that may be unfamiliar to listeners. You should plan so that you won't overexplain key concepts and ideas, because those who are really interested in a specific area of your talk can always ask questions later.

Try to keep your talk to about half an hour or less because the attention span of most people is about 20 to 30 minutes. If you have to give a longer talk, then you may want to mix your presentation with some humor or tell some interesting related story to keep your audience's attention. Maintain eye contact with all of your audience, not just one or two people. Don't ever have your back to the audience for too long! Use good visual aids. Use presentation software such as Power Point to prepare your talk. When possible, incorporate charts, graphs, animated drawings, short videos, or a model. When available, to demonstrate concepts for new products, you can make use of prototyping technology and have a prototype of the product on hand for your presentation. You may also want to have copies of the outline, along with notes on the important concepts and findings, ready to hand out to interested audience members. In summary, be organized, be well prepared, get right to the point when giving an oral presentation, and consider the needs and expectation of your listeners.

SUMMARY

Now that you have reached this point in the text

- You should understand the importance of fundamental dimensions in engineering analysis. You should also understand what is meant by an engineering system and an engineering component. You should also realize that physical laws are based on observation and experimentation.
- You should know the basic steps involved in the solution of engineering problems: What is the problem? What is to be determined? Can the analysis be simplified by making appropriate assumptions? What are the physical laws and principles that govern and predict the behavior of the given problem?
- You should realize that you must always find your own ways to verify your solutions to a problem. There are no answer books outside the classroom.
- You should realize that it is very important for engineers to know how to communicate well with others both orally and in written form. You should also be familiar with various ways of presenting your analysis of a problem, and the findings, by properly presenting your engineering work. This includes knowing the major sections of a technical engineering report.

PROBLEMS

1. For the following systems, identify the major components, and briefly explain the function or the role of each component: a dress shirt, pants, a skirt, shoes, a bicycle, roller blades.
2. For the following systems, identify the major components:
   a. a household refrigerator
   b. a computer
   c. the human body
   d. a building
   e. a hot water heater
   f. a toaster
   g. an airplane
3. Investigate what observations the following laws describe:
   a. Fourier's law
   b. Darcy's law
   c. Newton's law of viscosity
   d. Newton's law of cooling
   e. Coulomb's law
   f. Ohm's law
   g. the ideal gas law
   h. Hooke's law
   i. the first law of thermodynamics
   j. Fick's law
   k. Faraday's law
4. Investigate the operation of various turbines. Write a brief report explaining the operation of steam turbines, hydraulic turbines, gas turbines, and wind turbines.
5. In a brief report, discuss why we need various modes of transportation. How did they evolve? Discuss the role of public transportation, water transportation,

highway transportation, railroad transportation, and air transportation.

6. Identify the major components of a computer, and briefly explain the function or the role of each component.
7. Electronic communication is becoming increasingly important. In your own words, identify the various situations under which you should write a letter, send an e-mail, make a telephone call, or talk to someone in person. Explain why one particular form of communication is preferable to the others available.
8. Convert the information given in the accompanying table from SI units to U.S. Customary Units. Refer to the conversion tables on the front and back end sheets of this text.

Convert from SI Units	To U.S. Customary Units
120 km/hr	miles/h and ft/s
1000 W	Btu/h and hp
100 m^3	ft^3
80 kg	lb_m
1000 kg/m^3	lb_m/ft^3
900 N	lb_f
100 kPa	$lb_f/in.^2$
9.81 m/s^2	ft/s^2

9. The angle of twist for a shaft subjected to twisting torque can be expressed by the following equation:

$$\phi = \frac{TL}{JG}$$

where

ϕ = the angle of twist in radians
T = applied torque in N·m
L = length of the shaft in meter (m)
J = shaft's polar moment of inertia (measure of resistance to twisting)
G = shear modulus of the material in N/m^2

What is the appropriate unit for J, if the preceding equation is to be homogeneous in units?

10. Which one of the following equations is dimensionally homogeneous? Show your proof.
    i. $F = ma$
    ii. $F = m\frac{V^2}{R}$
    iii. $F(t_2 - t_1) = m(V_2 - V_1)$
    iv. $F = mV$
    v. $F = m\frac{(V_2 - V_1)}{(t_2 - t_1)}$

where

F = force (N)
m = mass (kg)
a = acceleration (m/s^2)
V = velocity (m/s)
R = radius (m)
t = time (s)

CHAPTER

4 Length and Length-Related Parameters

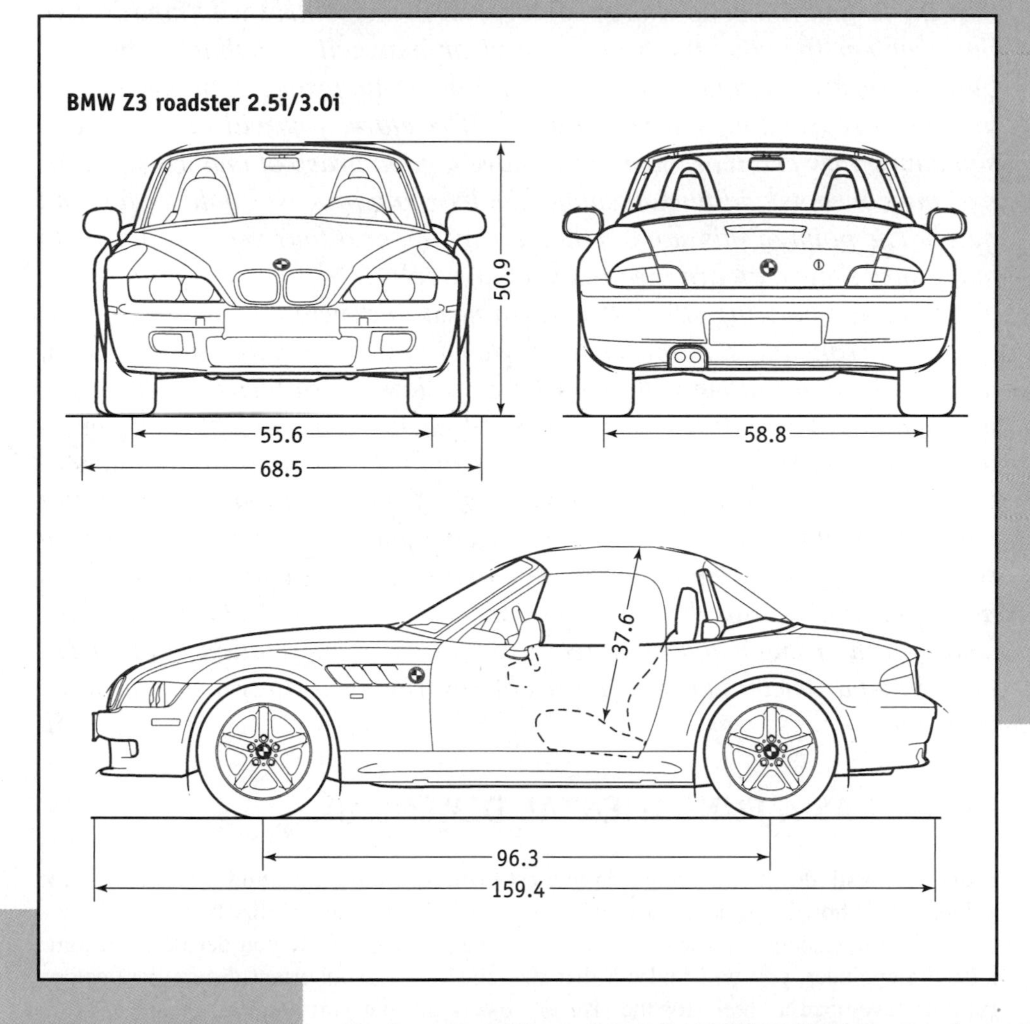

The important dimensions of a BMW Z3 Roadster are shown in the illustration. The fundamental dimension length and other length-related variables, such as area and volume (e.g., seating or trunk capacity), play important roles in engineering design. As a good engineer you will develop a "feel" for the relative magnitude of various length units, area units, and volume units. It is also important for you to know how to measure, how to caluclate, and how to approximate length, area, and volume.

Source: Z3 Roadster 2.5i/3.0i schematics reproduced with permission of BMW AG Munich.

When you become a practicing engineer, you will find out that you don't stop learning new things even after obtaining your engineering degree. For example, you may work on a project in which the noise of a machine is a concern, and you may be asked to come up with ways to reduce the level of noise. It may be the case that during the four or five years of your engineering education, you did not take a class in noise control. Considering your lack of understanding and background in noise reduction, you may first try to find someone who specializes in noise control who could solve the problem for you. But your supervisor may tell you that because of budget constraints and because this is a one-time project, you must come up with a reasonable solution yourself. Therefore, you will have to learn something new and learn fast. If you have a good grasp of underlying engineering concepts and fundamentals, the learning process could be fun and quick. The point of this story is that during the next four years you need to make sure that you learn the fundamentals well.

As an engineering student you also need to develop a keen awareness of your surroundings. In this chapter, we will investigate the role of length, area, and volume along with other length-related variables in engineering applications. You will learn how these physical variables affect engineering design decisions. The topics introduced in this chapter are fundamental in content, so developing a good grasp of them will make you a better engineer. You may see some of the concepts and ideas introduced here in some other form in the engineering classes that you will take later. The main purpose of introducing these concepts here is to help you become aware of their importance and learn to look for their relation to other engineering parameters in your future classes when you study a specific topic in detail. ■

4.1 LENGTH AS A FUNDAMENTAL DIMENSION

When you walk down a hallway, can you estimate how tall the ceiling is? Or how wide a door is? Or how long the hallway is? You should develop this ability because having a "feel" for dimensions will help you become a better engineer. If you decide to become a design engineer, you will find out that size and cost are important design parameters. Having developed a "feel" for the size of objects in your surroundings enables you to have a good idea about the acceptable range of values when you design something.

As you know, every physical object has a size. Some things are bigger than others. Some things are wider or taller than others. These are some common ways of expressing the relative size of objects. As discussed in Chapter 3, through their observation of nature people recognized the need for a physical quantity or a physical dimension (which today we call *length*) so that they could describe their surroundings better. They also realized that having a common definition for a physical quantity, such as length, makes communication easier. Earlier humans may have used a finger, arm, stride, stick, or rope to measure the size of an object. Chapter 3 also emphasized the need for having scales or divisions for the dimension length so that numbers could be kept simple and manageable. Today, we call these divisions or scales *systems of units*. In this chapter, we will focus our attention on length and such length-related derived quantities as area and volume.

Length is one of the seven fundamental or base dimensions that we use to properly express what we know of our natural world. In today's globally driven economy, where products are made in one place and assembled somewhere else, there exists an even greater need for a uniform and consistent way of communicating information about the fundamental dimension length and other related length variables so that parts manufactured in one place can easily be combined on an assembly line with parts made in other places. An automobile is a good example of this concept. It has literally thousands of parts that are manufactured by various companies in different parts of the world.

As explained in Chapter 3, we have learned from our surroundings and formulated our observations into laws and principles. We use these laws and physical principles to design, develop, and test products and services. Are you observing your surroundings carefully? Are you learning from your everyday observations? Here are some questions to consider: Have you thought about the size of a soda can? What are its dimensions? What do you think are the important design factors? Most of you drink a soda every day, so you know that it fits in your hand. You also know that the soda can is made from aluminum, so it is lightweight. What do you think are some of its other design factors? What are important considerations when designing signs for a highway? How wide should a hallway be? When designing a supermarket, how wide should an aisle be? Most of you have been going to class for at least 12 years, but have you thought about classroom seating arrangements? For example, how far apart are the desks? Or how far above the floor is the presentation board? For those of you who may take a bus to school, how wide are the seats in a bus? How wide is a highway lane? What do you think are the important factors when determining the size of a car seat? You can also look around home to think about the dimension length. Start with your bed: What are its dimensions, how far above the floor is it? What is a typical standard height for steps in a stairway? When you tell someone that you own a 20-inch TV, to what dimension are you referring? How high off the floor are a doorknob, showerhead, sink, light switch, and so on?

You are beginning to see that length is a very important fundamental dimension, and it is thus commonly used in engineering products. Coordinate systems are examples of another application where length plays an important role. Coordinate systems are used to locate things with respect to a known origin. In fact, you use coordinate systems every day, even though you don't think about it. When you go from your home to school or a grocery store or to meet a friend for lunch, you use coordinate systems. The use of coordinate systems is almost second nature to you. Let's say you live downtown, and your school is located on the northeast side of town. You know the exact location of the school with respect to your home. You know which streets to take for what distance and in which directions to move to get to school. You have been using a coordinate system to locate places and things most of your life, even if you did not know it. You also know the specific location of objects at home relative to other objects or to yourself. You know where your TV is located relative to your sofa or bedroom.

There are different types of coordinate systems such as rectangular, cylindrical, spherical, and so on, as shown in Figure 4.1. Based on the nature of a particular problem,

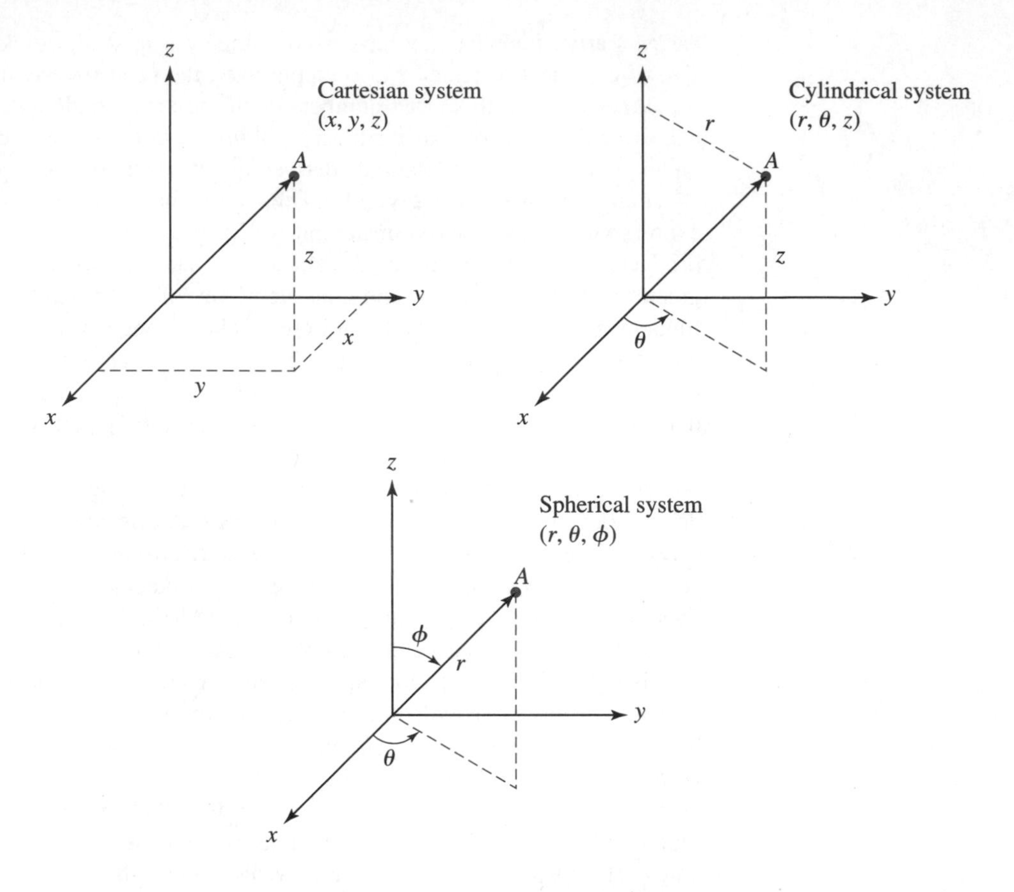

Figure 4.1
Examples of coordinate systems

we may use one or another. The most common coordinate system is the rectangular, or Cartesian, coordinate system (Figure 4.1). When you are going to school from home or meeting a friend for lunch you use the rectangular coordinate system. But you may not call it that; you may use the directions north, east, west, or south to get where you are going. You can think of the axes of a rectangular coordinate system as aligning with, for example, the east and north direction. People who are blind are expert users of rectangular coordinate systems. Because they cannot rely on their visual perception, people with a vision disability know how many steps to take and in which direction to move to go from one location to another. So to better understand coordinate systems, perform the following experiment. While at home, close your eyes for a few minutes and try to go from your bedroom to the bathroom. Note the number of steps you had to take and in which directions you had to move. Think about it!

All of us, at one time or another, have experienced not knowing where we are going. In other words, we were lost! The smarter ones among us use a map or stop and ask someone for directions and distances (x and y coordinates) to the desired place. An example of using a map to locate a place is shown in Figure 4.2. Coordinate systems are also integrated into software that drives computer numerically controlled (CNC) machines, such as a milling machine or a lathe that cuts materials into specific shapes.

Now that you understand the importance of the length dimension, let us look at its divisions or units. There are several systems of units in use in engineering today. We will

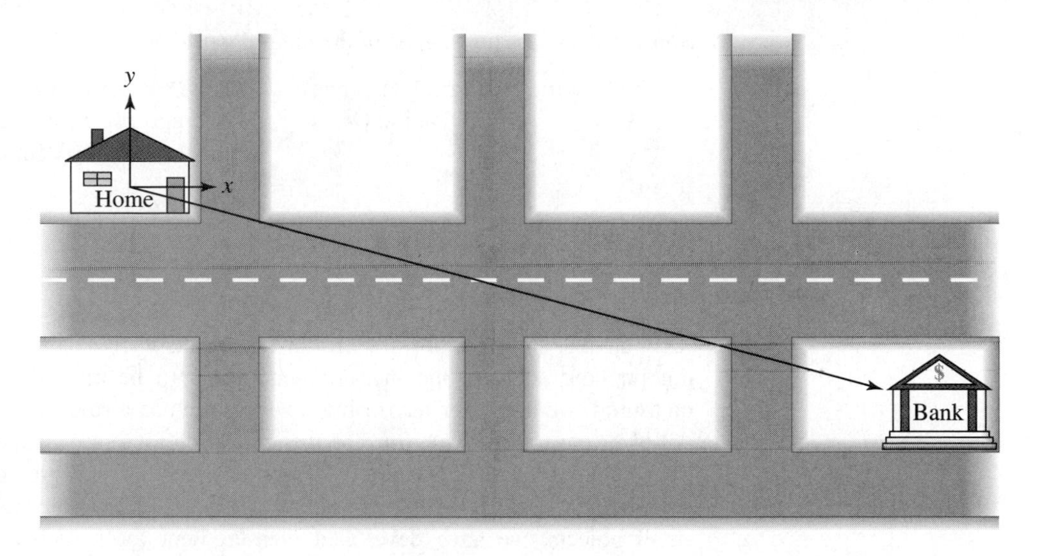

Figure 4.2 An example demonstrating the use of a coordinate system

focus on two of these systems: the International System of Units (SI) and the United States Customary Units. The unit of length in SI is the meter (m). We can use the multiples and fractions of this unit according to Table 3.2. Common multipliers of the meter are micrometer (μm), millimeter (mm), centimeter (cm), and kilometer (km). Recall from our discussion of units and multiplication prefixes in Chapter 3 that we use these multiplication prefixes to keep the numbers manageable. The International System of Units is used almost universally except in the United States, which is currently making the conversion from U.S. Customary to SI units. Soon everyone in the world will use SI units. The unit of length in the U.S. Customary system is foot (ft). The relation between foot and meter units is given by 1 ft = 0.3048 m. Table 4.1 shows other commonly used units and their equivalent values in an increasing order and includes both SI and U.S. Customary units to give you a sense of their relative magnitude.

TABLE 4.1 Units of Length and Their Equivalent Values

Units of Length in Increasing Order	Equivalent Value
1 angstrom	1×10^{-10} meter (m)
1 micrometer or 1 micron	1×10^{-6} meter (m)
1 mil	1/1000 inch $\approx 2.54 \times 10^{-5}$ meter (m)
1 point (printer's)	3.514598×10^{-4} (m)
1 millimeter	1/1000 meter (m)
1 pica (printer's)	4.217 millimeter (mm)
1 centimeter	1/100 meter (m)
1 inch	2.54 centimeter (cm)
1 foot	12 inches
1 yard	3 feet (ft)
1 meter	1.0936 yard $\approx$ 1.1 yard (yd)
1 kilometer	1000 meter (m)
1 mile	1.6093 kilometer (km) = 5280 feet (ft)

Some interesting dimensions in the natural world are

The highest mountain peak (Everest):	29,028 ft (8848 m)
Pacific Ocean:	Average depth: 4028 m (13,215 feet)
	Greatest known depth: 11,033 m (36,198 feet)

4.2 MEASUREMENT OF LENGTH

Early humans may have used finger length, arm span, stride length (step length), a stick, rope, chains, and so on to measure the size or displacement of an object. Today, depending on how accurate the measurement needs to be and the size of the object being measured, we use other measuring devices, such as a ruler, a yardstick, and a steel tape. All of us have used a ruler or tape measure to measure a distance or the size of an object. These devices are based on internationally defined and accepted units such as millimeters, centimeters, or meters or inches, feet, or yards. For more accurate measurements of small objects, we have developed measurement tools such as the micrometer or the Vernier caliper, which allow us to measure dimensions within 1/1000 of an inch. In fact, machinists use micrometers and Vernier calipers every day.

On a larger scale, you have seen the milepost markers along interstate highways. Some people actually use the mileposts to check the accuracy of their car's odometer. By measuring the time between two markers, you can also check the accuracy of the car's speedometer. In the last few decades, electronic distance measuring instruments (EDMI) have been developed that allow us to measure distances from a few feet to many miles with reasonable accuracy. These electronic distance measuring devices are used quite commonly for surveying purposes in civil engineering applications. The instrument sends out a light beam that is reflected by a system of reflectors located at the unknown distance.

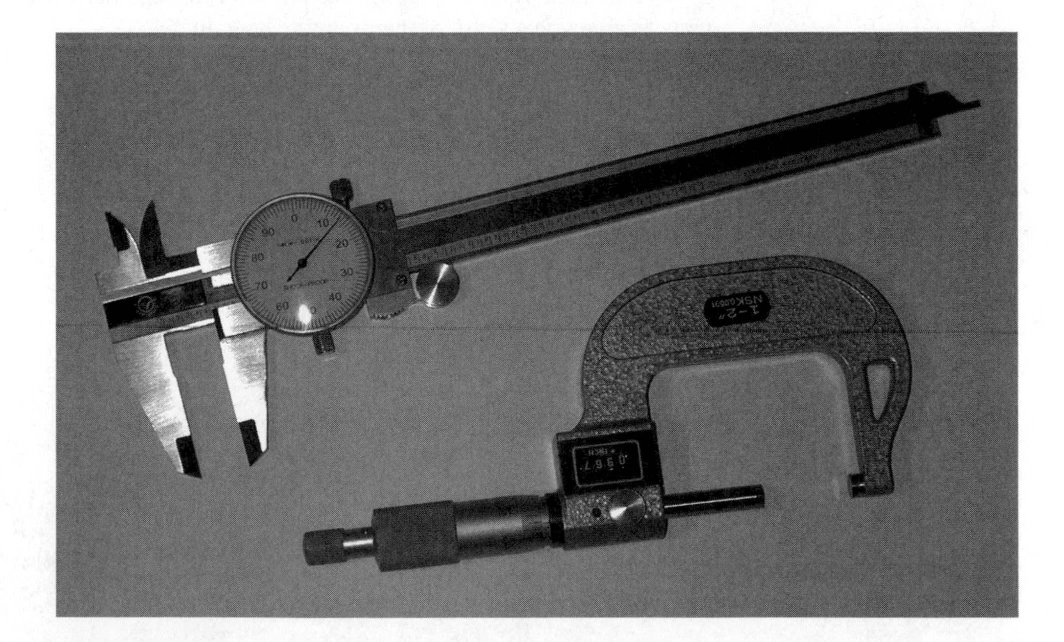

A Vernier caliper and a micrometer

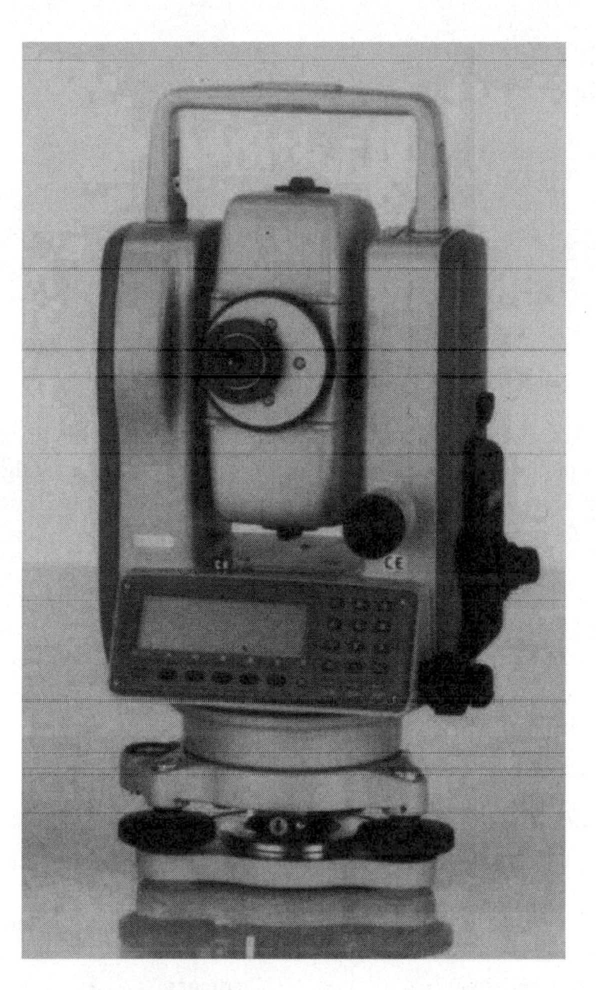

An example of an electronic distance measuring instrument used in surveying

The instrument and the reflector system are situated such that the reflected light beam is intercepted by the instrument. The instrument then interprets the information to determine the distance between the instrument and the reflector. The Global Positioning System (GPS) is another example of recent advances in locating objects on the surface of the earth with good accuracy. Currently, radio signals are sent from approximately 24 artificial satellites orbiting the earth. Tracking stations are located around the world to receive and interpret the signals sent from the satellites. Although originally the GPS was funded and controlled by the U.S. Department of Defense for military applications, it now has millions of users. GPS navigation receivers are now common in airplanes, automobiles, buses, and hand-held receivers used by hikers.

Sometimes, distances or dimensions are determined indirectly using trigonometric principles. For example, let us say that we want to determine the height of a building similar to the one shown in Figure 4.3. but do not have access to accurate measuring devices. With a drinking straw, a protractor, and a steel tape we can determine a reasonable value for the height of the building by measuring the angle α and the dimensions d and h_1. The

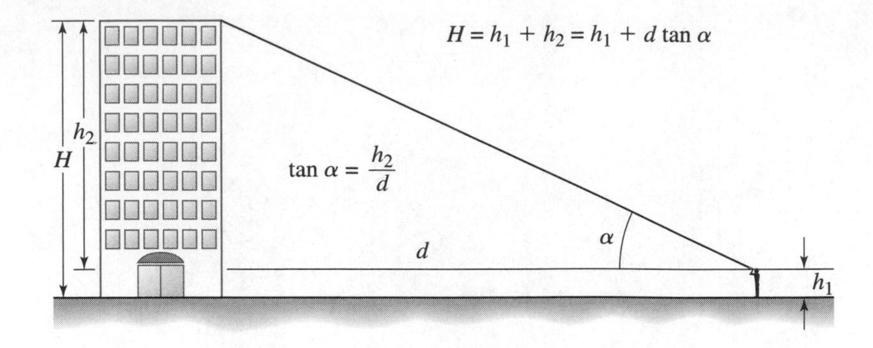

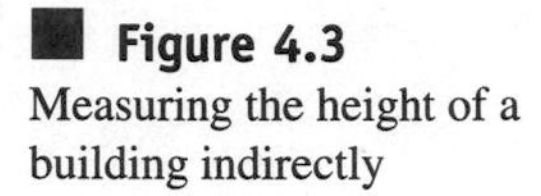

Figure 4.3
Measuring the height of a building indirectly

analysis is shown is Figure 4.3. Note that h_1 represents the distance from the ground to the eye of a person looking through the straw. The angle α is the angle between the straw, which is focused on the edge of the roof, and a horizontal line. The protractor is used to measure the angle that the straw makes with the horizontal line.

It is very likely that you have used trigonometric tools to analyze some problems in the past. It is also possible that you may have not used them recently. If that is the case, then the tools have been sitting in your mental toolbox for a while and quite possibly have collected some rust in your head! Or you may have forgotten altogether how to properly use the tools. Because of their importance, let us review some of these basic relationships and definitions. For a right triangle, the Pythagorean relation may be expressed by

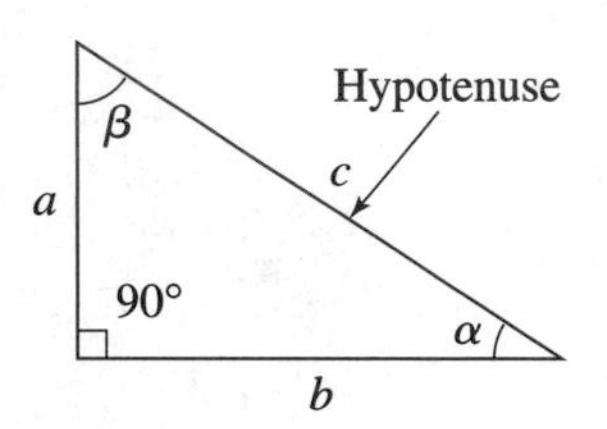

$$a^2 + b^2 = c^2$$

In the ***right triangle*** shown, the angle facing side a is denoted by α (alpha), and the angle facing side b by β (beta). The sine, cosine, and the tangent of an angle are defined by

$$\sin\alpha = \frac{\text{opposite}}{\text{hypotenuse}} = \frac{a}{c} \qquad \cos\alpha = \frac{\text{adjacent}}{\text{hypotenuse}} = \frac{b}{c} \qquad \tan\alpha = \frac{\sin\alpha}{\cos\alpha} = \frac{\text{opposite}}{\text{adjacent}} = \frac{a}{b}$$

$$\sin\beta = \frac{\text{opposite}}{\text{hypotenuse}} = \frac{b}{c} \qquad \cos\beta = \frac{\text{adjacent}}{\text{hypotenuse}} = \frac{a}{c} \qquad \tan\beta = \frac{\sin\beta}{\cos\beta} = \frac{\text{opposite}}{\text{adjacent}} = \frac{b}{a}$$

The sine and the cosine law (rule) for an ***arbitrarily shaped triangle*** is

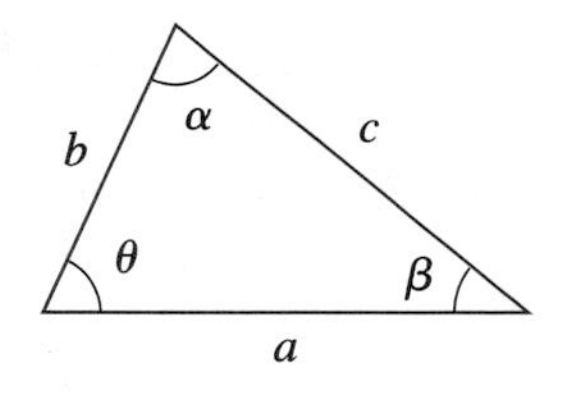

The sine rule: $$\frac{a}{\sin\alpha} = \frac{b}{\sin\beta} = \frac{c}{\sin\theta}$$

The cosine rule: $$a^2 = b^2 + c^2 - 2bc(\cos\alpha)$$

$$b^2 = a^2 + c^2 - 2ac(\cos\beta)$$

$$c^2 = a^2 + b^2 - 2bc(\cos\theta)$$

or $$\cos\alpha = \frac{b^2 + c^2 - a^2}{2bc}$$

$$\cos\beta = \frac{a^2 + c^2 - b^2}{2ac}$$

$$\cos\theta = \frac{a^2 + b^2 - c^2}{2bc}$$

Some other useful trigonometry identities are

$$\sin^2 \alpha + \cos^2 \alpha = 1$$

$$\sin 2\alpha = 2 \sin \alpha \cos \alpha$$

$$\cos 2\alpha = \cos^2 \alpha - \sin^2 \alpha = 2 \cos^2 \alpha - 1 = 1 - 2 \sin^2 \alpha$$

$$\sin(-\alpha) = -\sin \alpha$$

$$\cos(-\alpha) = \cos \alpha$$

$$\sin(\alpha + \beta) = \sin \alpha \cos \beta + \sin \beta \cos \alpha$$

$$\sin(\alpha - \beta) = \sin \alpha \cos \beta - \sin \beta \cos \alpha$$

$$\cos(\alpha + \beta) = \cos \alpha \cos \beta - \sin \alpha \sin \beta$$

$$\cos(\alpha - \beta) = \cos \alpha \cos \beta + \sin \alpha \sin \beta$$

4.3 NOMINAL SIZES VERSUS ACTUAL SIZES

You have all seen or used a 2×4 piece of lumber. If you were to measure the dimensions of the cross section of a 2×4, you would find that the actual width is less than 2 inches (approximately 1.5 inches) and the height is less than 4 inches (approximately 3.5 inches). Then why is it referred to as a "2 by 4"? Manufacturers of engineering parts use round numbers so that it is easier for people to remember the size and thus more easily refer to a specific part. The 2×4 is called the *nominal size* of the lumber. If you were to investigate other structural members, such as I-beams, you would also note that the nominal size tested by the manufacturer is different from the actual size. You will find a similar situation for pipes, tubes, screws, and many other engineering parts. Agreed-upon standards are followed by manufacturers when providing information about the size of the parts that they make. But manufacturers do provide actual sizes of parts in addition to nominal sizes as emphasized in Figure 4.4. This fact is important because, as you will learn in your future engineering classes, you need the actual size of parts for various engineering calculations. Examples of nominal sizes versus actual sizes of some engineering parts are given in Table 4.2.

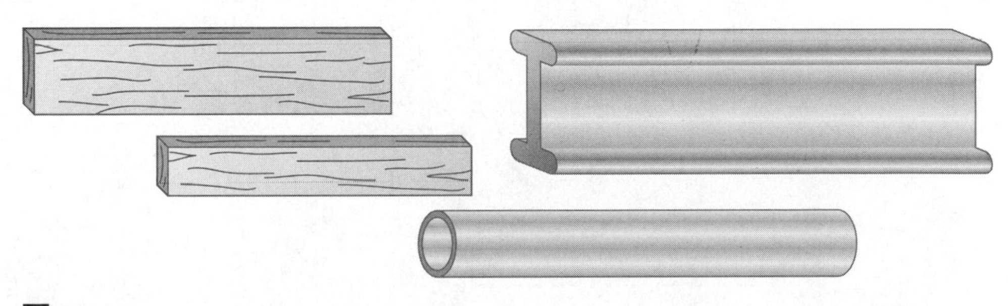

Figure 4.4 Manufacturers provide actual and nominal sizes for many parts and structural members.

TABLE 4.2 Examples of Nominal Size Versus Actual Size of Some Engineering Products

Dimension of Copper Tubing (Type K—for high pressure–temperature use)

Nominal Size (in.)	Outside Diameter (in.)	Inside Diameter (in.)	Wall Thickness (in.)	Flow Cross Section (in^2)
1/4	0.375	0.305	0.035	0.073
3/8	0.5	0.402	0.049	0.127
1/2	0.625	0.527	0.049	0.218
5/8	0.75	0.652	0.049	0.334
3/4	0.875	0.745	0.065	0.436
1	1.125	0.995	0.065	0.778
$1\frac{1}{4}$	1.375	1.245	0.065	1.22
$1\frac{1}{2}$	1.625	1.481	0.072	1.72
2	2.125	1.959	0.083	3.01
$2\frac{1}{2}$	2.625	2.435	0.095	4.66
3	3.125	2.907	0.109	6.64
$3\frac{1}{2}$	3.625	3.385	0.120	9.00
4	4.125	3.857	0.134	11.7
5	5.125	4.805	0.160	18.1

Dimension of Copper Tubing (Type L—for HVAC applications)

Nominal Size (in.)	Outside Diameter (in.)	Inside Diameter (in.)	Wall Thickness (in.)	Flow Cross Section (in^2)
1/4	0.375	0.315	0.030	0.078
3/8	0.5	0.44	0.035	0.145
1/2	0.625	0.545	0.040	0.233
5/8	0.75	0.666	0.042	0.334
3/4	0.875	0.785	0.045	0.484
1	1.125	1.025	0.065	0.825
$1\frac{1}{4}$	1.375	1.265	0.050	1.26
$1\frac{1}{2}$	1.625	1.505	0.055	1.78
2	2.125	1.985	0.060	3.09
$2\frac{1}{2}$	2.625	2.465	0.070	4.77
3	3.125	2.945	0.080	6.81
$3\frac{1}{2}$	3.625	3.425	0.090	9.21
4	4.125	3.905	0.100	11.0
5	5.125	4.875	0.110	18.1

Continued

Dimension of Copper Tubing (Type M—for HVAC and domestic water applications)

Nominal Size (in.)	Outside Diameter (in.)	Inside Diameter (in.)	Wall Thickness (in.)	Flow Cross Section (in^2)
3/8	0.5	0.450	0.025	0.159
1/2	0.625	0.569	0.028	0.254
3/4	0.875	0.811	0.032	0.517
1	1.125	1.055	0.035	0.874
$1\frac{1}{4}$	1.375	1.291	0.042	1.31
$1\frac{1}{2}$	1.625	1.527	0.049	1.83
2	2.125	2.009	0.058	3.17
$2\frac{1}{2}$	2.625	2.495	0.065	4.89
3	3.125	2.981	0.072	6.98
$3\frac{1}{2}$	3.625	3.459	0.083	9.40
4	4.125	3.935	0.095	12.2
5	5.125	4.907	0.109	18.9

Examples of Standard I-Beams (SI version)

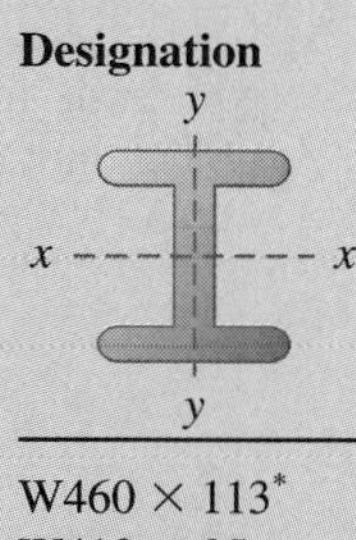

Designation	Depth (mm)	Width (mm)	Area (mm^2)
W460 × 113*	463	280	14,400
W410 × 85	417	181	10,800
W360 × 57	358	172	7,230
W200 × 46.1	203	203	5,890

*Nominal depth (mm) and mass (kg) per one meter length.

(English version)

Designation	Depth (in.)	Width (in.)	Area (in^2)
W18 × 76*	18.21	11.03	22.3
W16 × 57	16.43	7.12	16.8
W14 × 38	14.10	6.77	11.2
W8 × 31	8.00	7.95	9.1

*Nominal depth (in.) and weight (lb) per one foot length.

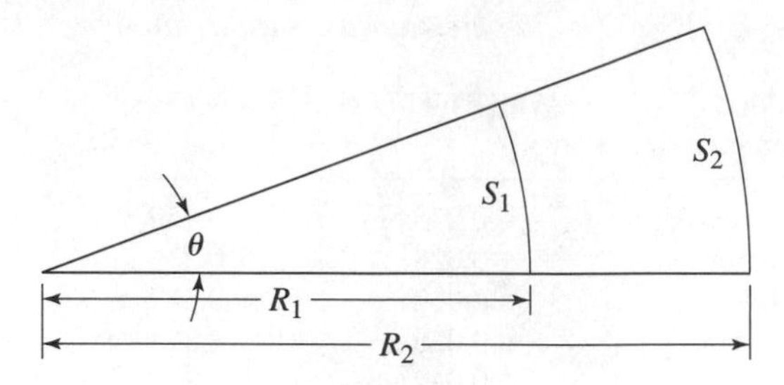

Figure 4.5 The relationship among arc length, radius, and angle

4.4 RADIANS AS A RATIO OF TWO LENGTHS

Consider the circular arc shown in Figure 4.5. The relationship among the arc length, S, radius of the arc, R, and the angle in radians, θ, is given by

$$\theta = \frac{S_1}{R_1} = \frac{S_2}{R_2} \tag{4.1}$$

Note that radians represents the ratio of two lengths and thus is unitless. As you will learn later in your physics and dynamics classes, you can use Equation (4.1) as the basis for establishing a relationship between the translational speed of a point on an object and its rotational speed.

4.5 STRAIN AS A RATIO OF TWO LENGTHS

When a material (e.g., in the shape of a rectangular bar) is subjected to a tensile load (pulling load), the material will deform. The deformation, ΔL, divided by the original length, L, is called *normal strain*, as shown in Figure 4.6. In Chapter 7, we will discuss the concepts of stress and strain in more detail. We will also explain how stress and strain information is used in engineering analysis. But for now, remember that strain is the ratio of deformation length to original length and thus is unitless.

L
ΔL
Load

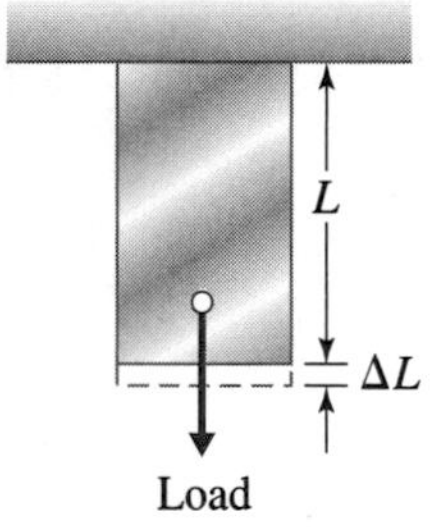

Figure 4.6 A bar subjected to a pulling load

$$\text{strain} = \frac{\Delta L}{L} \tag{4.2}$$

4.6 AREA

Area is a derived, or secondary, physical quantity. It plays a significant role in many engineering problems. For example, the rate of heat transfer from a surface is directly proportional to the exposed surface area. That is why a motorcycle engine or a lawn mower engine has extended surfaces, or fins as shown in Figure 4.7. If you look closely inside buildings around your campus, you will also see heat exchangers or radiators with

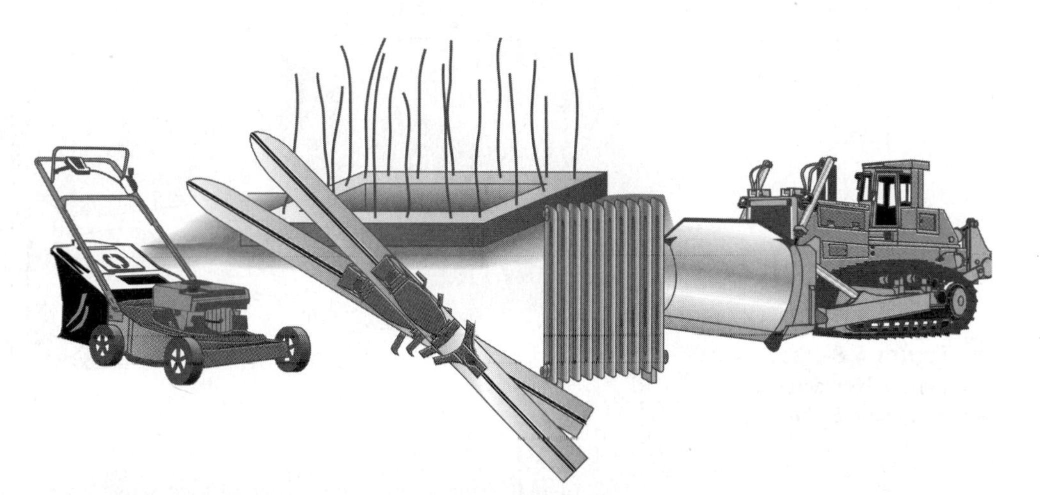

Figure 4.7 The important role of area in the design of various systems

extended surfaces under windows and against some walls. As you know, these heat exchangers or radiators supply heat to rooms and hallways during the winter. For another example, have you ever thought about why crushed ice cools a drink faster than ice cubes? It is because given the same amount of ice, the crushed ice has more surface exposed to the liquid. You may have also noticed that given the same amount of meat, it takes longer for a roast of beef to cook than it takes stew. Again, it is because the stew has more surface area exposed to the liquid in which it is being cooked. So next time you are planning to make some mashed potatoes, make sure you first cut the potatoes into smaller pieces. The smaller the pieces, the sooner they will cook. Of course, the reverse is also true. That is, if you want to reduce the heat loss from something, one way of doing this would be to reduce the exposed surface area. For example, when we feel cold we naturally try to curl up, which reduces the surface area exposed to the cold surroundings. You can tell by observing nature that trees take advantage of the effect and importance of surface area. Why do trees have lots of leaves rather than one big leaf? It is because they can absorb more solar radiation that way. Surface area is also important in engineering problems related to mass transfer. Your parents or grandparents perhaps recall when they used to hang out clothes to dry. The clothes were hung over the clothesline in such way that they had a maximum area exposed to the air. Bedsheets, for example, were stretched out fully across the clothesline.

Let us now investigate the relationship between a given volume and exposed surface area. Consider a 1 m × 1 m × 1 m cube. What is the volume? 1 m^3. What is the exposed surface area of this cube? 6 m^2. If we divide each dimension of this cube by half, we get 8 smaller cubes with the dimensions of 0.5 m × 0.5 m × 0.5 m, as shown in Figure 4.8. What is the total volume of the 8 smaller cubes? It is still 1 m^3. What is the total exposed surface area of the cubes? Each cube now has an exposed surface area of 1.5 m^2, which amounts to a total exposed surface area of 12 m^2.

Let us proceed with dividing the dimensions of our 1 m × 1 m × 1 m original cube into even smaller cubes with the dimensions of 0.25 m × 0.25 m × 0.25 m. We will now have 64 smaller cubes, and we note that the total volume of the cubes is still 1 m^3. However, the surface area of each cube is 0.375 m^2, leading to a total surface area of 24 m^2. Thus, the same cube divided into 64 smaller cubes now has an exposed surface area that is 4 times the original surface area.

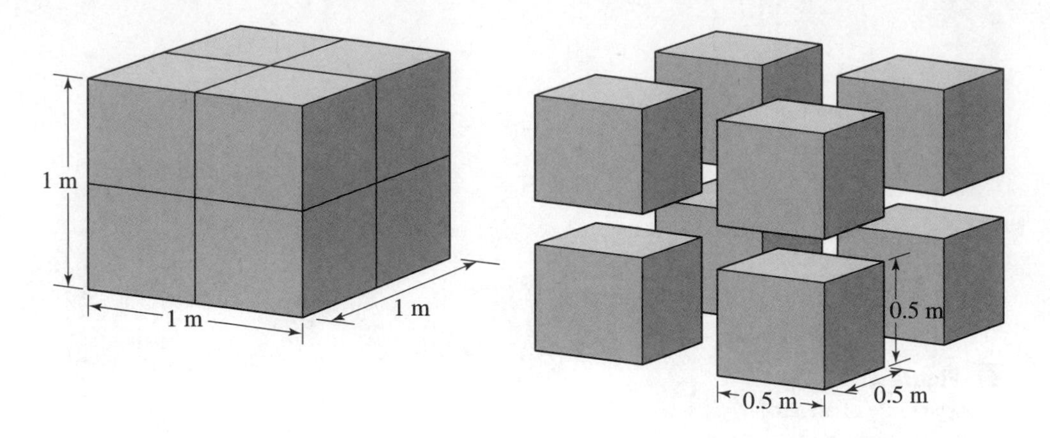

Figure 4.8 The relationship between the area and volume of a cube

This mental exercise gives you a good idea why, given the same amount of ice, the crushed ice cools a drink faster than ice cubes do. You may not know it, but you already own the best heat and mass exchanger in the world. Your lungs! Human lungs are the best heat and mass exchangers that we know of, with an approximate area density of 20,000 m^2/m^3.

Area plays an important role in aerodynamics as well. Air resistance to the motion of a vehicle is something that all of you are familiar with. As you may also know, the drag force acting on a vehicle is determined experimentally by placing it in a wind tunnel. The airspeed inside the tunnel is changed, and the drag force acting on the vehicle is measured. Engineers have learned that when designing new vehicles, the total exposed surface area and the frontal area are important factors in reducing air resistance. The experimental data is normally given by a single coefficient, which is called the *drag coefficient*. It is defined by the following relationship:

$$\text{drag coefficient} = \frac{\text{drag force}}{\frac{1}{2}(\text{air density})(\text{airspeed})^2(\text{frontal area})}$$

The frontal area represents the frontal projection of the vehicle's area and could be approximated simply by calculating 0.85 times the width and the height of a rectangle that outlines the front of the vehicle. This is the area that you see when you view the car or truck from a direction normal to the front grill. Later in your engineering education, some of you may take a class in the physics of flight, where you will learn that the lift force acting on the wings of a plane is proportional to the planform area of the wing. The planform area is the area that you would see if you were to look from above at the wing from a direction normal to the wing.

Cross-sectional area also plays an important role in distributing a force over an area. Foundations of buildings, hydraulic systems, and cutting tools (see Figure 4.9) are examples of objects for which the role of area is important. For example, have you ever thought about why the edge of a sharp knife cuts well? What do we mean by a "sharp" knife? A good sharp knife is one that has a cross-sectional area as small as possible along its cutting edge. The pressure along the cutting edge of a knife is simply determined by

$$\text{pressure at the cutting surface} = \frac{\text{force}}{\text{cross-sectional area at the cutting edge}} \tag{4.3}$$

Figure 4.9 Area plays an important role in the design of cutting tools.

You can see from Equation (4.3) that for the same force (push) on the knife, you can increase the cutting pressure by decreasing the cross-sectional area. As you can also tell from Equation (4.3), we can also reduce the pressure by increasing the area. In skiing, we use the area to our advantage and distribute our weight over a bigger surface so we won't sink into the snow. Next time you go skiing think about this relationship. Equation (4.3) also makes clear why high-heeled shoes are designed poorly as compared to walking shoes. The purpose of these examples is to make you realize that area is an important parameter in engineering design. During your engineering education, you will learn many new concepts and laws that are either directly or inversely proportional to the area. So keep a close watch out for area as you study various engineering topics.

TABLE 4.3 Units of Area and Their Equivalent Values

Units of Area in Increasing Order	Equivalent Value
1 mm^2	1×10^{-6} m^2
1 cm^2	1×10^{-4} m^2 = 100 mm^2
1 in^2	654.16 mm^2
1 ft^2	144 in^2
1 yd^2	9 ft^2
1 m^2	1.196 yd^2
1 acre	43,560 ft^2
1 km^2	1,000,000 m^2 = 247.1 acre
1 square mile	2.59 km^2 = 640 acres

Now that you understand the significance of area in engineering analysis, let us look at its units. The unit of area in SI is m^2. We can also use the multiples and the submultiples (see Table 3.2) of SI fundamental units to form other appropriate units for area, such as mm^2, cm^2, km^2, and so on. Remember, the reason we use these units is to keep numbers manageable. The common unit of area in the U.S. Customary system is ft^2. Table 4.3 shows other units commonly used in engineering practice today and their equivalent values.

Area Calculations

The areas of common shapes, such as a triangle, a circle, and a rectangle, can be obtained using the area formulas shown in Table 4.4. It is a common practice to refer to these simple areas as *primitive areas*. Many composite surfaces with regular boundaries can be divided into primitive areas. To determine the total area of a composite surface, such as the one shown in Figure 4.10, we first divide the surface into the simpler primitive areas that make it up, and then we sum the values of these areas to obtain the total area of the composite surface.

Examples of the more useful area formulas are shown in Table 4.4.

Approximation of Planar Areas

There are many practical engineering problems that require calculation of planar areas of irregular shapes. If the irregularities of the boundaries are such that they will not allow for the irregular shape to be represented by a sum of primitive shapes, then we need to resort

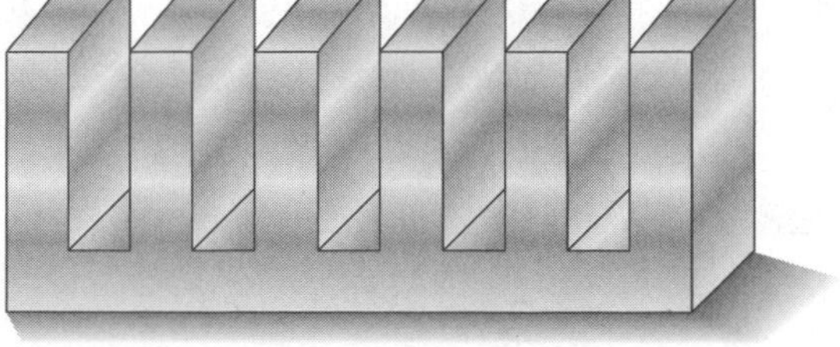

Figure 4.10 A composite surface (surface of a heat sink) that may be divided into primitive areas

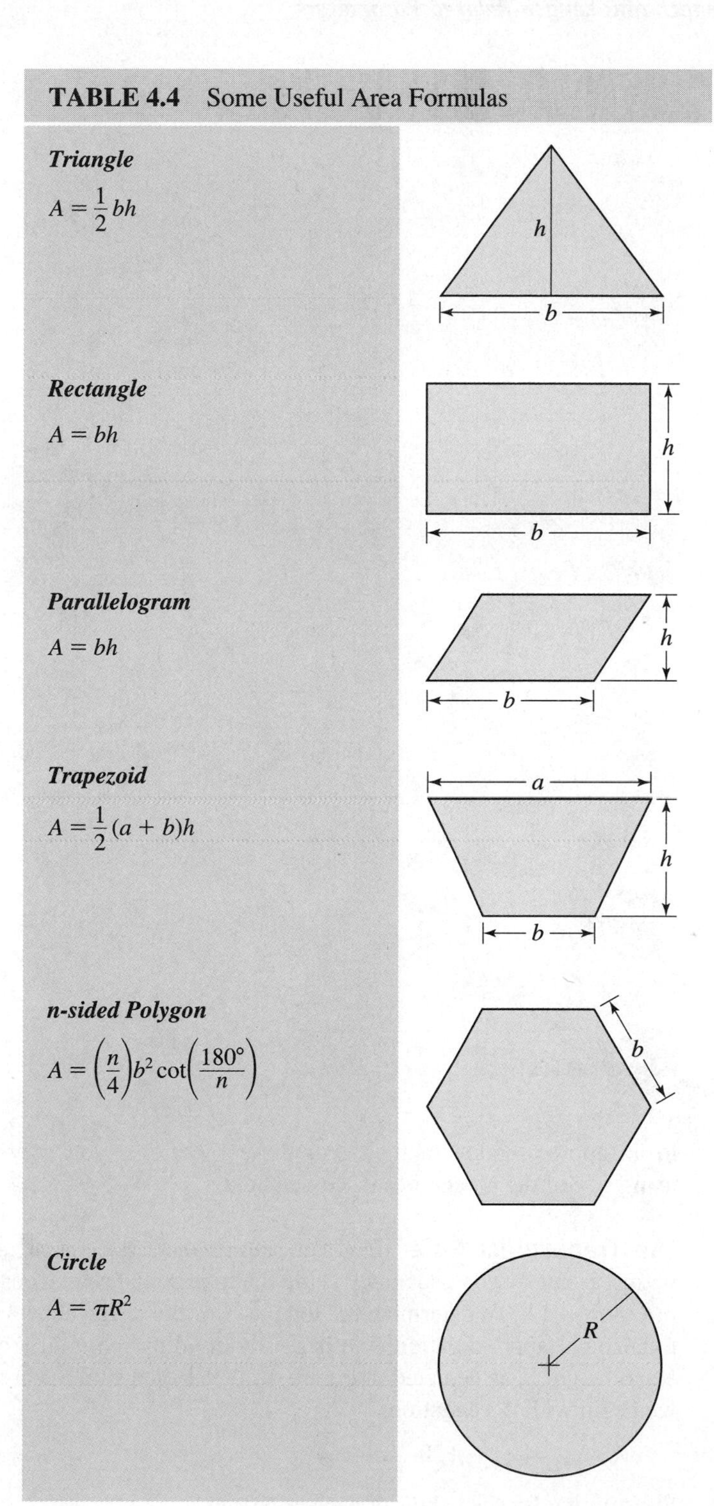

TABLE 4.4 Some Useful Area Formulas

Shape	Formula
Triangle	$A = \frac{1}{2}bh$
Rectangle	$A = bh$
Parallelogram	$A = bh$
Trapezoid	$A = \frac{1}{2}(a + b)h$
n-sided Polygon	$A = \left(\frac{n}{4}\right)b^2 \cot\left(\frac{180°}{n}\right)$
Circle	$A = \pi R^2$

Continued

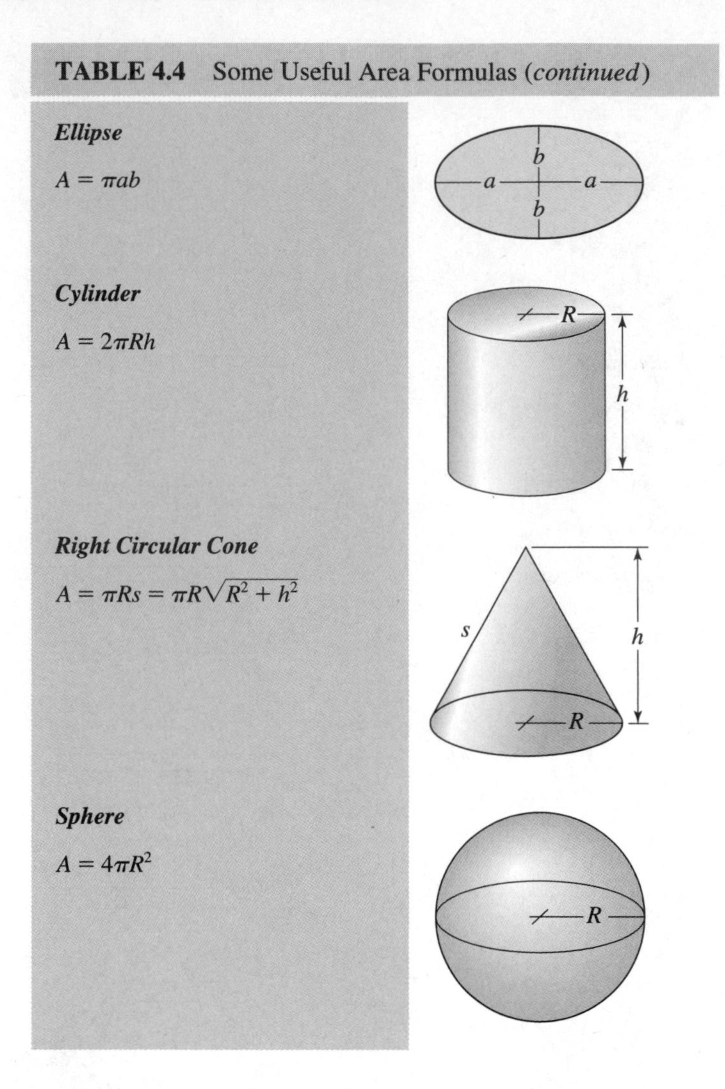

TABLE 4.4 Some Useful Area Formulas (*continued*)

Shape	Area formula
Ellipse	$A = \pi ab$
Cylinder	$A = 2\pi Rh$
Right Circular Cone	$A = \pi Rs = \pi R\sqrt{R^2 + h^2}$
Sphere	$A = 4\pi R^2$

to an approximation method. For these situations, you may approximate planar areas using any of the procedures discussed next.

The Trapezoidal Rule You can approximate the planar areas of an irregular shape with reasonably good accuracy using the trapezoidal rule. Consider the planar area shown in Figure 4.11. To determine the total area of the shape shown in Figure 4.11, we use the trapezoidal approximation. We begin by dividing the total area into small trapezoids of equal height h, as depicted in Figure 4.11. We then sum the areas of the trapezoids. Thus, we begin with the equation

$$A \approx A_1 + A_2 + A_3 + \cdots + A_n \qquad (4.4)$$

Substituting for the values of each trapezoid,

$$A \approx \frac{h}{2}(y_0 + y_1) + \frac{h}{2}(y_1 + y_2) + \frac{h}{2}(y_2 + y_3) + \cdots + \frac{h}{2}(y_{n-1} + y_n) \qquad (4.5)$$

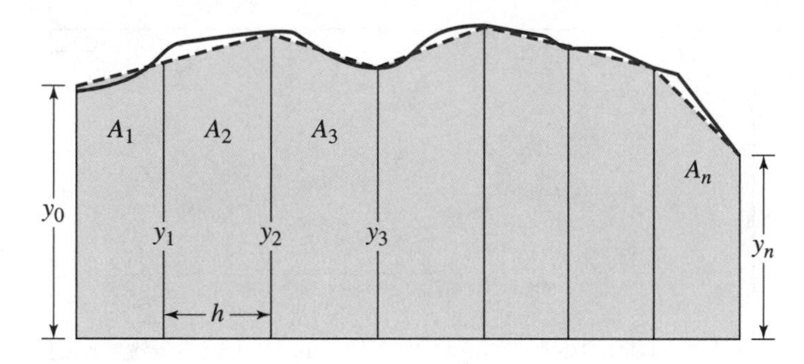

Figure 4.11
Approximation of a planar area by the trapezoidal rule

and simplifying Equation (4.5) leads to

$$A \approx h\left(\frac{1}{2}y_0 + y_1 + y_2 + \cdots + y_{n-2} + y_{n-1} + \frac{1}{2}y_n\right) \tag{4.6}$$

Equation (4.6) is known as the trapezoidal rule. Also note that the accuracy of the area approximation can be improved by using more trapezoids. This approach will reduce the value of h and thus improve the accuracy of the approximation.

Counting the Squares There are other ways to approximate the surface areas of irregular shapes. One such approach is to divide a given area into small squares of known size and then count the number of squares. This approach is depicted in Figure 4.12. You then need to add to the areas of the small squares the leftover areas, which you may approximate by the areas of small triangles.

Subtracting Unwanted Areas Sometimes, it may be advantageous to first fit large primitive area(s) around the unknown shape and then approximate and subtract the unwanted smaller areas. An example of such a situation is shown in Figure 4.13. Also keep in mind that for symmetrical areas you may make use of the symmetry of the shape. Approximate only 1/2 or 1/4 or 1/8 of the total area, and then multiply the answer by the appropriate factor.

Weighing the Area Another approximation procedure requires the use of an accurate analytical balance from a chemistry lab. Assuming the profile of the area to be determined can be drawn on a $8\frac{1}{2}$-×-11 sheet of paper, first weigh a blank $8\frac{1}{2}$-×-11 sheet of

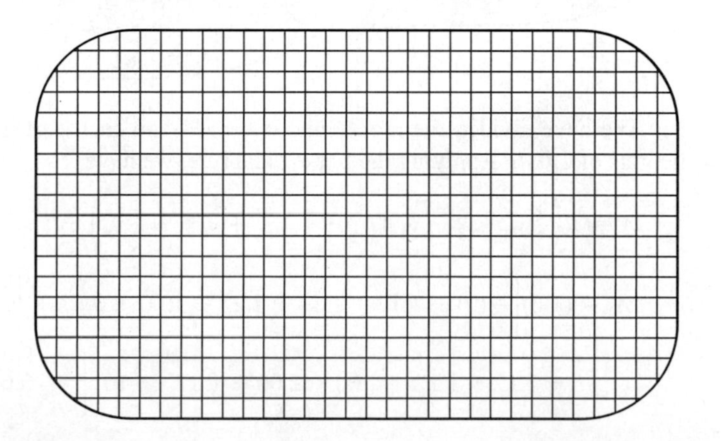

Figure 4.12
Approximation of a planar area using small squares

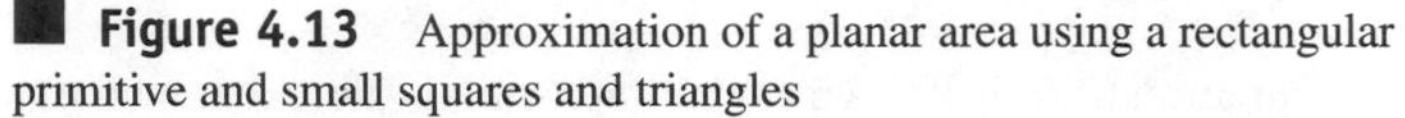

Figure 4.13 Approximation of a planar area using a rectangular primitive and small squares and triangles

paper, then using the analytical balance, weigh the sheet and record its weight. Next, draw the boundaries of the unknown area on the blank paper, and then cut around the boundary of that area. Determine the weight of the piece of paper that has the area drawn on it. Now, by comparing the weights of the blank sheet of paper to the weight of the paper with the profile, you can determine the area of the given profile. In using this approximation method, we assumed that the paper has uniform thickness and density.

EXAMPLE 4.1

Using the trapezoidal rule, determine the total ground-contact area of the athletic shoe shown in the accompanying figure.

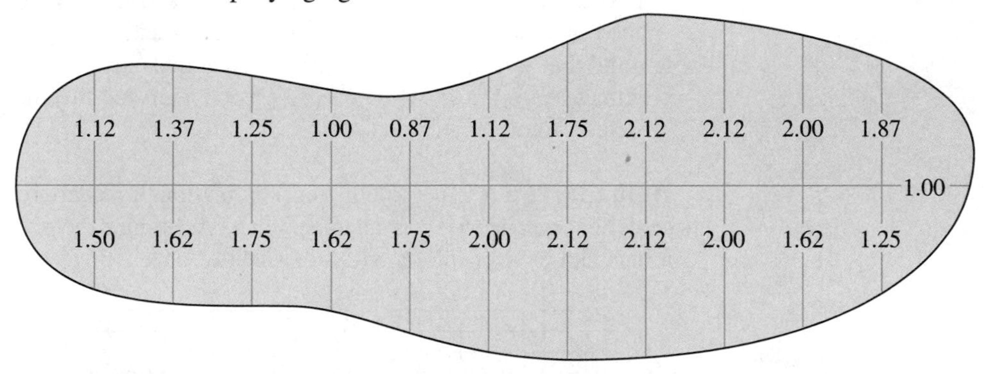

We have divided the profile into two parts and each part into 12 trapezoids of equal heights of 1.0 in. Applying the trapezoidal rule, we have

$$A \approx h\left(\frac{1}{2}y_0 + y_1 + y_2 + \cdots + y_{n-2} + y_{n-1} + \frac{1}{2}y_n\right)$$

$$A_1 \approx (1.0)\left[\frac{1}{2}(0) + 1.12 + 1.37 + 1.25 + 1.00 + 0.87 + 1.12 + 1.75 \right.$$

$$\left. + 2.12 + 2.12 + 2.00 + 2.00 + 1.87 + \frac{1}{2}(0)\right] \approx 16.6 \text{ in}^2$$

$$A_2 \approx (1.0)\left[\frac{1}{2}(0) + 1.50 + 1.62 + 1.75 + 1.62 + 1.75 + 2.00 + 2.12 + 2.12 + 2.00 + 1.62 + 1.25 + \frac{1}{2}(0)\right] \approx 19.3 \text{ in}^2$$

Then the total area is given by

$$A_{total} \approx 16.6 + 19.3 \approx 35.9 \text{ in}^2$$

4.7 VOLUME

Volume is another important physical quantity, or physical variable, that does not get enough respect. We live in a three-dimensional world, so it is only natural that volume would be an important player in how things are shaped or how things work. Let us begin by considering the role of volume in our daily lives. Today you may have treated yourself to a can of soda, which on average contains 12 fluid ounces or 355 milliliters of your favorite beverage. You may have driven a car whose engine size is rated in liters. For example, if you own a Buick Park Avenue, its engine size is 3.8 liters. Depending on the size of your car, it is also safe to say that in order to fill the gas tank you need to put in about 15 to 20 gallons (57 to 77 liters) of gasoline. We also express the gas consumption rate of a car in terms of so many miles per gallon of gasoline. Doctors tell us that we need to drink at least 8 glass of water (approximately 2.5 to 3 liters) a day. We breathe in oxygen at a rate of approximately 1.6 ft^3/h (0.0453 m^3/h). Of course, as you would expect, the volume of oxygen consumption or carbon dioxide production depends on the level of physical activity. The oxygen consumption, carbon dioxide production, and pulmonary ventilation for an average man is shown in Table 4.5.

We each consume on average about 20 to 40 gallons of water per day for personal grooming and cooking. Volume also plays an important role in food packaging and pharma-

TABLE 4.5 The Oxygen Consumption, Carbon Dioxide Production, and Pulmonary Ventilation for a Man

Level of Physical Activity	Oxygen Consumption (ft^3/h)	Carbon Dioxide Production (ft^3/h)	Rate of Breathing (ft^3/h)
Exhausting effort	6.6	5.7	146
Strenuous work or sports	4.44	3.8	97
Moderate exercise	2.96	2.5	64
Mild exercise; light work	1.84	1.55	40
Standing; desk work	1.10	0.93	24
Sedentary, at ease	0.74	0.62	16
Reclining, at rest	0.3	0.56	12

Source: American Society of Heating, Refrigerating, and Air-Conditioning Engineers, Inc.

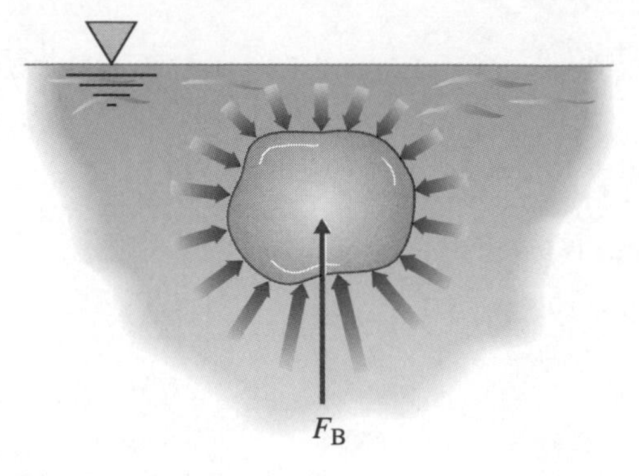

Figure 4.14 Buoyancy force acting on a submerged surface

ceutical applications. For example, a large milk container is designed to hold a gallon of milk. When administering drugs, the doctor may inject you with so many milliliters of some medicine. Many other materials are also packaged so that the package contains so many liters or gallons of something, for example, a gallon can of paint. Clearly, we use volume to express quantities of various fluids that we consume. Volume also plays a significant role in many other engineering concepts. For example, the density of a material represents how light or heavy a material is per unit volume. We will discuss density and other related mass and volume properties in Chapter 6. Buoyancy is another engineering principle where volume plays an important role. Buoyancy is the force that a fluid exerts on a submerged object. The net upward buoyancy force arises from the fact that the fluid exerts a higher pressure at the bottom surfaces of the object than it does on the top surfaces of the object, as shown in Figure 4.14. Thus, the net effect of fluid pressure distribution acting over the submerged surface of an object is the buoyancy force. The magnitude of the buoyancy force is equal to the weight of the volume of the fluid displaced. It is given by

$$F_B = \rho V g \tag{4.7}$$

where F_B is the buoyancy force (N), ρ represents the density of the fluid (kg/m^3), and g is acceleration due to gravity (9.81 m/s^2). If you were to fully submerge an object with a volume V, as shown in Figure 4.14, you would see that an equal volume of fluid has to be displaced to make room for the volume of the object. In fact, you can use this principle to measure the

TABLE 4.6 Units of Volume and Their Equivalent Values*

Volume Units in Increasing Capacity Order	Equivalent Value
1 milliliter	1/1000 liter
1 teaspoon (tsp)	4.928 milliliter
1 tablespoon (tbsp)	3 tsp
1 fluid ounce	2 tbsp $\approx$ 1/1000 ft^3
1 cup	8 ounces = 16 tbsp
1 pint	16 ounces = 2 cups
1 quart	2 pints
1 liter	1000 cm^3 $\approx$ 4.2 cups
1 gallon	4 quarts
1 cubic foot	7.4805 gallons
1 cubic meter	1000 liters $\approx$ 264 gallons $\approx$ 35.3 ft^3

*1 milliliter $<$ 1 teaspoon $<$ 1 tablespoon $<$ 1 fluid ounce $<$ 1 cup $<$ 1 pint $<$ 1 quart $<$ 1 liter $<$ 1 gallon $<$ 1 cubic foot $<$ 1 cubic meter.

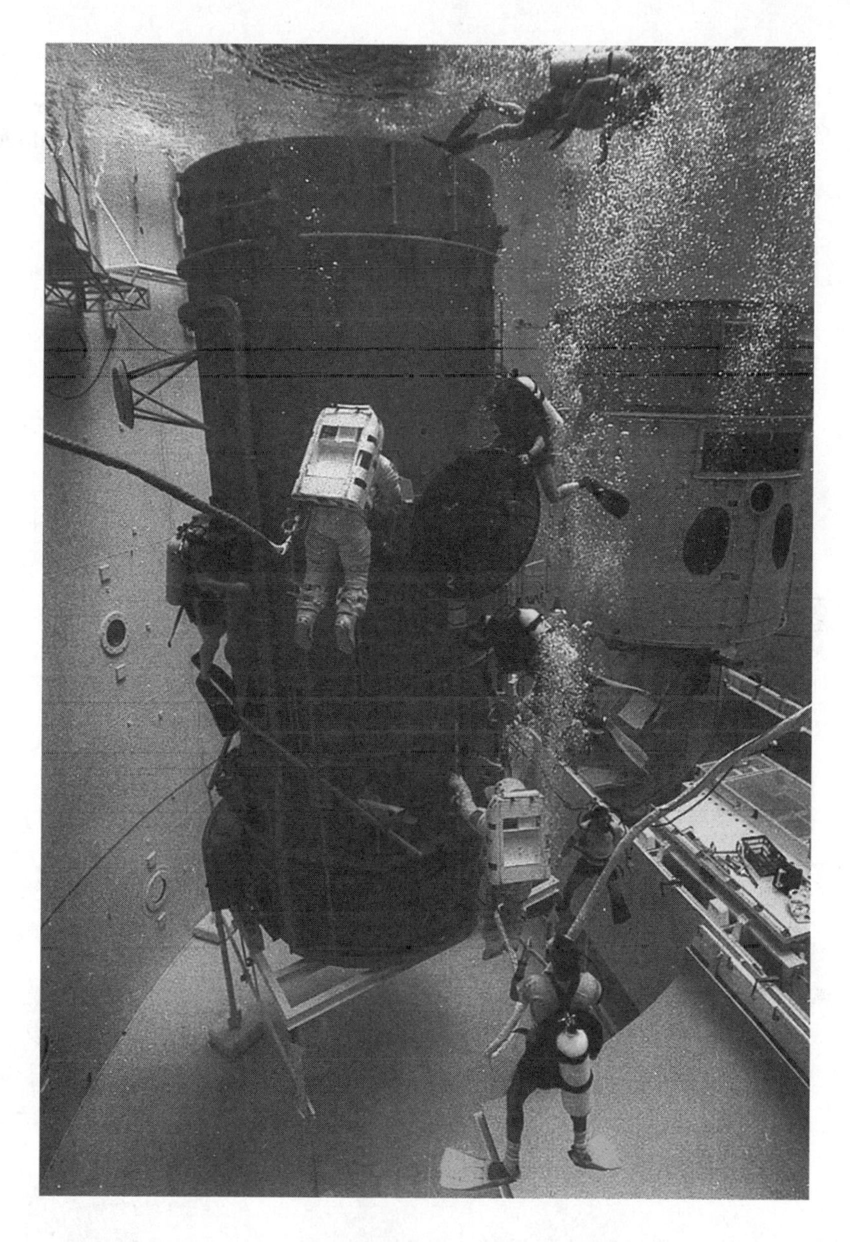

NASA Astronauts training in the Underwater Neutral Buoyancy Simulator to prepare for the in-orbit Hubbell Space Telescope repair

unknown volume of an object. This idea is demonstrated in Example 4.2. NASA astronauts also make use of buoyancy and train underwater to prepare for the in-orbit repair of satellites. This type of training takes place in the Underwater Neutral Buoyancy Simulator, shown in the accompanying photograph. The changes in the apparent weight of the astronaut allows him or her to prepare to work under near-zero-gravity (weightless) conditions.

Now that you understand the significance of volume in the analysis of engineering problems, let us look at some of the more common units in use. These units are shown in Table 4.6. As you go over Table 4.6, try to develop a "feel" for the order of volume quantity. For example, ask yourself whether a pint or a liter is the larger quantity, and so on.

Volume Calculations

The volume of simple shapes, such as a cylinder, a cone, or a sphere, may be obtained using volume formulas as shown in Table 4.7. Indirect and direct estimation of volumes of objects are demonstrated in Examples 4.2 and 4.3.

TABLE 4.7 Some Useful Volume Formulas

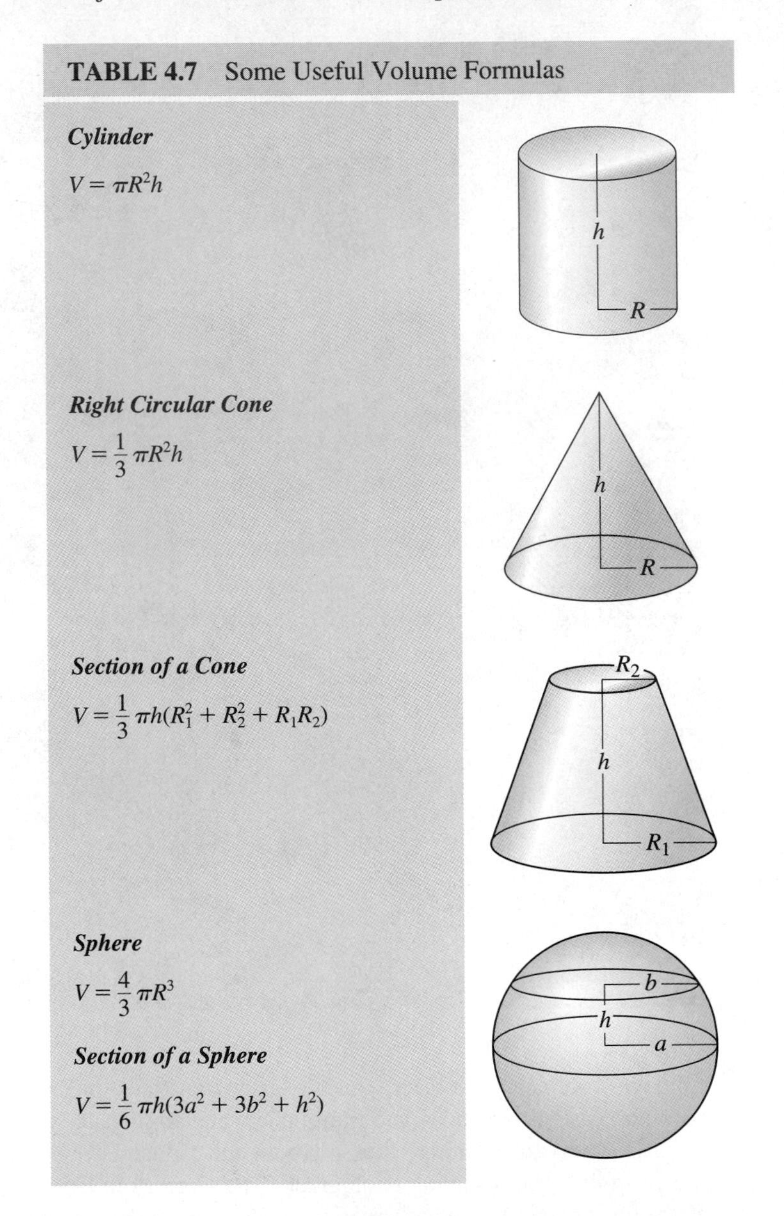

Cylinder

$$V = \pi R^2 h$$

Right Circular Cone

$$V = \frac{1}{3}\pi R^2 h$$

Section of a Cone

$$V = \frac{1}{3}\pi h(R_1^2 + R_2^2 + R_1 R_2)$$

Sphere

$$V = \frac{4}{3}\pi R^3$$

Section of a Sphere

$$V = \frac{1}{6}\pi h(3a^2 + 3b^2 + h^2)$$

EXAMPLE 4.2

Determine the exterior volume of the object shown in Figure 4.15.

For this example, we will use the buoyancy effect to measure the exterior volume of the object shown. We will consider two procedures. First, we obtain a large container that can accommodate the object. We will then fill the container completely to its rim with water and place the container inside a dry, empty tub. We next submerge the object with the unknown volume into the container until its top surface is just below the surface of the water. This will displace some volume of water, which is equal to the volume of the object. The water that overflowed and was collected in the tub can then be poured into a graduated cylinder to measure the volume of the object. This procedure is shown in Figure 4.16.

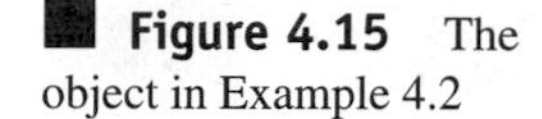
Figure 4.15 The object in Example 4.2

The second procedure makes direct use of the buoyancy force. We first suspend the object in air from a spring scale to obtain its weight. We then place the object, still suspended from the spring, into a container filled with water. Next, we record the apparent weight of the object. The difference between the actual weight of the object and the apparent weight of the object in water is the buoyancy force. Knowing the magnitude of the buoyancy force and using Equation (4.7), we can then determine the volume of the object. This procedure is depicted in Figure 4.17.

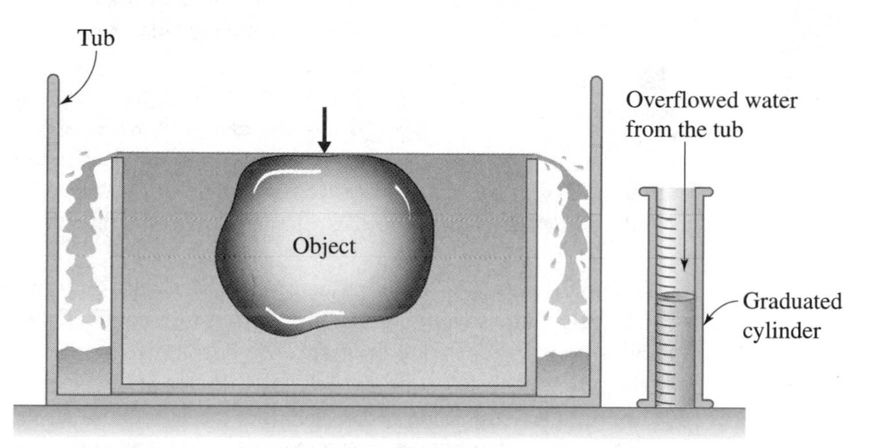

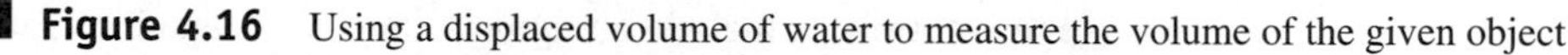
Figure 4.16 Using a displaced volume of water to measure the volume of the given object

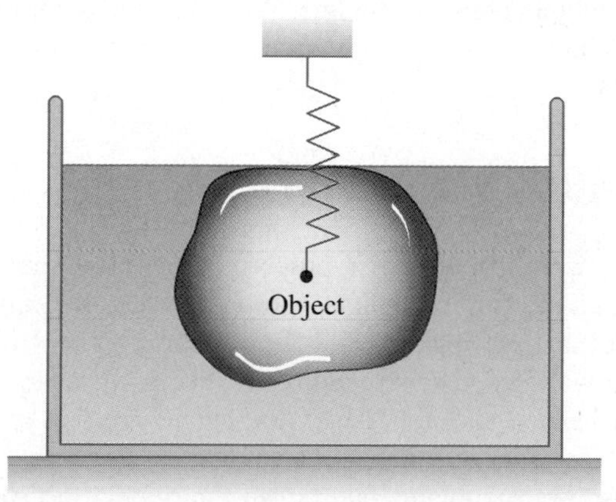

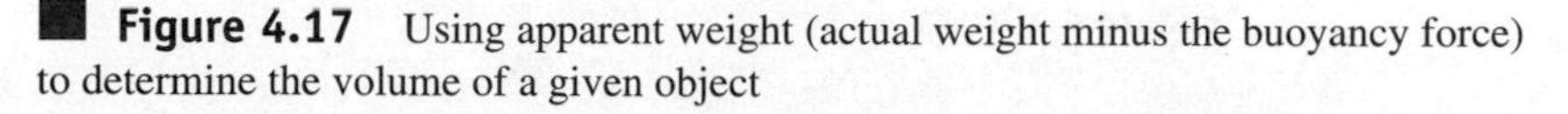
Figure 4.17 Using apparent weight (actual weight minus the buoyancy force) to determine the volume of a given object

EXAMPLE 4.3

Estimate the inside volume of a soda can. We have used a ruler to measure the height and the diameter of the can, as shown in Figure 4.18.

We may approximate the inside volume of the soda can using the volume of a cylinder of equal dimensions:

$$V = \pi R^2 h = (3.1415)\left(\frac{6.3\text{ cm}}{2}\right)^2(12.0\text{ cm}) = 374\text{ cm}^3 = 374\text{ mL}$$

Figure 4.18 Soda can in Example 4.3

Compared to the 355-mL value shown on a typical soda can, the approximated value seems reasonable. The difference between the approximated value and the indicated value may be explained in a number of ways. First, the soda container does not represent a perfect cylinder. If you were to look closely at the can you would note that the diameter of the can reduces at the top. This could explain our overestimation of volume. Second, we measured the outside diameter of the can, not the inside dimensions. However, this approach will introduce smaller inaccuracies because of the small thickness of the can.

We could have measured the inside volume of the can by filling the can with water and then pouring the water into a graduated cylinder or beaker to obtain a direct reading of the volume. As an alternative approach, we could have weighed the water and used the density of the water to obtain the inside volume of the can indirectly.

Finally, it is worth noting here that numerical solid modeling is an engineering topic that deals with computer generation of the surface areas and volumes of an actual object. Solid-modeling software programs are becoming quite common in engineering practice. Computer-generated solid models provide not only great visual images but also such information as magnitude of the area and the volume of the model. To generate numerical solid models of simple shapes, area and volume primitives are used. Other means of generating surfaces include dragging a line along a path or rotating a line about an axis, and, as with areas, you can also generate volumes by dragging or sweeping an area along a path or by rotating an area about a line. We will discuss computer solid modeling ideas in more detail in Chapter 15.

4.8 SECOND MOMENTS OF AREAS

In this section, we will consider a property of an area known as the *second moment of area*. The second moment of area, also known as the *area moment of inertia*, is an important property of an area that provides information on how hard it is to bend something. Next time you walk by a construction site, take a closer look at the cross-sectional area of the support beams, and notice how the beams are laid out. Pay close attention to the orientation of the cross-sectional area of an I-beam with respect to the directions of expected loads. Are the beams laid out in the orientation shown in Figure 4.19(a) or in Figure 4.19(b)?

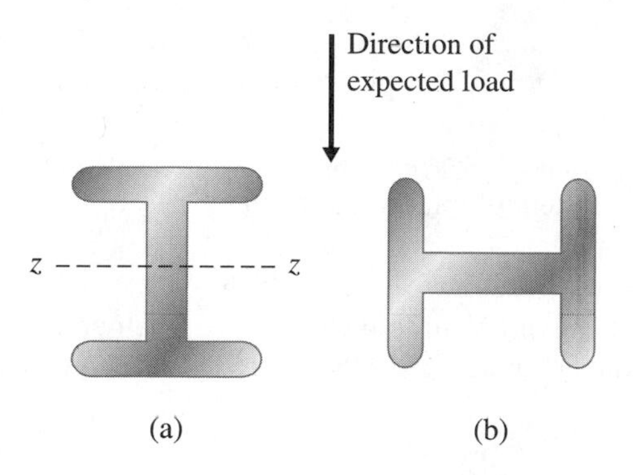

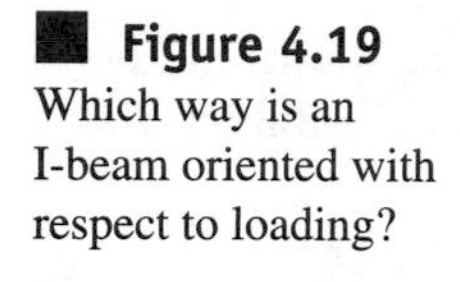
Figure 4.19 Which way is an I-beam oriented with respect to loading?

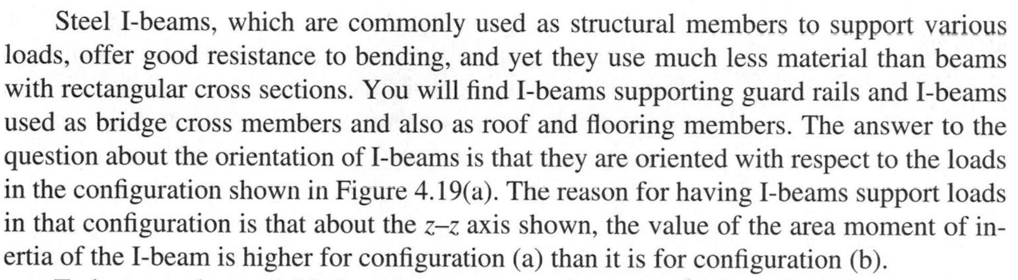

Steel I-beams, which are commonly used as structural members to support various loads, offer good resistance to bending, and yet they use much less material than beams with rectangular cross sections. You will find I-beams supporting guard rails and I-beams used as bridge cross members and also as roof and flooring members. The answer to the question about the orientation of I-beams is that they are oriented with respect to the loads in the configuration shown in Figure 4.19(a). The reason for having I-beams support loads in that configuration is that about the z–z axis shown, the value of the area moment of inertia of the I-beam is higher for configuration (a) than it is for configuration (b).

To better understand this important property of an area and the role of the second moment of inertia in offering a measure of resistance to bending, try the following experiment. Obtain a thin wooden rod and a yardstick. First try to bend the rod in the directions shown in Figure 4.20. If you were to report your findings, you would note that the circular cross section of the rod offers the same resistance to bending regardless of the direction of loading. This is because the circular cross section has the same distribution of area about an axis going through the center of the area. Note that we are concerned with bending a member, not twisting it! Now, try bending the yardstick in the directions shown in Figure 4.20. Which way is it harder to bend the yardstick? Of course, it is much harder to bend the yardstick in the direction shown in Figure 4.20(a). Again, that is because in the orientation shown in Figure 4.20(a), the second moment of area about the centroidal axis is higher.

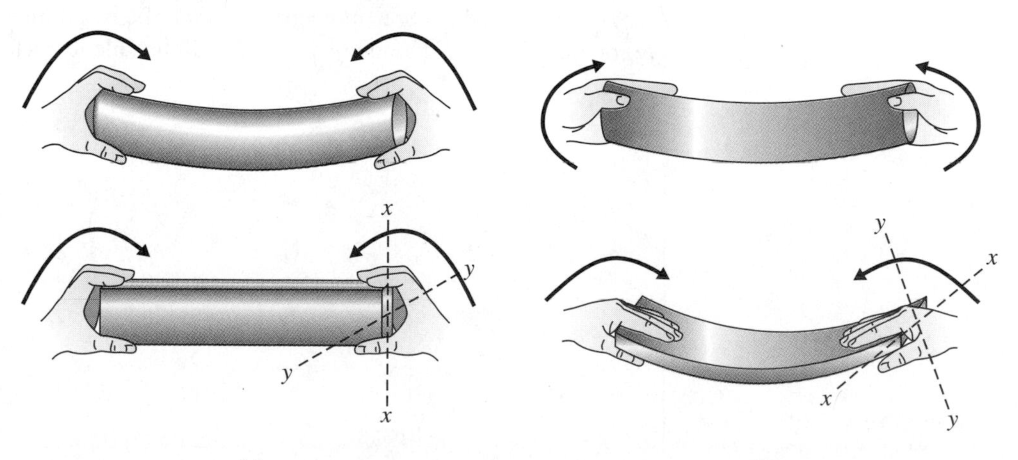

Figure 4.20 Bend the rod and yardstick in the directions shown.

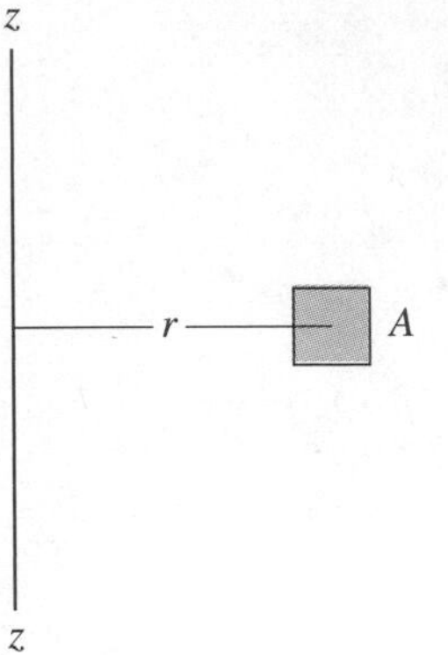

Figure 4.21 Small area element located at distance r from the z–z axis

Most of you will take a statics class, where you will learn more in depth about the formal definition and formulation of the second moment of area, or area moment of inertia, and its role in the design of structures. But for now, let us consider the simple situations shown in Figure 4.21. For a small area element dA, located at a distance r from the axis z–z, the area moment of inertia is defined by

$$I_{z-z} = r^2A \tag{4.8}$$

Now let us expand this problem to include more small area elements, as shown in Figure 4.22. The area moment of inertia for the system of discrete areas shown about the z–z axis is now

$$I_{z-z} = r_1^2A_1 + r_2^2A_2 + r_3^2A_3 \tag{4.9}$$

Similarly, we can obtain the second moment of area for a cross-sectional area such as a rectangle or a circle by summing the area moment of inertia of all the little area elements that make up the cross section. As you take calculus classes, you will learn that you can use integrals instead of summing the r^2A terms to evaluate the area moment of inertia of a continuous cross-sectional area. After all, the integral sign, $\int$, is nothing but a big S sign, indicating summation.

$$I_{z-z} = \int r^2\,dA \tag{4.10}$$

Also note that the reason this property of an area is called "second moment of area" is that the definition contains the product of *distance squared* and an area, hence the name "second moment of area." In Chapter 7, we will discuss the proper definition of a moment and how it is used in relation to the tendency of unbalanced forces to rotate things. As you will learn later, the magnitude of a moment of a force about a point is determined by the product of the perpendicular *distance* from the point about which the moment is taken to the line of action of the force and the magnitude of that force. You have to pay attention to what is meant by "a moment of a force about a point or an axis" and the way the term *moment* is incorporated into the name "the second moment of area" or "the area moment of inertia." Because the distance term is multiplied by another quantity (area), the word "moment" appears in the name of this property of an area.

You can obtain the area moment of inertia of any geometric shape by performing the integration given by Equation (4.10). You will be able to perform the integration and bet-

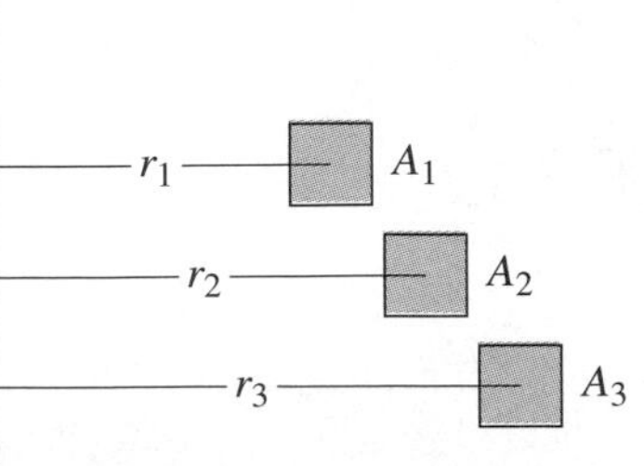

Figure 4.22 Second moment of area for three small area elements

ter understand what these terms mean in another semester or two. Keep a close watch for them in the upcoming semesters. For now, we will give you the formulas for area moment of inertia without proof. Examples of area moment of inertia formulas for some common geometric shapes are given next.

Rectangle

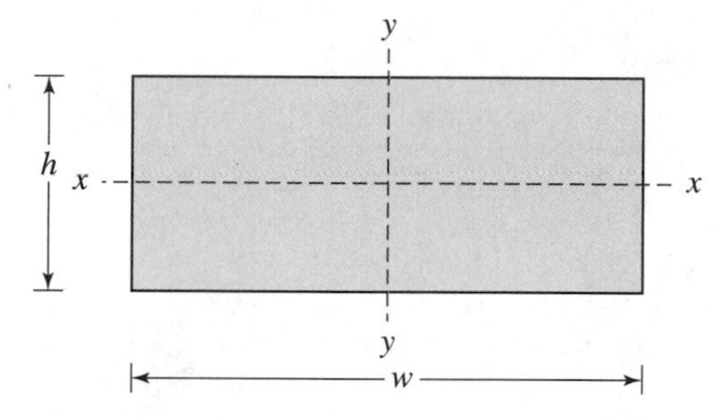

$$I_{x-x} = \frac{1}{12} wh^3 \tag{4.11}$$

$$I_{y-y} = \frac{1}{12} hw^3$$

Circle

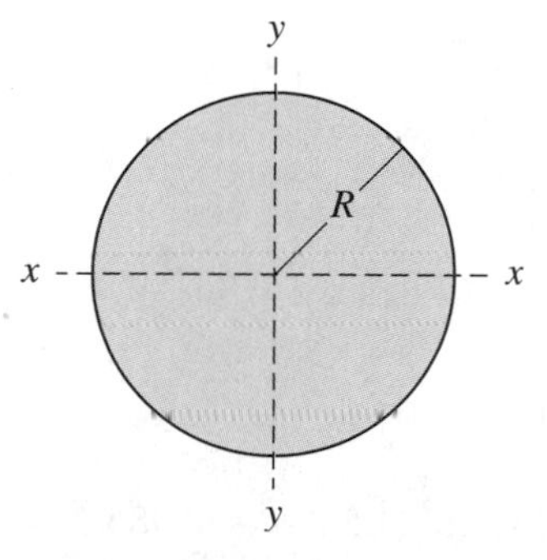

$$I_{x-x} = I_{y-y} = \frac{1}{4} \pi R^4 \tag{4.12}$$

The values of the second moment of area for standard I-beams are shown in Table 4.8.

In Chapter 6 we will look at another similarly defined property of an object, mass moment of inertia, which provides a measure of resistance to rotational motion.

TABLE 4.8 Examples of the Second Moment of Area of Standard I-Beams *(SI version)*

Designation	Depth (mm)	Width (mm)	Area (mm^2)	I_{x-x} (mm^4)	I_{y-y} (mm^4)
W460 × 113	463	280	14,400	554×10^6	63.3×10^6
W410 × 85	417	181	10,800	316×10^6	17.9×10^6
W360 × 57	358	172	7,230	160×10^6	11.1×10^6
W200 × 46.1	203	203	5,890	45.8×10^6	15.4×10^6

Table 4.8 *(English version)*

Designation	Depth (in.)	Width (in.)	Area (in²)	I_{x-x} (in⁴)	I_{y-y} (in⁴)
W18 × 76	18.21	11.03	22.3	1330	152
W16 × 57	16.43	7.12	16.8	758	43.1
W14 × 38	14.10	6.77	11.2	385	26.7
W8 × 31	8.00	7.95	9.1	110	37.1

EXAMPLE 4.4

Calculate the second moment of area for a 2 × 4 stud with the actual dimensions shown on the accompanying diagram.

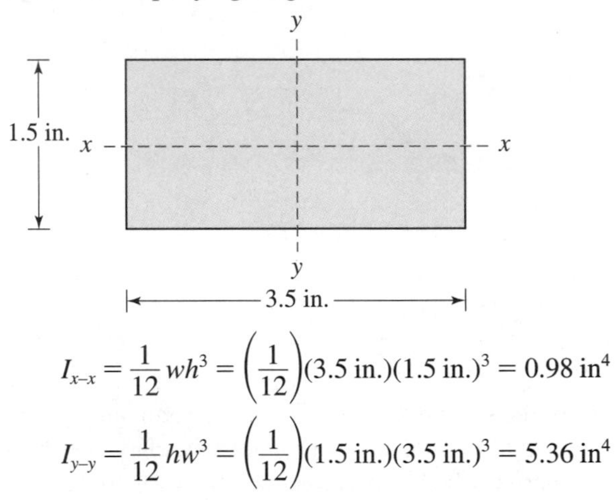

$$I_{x-x} = \frac{1}{12} wh^3 = \left(\frac{1}{12}\right)(3.5 \text{ in.})(1.5 \text{ in.})^3 = 0.98 \text{ in}^4$$

$$I_{y-y} = \frac{1}{12} hw^3 = \left(\frac{1}{12}\right)(1.5 \text{ in.})(3.5 \text{ in.})^3 = 5.36 \text{ in}^4$$

Note that the 2 × 4 lumber will show more than 5 times more resistance to bending about the *y–y* axis than it does about the *x–x* axis.

EXAMPLE 4.5

Calculate the second moment of area for a 5-cm-diameter shaft about the *x–x* and the *y–y* axes shown.

$$I_{x-x} = I_{y-y} = \frac{1}{4}\pi R^4 = \frac{1}{4}(3.1415)\left(\frac{5 \text{ cm}}{2}\right)^4 = 30.7 \text{ cm}^4$$

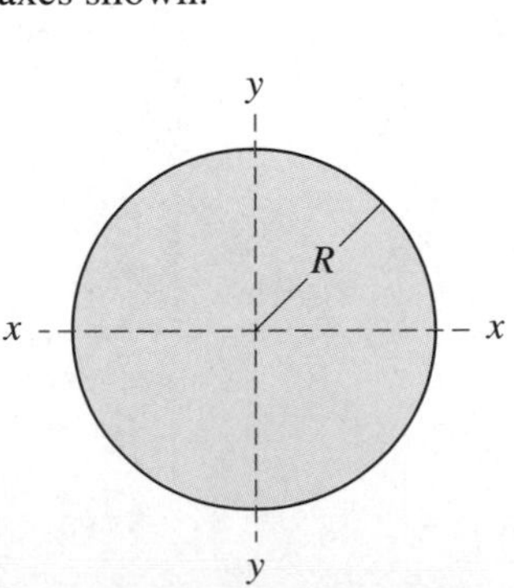

Finally, it is important to emphasize the fact that all physical variables discussed in this chapter are based on the fundamental dimension length. For example, area has a dimension of $(\text{length})^2$, volume has a dimension of $(\text{length})^3$, and the second moment of area has a dimension of $(\text{length})^4$. In Chapter 5, we will look at time- and length-related parameters in engineering.

SUMMARY

Now that you have reached this point in the text

- You should understand the significant role the fundamental dimension length plays in engineering problems. You should also realize the importance of area and volume in engineering applications.
- You should have developed a "feel" for the relative magnitude of various length units, area units, and volume units.
- You should know the difference between the actual size and a nominal size of objects.
- You should know how to measure areas and to approximate planar surfaces using the trapezoidal rule.
- You should have a good understanding of the physical significance of the second moment of area in structural analysis.

PROBLEMS

1. Seasoned engineers are good at estimating physical values without using tools. Therefore, you need to begin developing a "feel" for the sizes of various physical quantities. This exercise is intended to help you develop this ability. Using the table below, first estimate the dimensions of the given objects. Next measure, or look up, the actual dimensions of the objects, and compare them to your estimated values. How close are your estimations? Do you have a "feel" for units of length yet?

	Estimated Values		Measured Values		Difference	
Object	(cm)	(in.)	(cm)	(in.)	(cm)	(in.)
This book						
A pen or a pencil						
A laptop computer (closed)						
A 12-fl.-oz. soda can						
The distance from home to school. Use your car's odometer to measure the actual distance.	(km)	(miles)	(km)	(miles)	(km)	(miles)
A dollar bill						
The height of your engineering building						
Wingspan of a Boeing 747	(m)	(ft)	(m)	(ft)	(m)	(ft)

2. The following exercises are designed to help you become aware of the significance of various dimensions around you. You see these dimensions every day, but perhaps you never looked at them with the eyes of an engineer. Measure and discuss the significance of the dimensions of the following items
   a. The dimensions of your bedroom or living room
   b. The dimensions of the hallway
   c. The window dimensions
   d. The width, height, and thickness of your apartment doors or dormitory doors
   e. The distance from the floor to the doorknob
   f. The distance from the floor to the light switches
   g. The dimensions of your desk
   h. The dimensions of your bed
   i. The distance from the floor to the bathroom sink
   j. The distance from the tub surface to the showerhead

3. In this exercise, you are to explore the size of your classroom, the seating arrangements, the location of the chalkboard (or the whiteboard) with respect to the classroom's main entrance, and the size of the board relative to classroom size. Discuss your findings in a brief report to your instructor.
   a. What are the dimensions of the classroom?
   b. How far apart are the seats placed? Is this a comfortable arrangement? Why or why not?
   c. How far above the floor is the chalkboard (or whiteboard) placed? How wide is it? How tall is it? What is its relative placement in the classroom? Can someone sitting in the back corner of the room see the board without too much difficulty?

4. This is a sports-related assignment. First look up the dimensions, and then show the dimensions on a diagram. You may need to do some research to obtain the information required here.
   a. A basketball court
   b. A tennis court
   c. A football field
   d. A soccer field
   e. A volleyball court

5. These dimensions concern transportation systems. Look up the dimensions of the following vehicles; give the model year and the source of your information.
   a. Buick LeSabre or a car of your choice
   b. Ford Taurus
   c. Honda Accord
   d. How wide is the driver's seat in each car mentioned in (a)–(c)?
   e. A city bus
   f. How wide are the bus seats?
   g. A high-speed passenger train
   h. How wide are the train seats in the coach section?

6. This is a bioengineering assignment. Investigate the following dimensions and write a brief report to your instructor about your findings.
   a. What is the average diameter of a healthy red blood cell?
   b. What is the average diameter of a white blood cell?
   c. Measure and record the length of each finger of ten male adults in your class. Also, measure and record the length of each finger of ten female adults. Compute the averages for males and females, and compare the female results to the male results. Present the results in both SI and English units.
   d. Measure the average surface area of an adult male arm by covering the arm with paper and then measuring the area of the paper. Measure the surface area of an adult male leg. How much plaster would be required for a plaster cast around an average adult male leg, assuming 2-mm thickness? What is the volume of the plaster needed?
   e. Estimate the average surface (skin) area of an adult male.
   f. What are the lengths of human small and large intestines?

7. This assignment is related to civil engineering.
   a. How wide is each lane of a street in your neighborhood? Talk to your city engineer to obtain information.
   b. Visit the U.S. Department of Transportation Web site to find out how wide each lane of an interstate highway is. Do all interstate highway lanes have the same width?
   c. Find out how tall the Hoover Dam is. How wide is the Hoover Dam? What is the area of the dam exposed to upstream water? Discuss how and why the thickness of the dam varies with its height.
   d. How tall are the interstate bridges so that an average truck can go under them?
   f. How far above the road are the highway signs placed?
   g. What is the average height of a tunnel? What is the area of a tunnel at the entrance?

Write a brief report discussing your findings.

8. Investigate the size of the main waterline in your neighborhood. What are the inside and outside diameters? What is the nominal size of the pipe? What is the cross-sectional area of water flow? Investigate the size of piping used in your home. Write a brief memo to your instructor discussing your findings.
9. Look up the length of the Alaska pipeline. What are the inside and outside diameters of the pipes used in transporting oil? How far apart are the booster pump stations? How thick is the pipeline? What is the cross-sectional area of the pipe? What is the nominal size of the pipe? Write a brief memo to your instructor discussing your findings.
10. Investigate the size of pipes used in transporting natural gas to your state. What are the inside and outside diameters of the pipe? What is the distance between boosting stations? What is the cross-sectional area of the pipe? Write a brief memo to your instructor discussing your findings.
11. Investigate the diameter of the electrical wire used in your home. How thick are the interstate transmission lines, and what is their cross-sectional area? Write a brief memo to your instructor discussing your findings.
12. What is the operating wavelength range for the following items?
    a. A cellular phone
    b. FM radio transmissions
    c. Satellite TV broadcasting

    Write a brief memo to your instructor discussing your findings.
13. Derive the formula given for the area of a trapezoid. Start by dividing the area into two triangular areas and one rectangular area.
14. Trace on a white sheet of paper the boundaries of the area of the United States shown in the accompanying diagram.
    a. Use the trapezoidal rule to determine the total area.
    b. Approximate the total area by breaking it into small squares. Count the number of total squares and add what you think is an appropriate value for the remaining area. Compare your findings to part (a).
    c. Use an analytical balance from a chemistry lab to weigh an a $8\frac{1}{2}$-×-11 sheet of paper. Record the dimensions of the paper. Draw the boundaries of the area shown on the accompanying figure and cut around the boundary of the area. Determine the weight of the piece of paper that has the area drawn on it. Compare the weights and determine the area of the given profiles. What assumptions did you make to arrive at your solution? Compare the area computed in this manner to your results in parts (a) and (b). Are there any other ways that you could have determined the area of the profile? Explain.

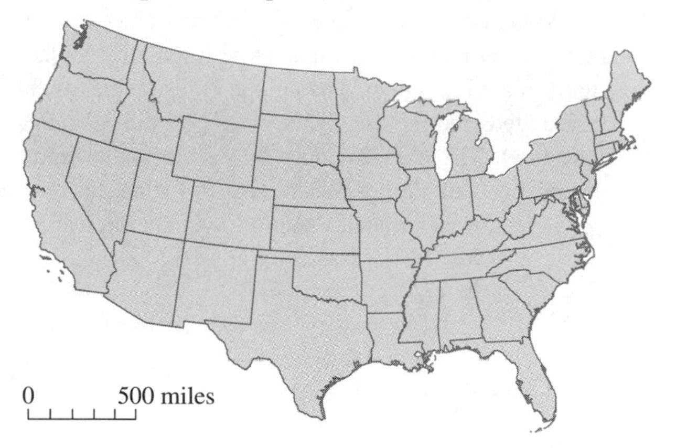

15. Obtain a woman's high-heeled dress shoe and a woman's athletic walking shoe. For each shoe, make imprints of the floor contact surfaces. Determine the total area for each shoe, and assuming the woman who wears these shoes weighs 120 lb, calculate the average pressure at the bottom of the shoe for each shoe style. What are your findings? What are your recommendations? Recall that

$$\text{pressure} = \frac{\text{force}}{\text{area}}$$

16. Obtain two different brands of cross-country skis. Make imprints of the expected contact surface of the skies with snow. Obtain a value for the average pressure for each ski, assuming the person wearing the ski weighs 180 lb.
17. Estimate the average pressure exerted on the road by the following.
    a. A family car
    b. A truck
    c. A bulldozer

    Discuss your findings.
18. Using area as your variable, suggest ways to cool freshly baked cookies faster.

19. Estimate how much material is needed to make 100,000 stop signs. Investigate the size of one side of a stop sign and the kind of material it is made from. Write a brief memo to your instructor discussing your findings.

20. Estimate how much material is needed to make 100,000 traffic yield signs. Investigate the size of one side of a yield sign and the kind of material it is made from. Write a brief memo to your instructor discussing your findings.

21. As explained in the chapter, the air resistance to motion of a vehicle is determined experimentally by placing it in a wind tunnel. The air speed inside the tunnel is changed, and the drag force acting on the vehicle is measured. The experimental data is normally represented by a single coefficient that is called the *drag coefficient*. It is defined by the following relationship:

$$\text{drag coefficient} = \frac{\text{drag force}}{\frac{1}{2}(\text{air density})(\text{air speed})^2(\text{frontal area})}$$

or, in a mathematical form,

$$C_d = \frac{F_d}{\frac{1}{2}\rho V^2 A}$$

It was also explained in this chapter that the frontal area A is the frontal projection of the area, and could be approximated simply by multiplying 0.85 times the width and the height of a rectangle that outlines the front of a vehicle when you view it from a direction normal to the front windshield. The 0.85 factor is to adjust for rounded corners, open space below the bumper, and so on. Typical drag coefficient values for sports cars are between 0.27 and 0.38, and for sedans the values are between 0.34 and 0.5. This assignment requires you to actually measure the frontal area of a car. Tape a ruler or a yardstick to the bumper of your car. The ruler will serve as a scale. Take a photograph of the car and use any of the methods discussed in this chapter to compute the frontal area of the car.

22. A machinist in an engineering machine shop has ordered a sheet of plastic with dimensions of 10 ft × 12 ft × 1 in. width, length, and thickness, respectively. Can the machinist get the sheet of plastic inside the machine shop provided that the door dimensions are 6 ft × 8 ft? Give the maximum dimensions of a sheet that can be moved inside the shop and explain how.

23. Investigate the volume capacity of a barrel of oil in gallons, cubic feet, and cubic meters. Also, determine the volume capacity of a bushel of agricultural products in cubic inches, cubic feet, and cubic meters. Write a brief memo to your instructor discussing your findings.

24. Measure the outside diameter of a flagpole or a streetlight pole by first wrapping a piece of string around the object to determine its circumference. Why do you think the flagpole or the streetlight pole is designed to be thicker at the bottom near the ground than at the top? Explain your answer.

25. Measure the width and the length of a treadmill machine. Discuss what some of the design factors are that determine the appropriateness of the values you measured.

26. Collect information on the standard sizes of automobile tires. Create a list of these dimensions. What do the numbers indicated on a tire mean? Write a brief memo to your instructor discussing your findings.

27. Visit a hardware store and obtain information on the standard size of the following items.

a. Screws
b. Plywood sheets
c. PVC pipes
d. Lumber

Create a list with both the actual and the nominal sizes.

28. Using just a yardstick or a measuring tape, open the classroom door 35 degrees. *You cannot use a protractor.*

29. This is a group project. Determine the area of each class member's right hand. You can do that by tracing the profile of everyone's fingers on a white sheet of paper and using any of the techniques discussed in this chapter to compute the area of each hand. Compile a data bank containing the areas of the right hands of each person in class. Based on some average, classify the data into small, medium, and large. Estimate how much material (outer leather and inner lining) should be ordered to make 100,000 gloves. How many spools of thread should be ordered? Describe and/or draw a diagram of the best way to cut the hand profiles from sheets of leather to minimize wasted materials.

30. This is a group project. Given a square kilometer of land, how many cars can be parked safely there? Determine the appropriate spacing between cars and the width of drive lanes. Prepare a diagram showing your solution, and write a brief report to your instructor discussing your assumptions and findings.
31. Estimate the length of tubing used to make a bicycle rack on your campus.
32. Determine the base area of an electric steam iron.
33. Determine the area of an incandescent 100-W light bulb.
34. Obtain a trash bag and verify its listed volume.
35. Discuss the size of a facial tissue, a paper towel, and a sheet of toilet paper.
36. Estimate how much silver is needed to make a four-piece set of silverware containing teaspoons, tablespoons, forks, and knives (16 pieces in all). Discuss your assumptions and findings.
37. Calculate the second moment of area of a 2-in. diameter shaft about the x–x and y–y axes, as shown.

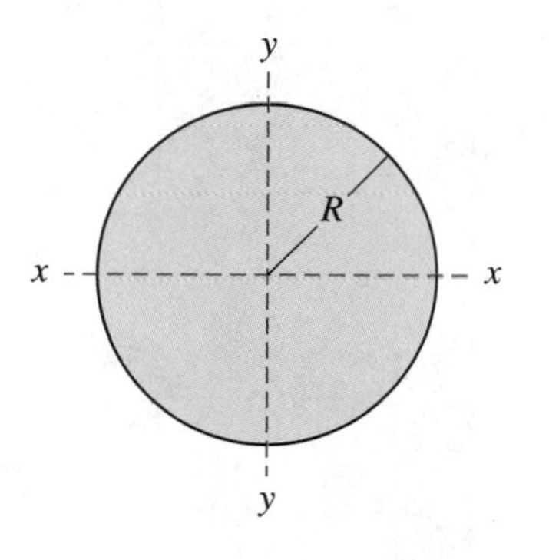

38. Calculate the second moment of area for a 2×6 piece of lumber about the x–x and y–y axes, as shown. Visit a hardware store and measure the actual dimensions of a typical 2×6 piece of lumber.

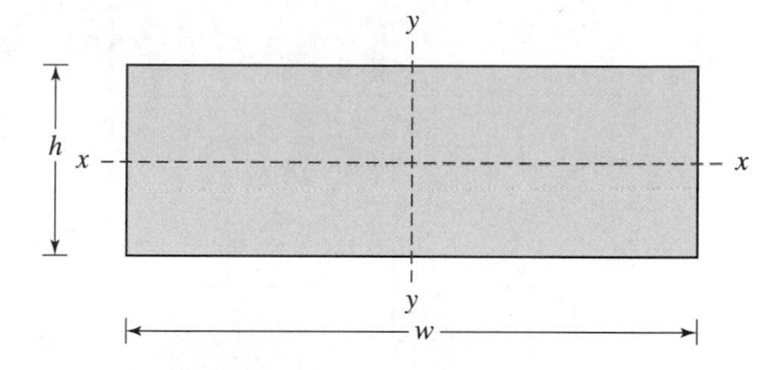

The New York City Water Tunnel No. 3*

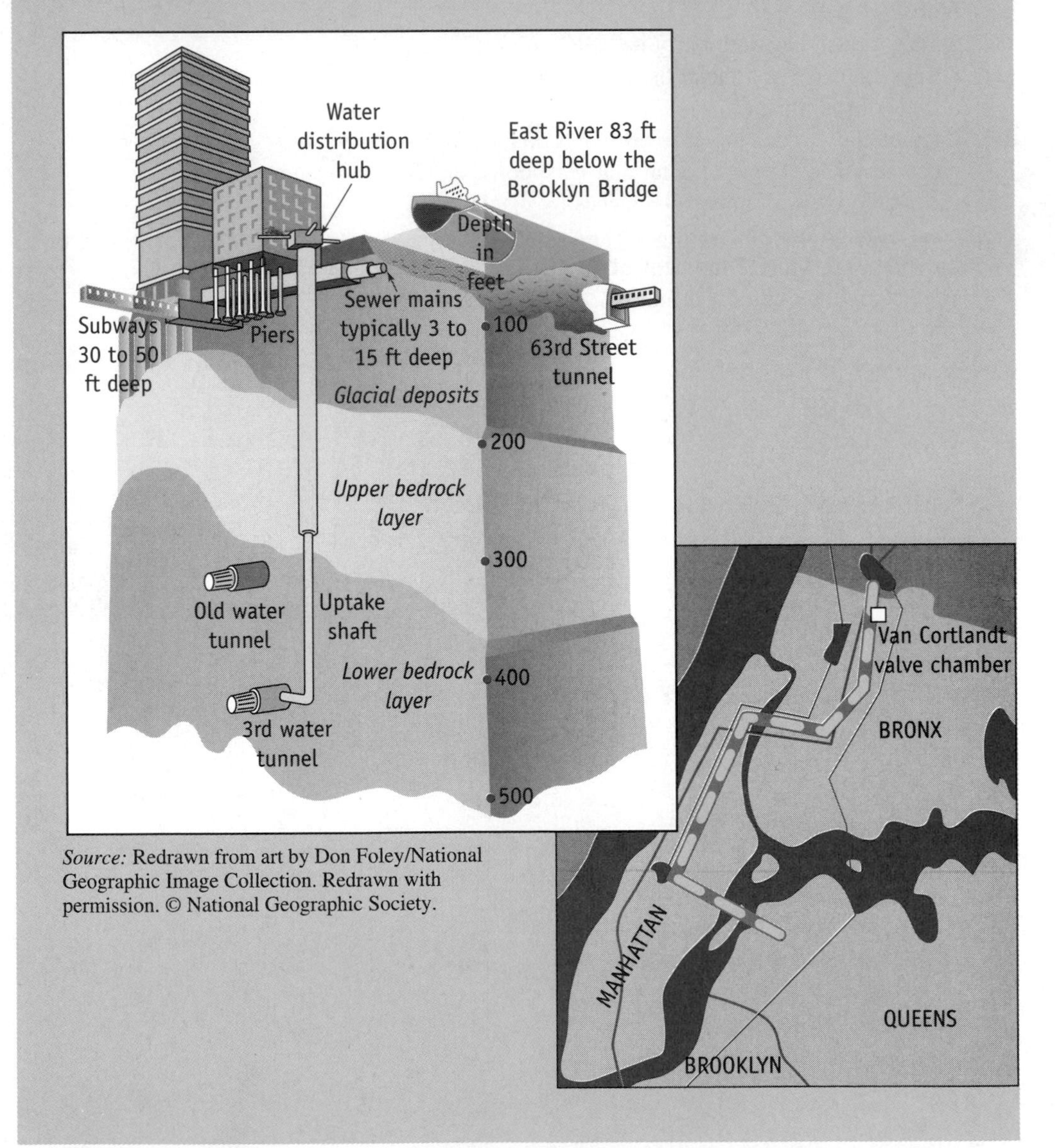

Source: Redrawn from art by Don Foley/National Geographic Image Collection. Redrawn with permission. © National Geographic Society.

* Materials were adapted with permission of the New York City Department of Environmental Protection.

In 1954, New York City recognized the need for a third water tunnel to meet the growing demand on its more than 150-year-old water supply system. Planning for City Tunnel No. 3 began in the early 1960s, and actual construction commenced nearly a decade later in 1970. The New York City Water Tunnel No. 3, because of its size, length, controlling devices, and depth of excavation, represents one of the most complex and challenging engineering projects in the world today. The New York City Water Tunnel No. 3 also represents the largest capital improvement project in the city's history. Tunnel No. 3, when complete, is projected to cost nearly $6 billion dollars. Constructed by the New York City Department of Environmental Protection (DEP) employees, engineers, and underground construction workers (known as sandhogs), the tunnel will eventually span more than 60 miles. The tunnel is expected to be complete in 2020. Since 1970, when construction on the tunnel began, a total of 24 people have died in construction-related accidents.

Although City Tunnel No. 3 will not replace City Tunnels No. 1 and No. 2, it will enhance and improve the adequacy and dependability of the entire water supply system as well as improve service and pressure to outlying areas of the city. It will also allow the DEP to inspect and repair City Tunnels No. 1 and No. 2 for the first time since they were put into operation in 1917 and 1936, respectively.

The City Tunnel No. 3 project is to be completed in four stages. Stage 1 of the project has already been completed. Similar to City Tunnels No. 1 and 2, stage 1 of City Tunnel No. 3 begins at the Hillview Reservoir in Yonkers. It is constructed in bedrock 250 to 800 ft below the surface and runs 13 mi, extending across Central Park until about Fifth Avenue and 78th Street, then stretching eastward under the East River and Roosevelt Island into Astoria, Queens. The first stage of the tunnel, which cost $1 billion to construct, consists of a 24-ft-diameter concrete-lined pressure tunnel that steps down in diameter to 20 ft. Water travels along this route and rises from the tunnel via 14 supply shafts, or "risers," and feeds into the city's water distribution system. Currently, City Tunnel No. 3 is serving the Upper East and Upper West Sides of Manhattan, Roosevelt Island, and many neighborhoods in the Bronx west of the Bronx River.

There are four unique valve chambers that will allow stage 1 to connect to future portions of the tunnel without disrupting the flow of water service. Each of the existing valve chambers contains a series of 96-in.-diameter conduits with valves and flowmeters to direct, control, and measure the flow of water in sections of the tunnel. The largest of the valve chambers is in Van Cortlandt Park in the Bronx. Built 250 ft below the surface, this valve chamber will control the daily flow of water from the Catskill and Delaware water supply systems into Tunnel No. 3. In a design departure from the two existing tunnels, valves that control the flow of water in Tunnel No. 3 will be housed in large underground valve chambers, making them accessible for maintenance and repair. The valves for City Tunnels No. 1 and 2 are at the tunnel level and thus inaccessible when the tunnels are in service. Three of these four unique subsurface valve chambers have already been built to allow the connection of future stages of the tunnel without removing the water or taking any other stage of the tunnel out of service. These three valve chambers are located in the Bronx at Van Cortlandt Park (Shaft 2B), in Manhattan at Central Park (Shaft 13B), and Roosevelt Island (Shaft 15B).

Stage 2 of City Tunnel No. 3 construction is comprised of a two-leg Brooklyn/Queens section and a section in Manhattan. Stage 2 is scheduled to be completed in 2008 at an approximate cost of $1.5 billion. The combination of stages 1 and 2 will provide the system with the ability to bypass one or both of City Tunnels No. 1 and 2.

Stage 3 involves the construction of a 16-mi-long section extending from the Van Cortlandt Park Valve Chamber to the Kensico Reservoir, which contains water from the Catskill and Delaware systems. When stage 3 is completed, City Tunnel No. 3 will operate at greater pressure, induced by the high elevation of Kensico Reservoir. It will also provide an additional aqueduct to supply water to the city that will parallel the Delaware and Catskill aqueducts. Stage 4, which will be 14 mi long, will travel from Van Cortlandt Park under the East River into Woodside, Queens.

Construction of the remaining three stages of City Tunnel No. 3 is being accelerated by the use of a mechanical rock excavator called a tunnel-boring machine (TBM). This machine, which is lowered in sections and assembled on the tunnel floor, chips off sections of bedrock through the continuous rotation of a series of steel cutting teeth. The TBM, which replaces the conventional drilling and blasting methods used during the construction of stage 1, allows for faster and safer excavation.

A photo summary illustrating various aspects of Tunnel No. 3 is shown in the accompanying photos.*

Excavation of stage 1 of City Water Tunnel No. 3 looked like this in 1972. Today, water filling this portion of the tunnel is serving areas of the Bronx and Manhattan.

*Photos courtesy of New York City Department of Environmental Protection.

A form is constructed to prepare the tunnel to be lined with concrete.

This finished section of the tunnel, which is 20 feet in diameter, is under Shaft 13B, near the Central Park Reservoir.

The valve chamber in Van Cortlandt Park was completed in the early 1990s.

The tunnel boring machine (TBM), which was lowered in sections and assembled on the tunnel floor, chips off sections of bedrock in stage 2 of City Water Tunnel No. 3. The TBM's rotating steel cutting teeth replace the dynamite needed for earlier excavation.

A close-up view of the tunnel boring machine (TBM).

PROBLEM

1. Estimate the volume of the earth that has to be removed to make room for the 60-mi-long, 24-ft-diameter, concrete-lined water tunnel. Also, investigate the capacity of typical dump trucks used in removal of earth materials.

CHAPTER 5 Time and Time-Related Parameters

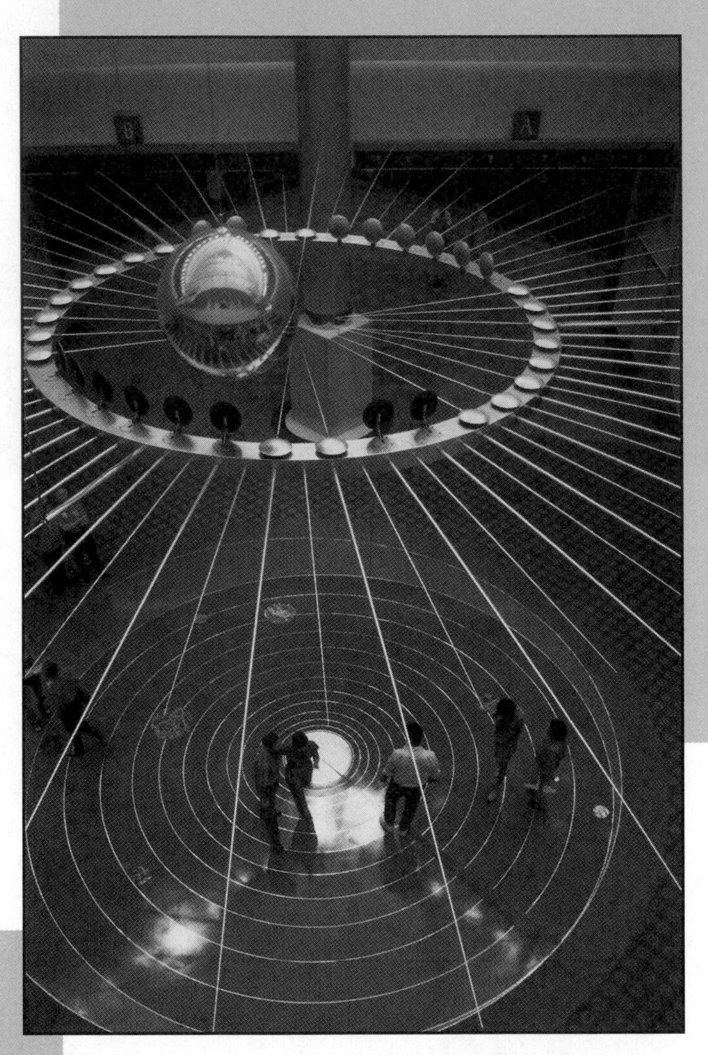

Good engineers recognize the role of time in their lives and in calculating speed, acceleration, and flow of traffic, as well as flow of materials and substances. They know what is meant by *frequency* and *period* and understand the difference between a steady and a transient process. Good engineers have a comfortable grasp of rotational motion and understand how it differs from translational motion.

In the previous chapter, we considered the role of length and length-related parameters in engineering. In this chapter, we will investigate the role of time as a fundamental dimension and other time-related engineering parameters, such as frequency and period. We will first discuss why the time variable is needed to describe events, processes, and other occurrences in our physical surroundings. We will then explain the role of periods and frequency in recurring, or periodic, events (events that repeat themselves). A brief introduction to parameters describing traffic flow is also given in this chapter. Finally, we will look at engineering variables that involve length and time, including linear velocities and accelerations and volume flow rate of fluids. Rotational motion is also introduced at the end of this chapter. ■

5.1 TIME AS A FUNDAMENTAL DIMENSION

We live in a dynamic world. Everything in the universe is in a constant state of motion. Think about it! Everything in the universe is continuously moving. The earth and everything associated with it moves around the sun. All of the solar planets, and everything that comprises them, are moving around the sun. We know that everything outside our solar system is moving too. From our everyday observation we also know that some things move faster than others. For example, people can move faster than ants, or a rabbit can move faster than a turtle. A jet plane in flight moves faster than a car on a highway.

Time is an important parameter in describing motion. How long does it take to cover a certain distance? A long time ago, humans learned that by defining a parameter called *time* they could use it to describe the occurrences of events in their surroundings. Think about the questions frequently asked in your everyday life: How old are you? How long does it take to go from here to there? How long does it take to cook this food? How late are you open? How long is your vacation? How long will you be gone? How long is the Christmas break? How long have you two known each other? We use the parameter of time to express how old we are. We have also associated time with natural occurrences in our lives, for example, to express the relative position of the earth with respect to the sun, we use day, night, 2:00 A.M. or 3:00 P.M., or May 30. The parameter time has been conveniently divided into smaller and larger intervals, such as seconds, minutes, hours, days, months, years, centuries, and millennia. We are continuously learning more and more about our surroundings in terms of how nature was put together and how it works, thus we need shorter and shorter time divisions, such as microseconds and nanoseconds. For example, with the advent of high-speed communication lines, the time that it takes electrons to move between short distances can be measured in nanoseconds.

We have also learned from our observation of the world around us that we can combine the parameter time with the parameter length to describe how fast something is moving. When we ask how fast, we should be careful to state with respect to what. Remember, everything in the universe is moving.

Before discussing the role of time in engineering analysis, let us focus on the role of time in our lives—our limited time budget. Today we can safely assume that the average life expectancy of a person living in the Western world is around 75 years. Let us use this number and perform some simple arithmetic operations to illustrate some interesting points. Converting the 75 years to hours, we have:

$$(75 \text{ years})(365.25 \text{ days/year})(24 \text{ hours/day}) = 657{,}450 \text{ hours}$$

On an average basis, we spend about 1/3 of our lives sleeping; this leaves us with 438,300 waking hours. Considering that traditional college freshmen are 18 years old, you have 333,108 waking hours still available to you if you live to the age of 75 years. Think about this for a while. If you were given only $333,108 for the rest of your life, would you throw away a dollar here and a dollar there as you were strolling through life? Perhaps not, especially knowing that you will not get any more money. Life is short! Make good use of your time, and at the same time enjoy your life.

Now let us look at the role of time in engineering problems and solutions. Most engineering problems may be divided into two broad areas: *steady* and *unsteady* analysis. The problem is said to be *steady* when the value of a physical quantity under investigation does not change over time. If the value of a physical quantity changes with time, then the problem is said to be *unsteady* or *transient*.

A good example of an unsteady situation that is familiar to everyone is the rate of a person's physical growth. From the time you were born until you reached your late teens or early twenties, your physical dimensions changed with time. You became taller, your arms grew longer, your shoulders got wider. Then, from 20 years and older, the situation becomes steady. Of course, in this example not only the dimension length changes with time but also your mass changes with time. Another common example of an unsteady event that you are familiar with deals with the physical variable temperature. When you lay freshly baked cookies out to cool, the temperature of the cookies decreases with time until they reach the air temperature in the kitchen. There are many engineering problems

There is still time for you to do some good things in life.

Once I reach the age of 20, does my physical dimension really become steady? What if I were to gain mass?

that are also unsteady. You will find examples of unsteady processes in materials casting, materials processing, plastic forming, heating or cooling applications, and in food processing. The transient response of a mechanical or structural system to a suddenly applied force is another instance of an unsteady engineering problem; for example, the response of a car's suspension system as you drive through a pothole, or the response of a building to an earthquake.

5.2 MEASUREMENT OF TIME

Early humans relied on the relative position of the earth with respect to the sun, moon, stars, or other planets to keep track of time. The lunar calendar was used by many early civilizations. These celestial calendars were useful in keeping track of long periods of time, but humans needed to devise a means to keep track of shorter time intervals, such as what today we call an hour. This need led to the development of clocks (see Figure 5.1). Sun clocks, also known as shadow clocks or sundials, were used to divide a given day into smaller periods. The moving shadow of the dial marked the time intervals. Like other human-invented instruments, the sundial evolved over time into elaborate instruments that accounted for the shortness of the day during the winter as compared to the summer to provide for a better year-round accuracy. Sand glasses (glass containers filled with sand) and water clocks were among the first time-measuring devices that did not make use of the relative position of the earth with respect to the sun or other celestial bodies. Most of you have seen a sand glass (sometimes referred to as an hourglass); the water clocks were basically made of a graduated container with a small hole near the bottom. The container held water and was tilted so that the water would drip out of the hole slowly. Graduated cylindrical containers, into which water dripped at a constant rate, were also used to measure the passage of an hour. Over the years, the design of water clocks was also modified. The next revolution in timekeeping came during the 14th century when weight-driven mechanical

Figure 5.1 Are we really measuring time?

clocks were used in Europe. Later, during the 16th century, came spring-loaded clocks. The spring mechanism design eventually led to smaller clocks and to watches. The oscillation of a pendulum was the next advancement in the design of clocks.

Quartz clocks eventually replaced the mechanical clocks around the middle of the 20th century. A quartz clock or watch makes use of the piezoelectric property of quartz crystal. A quartz crystal, when subjected to a mechanical pressure, creates an electric field. The inverse is also true—that is to say, the shape of the crystal changes when it is subjected to an electric field. These principles are used to design clocks that make the crystal vibrate and generate an electric signal of constant frequency.

As we stated in Chapter 3, the natural frequency of the cesium atom was adopted as the new standard unit of time. The unit of a second is now formally defined as the duration of 9,192,631,770 periods of the radiation corresponding to the transition between the two hyperfine levels of the ground state of the cesium 133 atom.

The Need for Time Zones

You know that the earth rotates about an axis that runs from the South Pole to the North Pole, and it takes the earth 24 hours to complete one revolution about this axis. Moreover, from studying globes and maps, you may have noticed that the earth is divided into 360 circular arcs that are equally spaced from east to west; these arcs are called *longitudes*. The zero longitude was arbitrarily assigned to the arc that passes through Greenwich, England. Because it takes the earth 24 hours to complete one revolution about its axis, every 15 degrees longitude corresponds to 1 hour (360 degrees/24 hours = 15 degrees per one hour). For example, someone exactly 15 degrees west of Chicago will see the sun in the same exact position one hour later as it was observed by another person in Chicago one hour before. The earth is also divided into latitudes, which measure the angle formed by the line connecting the center of the earth to the specific location on the surface of the earth and the equatorial plane, as shown in Figure 5.2. The latitude varies

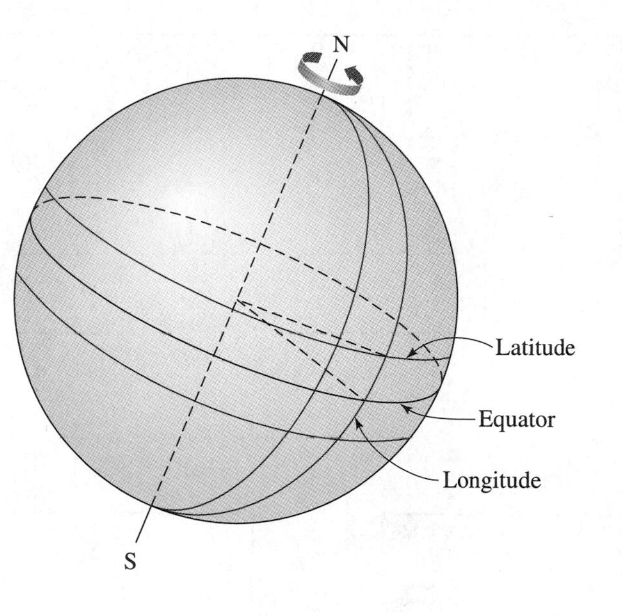

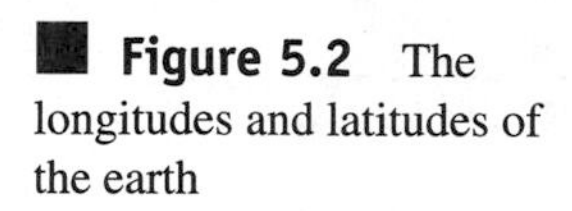

Figure 5.2 The longitudes and latitudes of the earth

from 0 (equatorial plane) degrees to 90 degrees north (North Pole) and from 0 degrees to 90 degrees south (South Pole).

The need for time zones was not realized until the latter part of the 19th century when the railroad companies were expanding. The railroad companies realized a need for standardizing their schedules. After all, 8:00 A.M. in New York City did not correspond to 8:00 A.M. in Denver, Colorado. Thus, a need for a uniform means to keep track of time and its relationship to other locations on the earth was born. It was railroad scheduling and commerce that eventually brought nations together to define time zones. The standard time zones are shown in Figure 5.3.

Daylight Saving Time

Daylight saving time was originally put into place to save fuel (energy) during hard times such as World War I and World War II. The idea is simple; by setting the clock forward in the spring and keeping it that way during the summer and the early fall, we extend the daylight hours and, consequently, we save energy. For example, on a certain day in the spring, without daylight saving time, it would get dark at 8:00 P.M., but with the clocks set forward by one hour, it would then get dark at 9:00 P.M. So we turn our lights on an hour later. According to a U.S. Department of Transportation study, daylight saving time saves energy because we tend to spend more time outside of our homes engaging in outdoor activities. Moreover, because more people drive during the daylight hours, daylight saving could also reduce the number of automobile accidents, consequently saving many lives.

In 1966, the U.S. Congress passed the Uniform Time Act to establish a system of uniform time. Moreover, in 1986, Congress changed the initiation of daylight saving time from the last Sunday in April to the first Sunday in April. Today, most countries around the world follow some type of daylight saving schedule. In the European Union countries,

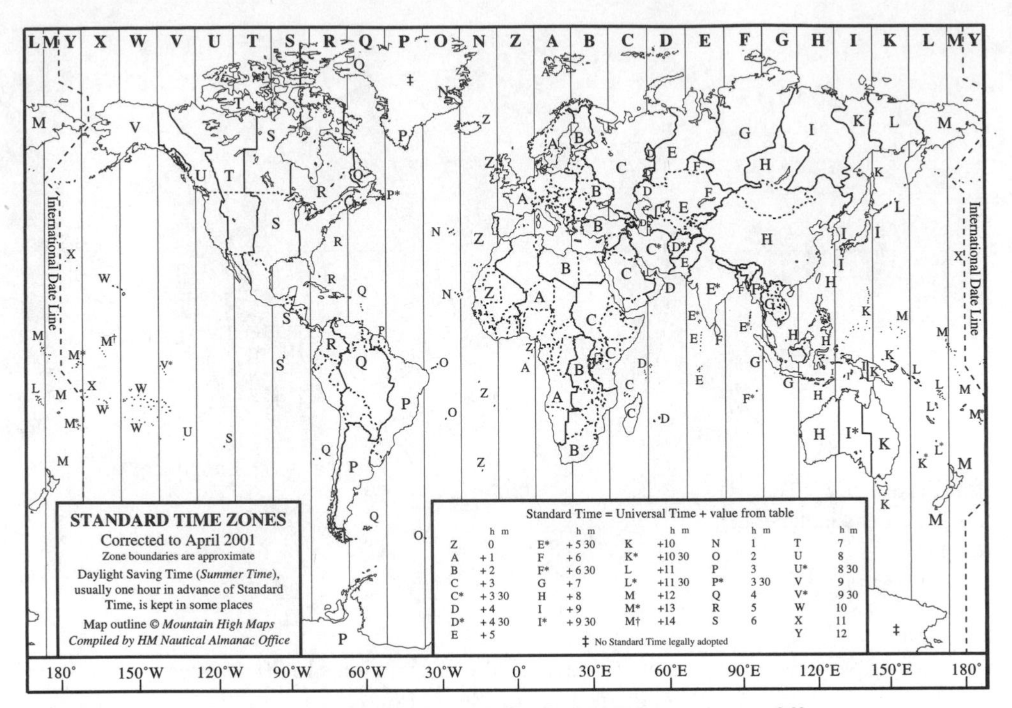

Figure 5.3 Standard time zones

Source: Map outline © Mountain High Maps. Data supplied by HM Nautical Almanac Office.

the daylight saving begins on the last Sunday in March and continues through the last Sunday in October.

5.3 PERIODS AND FREQUENCIES

For periodic events, a ***period*** is the time that it takes for the event to repeat itself. For example, every 365.24 days the earth lines up in exactly the same position with respect to the sun. The orbit of the earth around the sun is said to be periodic because this event repeats itself. The inverse of a period is called a ***frequency***. For example, the frequency at which the earth goes around the sun is once a year.

Let us now use other simple examples to explain the difference between period and frequency. It is safe to assume that most of you do laundry once a week or buy groceries once a week. In this case, the frequency of your doing laundry or buying groceries is once a week. Or you may go see your dentist once every six months. That is the frequency of your dentist visits. Therefore, frequency is a measure of how frequently an event or a process occurs, and period is the time that it takes for that event to complete one cycle. Some engineering examples that include periodic motion are oscillatory systems such as shakers, mixers, and vibrators. The piston inside a car's engine cylinder is another good example of periodic motion. Your car's suspension system, the wings of a plane in a turbulent flight, or a building being shaken by strong winds may also show some component of periodic motion.

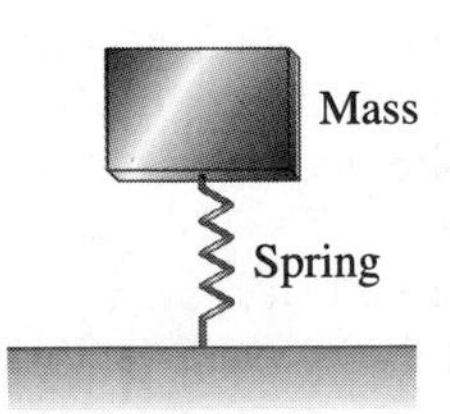

Figure 5.4 A spring–mass system

Consider a simple spring–mass system, as shown in Figure 5.4. The spring–mass system shown could represent a very simple model for a vibratory system, such as a shaker

TABLE 5.1 Examples of Frequencies of Various Electrical and Electronic Systems

Application	Frequency
Alternating current (USA)	60 Hz
AM radio	540 kHz–1.6 MHz
Television (channels 2–6)	54–88 MHz
FM radio	88–108 MHz
Emergency, fire, police	153–159 MHz
Television (channels 7–13)	174–216 MHz
Personal computer clocks (as of year 2000)	up to 1 GHz
Cordless phone (as of year 2000)	900 MHz

or a vibrator. What happens if you were to push down on the mass and then let go of it? The mass will oscillate in a manner that manifests itself by an up-and-down motion. If you study mechanical vibration, you will learn that the natural undamped frequency of the system is given by

$$f_n = \frac{1}{2\pi}\sqrt{\frac{k}{m}} \tag{5.1}$$

where f_n is the natural frequency of the system in cycles per second, or Hertz, k represents the stiffness of the spring or an elastic member (N/m), and m is the mass of the system (kg). The period of oscillation, T, for the given system—or in other words, the time that it takes for the mass to complete one revolution—is given by

$$T = \frac{1}{f_n} \tag{5.2}$$

Most of you have seen oscillating pendulums in clocks. The pendulum is another good example of a periodic system. The period of oscillation for a pendulum is given by

$$T = 2\pi\sqrt{\frac{L}{g}} \tag{5.3}$$

where L is the length of the pendulum (m) and g is the acceleration due to gravity (m/s^2). Note that the period of oscillation is independent of the mass of the pendulum. Not too long ago, oil companies used measured changes in the period of an oscillating pendulum to detect variations in acceleration due to gravity that could indicate an underground oil reservoir.

An understanding of periods and frequencies is also important in the design of electrical and electronic components. In general, excited mechanical systems have much lower frequencies than electrical/electronic systems. Examples of frequencies of various electrical and electronic systems are given in Table 5.1.

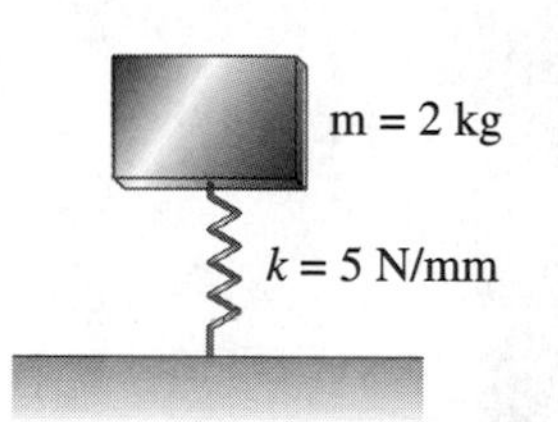

Figure 5.5 A simple spring–mass system

EXAMPLE 5.1

Determine the natural frequency of the simple spring–mass system shown in Figure 5.5.

Using Equation (5.1), we have

$$f_n = \frac{1}{2\pi}\sqrt{\frac{k}{m}} = \frac{1}{2\pi}\sqrt{\frac{5000\text{ N/m}}{2\text{ kg}}} \approx 8\text{ Hz}$$

5.4 FLOW OF TRAFFIC

Those of you who live in a big city know what we mean by traffic. A branch of civil engineering deals with the design and layout of highways, roads, and streets and the location and timing of traffic control devices that move vehicles efficiently. In this section, we will provide a brief overview of some elementary concepts related to traffic engineering. These variables are time related. Let us begin by defining what we mean by traffic flow. In civil engineering, traffic flow is formally defined by the following relationship:

$$q = \frac{3600n}{T} \tag{5.4}$$

In Equation (5.4), q represents the traffic flow in terms of number of vehicles per hour, n is the number of vehicles passing a known location during a time duration T in seconds. Another useful variable of traffic information is density—how many cars occupy a stretch of a highway. Density is defined by

$$k = \frac{1000n}{d} \tag{5.5}$$

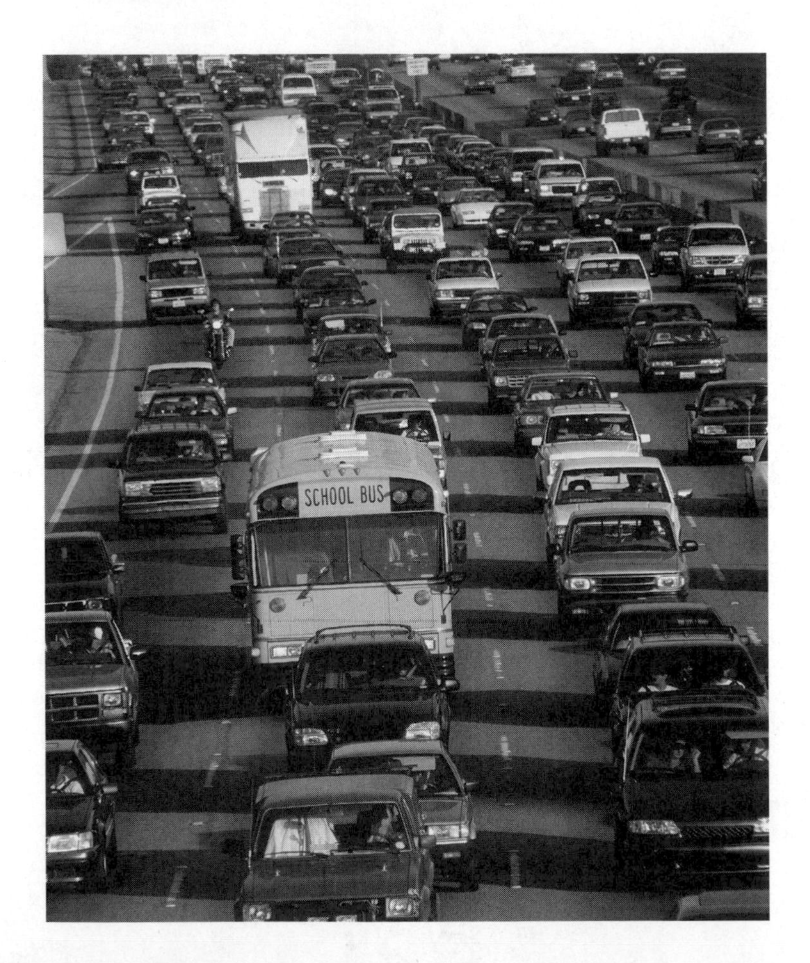

A congested flow!

where k is density and represents the number of vehicles per kilometer, and n is the number of vehicles on a stretch of highway, d, measured in meters.

Knowing the average speed of cars also provides valuable information for the design of road layouts and the location and timing of traffic control devices. The average speed of cars is determined from

$$\bar{u} = \frac{1}{n}\sum_{i=1}^{n} u_i \tag{5.6}$$

In Equation (5.6), u_i is the speed of individual cars (we will discuss the definition of speed in more detail in Section 5.5.), and n represents the total number of cars. There is a relationship among the traffic parameters—namely, the flow of traffic, density, and the average speed—according to

$$q = k\bar{u} \tag{5.7}$$

where q, k, and u were defined earlier by Equations (5.4) through (5.6).

The relationship among flow, density, and average speed is shown in Figures 5.6 and 5.7. Figure 5.6 shows that when the average speed of moving cars is high, the value of density is near zero, implying that there are not that many cars on that stretch of highway. As you may expect, as the value of density increases (the number of vehicles per kilometer), the average speed of vehicles decreases and will eventually reach a zero value, meaning bumper-to-bumper traffic. Figure 5.7 shows the relationship between the average speed of vehicles and the flow of traffic. For bumper-to-bumper traffic, with no cars moving, the flow of traffic stops, and thus q has a value of zero. This is the beginning of the congested region shown in Figure 5.7. As the average speed of the vehicles increases, so does the flow of traffic, eventually reaching a maximum value. As the average speed of vehicles continues to increase, the flow of traffic (the number of vehicles per hour) decreases. This region is marked by an uncongested flow region.

Finally, traffic engineers use various measurement devices and techniques to obtain real-time data on the flow of traffic. They use the collected information to make improvements to move vehicles more efficiently. You may have seen examples of traffic measurement devices such as pneumatic road tubes and counters. Other common traffic measurement devices include magnetic induction loops and speed radar.

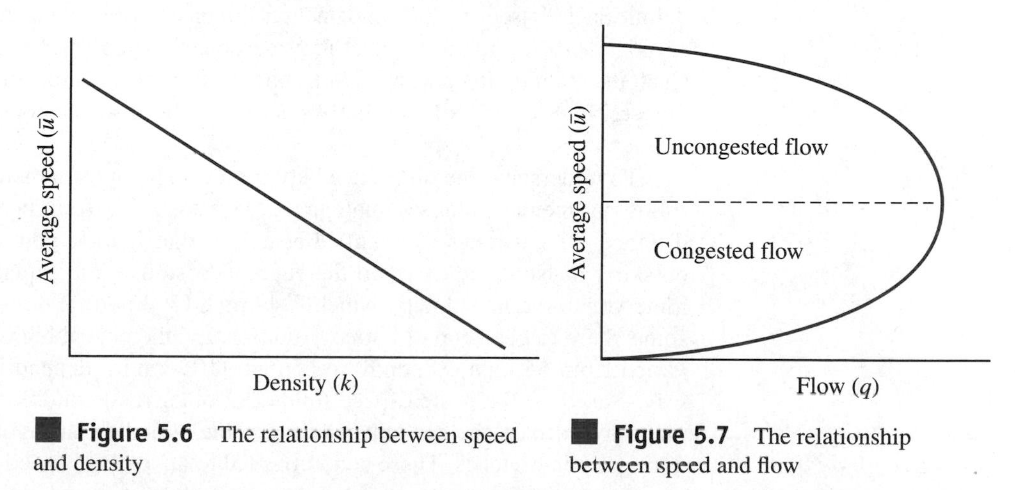

Figure 5.6 The relationship between speed and density

Figure 5.7 The relationship between speed and flow

EXAMPLE 5.2

Show that Equation (5.7) is dimensionally homogeneous by carrying out the appropriate units for each term in that equation.

$$q\left(\frac{\text{vehicles}}{\text{hour}}\right) = k\left(\frac{\text{vehicles}}{\text{kilometer}}\right)\bar{u}\left(\frac{\text{kilometer}}{\text{hour}}\right)$$

5.5 PARAMETERS INVOLVING LENGTH AND TIME

In this section, we will consider *derived physical quantities* that are based on the fundamental dimensions of length and time. We will first discuss the concepts of linear speed and acceleration and then define volumetric flow rate.

Linear Velocities

We will begin by explaining the concept of linear speed. Knowledge of linear speed and acceleration is important to engineers when designing conveyer belts that are used to load luggage into airplanes and product assembly lines, treadmills, elevators, automatic walkways, escalators, water or gas flow inside pipes, space probes, roller coasters, transportation systems (such as cars, boats, airplanes, and rockets), snow removal equipment, backup computer tape drives, and so on. Civil engineers are also concerned with velocities, particularly wind velocities, when designing structures. They need to account for the wind speed and its direction in their calculations when sizing structural members.

Let us now take a close look at what we mean by linear speed and linear velocity. All of you are familiar with a car speedometer. It shows the instantaneous speed of a car. Before we explain in more detail what we mean by the term *instantaneous speed*, let us define a physical variable that is more easily understood, the *average speed*, which is defined as

$$\text{average speed} = \frac{\text{change in the position of the moving object}}{\text{time}} = \frac{\text{distance traveled}}{\text{time}} \qquad (5.8)$$

Note that the fundamental (base) dimensions of length and time are used in the definition of the average speed. The average speed is called a *derived physical quantity* because its definition is based on the fundamental dimensions of length and time. The SI unit for speed is m/s, although for fast-moving objects km/h is also commonly used. In U.S. Customary units, ft/s and miles/h (mph) are used to quantify the magnitude of a moving object. Now let us look at what we mean by the instantaneous speed and see how it is related to the average speed.

To understand the difference between average and instantaneous speed, consider the following mental exercise. Imagine that you are going from New York City to Boston, a distance of 220 miles (354 km). Let us say that it took you 4.5 hours to go from the outskirts of New York City to the edge of Boston. From Equation (5.8), you can determine your average velocity, which is 49 mph (79 km/h). You may have made a rest stop somewhere to get a cup of coffee. Additionally, the posted highway speed limit may have varied from 55 mph (88 km/h) to 65 mph (105 km/h), depending on the stretch of highway. Based on the posted speed limits and other road conditions, and how you felt, you may have driven the car faster during some stretches, and you may have gone slower during other stretches. These conditions led to an average speed of 49 mph (79 km/h). Let us also imagine that you recorded the speed of your car as indicated by the speedometer

TABLE 5.2 Examples of Some Speeds

Situation	m/s	km/h	ft/s	mph
Average speed of a person walking	1.3	4.7	4.3	2.9
The fastest runner in the world (100 m)	10.2	36.4	33.5	22.8
Professional tennis player serving a ball	58	209	190	130
Top speed of a sports car	67	241	220	150
777 Boeing airplane (cruise speed)	248	893	814	555
Orbital speed of the space shuttle	7741	27,869	25,398	17,316
Average orbital speed of the earth around the sun	29,000	104,400	95,144	64,870

every second. The actual speed of the car at any given instant while you were driving it is called the ***instantaneous velocity***.

To better understand the difference between the average speed and the instantaneous speed, ask yourself the following question. If you needed to locate someone, would you be able to locate the car knowing just the average speed of the car? The knowledge of the average speed of the car would not be sufficient. To know where the car is at all times, you need more information, such as the instantaneous speed of the car and the direction in which it is traveling. This means you must know the instantaneous velocity of the car. Note that when we say velocity of a car, we not only refer to the speed of the car but also the direction in which it moves.

Physical quantities that possess both a magnitude and a direction are called ***vectors***. You will learn more about vectors in your calculus, physics, and mechanics classes. For now, just remember the simple definition of a vector quantity—a quantity that has both magnitude and direction. A physical quantity that is described only by a magnitude is called a ***scalar*** quantity. Examples of scalar quantities include temperature, volume, and mass. Examples of the range of speed of various objects are given in Table 5.2.

Linear Accelerations

Acceleration provides a measure of how velocity changes with time. Something that moves with a constant velocity has a zero acceleration. *Because velocity is a vector quantity and has both magnitude and direction, any change in either the direction or the magnitude of acceleration can create acceleration.* For example, a car moving at a constant speed following a circular path has an acceleration component due to the change in the direction of the velocity vector, as shown in Figure 5.8. Here we will focus on an object moving along a straight line. The average acceleration is defined as

$$\text{average acceleration} = \frac{\text{change in velocity}}{\text{time}} \tag{5.9}$$

Again note that acceleration uses only the dimensions of length and time. Acceleration represents the rate at which the velocity of a moving object changes with time. Therefore, acceleration is the time rate of change of velocity. The SI unit for acceleration is m/s^2 and in U.S. Customary units ft/s^2.

The difference between instantaneous acceleration and average acceleration is similar to the difference between instantaneous velocity and average velocity. The instantaneous

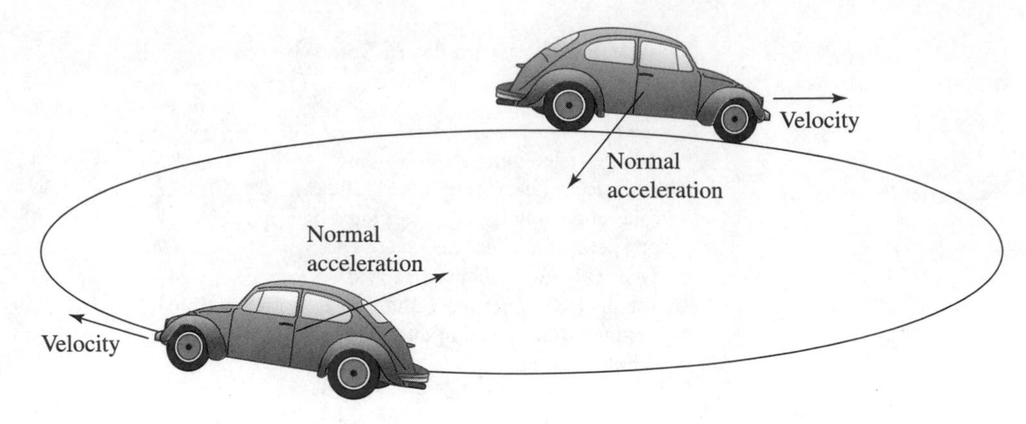

Figure 5.8 The acceleration of a car moving at a constant speed following a circular path

acceleration can be obtained from Equation (5.9) by making the time interval smaller and smaller. That is to say, instantaneous acceleration shows how the velocity of a moving object changes at any instant.

Let us now turn our attention to acceleration due to gravity. Acceleration due to gravity plays an important role in our everyday lives in the weight of objects and in the design of projectiles. What happens when you let go of an object in your hand? It falls to the ground. Sir Isaac Newton discovered that two masses attract each other according to

$$F = \frac{Gm_1m_2}{r^2} \tag{5.10}$$

where F is the attractive force between the masses (N), G represents the universal gravitational attraction and is equal to 6.7×10^{-11} m³/kg·s², m_1 and m_2 are the mass of each particle (kg), and r denotes the distance between the center of each particle. At the surface of the earth, the attractive force is called the *weight* of an object, and the acceleration created by the force is referred to as the *acceleration due to gravity*. At the surface of the earth, the value of g is equal to 9.81 m/s², or 32.2 ft/s². In general, acceleration due to gravity is a function of latitude and longitude, but for most practical engineering applications near the surface of the earth the values of 9.81 m/s² or 32.2 ft/s² are sufficient.

TABLE 5.3 Speed, Acceleration, and Distance Traveled by a Falling Object Neglecting the Air Resistance

Time (seconds)	Acceleration of the Object (m/s²)	Speed of the Falling Object (m/s) ($V = gt$)	Distance Traveled (m) $\left(d = \frac{1}{2}gt^2\right)$
0	9.81	0	0
1	9.81	9.81	4.90
2	9.81	19.62	19.62
3	9.81	29.43	44.14
4	9.81	39.24	78.48
5	9.81	49.05	122.62
6	9.81	58.86	176.58

Table 5.3 shows speed, acceleration, and the distance traveled by a falling object from the roof of a high-rise building. We have neglected air resistance in our calculations. Note how the distance traveled by the falling object and its velocity change with time.

EXAMPLE 5.3

A car starts from rest and accelerates to a speed of 100 km/h in 15 s. The acceleration during this period is constant. For the next 30 minutes the car moves with a constant speed of 100 km/h. At this time the driver of the car applies the brake and the car decelerates to a full stop in 10 s. The variation of the speed of the car with time is shown in the accompanying diagram. We are interested in determining the total distance traveled by the car and its average speed over this distance.

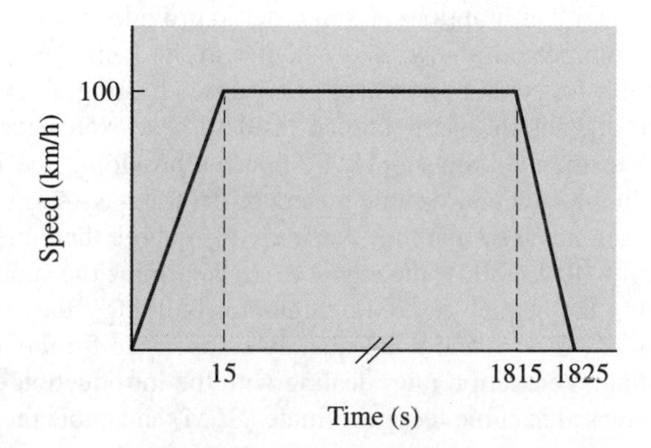

During the initial 15 seconds, the speed of the car increases linearly from a value of zero to 100 km/h. Therefore, the average speed of the car is 50 km/h during this period. The distance traveled during this period is

$$d_1 = \text{(time)(average speed)} = (15\ \text{s})\left(50\,\frac{\text{km}}{\text{h}}\right)\left(\frac{1\ \text{h}}{3600\ \text{s}}\right)\left(\frac{1000\ \text{m}}{1\ \text{km}}\right) = 208.3\ \text{m}$$

During the next 30 minutes the car moves with a constant speed of 100 km/h and the distance traveled during this period is

$$d_2 = (1800\ \text{s})\left(100\,\frac{\text{km}}{\text{h}}\right)\left(\frac{1\ \text{h}}{3600\ \text{s}}\right)\left(\frac{1000\ \text{m}}{1\ \text{km}}\right) = 50000\ \text{m}$$

Because the car decelerates at a constant rate from a speed of 100 km/h to 0, the average speed of the car during the last 10 seconds is also 50 km/h, and the distance traveled during this period is

$$d_3 = (10\ \text{s})\left(50\,\frac{\text{km}}{\text{h}}\right)\left(\frac{1\ \text{h}}{3600\ \text{s}}\right)\left(\frac{1000\ \text{m}}{1\ \text{km}}\right) = 138.9\ \text{m}$$

The total distance traveled by the car is

$$d = d_1 + d_2 + d_3 = 208.3 + 50000 + 138.9 = 50347.2\ \text{m} = 50.3472\ \text{km}$$

And finally, the average speed of the car for the entire duration of travel is

$$V_{\text{average}} = \frac{\text{distance traveled}}{\text{time}} = \frac{50347.2}{1825} = 27.56\ \text{m/s} = 99.2\ \text{km/h}$$

Volume Flow Rate

Engineers design flow-measuring devices to determine the amount of a material or a substance flowing through a pipeline in a processing plant. Volume flow rate measurements are necessary in many industrial processes to keep track of the amount of material being transported from one point to the next point in a plant. Additionally, knowing the flow rate of a material, engineers can determine the consumption rate so that they can provide the necessary supply for a steady state operation. Flow-measuring devices are also employed to determine the amount of water or natural gas being used by us in our homes during a specific period of time. City engineers need to know the daily or monthly volumetric water consumption rates in order to provide an adequate supply of water to our homes and commercial plants. Many homes in the United States, and around the world, use natural gas for cooking purposes or for space heating. For example, for a gas furnace to heat cold air, natural gas is burned inside a heat exchanger that transfers the heat of combustion to the cold air supply. Companies providing the natural gas need to know how much fuel—how many cubic meters of natural gas—are burned every day, or every month, by each home so that they can correctly charge their customers. Knowing the fuel consumption will also allow the engineers to determine the supply rate for a steady operation. In sizing the heating or cooling units for buildings, the volumetric flow of warm or cool air must be determined to adequately compensate for the heat loss or heat gain for a given building. Ventilation rates dealing with the introduction of fresh air into a building are also expressed in cubic feet per minute (CFM) and cubic meter per minute or per hour.

Another example that you may be familiar with is the flow rate of a water–antifreeze mixture necessary to cool your car's engine. Engineers who design the cooling systems for a car engine need to determine the volumetric flow rate of the water–antifreeze mixture (gallons of water–antifreeze mixture per minute) through the car's engine block and through the car's radiator to keep the engine block cool at safe temperatures.

Now that you have an idea what we mean by the term *volumetric flow rate*, let us formally define it. The volume flow rate is simply defined by the volume of a given substance that flows through something per unit time.

$$\text{volume flow rate} = \frac{\text{volume}}{\text{time}} \tag{5.11}$$

Some of the more common units for volume flow rate include, m^3/s, m^3/h, L/s, ml/s, ft^3/s, or gal/min (gpm). Note that in the definition given by Equation (5.11), the fundamental dimensions of length (length cubed) and time are used.

For fluids flowing through pipes, conduits, or nozzles, there exists a relationship between the volumetric flow rate, the average velocity of the flowing fluid, and the cross-sectional area of the flow, according to

$$\text{volumetric flow rate} = (\text{average velocity})\,(\text{cross-sectional area of the flow}) \tag{5.12}$$

In fact, some flow-measuring devices make use of Equation (5.12) to determine the volume flow rate of a fluid by first measuring the fluid's average velocity and the cross-sectional area of the flow. Finally, note that in Chapter 6 we will explain another closely defined variable, mass flow rate of a flowing material, which provides a measure of time rate of mass flow through pipes or other carrying conduits.

EXAMPLE 5.4

Consider the piping system shown in the accompanying figure. The average speed of water flowing through the 12-inch-diameter section of the piping system in 5 ft/s. What is the volume flow rate of water in the piping system? Express the volume flow rate in ft^3/s, gpm, and L/s. For the case of steady flow of water through the piping system, what is the speed of water in the 6-inch-diameter section of the system?

We can determine the volume flow rate of water through the piping system using Equation (5.12).

$$\text{volume flow rate} = (\text{average velocity})(\text{cross-sectional area of flow})$$

$$Q = \text{volume flow rate} = (5\ \text{ft/s})\left(\frac{\pi}{4}\right)(1\ \text{ft})^2 = 3.926\ \text{ft}^3/\text{s}$$

$$Q = (3.926\ \text{ft}^3/\text{s})\left(\frac{7.48\ \text{gal}}{1\ \text{ft}^3}\right)\left(\frac{60\ \text{s}}{1\ \text{min}}\right) = 1762\ \text{gpm}$$

$$Q = (3.926\ \text{ft}^3/\text{s})\left(\frac{28.3168\ \text{L}}{1\ \text{ft}^3}\right) = 111.2\ \text{L/s}$$

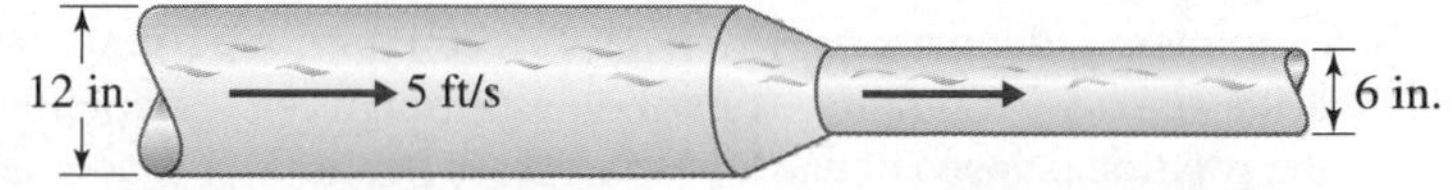

For the steady flow of water through the piping system the volume flow rate is constant. This fact allows us to calculate the speed of water in the 6-inch section of the pipe.

$$Q = 3.926 \text{ ft}^3/\text{s} = (\text{average velocity})(\text{cross-sectional area of flow})$$

$$3.926 \text{ ft}^3/\text{s} = (\text{average speed})\left(\frac{\pi}{4}\right)(0.5 \text{ ft})^2$$

$$\text{average speed} = 20 \text{ ft/s}$$

From the results of this example you can see that when you reduce the pipe diameter by a factor of 2, the water speed in the reduced section of the pipe increases by a factor of 4.

5.6 ANGULAR MOTION

In the next two sections, we will discuss angular motion, including angular velocities and accelerations of rotating objects.

Angular (Rotational) Velocities

In the previous section, we explained the concept of linear velocity and acceleration. Now we will consider variables that define angular motion. Rotational motion is also quite common in engineering applications. Examples of engineering components with rotational motion include shafts, wheels, gears, drills, pulleys, fan or pump impellers, helicopter blades, hard drives, CD drives, Zip drives, and so on.

The average angular speed of a line segment located on a rotating object such as a shaft is defined as the change in its angular position (angular displacement) over the time that it took the line to go through the angular displacement.

$$\omega = \frac{\Delta\theta}{\Delta t} \tag{5.13}$$

In Equation (5.13), ω represents the average angular speed in radians per second, $\Delta\theta$ is the angular displacement (radians), and Δt is the time interval in seconds. Similar to the definition of instantaneous velocity given earlier, the instantaneous angular speed is defined by making the time increment become smaller and smaller. Again, when we speak of angular velocity, we not only refer to the speed of rotation but also to the direction of rotation. It is a common practice to express the angular speed of rotating objects in revolutions per minute (rpm) instead of radians per second (rad/s). For example, the rotational speed of a pump impeller may be expressed as 1600 rpm. To convert the angular speed value from rpm to rad/s, we make the appropriate conversion substitutions.

$$1600\left(\frac{\text{revolutions}}{\text{minutes}}\right)\left(\frac{2\pi \text{ radians}}{1 \text{ revolution}}\right)\left(\frac{1 \text{ minute}}{60 \text{ seconds}}\right) = 167.5\,\frac{\text{rad}}{\text{s}}$$

In practice, the angular speed of rotating objects is measured using a stroboscope or a tachometer.

To get some idea how fast some common objects rotate, consider the following examples: a dentist's drill runs at 400,000 rpm; a current state-of-the-art computer hard drive runs at 7200 rpm; the earth goes through one complete revolution in 24 hours, thus the rotational speed of earth is 15 degrees per hour or 1 degree every 4 minutes.

There exists a relationship between linear and angular velocities of objects that not only rotate but also translate as well. For example, a car wheel, when not slipping, will not only rotate but also translate. To establish the relationship between rotational speed and the translational speed we begin with

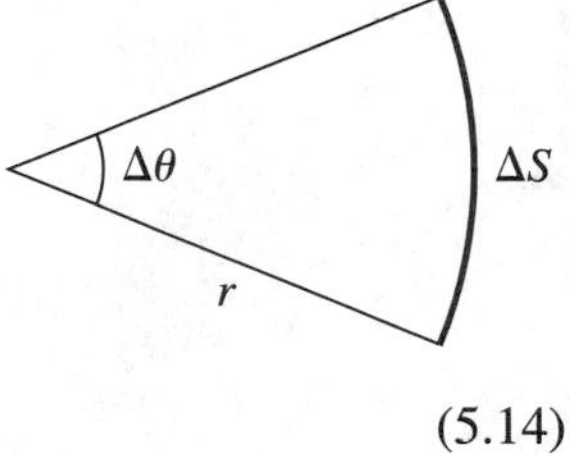

$$\Delta S = r\Delta\theta \tag{5.14}$$

dividing both sides by time increment Δt

$$\frac{\Delta S}{\Delta t} = r\frac{\Delta\theta}{\Delta t} \tag{5.15}$$

and making use of the definitions of linear and angular velocities, we have

$$V = r\omega \tag{5.16}$$

For example, the actual linear velocity of a particle located 0.1 m away from the center of shaft that is rotating at an angular speed of 1000 rpm (104.7 rad/s) is approximately 10.5 m/s.

EXAMPLE 5.5

Determine the rotational speed of a car wheel if the car is translating along at a speed of 55 mph. The radius of the wheel is 12.5 in.

Using Equation (5.15), we have

$$\omega = \frac{V}{r} = \frac{\left(55\,\frac{\text{miles}}{\text{h}}\right)\left(\frac{1\text{ h}}{3600\text{ s}}\right)\left(\frac{5280\text{ ft}}{1\text{ mile}}\right)}{(12.5\text{ in.})\left(\frac{1\text{ ft}}{12\text{ in.}}\right)} = 77.4\text{ rad/s} = 739\text{ rpm}$$

Angular Accelerations

In the previous section, we defined angular velocity and its importance in engineering applications. We now define angular acceleration in terms of the rate of change of angular velocity. Angular acceleration is a vector quantity. Note also that similar to the relationship between the average acceleration and instantaneous acceleration, we could first define an average angular acceleration as

$$\text{angular acceleration} = \frac{\text{change in angular speed}}{\text{time}} \tag{5.17}$$

and then define the instantaneous angular acceleration by making the time interval smaller and smaller in Equation (5.17).

EXAMPLE 5.6

It takes 5 s for a shaft of a motor to go from zero to 1600 rpm. Assuming constant angular acceleration, what is the value of the angular acceleration of the shaft?

First, we convert the final angular speed of the shaft from rpm to rad/s.

$$1600\left(\frac{\text{revolutions}}{\text{minutes}}\right)\left(\frac{2\pi\text{ radians}}{1\text{ revolution}}\right)\left(\frac{1\text{ minute}}{60\text{ seconds}}\right) = 167.5\,\frac{\text{rad}}{\text{s}}$$

Then we use Equation (5.17) to calculate the angular acceleration.

$$\text{angular acceleration} = \frac{\text{changes in angular speed}}{\text{time}} = \frac{(167.5 - 0)\text{ rad/s}}{5\text{ s}} = 33.5\,\frac{\text{rad}}{\text{s}^2}$$

Finally, remember that all the parameters discussed in this chapter involved time, or length and time.

SUMMARY

Now that you have reached this point in the text

- You should have a good grasp of yet another fundamental dimension in engineering, namely time, and its role in engineering analysis. You should also recognize the role of time in calculating speed, acceleration, and flow of traffic, as well as flow of materials and substances.
- You should know what we mean by frequency and a period, and be able to give some examples of mechanical and electrical systems with frequency and periods.
- You should know how to define average and instantaneous velocity. You should also be able to define average acceleration and instantaneous acceleration.
- You should have a comfortable grasp of a rotational motion and how it differs from a translation motion. You should know how to define angular velocity and acceleration.
- You should understand the significance of volume flow rate in our everyday lives and in engineering applications. You should also have a good grasp of what is meant by volume flow rate.

PROBLEMS

1. Create a calendar showing the beginning and the end of daylight saving time for the years 2004–2011.

Year	Daylight Saving Time Begins at 2:00 A.M.	Daylight Saving Time Ends at 2:00 A.M.
2004		
2005		
2006		
2007		
2008		
2009		
2010		
2011		

2. According to the Department of Transportation analysis of energy consumption figures for 1974 and 1975, observation of daylight saving time in the month of April in the years of 1974 and 1975 saved the country an estimated energy equivalent of 10,000 barrels of oil each day. Estimate how much energy is saved in the United States or your country due to observing daylight saving time in the current year.
3. Is there a need for a country near the equator to observe daylight saving time? Explain.
4. Besides energy savings, what are other advantages that observation of daylight saving time may bring?
5. Every year all around the world we celebrate certain cultural events dealing with our past. For example, in Christianity, Easter is celebrated in the spring

and Christmas is celebrated in December; in the Jewish calendar, Yom Kippur is celebrated in October; and Ramadan, the time of fasting, is celebrated by Muslims according to a lunar calendar. Briefly discuss the basis of the Christian, Jewish, Muslim, and Chinese calendars.

6. In this problem you are asked to investigate how much water a leaky faucet wastes in one week, one month, and one year. Perform an experiment by placing a container under a leaky faucet and actually measure the amount of water accumulated in an hour (you can simulate a leaky faucet by just partially closing the faucet). You are to design the experiment. Think about the parameters that you need to measure. Express and project your findings in gallons/day, gallons/week, gallons/month, and gallons/per year. At this rate, how much water is wasted by 10,000,000 households with leaky faucets. Write a brief report to discuss your findings.
7. Next time you are putting gasoline in your car, determine the volumetric flow rate of the gasoline at the pumping station. Record the time that it takes to pump a known volume of gasoline into your car's gas tank. The flow meter at the pump will give you the volume in gallons, so all you have to do is to measure the time. Investigate the size of the storage tanks in your neighborhood gas station. Estimate how often the storage tank needs to be refilled. State your assumptions.
8. You are to investigate the water consumption in your house, apartment, or dormitory—whichever is applicable. For example, to determine bathroom water consumption, time how long it normally takes you to shower. Then place a bucket under the showerhead for a known period of time. Determine the total volume of water used when you take a typical shower. Estimate how much water you use during the course of a year just by showering. Identify other activities where you use water and estimate your amount of use.
9. Using the concepts discussed in this chapter, measure the volumetric flow rate of water out of a drinking fountain.
10. Convert the following speed limits from miles per hour to kilometers per hour and from feet per second to meters per second. Think about the relative magnitude of values as you go from mph to ft/s and as you go from km/h to m/s. You may use Microsoft® Excel to solve this problem.

Speed Limit (mph)	Speed Limit (km/h)	Speed Limit (ft/s)	Speed Limit (m/s)
15			
25			
30			
35			
40			
45			
55			
65			
70			

11. Most car owners drive their cars an average of 12,000 miles a year. Assuming a 20 miles/gallon gas consumption rate, determine the amount of fuel consumed by 150 million car owners on the following time basis:
    a. average daily basis
    b. average weekly basis
    c. average monthly basis
    d. average yearly basis
    e. over a period of ten years

    Express your results in gallons and liters.
12. Calculate the speed of sound for the U.S. standard atmosphere using $c = \sqrt{kRT}$, where c represents the speed of sound in m/s, k is the specific heat ratio for air ($k = 1.4$), and R is the gas constant for the air ($R = 286.9$ J/kg·K) and T represents the temperature of the air in Kelvin. The speed sound in atmosphere is the speed that sound propagates through the air. You may use Excel to solve this problem.

Altitude (m)	Air Temperature (K)	Speed of Sound (m/s)	Speed of Sound (km/h)
500	284.9		
1000	281.7		
2000	275.2		
5000	255.7		
10,000	223.3		
15,000	216.7		
20,000	216.7		
40,000	250.4		
50,000	270.7		

13. Express Equation (5.5), the traffic density, in terms of number of vehicles per mile.

14. Express the angular speed of the earth in rad/s and rpm.

15. What is the magnitude of the speed of a person at the equator due to rotational speed of the earth.

16. Calculate the average speed of the gasoline exiting a nozzle at a gas station. Next time you go to a gas station, measure the volumetric flow rate of the gas first, and then measure the diameter of the nozzle. Use Equation (5.12) to calculate the average speed of the gasoline coming out of the nozzle. *Hint:* First measure the time that it takes to put so many gallons of gasoline into the gas tank!

17. Measure the volumetric flow rate of water coming out a of faucet. Also determine the average velocity of water leaving the faucet.

18. Determine the speed of a point on the earth's surface in ft/s, m/s, mph, km/h.

19. Into how many time zones are the United States and its territories divided?

20. Estimate the rotational speed of your car wheels when you are traveling at 60 mph.

21. Determine the natural frequency of a pendulum whose length is 5 m.

22. Determine the spring constant for Example 5.1 if the system is to oscillate with a natural frequency of 5 Hz.

23. Determine the traffic flow if 100 cars pass a known location during 10 s.

24. A conveyer belt runs on 3-in. drums that are driven by a motor. If it takes 6 seconds for the belt to go from zero to the speed of 3 ft/s, calculate the final angular speed of the drum and its angular acceleration. Assume constant acceleration.

25. Chinook, a military helicopter, has two three-blade rotor systems, each turning in opposite direction. Each blade has a diameter of approximaterly 41 ft. The blades can spin at angular speeds of up to 225 rpm. Determine the translational speed of a particle located at the tip of a blade. Express your answer in ft/s, mph, m/s, and km/h.

26. Consider the piping system shown in the accompanying figure. The speed of water flowing through the 4-inch-diameter section of the piping system is 3 ft/s. What is the volume flow rate of water in the piping system? Express the volume flow rate in ft^3/s, gpm, and L/s. For the case of steady flow of water through the piping system, what is the speed of water in the 3-inch-diameter section of the system?

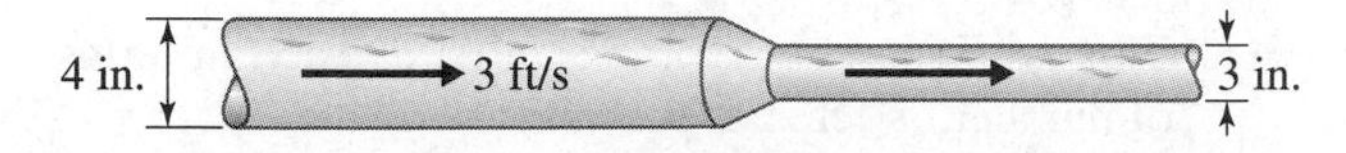

27. Consider the duct system shown in the accompanying figure. Air flows through two 8-inch-by-10-inch ducts that merge into a 18-inch-by-14-inch duct. The average speed of the air in each of the 8 × 10 ducts is 30 ft/s. What is the volume flow rate of air in the 18 × 14 duct? Express the volume flow rate in ft^3/s, ft^3/min, and m^3/s. What is the average speed of air in the 18 × 14 duct?

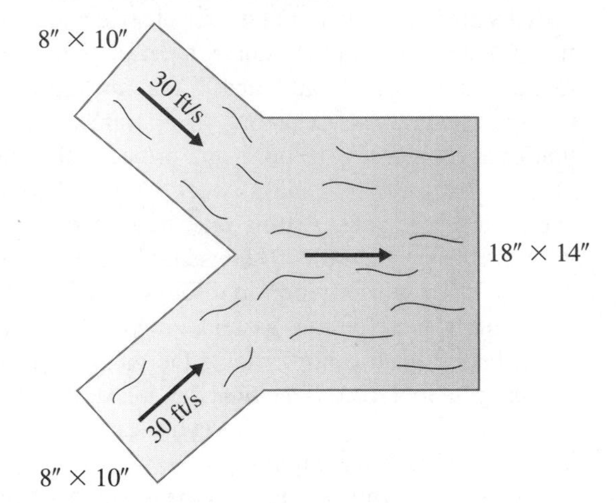

28. A car starts from rest and accelerates to a speed of 60 mph in 20 s. The acceleration during this period is constant. For the next 20 minutes the car moves with the constant speed of 60 mph. At this time the driver of the car applies the brake and the car decelerates to a full stop in 10 s. The variation of the speed of the car with time is shown in the accompanying diagram. Determine the total distance traveled by the car and the average speed of the car over this distance. Also plot the acceleration of the car as a function of time.

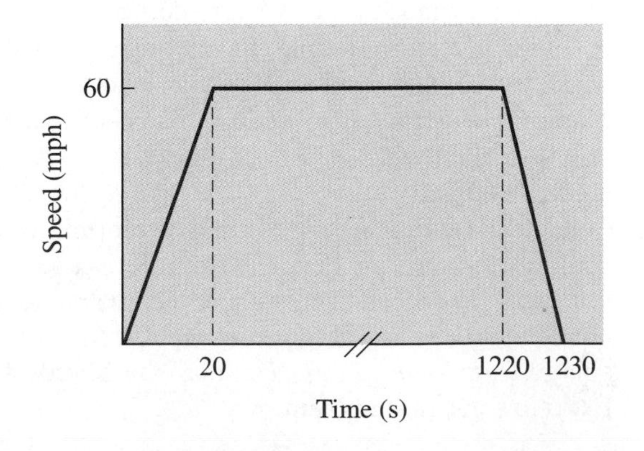

6 Mass and Mass-Related Parameters

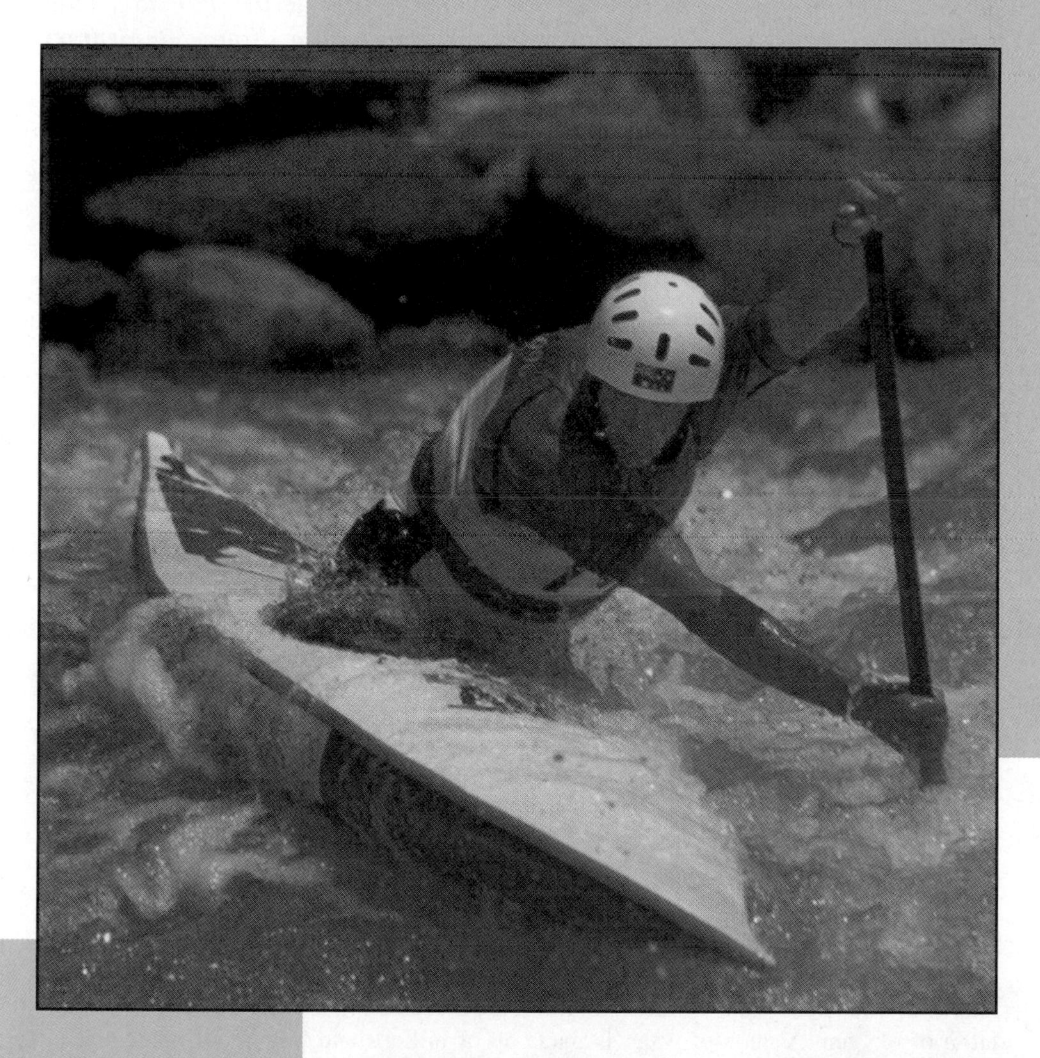

A kayaker runs through a gate in a whitewater kayaking slalom event. Mass is another important fundamental dimension that plays an important role in engineering analysis and design. Mass provides a measure of resistance to translational motion. Knowledge of mass is important in determining the momentum of moving objects. In engineering, to represent how light or heavy materials are, we use properties such as density and specific gravity.

The objective of this chapter is to introduce the concept of mass and mass-related quantities encountered in engineering. We will begin by discussing the building blocks of all matter: atoms and molecules. We will then introduce the concept of mass in terms of a quantitative measure of the amount of atoms possessed by a substance. We will then define and discuss other mass-related engineering quantities, such as density, specific gravity, momentum, and mass flow rate. In this chapter, we will also consider conservation of mass and its application in engineering. Some elementary statistical information, such as mean, variance, and standard deviation, is discussed at the end of the chapter. It is common in engineering to represent the tendencies of data by a single number, such as an arithmetic mean of data or the spread of data by a measure that is called standard deviation. *We will use a set of measured data related to the density of water to explain these elementary statistical ideas.* ■

6.1 MASS AS A FUNDAMENTAL DIMENSION

As we discussed in Chapter 3, from their day-to-day observations humans noticed that some things were heavier than others and thus recognized the need for a physical quantity to describe that observation. Early humans did not fully understand the concept of gravity; consequently, the correct distinction between mass and weight was made later. Let us now look more carefully at mass as a physical variable. Consider the following. When you look around at your surroundings, you will find that matter exists in various forms and shapes. You will also notice that matter can change shape when its condition or its surroundings are changed. All solid objects and living things are made of matter, and matter itself is made of atoms, or chemical elements. There are 106 known chemical elements to date. Atoms of similar characteristics are grouped together in a table, which is called the *periodic table of chemical elements*. An example of the chemical periodic table is shown in Figure 6.1.

Atoms are made up of even smaller particles we call *electrons*, *protons*, and *neutrons*. In your first chemistry class you will study these ideas in more detail, if you have not yet done so. Some of you may decide to study chemical engineering, in which case you will spend much more time studying chemistry. But for now, remember that atoms are the basic building block of all matter.

Atoms are combined naturally, or in a laboratory setting, to create molecules. For example, as you already know, water molecules are made of two atoms of hydrogen and one atom of oxygen. A glass of water is made up of billions and billions of homogeneous water molecules. Molecules are the smallest portion of a given matter that still possesses its characteristic properties. Matter can exist in four states, depending on its own and the surrounding conditions: solid, liquid, gaseous, or plasma. Let us consider the water that we drink every day. As you already know, under certain conditions, water exists in a solid form that we call ice. At a standard atmospheric pressure, water exists in a solid form as long its temperature is kept under 0°C. Under standard atmospheric pressure, if you were

	IA	IIA	IIIB	IVB	VB	VIB	VIIB	VIIIB	VIIIB	VIIIB	IB	IIB	IIIA	IVA	VA	VIA	VIIA	VIIIA
1	1 H 1.0079																	2 He 4.003
2	3 Li 6.941	4 Be 9.012											5 B 10.811	6 C 12.011	7 N 14.007	8 O 15.999	9 F 18.998	10 Ne 20.180
3	11 Na 22.990	12 Mg 24.305											13 Al 26.982	14 Si 28.086	15 P 30.974	16 S 32.066	17 Cl 35.453	18 Ar 39.948
4	19 K 39.098	20 Ca 40.078	21 Sc 44.956	22 Ti 47.88	23 V 50.942	24 Cr 51.996	25 Mn 54.938	26 Fe 55.845	27 Co 58.933	28 Ni 58.69	29 Cu 63.546	30 Zn 65.39	31 Ga 69.723	32 Ge 72.61	33 As 74.922	34 Se 78.96	35 Br 79.904	36 Kr 83.8
5	37 Rb 85.468	38 Sr 87.62	39 Y 88.906	40 Zr 91.224	41 Nb 92.906	42 Mo 95.94	43 Tc 98	44 Ru 101.07	45 Rh 102.906	46 Pd 106.42	47 Ag 107.868	48 Cd 112.411	49 In 114.82	50 Sn 118.71	51 Sb 121.76	52 Te 127.60	53 I 126.905	54 Xe 131.29
6	55 Cs 132.905	56 Ba 137.327	57 La 138.906	72 Hf 178.49	73 Ta 180.948	74 W 183.84	75 Re 186.207	76 Os 190.23	77 Ir 192.22	78 Pt 195.08	79 Au 196.967	80 Hg 200.59	81 Tl 204.383	82 Pb 207.2	83 Bi 208.980	84 Po 209	85 At 210	86 Rn 222
7	87 Fr 223	88 Ra 226.025	89 Ac 227.028	104 Rf 261	105 Db 262	106 Sg 263	107 Bh 262	108 Hs 265	109 Mt 266	110 Uun 269	111 Uuu 272	112 Uub 277		114		116		118

Lanthanide series	58 Ce 140.115	59 Pr 140.908	60 Nd 144.24	61 Pm 145	62 Sm 150.36	63 Eu 151.964	64 Gd 157.25	65 Tb 158.925	66 Dy 162.5	67 Ho 164.93	68 Er 167.26	69 Tm 168.934	70 Yb 173.04	71 Lu 174.967
Actinide series	90 Th 232.038	91 Pa 231.036	92 U 238.029	93 Np 237.048	94 Pu 244	95 Am 243	96 Cm 247	97 Bk 247	98 Cf 251	99 Es 252	100 Fm 257	101 Md 258	102 No 259	103 Lr 262

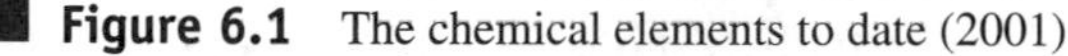

Figure 6.1 The chemical elements to date (2001)

to heat the ice and consequently change its temperature, the ice would melt and change into a liquid form. Under standard pressure at sea level, the water remains liquid up to a temperature of 100°C as you continue heating the water. If you were to carry out this experiment further by adding more heat to the liquid water, eventually the liquid water changes its phase from a liquid into a gas. This phase of water we commonly refer to as steam. If you had the means to heat the water to even higher temperatures, temperatures exceeding 2000°C, you would find that you can break up the water molecules into their atoms, and eventually the atoms break up into free electrons and nuclei that we call plasma.

Well, what does all this have to do with mass? Mass provides a quantitative measure of how many molecules or atoms are in a given object. The matter may change its phase, but its mass remains constant. Some of you will take a class in dynamics where you will learn that on a macroscopic scale mass also serves as a measure of resistance to motion. You already know this from your daily observations. Which is harder to push, a motorcycle or a truck? As you know, it takes more effort to push a truck. When you want to rotate something, the distribution of the mass about the center of rotation also plays a significant role. The further away the mass is located from the center of rotation, the harder it will be to rotate the mass about that axis. A measure of how hard it is to rotate something with respect to center of rotation is called *mass moment of inertia*. We will discuss this in detail in Section 6.5.

The other mass-related parameter that we will investigate in this chapter is momentum. Consider two objects with differing masses moving with the same velocity. Which object is harder to bring to a stop, the one with the small mass or the one with the larger mass? Again, you already know the answer to this question. The object with the bigger mass is harder to stop. You see that in the game of football, too. If two players of differing mass are running at the same speed, it is harder to bring down the bigger player. These observations lead to the concept of momentum, which we will explain in Section 6.6.

Mass also plays an important role in storing thermal energy. The more massive something is, the more thermal energy you can store within it. Some materials are better than

others at storing thermal energy. For example, water is better at storing thermal energy than air is. In fact, the idea of storing thermal energy within a massive medium is fully utilized in the design of passive solar houses. There are massive brick or concrete floors in the sunrooms. Some people even place big barrels of water in the sunrooms to absorb the available daily solar radiation and store the thermal energy in the water to be used overnight. We will discuss heat capacities of materials in more detail in Chapter 8.

6.2 MEASUREMENT OF MASS

The ***kilogram*** is the unit of mass in SI; it is equal to the mass of the international prototype of the kilogram. In practice, the mass of an object is measured indirectly by using how much something weighs. The weight of an object on earth is the force that is exerted on the mass due to the gravitational pull of the earth. You are familiar with spring scales that measure the weight of goods at a supermarket or bathroom scales at home. Force due to gravity acting on the unknown mass will make the spring stretch or compress. By measuring the deflection of the spring, one can determine the weight and consequently the mass of the object that created that deflection. Some weight scales use force transducers consisting of a metallic member that behaves like a spring except the deflection of the metal member is measured electronically using strain gauges. You should understand the difference between weight and mass, and be careful how you use them in engineering analysis. We will discuss the concept of weight in more detail in Chapter 7.

6.3 DENSITY, SPECIFIC VOLUME, SPECIFIC GRAVITY

In engineering practice, to represent how light or how heavy materials are we often define properties that are based on a unit volume; in other words, how massive something is per unit volume. Given 1 cubic foot of wood and 1 cubic foot of steel, which one has more mass? The steel of course! The ***density*** of any substance is defined as the ratio of the mass to the volume that it occupies, according to

$$\text{density} = \frac{\text{mass}}{\text{volume}} \tag{6.1}$$

Density provides a measure of how compact the material is for a given volume. Materials such as mercury or gold with relatively high values of density have more mass per 1 ft^3 volume or 1 m^3 volume than those with lower density values, such as water. It is important to note that the density of matter changes with temperature and could also change with pressure. The SI unit for density is kg/m^3, and in U.S. Customary system the density is expressed in slugs/ft^3.

Specific volume, which is the inverse of density, is defined by

$$\text{specific volume} = \frac{\text{volume}}{\text{mass}} \tag{6.2}$$

Specific volume is commonly used in the study of thermodynamics. The SI unit for specific volume is m^3/kg. Another common way to represent the heaviness or lightness of some material is by comparing its density to the density of water. This comparison is called the ***specific gravity*** of a material and is formally defined by

$$\text{specific gravity} = \frac{\text{density of a material}}{\text{density of water@4°C}} \tag{6.3}$$

It is important to note that specific gravity is unitless because it is the ratio of the value of two densities. Therefore, it does not matter which system of units is used to compute the specific gravity of a substance.

Specific weight is another way to measure how truly heavy or light a material is for a given volume. Specific weight is defined as the ratio of the *weight* of the material by the volume that it occupies, according to

$$\text{specific weight} = \frac{\text{weight}}{\text{volume}} \tag{6.4}$$

Again, the weight of something on earth is the force that is exerted on its mass due to gravity. We will discuss the concept of weight in more detail in Chapter 7. In this chapter,

TABLE 6.1 Density, Specific Gravity, and Specific Weight of Some Materials (at room temperature or at the specified temperature)

Material	Density (kg/m³)	Specific Gravity	Specific Weight (N/m³)
Aluminum	2740	2.74	26,880
Asphalt	2110	2.11	20,700
Cement	1920	1.92	18,840
Clay	1000	1.00	9810
Fireclay brick	1790@100°C	1.79	17,560
Glass (soda lime)	2470	2.47	24,230
Glass (lead)	4280	4.28	41,990
Glass (Pyrex)	2230	2.23	21,880
Iron (cast)	7210	7.21	70,730
Iron (wrought)	7700@100°C	7.70	75,540
Paper	930	0.93	9120
Steel (mild)	7830	7.83	76,810
Steel (stainless 304)	7860	7.86	77,110
Wood (ash)	690	0.69	6770
Wood (mahogany)	550	0.55	5400
Wood (oak)	750	0.75	7360
Wood (pine)	430	0.43	4220
Fluids			
Standard air	1.225	0.0012	12
Gasoline	720	0.72	7060
Glycerin	1260	1.26	12,360
Mercury	13,550	13.55	132,930
SAE 10W oil	920	0.92	9030
Water	1000@4°C	1.0	9810

it is left as an exercise (see Problem 4) for you to show the relationship between density and specific weight:

specific weight = (density)(acceleration due to gravity)

The density, specific gravity, and the specific weight of some materials are shown in Table 6.1.

6.4 MASS FLOW RATE

In Chapter 5, we discussed the significance of volume flow rate. The mass flow rate is another closely related parameter that plays an important role in many engineering applications. There are many industrial processes that depend on the measurement of fluid flow. Mass flow rate tells engineers how much material is being used or moved over a period of time so that they can replenish the supply of the material. Engineers use flowmeters to measure volume or mass flow rates of water, oil, gas, chemical fluids, and food products. The design of any flow measuring device is based on some known engineering principles. Mass and volume flow measurements are also necessary and common in our everyday life. For example, when you go to a gasoline service station, you need to know how many gallons of gas are pumped into the tank of your car. Another example is measuring the amount of domestic water used or consumed. There are over 100 types of commercially available flowmeters to measure mass or volume flow rates. The selection of a proper flowmeter will depend on the application type and other variables, such as accuracy, cost, range, ease of use, service life, and type of fluid: gas or liquid, dirty or slurry or corrosive, for example.

The ***mass flow rate*** is simply defined by the amount of mass that flows through something per unit of time.

$$\text{mass flow rate} = \frac{\text{mass}}{\text{time}} \tag{6.5}$$

Some of the more common units for mass flow rate include kg/s, kg/min, kg/h or slugs/s or lbm/s (pounds mass/second). How would you measure the mass flow rate of water coming out of a faucet or a drinking fountain? Place a cup under a drinking fountain and measure the time that it takes to fill the cup. Also, measure the total mass of the cup and the water and then subtract the mass of the cup from the total to obtain the mass of the water. Divide the mass of the water by the time interval it took to fill the cup.

We can relate the volume flow rate of something to its mass flow rate provided that we know the density of the flowing fluid or flowing material. The relationship between the mass flow rate and the volume flow rate is given by

$$\begin{aligned}\text{mass flow rate} &= \frac{\text{mass}}{\text{time}} = \frac{(\text{density})(\text{volume})}{\text{time}} = (\text{density})\left(\frac{\text{volume}}{\text{time}}\right) \\ &= (\text{density})(\text{volume flow rate})\end{aligned} \tag{6.6}$$

The mass flow rate calculation is also important in excavation or tunnel-digging projects in determining how much soil can be removed in one day or one week, taking into consideration the parameter of the digging and transport machines.

6.5 MASS MOMENT OF INERTIA

As mentioned earlier in this chapter, when it comes to the rotation of objects, the distribution of mass about the center of rotation plays a significant role. The further away the mass is located from the center of rotation, the harder it is to rotate the mass about the given center of rotation. A measure of how hard it is to rotate something with respect to center of rotation is called ***mass moment of inertia***. All of you will take a class in physics and some of you may even take a dynamics class where you will learn in more depth about the formal definition and formulation of mass moment of inertia. But for now, let us consider the following simple situations, as shown in Figure 6.2. For a single mass particle m, located at a distance r from the axis of rotation z–z, the mass moment of inertia is defined by

$$I_{z-z} = r^2 m \tag{6.7}$$

Now let us expand this problem to include a system of mass particles, as shown in Figure 6.3. The mass moment of inertia for the system of masses shown about the z–z axis is now

$$I_{z-z} = r_1^2 m_1 + r_2^2 m_2 + r_3^2 m_3 \tag{6.8}$$

Similarly, we can obtain the mass moment of inertia for a body, such as a wheel or a shaft, by summing the mass moment of inertia of each mass particle that makes up the body. As you take calculus classes you will learn that you can use integrals instead of summations to evaluate the mass moment of inertia of continuous objects. After all, the integral sign $\int$ is nothing but a big S sign, indicating summation.

$$I_{z-z} = \int r^2\, dm \tag{6.9}$$

The mass moment of inertia of objects with various shapes can be determined from Equation (6.9). You will be able to perform this integration in another semester or two. Examples of mass moment of inertia formulas for some typical bodies such as a cylinder, disk, sphere, and a thin rectangular plate are given on the following page.

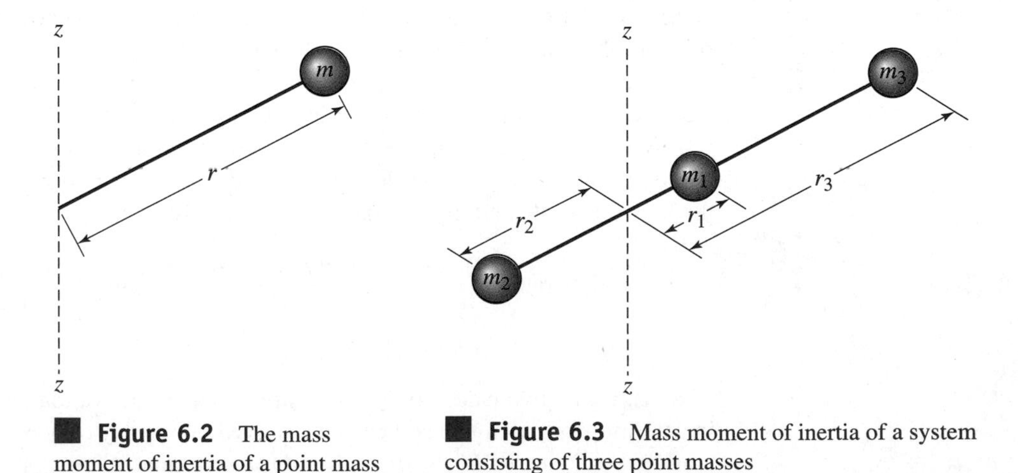

Figure 6.2 The mass moment of inertia of a point mass

Figure 6.3 Mass moment of inertia of a system consisting of three point masses

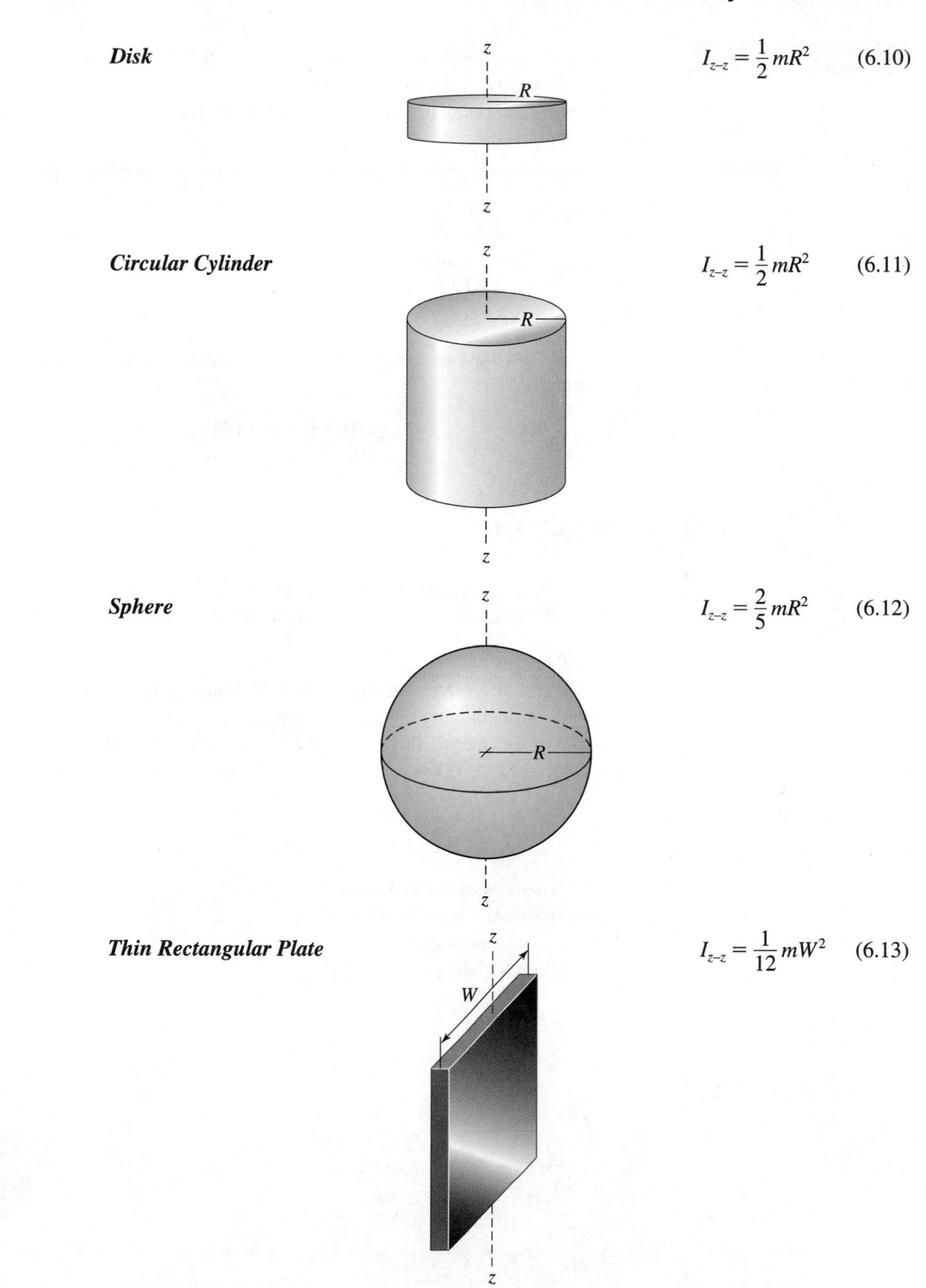

Disk
z
R
z
$I_{z-z} = \frac{1}{2} mR^2$ (6.10)
Circular Cylinder
z
R
z
$I_{z-z} = \frac{1}{2} mR^2$ (6.11)
Sphere
z
R
z
$I_{z-z} = \frac{2}{5} mR^2$ (6.12)
Thin Rectangular Plate
z
W
z
$I_{z-z} = \frac{1}{12} mW^2$ (6.13)

EXAMPLE 6.1

Determine the mass moment inertia of a steel shaft that is 2 m long and has a diameter of 10 cm. The density of steel is 7860 kg/m^3.

We will first calculate the mass of the shaft using Equation (6.1). The volume of a cylinder is given by

$$\text{volume} = (\pi/4)(d^2)(\text{length}) = (\pi/4)(0.1\text{ m})^2(2\text{ m}) = 0.01571\text{ m}^3$$

$$\text{density} = \frac{\text{mass}}{\text{volume}}$$

$$7860\text{ kg/m}^3 = \frac{\text{mass}}{0.01571\text{ m}^3}$$

$$\text{mass} = 123.5\text{ kg}$$

To calculate the mass moment of inertia of the shaft about its longitudinal axis, we use Equation (6.11).

$$I_{z\text{-}z} = \frac{1}{2}mR^2 = \frac{1}{2}(123.5\text{ kg})(0.05\text{ m})^2 = 0.154\text{ kg}\cdot\text{m}^2$$

6.6 MOMENTUM

Mass also plays an important role in problems dealing with moving objects. ***Momentum*** is a physical variable that is defined as the product of mass and velocity.

$$\vec{L} = m\vec{V} \tag{6.14}$$

In equation (6.14), L represents momentum, m is mass, and V is the velocity vector. Because the velocity of the moving object has a direction, we associate a direction with momentum as well. The momentum's direction is the same as the direction of the velocity vector or the moving object. So a 1000-kg car moving north at a rate of 20 m/s has a momentum with a

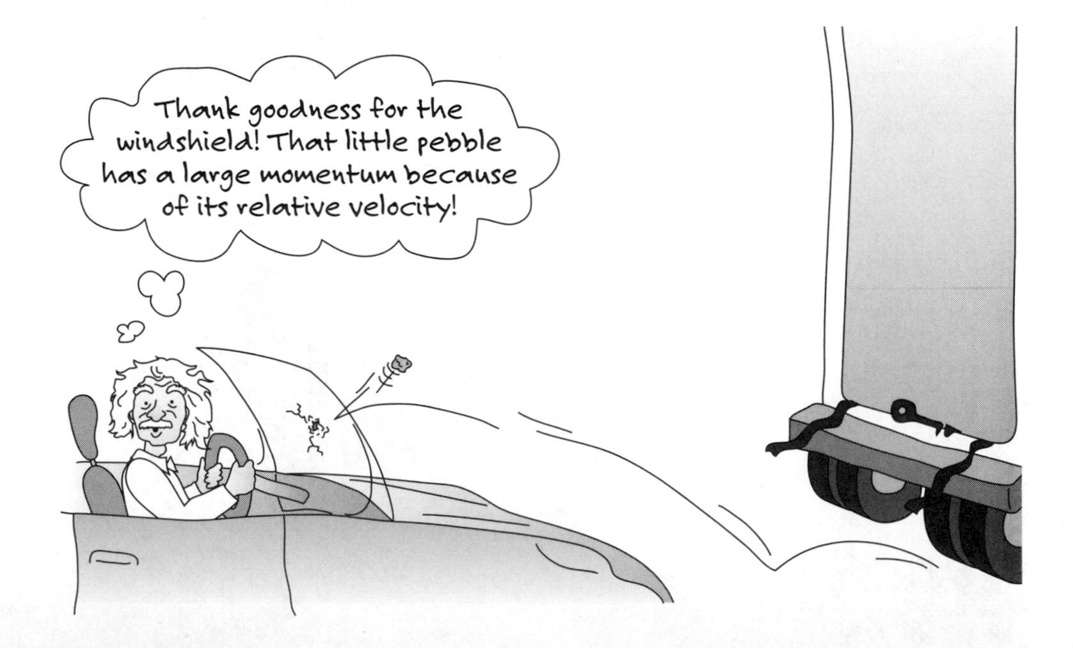

magnitude of 20,000 kg·m/s in the north direction. Momentum is one of those physical concepts that are commonly abused by sports broadcasters when discussing sporting events. During a timeout period when athletes are standing still and listening to their coaches, the broadcaster may say, "Bob, clearly the momentum has shifted, and team B has more momentum now going into the fourth quarter." Using the preceding definition of momentum, now you know that relative to earth, the momentum of an object or a person at rest is zero!

Because the magnitude of linear momentum is simply mass times velocity, something with relatively small mass could have a large momentum value, depending on its velocity. For example, a bullet with a relatively small mass shot out of a gun can do lots of harm and penetrate a surface because of its high velocity. The magnitude of the momentum associated with the bullet could be relatively large. You have seen stunt actors falling from the top of tall buildings. As the stunt performer approaches the ground, he or she has a relatively large momentum, so how are injuries avoided? Of course, the stunt performer falls onto an air bag and some soft materials that increase the time of contact to reduce the forces that act on his or her body. We will discuss the relationship between linear momentum and linear impulse in Chapter 7 after we discuss the concept of force.

EXAMPLE 6.2

Determine the linear momentum of a person whose mass is 80 kg and who is running at a rate of 3 m/s. Compare it to the momentum of a car that has a mass of 2000 kg and is moving at a rate of 30 m/s.

We will use Equation (6.14) to answer the questions.

$$\text{Person:}\quad \vec{L} = m\vec{V} = (80\text{ kg})(3\text{ m/s}) = 240\text{ kg}\cdot\text{m/s}$$

$$\text{Car:}\quad \vec{L} = m\vec{V} = (2000\text{ kg})(30\text{ m/s}) = 60{,}000\text{ kg}\cdot\text{m/s}$$

6.7 CONSERVATION OF MASS

Recall that in Chapter 3 we mentioned that engineers are good bookkeepers. In the analysis of engineering work we need to keep track of physical quantities such as mass, energy, momentum, and so on. Let us now look at how engineers may go about keeping track of mass and the associated bookkeeping procedure (see Figure 6.4). Simply stated, the conservation of mass says that we cannot create or destroy mass. Consider the following example. You are taking a shower in your bathtub. You turn on the water and begin to wash yourself. Let us focus our attention on the tub. With the drain open and clear of any hair and dead skin tissue, the rate at which water comes to the tub from the showerhead is equal to the rate at which water leaves the tub. This is a statement of conservation of mass applied to the water inside the tube. Now what happens if the drain becomes plugged up by hair or dead skin tissue? Many of us have experienced this at one time or another. Liquid Drāno time! The rate at which water comes to the tub now is not exactly equal to the rate at which water leaves the tub, which is why the water level in the tub begins to rise. How would you use the conservation of mass principle to describe this situation? You can express it this way: *The rate at which water comes to the tub is equal to the rate at which the water leaves the tub plus the time rate of accumulation of the mass of water within the tub.*

What happens if you were taking a bath and you had the tub filled with water? Let us say that after you are done taking the bath you open the drain, and the water begins to leave. Being the impatient person that you are, you slowly turn on the shower as the tub is

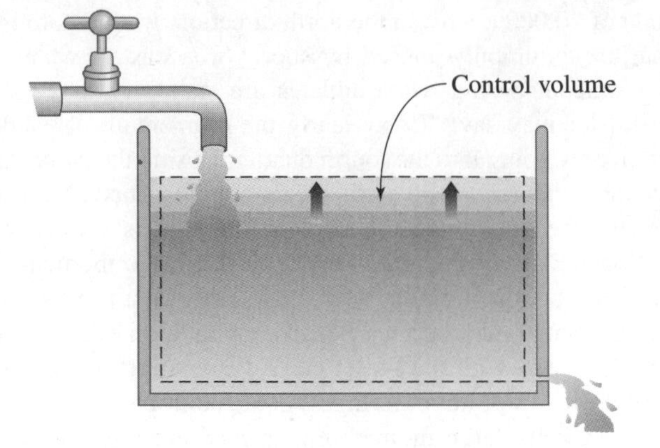

Figure 6.4 The rate at which water enters the container minus the rate at which water leaves the container should be equal to the rate of accumulation or depletion of the mass of water within the container.

being drained. Now you notice that the water level is going down but at a slow rate. The rate the water is coming into the tub minus the rate at which water is leaving the tub is equal to the rate of water depletion (reduction) within the tub.

Let us now turn our attention to an engineering presentation of conservation of mass. In engineering we refer to the tub as a control volume, because we focus our attention on a specific object occupying a certain volume in space. The control volume could represent the flow boundaries in a pump, or a section of a pipe, or the inside volume of a tank, the flow passage in a compressor or water heater or the boundaries of a river. We can use the bathtub example to formulate a general statement for conservation of mass, which states: *The rate at which a fluid enters a control volume minus the rate at which the fluid leaves*

I think I'll invite my class here tomorrow to demonstrate the conservation of mass!

the control volume should be equal to the rate of accumulation or depletion of the mass of fluid within the given control volume. There is also a broad area in mathematics, operations research, and engineering management called *queuing*. It is a study of "queues" of people waiting in service lines, or products waiting in assembly lines, or digital information waiting to move through computer networks. What happens during busy hours at banks, gas stations, or supermarket checkout counters? The lines formed by people or cars grow longer. The flow into a line is not equal to the flow out of the line and thus the line gets longer. You can think of this in terms of conservation of mass. During busy hours, the rate at which people enter a line is not equal to the rate at which people leave the line and thus there is a rate of accumulation at the line. This analogy is meant to make you think about other issues closely related to the flow of things in everyday life.

EXAMPLE 6.3

How much water is stored after 5 minutes in each of the tanks shown in the accompanying figure? How long will it take to fill the tanks completely provided that the volume of each tank is 12 m^3? Assume the density of water is 1000 kg/m^3.

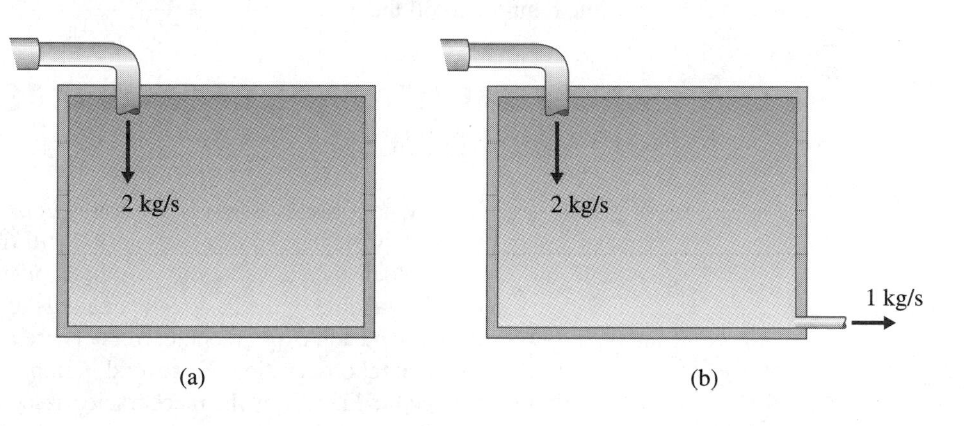

We will use the conservation of mass statement to solve this problem.

(the rate at which fluid enters a control volume) − (the rate at which the fluid leaves the control volume) = (the rate of accumulation or depletion of the mass of fluid within the given control volume)

Realizing that no water leaves tank (a) we have

$$(2\ \text{kg/s}) - (0) = \frac{\text{changes of mass inside the control volume}}{\text{change in time}}$$

After 5 minutes,

$$\text{change of mass inside the control volume} = (2\ \text{kg/s})(5\ \text{min})\left(\frac{60\ \text{s}}{1\ \text{min}}\right) = 600\ \text{kg}$$

To determine how long it will take to fill the tank, first we will make use of the relationship among mass, density, and volume—mass = (density)(volume)—to compute how much mass each tank can hold.

$$\text{mass} = (1000\ \text{kg/m}^3)(12\ \text{m}^3) = 12000\ \text{kg}$$

By rearranging terms in the conservation of mass equation we can now solve for the time that is required to fill the tank.

$$\text{time required to fill the tank} = \frac{12000\ \text{kg}}{(2\ \text{kg/s})} = 6000\ \text{s} = 100\ \text{min}$$

Water enters tank (b) at 2 kg/s and leaves the tank at 1 kg/s. Applying conservation of mass to tank (b) we have

$$(2\ \text{kg/s}) - (1\ \text{kg/s}) = \frac{\text{changes of mass inside the control volume}}{\text{change in time}}$$

After 5 minutes,

$$\text{change of mass inside the control volume} = [(2\ \text{kg/s}) - (1\ \text{kg/s})](5\ \text{min})\left(\frac{60\ \text{s}}{1\ \text{min}}\right) = 300\ \text{kg}$$

and the time that is required to fill the tank (b)

$$\text{time required to fill the tank} = \frac{12000\ \text{kg}}{[(2\ \text{kg/s}) - (1\ \text{kg/s})]} = 12000\ \text{s} = 200\ \text{min}$$

6.8 STATISTICAL MEASURES OF CENTRAL TENDENCY AND VARIATION

As you have noticed by now, some assignments in this book require the actual measurement of various physical variables. In this section, we will discuss some simple ways to examine the central tendency and variations within a given data set. Every engineer should have some understanding of the basic fundamentals of statistics and probability for analyzing experimental data and experimental errors. There are always inaccuracies associated with all experimental observations. If several variables are measured to compute a final result, then we need to know how the inaccuracies associated with these intermediate measurements will influence the accuracy of the final result. There are basically two types of observation errors: systematic errors and random errors. Suppose you were to measure the boiling temperature of pure water at sea level and standard pressure with a thermometer that reads 104°C. If readings from this thermometer are used in an experiment, it will result in systematic errors. Therefore, ***systematic errors***, sometimes called ***fixed errors***, are errors associated with using an inaccurate instrument. These errors can be detected and avoided by properly calibrating instruments. On the other hand, ***random errors*** are generated by a number of unpredictable variations in a given measurement situation. Mechanical vibrations of instruments or variations in line voltage friction or humidity could lead to fluctuations in experimental observations. These are examples of random errors.

Suppose two groups of students in an engineering class measured the density of water at 20°C. Each group consisted of ten students. They reported the results shown in Table 6.2. We would like to know if any of the reported data is in error.

Let us first consider the ***mean*** (arithmetic average) for each group's findings. The mean of densities reported by each group is 1000 kg/m^3. The mean alone cannot tell us whether any student or which student(s) in each group may have made a mistake. What we need is a way of defining the dispersion of the reported data. There are a number of ways to do this. Let us compute how much each reported density deviates from the mean, add up

TABLE 6.2 Reported Densities of Water at 20°C

Group A Findings	Group B Findings
$\rho(\text{kg/m}^3)$	$\rho(\text{kg/m}^3)$
1020	950
1015	940
990	890
1060	1080
1030	1120
950	900
975	1040
1020	1150
980	910
960	1020
$\rho_{avg} = 1000$	$\rho_{avg} = 1000$

all the deviations, and then take their average. Table 6.3 shows the deviation from the mean for each reported density. As one can see, the sum of the deviations is zero for both groups. This is not a coincidence. In fact, the sum of deviations from the mean for any given sample is always zero. This can be readily verified by considering the following:

$$\bar{x} = \frac{1}{n}\sum_{i=1}^{n} x_i \tag{6.15}$$

$$d_i = (x_i - \bar{x}) \tag{6.16}$$

$$\sum_{i=1}^{n} d_i = \sum_{i=1}^{n} x_i - \sum_{i=1}^{n} \bar{x} \tag{6.17}$$

$$\sum_{i=1}^{n} d_i = n\bar{x} - n\bar{x} = 0 \tag{6.18}$$

TABLE 6.3 Deviations from the Mean

ρ	$(\rho - \rho_{avg})$	$\lvert(\rho - \rho_{avg})\rvert$	ρ	$(\rho - \rho_{avg})$	$\lvert(\rho - \rho_{avg})\rvert$
1020	+20	20	950	−50	50
1015	+15	15	940	−60	60
990	−10	10	890	−110	110
1060	+60	60	1080	+80	80
1030	+30	30	1120	+120	120
950	−50	50	900	−100	100
975	−25	25	1040	+40	40
1020	+20	20	1150	+150	150
980	−20	20	910	−90	90
960	−40	40	1020	+20	20
	$\sum = 0$	$\sum = 290$		$\sum = 0$	$\sum = 820$

Therefore, the average of the deviations from the mean of the data set cannot be used to measure the spread of a given data set. What if one considers the absolute value of each deviation from the mean? We can then calculate the average of the absolute values of deviations. The result of this approach is shown in the third column of Table 6.3. For group A, the mean deviation is 29, whereas for group B the mean deviation is 82. It is clear that the result provided by group B is more scattered than the group A data. Another common way of measuring the dispersion of data is by calculating the ***variance***. Instead of taking the absolute values of each deviation, one may simply square the deviations and compute their averages:

$$v = \frac{\sum_{i=1}^{n}(x_i - \bar{x})^2}{n - 1} \tag{6.19}$$

Notice, however, the variance yields units that are $(\mathrm{kg/m^3})^2$ for the given example. To remedy this problem, we can take the square root of the variance, which results in a number that is called ***standard deviation***.

$$s = \sqrt{\frac{\sum_{i=1}^{n}(x_i - \bar{x})^2}{n - 1}} \tag{6.20}$$

This may be an appropriate place to say a few words about why we use $n - 1$ rather then n to obtain the standard deviation. This is done to obtain conservative values because generally the number of experimental trials are few and limited. Let us turn our attention to the standard deviations computed for each group of densities in Table 6.4. Group A has a standard deviation that is smaller than group B's. This shows the densities reported by group A are bunched near the mean ($\rho = 1000\ \mathrm{kg/m^3}$), whereas the results reported by

TABLE 6.4 Standard Deviation Calculation for Each Group

Group A	Group B
$(\rho - \rho_{avg})^2$	$(\rho - \rho_{avg})^2$
400	2500
225	3600
100	12,100
3600	6400
900	14,400
2500	10,000
625	1600
400	22,500
400	8100
1600	400
$\sum = 10{,}750$	$\sum = 81{,}600$
$s = 34.56\ (\mathrm{kg/m^3})$	$s = 95.22\ (\mathrm{kg/m^3})$

group B are more spread out. The standard deviation can also provide information about the frequency of a given data set. For normal distribution of a data set, it can be shown that approximately 68% of the data will fall in the interval of (mean − s) to (mean + s), about 95% of the data should fall between (mean − $2s$) to (mean + $2s$), and almost all data points must lie between (mean − $3s$) to (mean + $3s$).

SUMMARY

Now that you have reached this point in the text

- You should have a good understanding of what is meant by mass and know about the important roles of mass in engineering applications and analysis.
- You should know that in engineering to show how light or heavy materials are, we use properties such as density, specific volume, specific weight, and specific gravity. You should also know the formal definition of these properties.
- You should have a good understanding of what is meant by mass flow rate and how it is related to volume flow rate.
- You should realize that mass provides a measure of resistance to translational motion. You should also know that if you want to rotate something, the distribution of mass about the center of rotation also plays an important role. The further the mass from a center of rotation, the more resistance mass offers to rotational motion.
- You should know how we define momentum for a moving object. You should also know that something with a relatively small mass could have a relatively large momentum.
- You should know how to compute basic statistical information such as mean, variance, and standard deviation.

PROBLEMS

1. Look up the mass of the following objects:
   a. recent model of an automobile
   b. the earth
   c. a Boeing 777 fully loaded
   d. an ant

2. The density of standard air is a function of temperature and may be approximated using the ideal gas law, according to

$$\rho = \frac{P}{RT}$$

where

ρ = density (kg/m^3)
P = standard atmospheric pressure (101.3 kPa)
R = gas constant; its value for air is $286.9 \left(\frac{\text{J}}{\text{kg}\cdot\text{K}}\right)$
T = air temperature in Kelvin

Create a table that shows the density of air as a function of temperature in the range of 0°C (273.15 K) to 50°C (323.15 K) in increments of 5°C. Also, create a graph showing the value of density as a function of temperature.

3. Determine the specific gravity of the following materials: gold (ρ = 1208 lb/ft^3), platinum (ρ = 1340 lb/ft^3), silver (ρ = 654 lb/ft^3), sand (ρ = 94.6 lb/ft^3), freshly fallen snow (ρ = 31 lb/ft^3), tar (ρ = 75 lb/ft^3), hard rubber (ρ = 74.4 lb/ft^3).

4. Show that the specific weight and density are related according to specific weight = (density)(acceleration due to gravity).

5. Compute the values of momentum for the following situations:
   a. a 90-kg football player running at 6 m/s
   b. a 1500-kg car moving at a rate of 100 km/h

c. a 200,000-kg Boeing 777 moving at a speed of 500 km/h
d. a bullet with a mass of 15 g traveling at a speed of 500 m/s
e. a 140-g baseball traveling at 120 km/h
f. an 80-kg stunt performer falling off a ten story building reaching a speed of 30 m/s

6. Investigate the mass flow rate of blood through your heart, and write a brief report to your instructor discussing your findings.

7. Investigate the mass flow rate of oil inside the Alaskan pipeline, and write a brief report to your instructor discussing your findings.

8. Investigate the mass flow rate of water through the Mississippi River during a normal year, and write a brief report discussing your findings. Explain flooding using the conservation of mass statement.

9. Determine the mass flow rate of fuel from the gasoline tank to the car's fuel injection system. Assume that the gasoline consumption of the car is 20 mpg when the car is moving at 60 mph. Use the specific gravity value of 0.72 for the gasoline.

10. Determine the average, variance, and standard deviation for the following parts. The measured values are given in the accompanying table.

Screw Length (cm)	Pipe Diameter (in.)	2 × 4 Lumber Width (in.)	Steel Spherical Balls (cm)
2.55	1.25	3.50	1.00
2.45	1.18	3.55	0.95
2.55	1.22	3.45	1.05
2.35	1.15	3.60	1.10
2.60	1.17	3.55	1.00
2.40	1.19	3.40	0.90
2.30	1.22	3.40	0.85
2.40	1.18	3.65	1.05
2.50	1.17	3.35	0.95
2.50	1.25	3.60	0.90

11. Next time you make a trip to a supermarket ask the manager if you can measure the mass of at least ten cereal boxes of your choice. Choose the same brand and the same size boxes. Tell the manager this is an assignment for a class. Report the average mass, variance, and standard deviation for the cereal boxes. Does the manufacturer's information noted on the box fall within your measurement?

12. Repeat Problem 11 using three other products such as cans of soup, tuna, or peanuts.

13. Obtain the height, age, and mass of players on your favorite professional basketball team. Determine the average, variance, and standard deviation for the height and age and the mass. Discuss your findings. If you do not like basketball, perform the experiment using data from a soccer team, football team, or a sports team of your choice.

14. Determine the mass moment of inertia for the following objects: a thin disk, circular cylinder, and a sphere. Refer to Equations (6.10) through (6.13) for appropriate relationships. If you were to place these objects alongside of each other on an inclined surface, which one of the objects would get to the bottom first, provided that they all have the same mass and diameter?

15. The use of ceiling fans to circulate air has become quite common. Suggest ways to correct the rotation of a wobbling fan using the concept of mass moment of inertia. As a starting point, you can use pocket change such as dimes, nickels, and quarters and chewing gum. How would you stop the fan from wobbling? Exercise caution!

16. Determine the mass moment of inertia of a steel shaft that is 1 m long and has a diameter of 5 cm. Determine the mass of the shaft using the density information provided in Table 6.1.

17. Determine the mass moment of inertia of steel balls used in ball bearings. Use a diameter of 2 cm.

18. Determine the mass moment of inertia of the earth about its axis of rotation, going through the poles. Assume the shape of the earth to be spherical. Look up information such as the mass of the earth and the radius of the earth at the equatorial plane.

19. Next time you put gasoline in your car, measure the mass flow rate (kg/s) of gasoline at the pumping station. Record the amount of gas in gallons (or liters if you are doing the experiment outside the United States) that you placed in your car's gas tank and the time that it took to do so. Make the appropriate conversion from volume flow rate to mass flow rate using the density of gasoline.

20. Measure the mass flow rate of water coming out of a drinking fountain by placing a cup under the running water and by measuring the time that it took to fill the cup. Measure the mass of the water by subtracting the total mass from the mass of the cup.

21. Obtain a graduated beaker and accurate scale from a chemistry lab and measure the density of the following liquids:
    a. a cooking oil
    b. SAE 10W-40 engine oil
    c. water
    d. milk
    e. ethylene glycol, antifreeze

    Express your findings in kg/m^3. Also determine the specific gravity of each liquid.
22. Obtain pieces of steel, wood, and concrete of known volume. Find pieces with simple shapes so that you can measure the dimensions and calculate the volume quickly. Determine their mass by placing each on an accurate scale, and calculate their densities.
23. Take a 500-sheet ream of computer paper as it comes wrapped. Unwrap it and measure the height, width, and length of the stack. Determine the volume, and measure the mass of the ream, and obtain the density. Determine how many reams come in a standard box. Estimate the total mass of the box. Discuss your assumptions and estimation procedure.
24. In this assignment you will investigate how much water may be wasted by a leaky faucet. Place a large cup under a leaky faucet. If you don't have a leaky faucet at home, open the faucet so that it just drips into the cup. Record the time that you started the experiment. Allow the water to drip into the cup for about an hour or two. Record the time when you remove the cup from under the faucet. Determine the mass flow rate. Estimate the water wasted by 100,000 people with leaky faucets during a period of one year.
25. Obtain a graduated beaker and accurate scale from a chemistry lab and measure the density of SAE 10W-40 engine oil. Repeat the experiment ten times. Determine the mean, variance, and standard deviation for your density measurement.
26. *A hydrometer* is a device which uses the principle of buoyancy to measure the Specific Gravity (S. G.) of a liquid. Design a pocket-sized hydrometer that can be used to measure S. G. for liquids having $1.2 < \text{S. G.} < 1.5$. Specify important dimensions and the mass of the device. Write a brief report to your instructor explaining how you arrived at the final design. Also include a sketch showing the calibrated scale from which one would read S. G. In addition, construct a model and attach the scale to the working model. To keep things simple, assume that the hydrometer has a cylindrical shape.
27. Referring to diagrams given in Example 6.3, how much water is stored after 20 minutes in each of the tanks? How long will it take to fill the tanks completely provided that the volume of tank (a) is 24 m^3 and tank (b) has a volume of 36 m^3? Assume the density of water is 1000 kg/m^3.
28. Investigate the size of the storage tank in a gas station. Apply the conservation of mass statement to the gasoline flow in the gas station. Draw a control volume showing appropriate components of the gasoline flow system. Estimate how much gasoline is removed from the storage tank per day. How often does the storage tank need to be filled? Is this a steady process?

CHAPTER

7 Force and Force-Related Parameters

To surface, air is blown into the blast tank of a submarine to push water out of the tank to make the submarine lighter—the submarine floats to surface. To dive, the blast tank is opened to let water in and to push out the air. The submarine sinks below the surface. The diving and surfacing of the submarine demonstrate the relationship between the weight of the submarine and the buoyancy force acting on it.

The objectives of this chapter are to introduce the concept of force, its various types, and other force related variables such as pressure and stress. We will discuss Newton's laws in this chapter. Newton's laws form the foundation of mechanics and analysis and design of many engineering problems including structures, airplane airframes (fuselage and wings), car frames, medical implants for hips and other joint replacements, machine parts, and orbit of satellites. We will also explain the tendencies of unbalanced mechanical forces, which are to translate and rotate objects. Moreover, we will consider some of the mechanical properties of materials that show how stiff or flexible a material is when subjected to a force. We will then explain the effect of a force acting at a distance in terms of creating a moment about a point; the effect of a force acting over a distance, what is formally defined as mechanical work; and the effect of a force acting over a period of time in terms of what is generally referred to as linear impulse. ■

7.1 WHAT WE MEAN BY FORCE

What is force? The simplest form of a force that represents the interaction of two objects is a push or a pull. When you push or pull on a lawn mower or on a vacuum cleaner, that interaction between your hand and the lawn mower or the vacuum cleaner is called *force*. When an automobile pulls a U-Haul trailer, a force is exerted by the bumper hitch on the trailer. The interaction representing the trailer being pulled by the bumper hitch is represented by a force. In these examples, the force is exerted by one body on another body by direct contact. Not all forces result from direct contact. For example, gravitational and magnetic forces are not exerted by direct contact. If you hold your book, say, 3 feet above the ground and let it go, what happens? It falls; that is due to gravitational force which is exerted by the earth on the book. The gravitational attractive forces act at a distance. A satellite orbiting the earth is continuously being pulled by the earth toward the center of the earth, and this allows the satellite to maintain its orbit. We will discuss the universal law of gravitational attraction in detail later. All forces, whether they represent the interaction of two bodies in direct contact or the interaction of two bodies at a distance (gravitational force), are defined by their magnitudes, their directions, and the points of application. The natural tendency of a force acting on an object, if unbalanced, will be to translate the object (move it along) and to rotate it. Moreover, forces acting on an object could squeeze or shorten, elongate, bend, or twist the object. The amount by which the object will elongate, shorten, bend, or twist will depend on the type of material the object is made from. Machine components, tools, parts of the human body, and structural members are generally subject to push–pull, bending, or twisting types of loading. Those of you who are studying to become aerospace, civil, manufacturing, or mechanical engineers will take classes in basic mechanics and mechanics of materials in which you will explore force, stresses, and materials behavior in more depth.

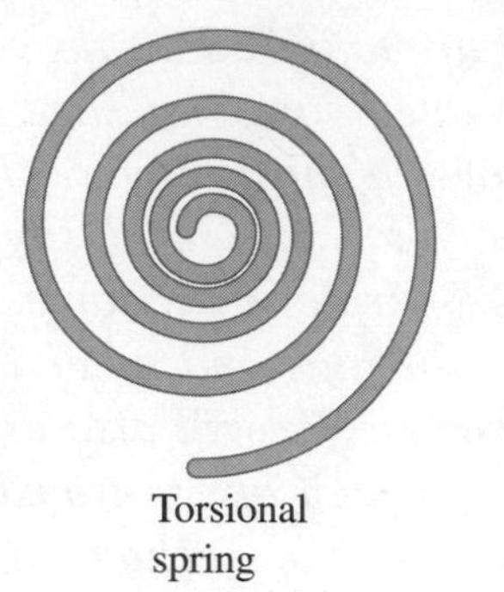
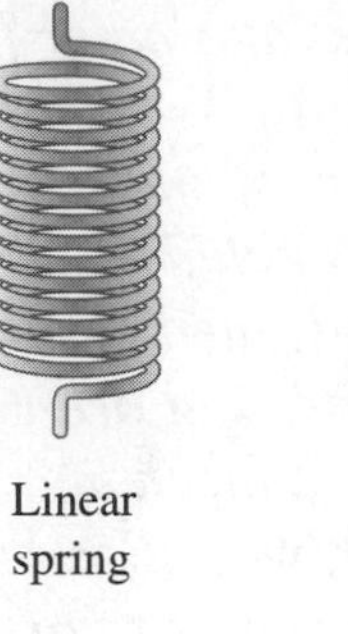
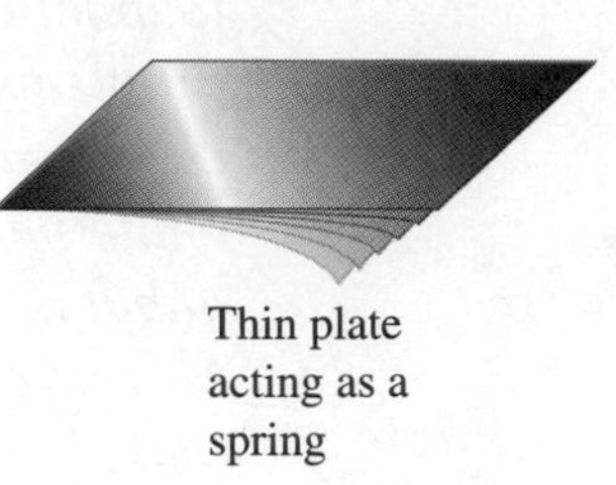

Figure 7.1 Examples of different types of springs

Spring Forces and Hooke's Law

Most of you have seen springs that are used in cars, locomotives, clothespins, weighing scales, and clips for potato chip bags. Springs are also used in medical equipment, electronics equipment such as printers and copiers, and in many restoring mechanisms (a mechanism that returns a component to its original position) covering a wide range of applications. There are different types of springs, including extension or compression and torsional springs. Examples of different types of springs are shown in Figure 7.1.

Hooke's law (named after Robert Hooke, an English scientist who proposed the law) states that over the elastic range the deformation of a spring is directly proportional to the applied force, according to

$$F = kx \tag{7.1}$$

where

F = applied force (N or lb)

k = spring constant (N/mm or N/cm or lb/in.)

x = deformation of the spring (mm or cm or in.—use units that are consistent with k)

By *elastic range* we mean the range over which if the applied force is removed the spring will return to its original unstretched shape and size and do so with no permanent deformation. The value of the spring constant depends on the type of material used to make the spring. Moreover, the shape and winding of the spring will also affect its k value. The spring constant can be determined experimentally.

EXAMPLE 7.1

For a given spring, in order to determine the value of the spring constant, we have attached dead weights to one end of the spring, as shown in Figure 7.2. We have measured and recorded the deflection caused by the corresponding weights as given in Table 7.1. What is the value of the spring constant? We have plotted the results of the experiment using Excel (shown in Figure 7.3).

The spring constant k is determined by calculating the slope of a force-deflection line (recall that the slope of a line is determined from slope = rise/run, and for this problem, the slope = change in force/change in deflection). This approach leads to a value of k = 0.54 N/mm.

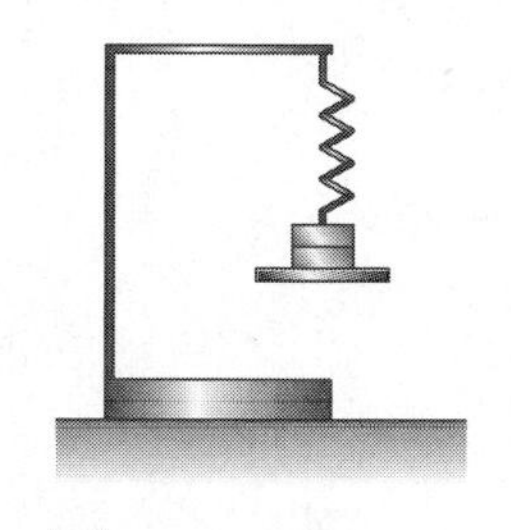

Figure 7.2 The spring setup

TABLE 7.1 The Results of the Experiment for Example 7.1

Weight (N)	The Deflection of the Spring (mm)
4.9	9
9.8	18
14.7	27
19.6	36

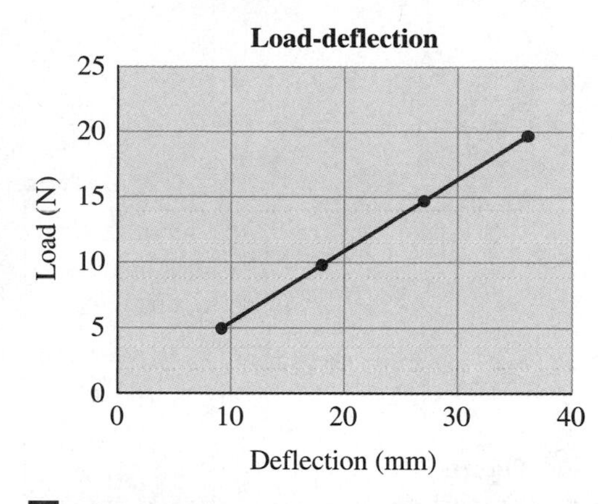

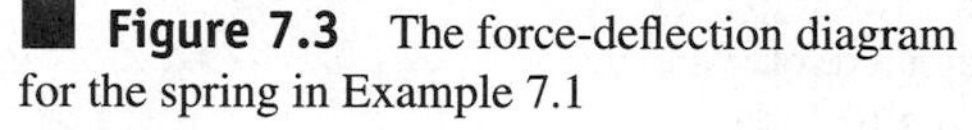

Figure 7.3 The force-deflection diagram for the spring in Example 7.1

Often, when connecting experimental force-deflection points, you may not obtain a straight line that goes through each experimental point. In this case, you will try to come up with the best fit to the data points. There are mathematical procedures (including least-squares techniques) that allow you to find the best fit to a set of data points, but for now just draw a line that you think best fits the data points. As an example, we have shown a set of data points in Table 7.1(a) and a good corresponding fit in Figure 7.4.

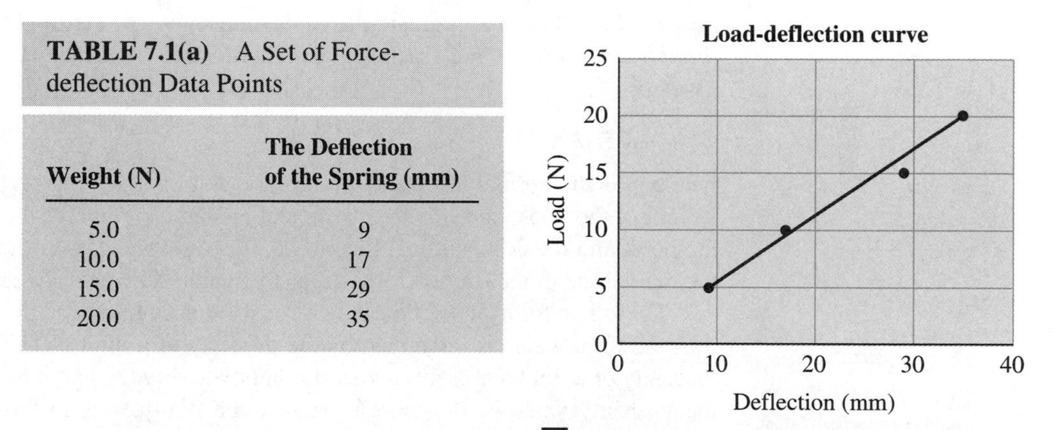

TABLE 7.1(a) A Set of Force-deflection Data Points

Weight (N)	The Deflection of the Spring (mm)
5.0	9
10.0	17
15.0	29
20.0	35

Figure 7.4 A good fit to a set of force-deflection data points

Friction Forces—Dry Friction and Viscous Friction

There are basically two types of frictional forces that are important in engineering design: ***dry frictional forces*** and ***viscous friction*** (or the fluid friction). Let us first take a closer look at dry friction, which allows us to walk or to drive our cars. Remember what happens when frictional forces get relatively small, as is the case when you try to walk on ice or

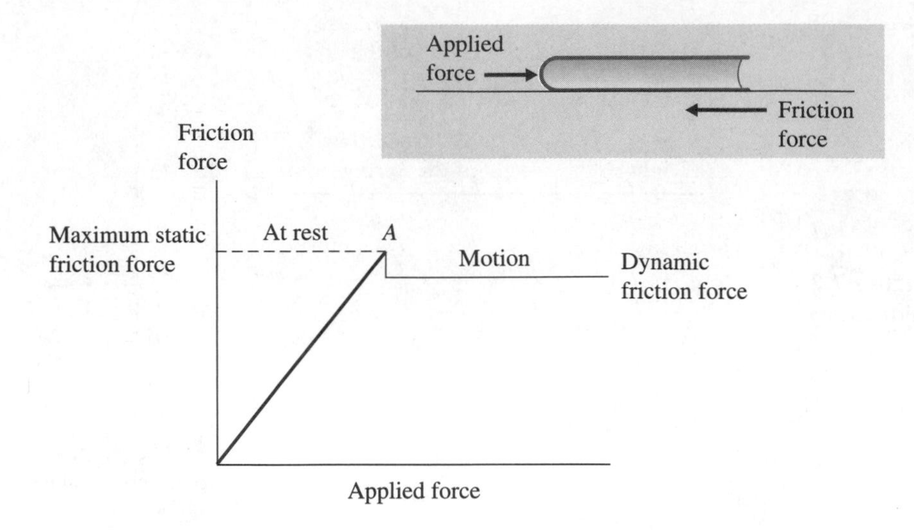

Figure 7.5 The relationship between applied force and the friction force

drive your car on a sheet of ice. Stated in a simple way, dry friction exists because of irregularities between surfaces in contact. To better understand how frictional forces build up, imagine you were to perform the following experiment. Place a book on a table and begin by pushing the book gently. You will notice that the book will not move. The applied force is balanced by the friction force generated at the contact surface. Now push on the book harder, you will notice that the book will not move until the frictional force is not great enough to prevent the book from moving. The results of your experiment may be plotted in a diagram similar to the one shown in Figure 7.5. Note that the *friction force is not constant* and reaches a maximum value (see point A in Figure 7.5), which is given by

$$F_{\text{max}} = \mu \text{N} \tag{7.2}$$

where N is the normal force and, in the case of a book resting on the table is equal to the weight of the book, and μ is the coefficient of static friction for the two surfaces involved, the book and the desk surface. Figure 7.5 also shows that once the book is set in motion, the magnitude of the friction force drops to a value called the *dynamic*, or kinetic, *friction*.

The other form of friction which must be accounted for in engineering analysis is fluid friction, which is quantified by the property of a fluid called *viscosity*. The value of viscosity of a fluid represents a measure of how easily the given fluid can flow. The higher the viscosity value is, the more resistance the fluid offers to flow. For example, if you were to pour water and honey side by side on an inclined surface, which of the two liquids will flow easier? You know the answer: The water will flow down the inclined surface faster because it has less viscosity. Fluids with relatively smaller viscosity require less energy to be transported in pipelines. For example, it would require less energy to transport water in a pipe than it does to transport motor oil or glycerin. The viscosity of fluids is a function of temperature. In general, the viscosity of gases increases with increasing temperature, and the viscosity of liquids generally decreases with increasing temperature. Knowledge of the viscosity of a liquid and how it may change with temperature also helps when selecting lubricants to reduce wear and friction between moving parts.

EXAMPLE 7.2

The coefficient of static friction between a book and a desk surface is 0.6. The book weighs 20 N. If a horizontal force of 10 N is applied to the book, would the book move? And if not, what is the magnitude of the friction force? What should be the magnitude of the horizontal force to set the book in motion?

The maximum friction force is given by

$$F_{max} = \mu N = (0.6)(20) = 12 \text{ newtons}$$

Since the magnitude of the horizontal force (10 N) is less than the maximum friction force (12 N), the book will not move, and hence the friction force will equal the applied force. A horizontal force having a magnitude greater than 12 N will be required to set the book in motion.

7.2 NEWTON'S LAWS IN MECHANICS

As we mentioned in Chapter 3, physical laws are based on observations. In this section, we will briefly discuss Newton's laws, which form the foundation of mechanics. The design and analysis of many engineering problems, including structures, machine parts, and the orbit of satellites, begins with the application of Newton's laws. Most of you will have the opportunity to take a physics, statics, or dynamics class that will explore Newton's laws and their applications further.

Newton's First Law

If a given object is at rest, and if there are no unbalanced forces acting on it, the object will then remain at rest. If the object is moving with a constant speed in a certain direction, and if there are no unbalanced forces acting on it, the object will continue to move with its constant speed and in the same direction. Newton's first law is quite obvious and should be intuitive. For example, you know from your everyday experiences that if a book is resting on a table and you don't push, pull or lift it, then the book will lie on that table in that position until it is disturbed with an unbalanced force.

Newton's Second Law

We briefly explained this law in Chapter 3, where we said that if you place a book on a smooth table and push it hard enough, it will move. Newton observed this and formulated his observation into what is called Newton's second law of motion. Newton observed that as he increased the mass of the object being pushed, while keeping the magnitude of the force constant, the object did not move as quickly. Thus, Newton noticed that there was a direct relationship between the push (the force), the mass of the object being pushed, and the acceleration of the object. He also noticed that there was a direct relationship between the direction of the force and the direction of the acceleration. Newton's second law of motion simply states that the net effect of unbalanced forces is equal to mass times acceleration of the object, which is given by the expression

$$\sum F = ma \tag{7.3}$$

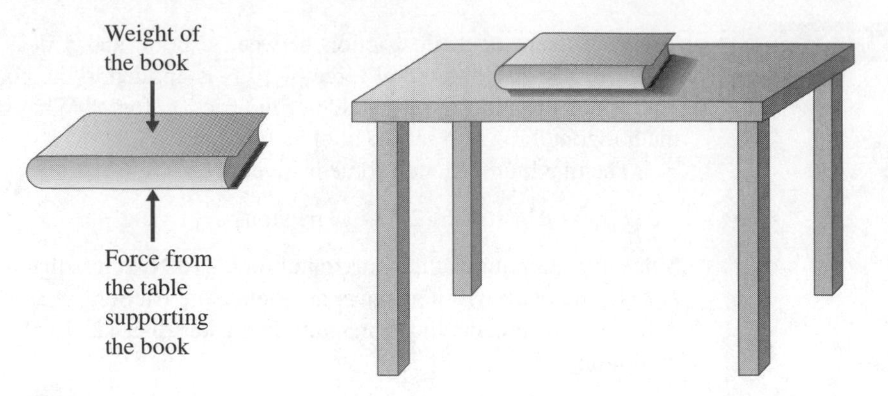

Figure 7.6 The forces of action and reaction; they have the same magnitude and the same line of action, but they act in opposite directions.

where the symbol Σ (sigma) means summation, F represents the forces in units of newton, m is the mass of the object in kg, and a is the resulting acceleration of the mass center of the object in m/s^2. In Equation (7.3), a summation of forces is used to allow for application of more than one force on an object.

Newton's Third Law

Returning to our example of a book resting on a desk, as shown in Figure 7.6, because the weight of the book is pushing down on the desk, simultaneously the desk also pushes up on the book. Otherwise, the desk won't be supporting the book. Newton's third law states that for every action there exists a reaction, and the forces of action and reaction have the same magnitude and act along the same line, but they have opposite directions.

Newton's Law of Gravitation

The weight of an object is the force that is exerted on the mass of the object by the earth's gravity. Newton discovered that any two masses, m_1 and m_2, attract each other with a force that is equal in magnitude and acts in the opposite direction, as shown in Figure 7.7, according to the following relationship:

$$F = \frac{Gm_1m_2}{r^2} \tag{7.4}$$

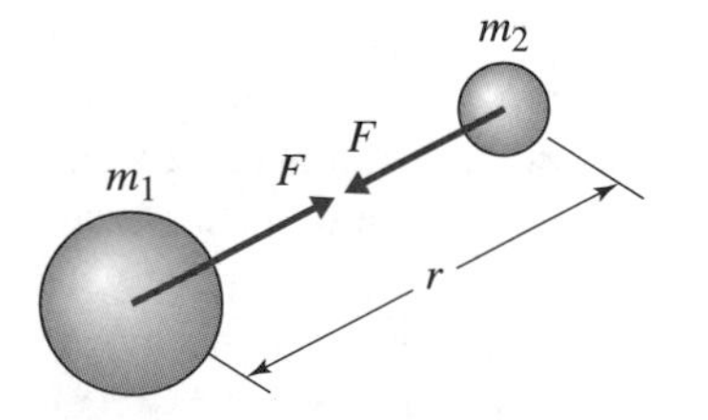

Figure 7.7 The gravitational attraction of two masses

where

F = attractive force (N)

G = universal gravitational constant (6.673×10^{-11} m^3/kg·s^2)

m_1 = mass of particle 1 (kg, see Figure 7.7)

m_2 = mass of particle 2 (kg, see Figure 7.7)

r = distance between the center of each particle (m)

Using Equation (7.4), we can determine the weight of an object having a mass m, on the earth, by substituting m for the mass of particle 1 and substituting for the mass of particle 2 the mass of the earth (M_{earth}), and using the radius of the earth as the distance between the center of each particle. Note that the radius of the earth is much larger than the physical dimension of any object on earth, and, thus, for an object resting on or near the surface of the earth, r could be replaced with R_{earth} as a good approximation. Thus, the weight (W) of an object on the surface of the earth having a mass m is given by

$$W = \frac{GM_{\text{earth}}m}{R^2_{\text{earth}}}$$

And after letting $g = \dfrac{GM_{\text{earth}}}{R^2_{\text{earth}}}$, we can write

$$W = mg \tag{7.5}$$

Equation (7.5) shows the relationship among the weight of an object, its mass, and the local acceleration due to gravity. Make sure you fully understand this simple relationship. Because the earth is not truly spherical in shape, the value of g varies with latitude and longitude. However, for most engineering applications, g = 9.81 m/s^2 or g = 32.2 ft/s^2 is used. Also note that the value of g decreases as you get further away from the surface of the earth. This fact is evident from examining Equations (7.4) and (7.5).

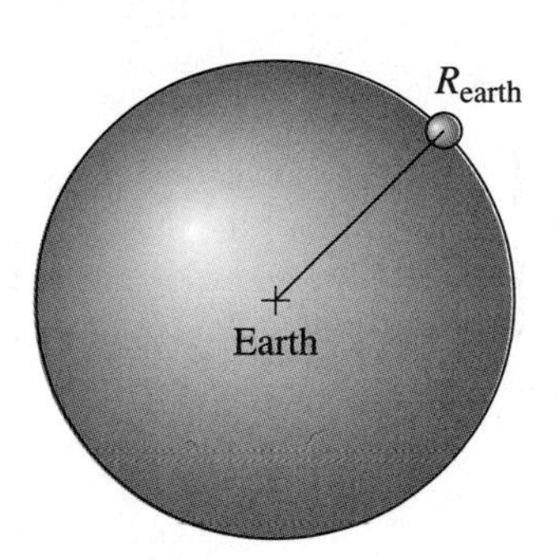

EXAMPLE 7.3

Determine the weight of an exploration vehicle whose mass is 250 kg on earth. What is the mass of the vehicle on the moon (g_{Moon} = 1.6 m/s^2) and the planet Mars (g_{Mars} = 3.7 m/s^2)? What is the weight of the vehicle on the moon and Mars?

The mass of the vehicle is 250 kg on the moon and on the planet Mars. The weight of the vehicle is determined from $W = mg$:

On earth: $W = (250\text{ kg})\left(9.8\ \frac{\text{m}}{\text{s}^2}\right) = 2450\text{ N}$

On moon: $W = (250\text{ kg})\left(1.6\ \frac{\text{m}}{\text{s}^2}\right) = 400\text{ N}$

On Mars: $W = (250\text{ kg})\left(3.7\ \frac{\text{m}}{\text{s}^2}\right) = 925\text{ N}$

EXAMPLE 7.4

The space shuttle orbits the earth at altitudes from as low as 250 km (155 miles) to as high as 965 km (600 miles), based on its missions. Determine the value of g for an astronaut in a space shuttle. If an astronaut has a mass of 70 kg on the surface of the earth, what is her weight when in orbit around the earth?

When the space shuttle is orbiting at an altitude of 250 km above the earth surface:

$$g = \frac{GM_{\text{earth}}}{R^2} = \frac{\left(6.673 \times 10^{-11} \frac{\text{m}^3}{\text{kg}\cdot\text{s}^2}\right)(5.97 \times 10^{24}\,\text{kg})}{[(6378 \times 10^3 + 250 \times 10^3)\,\text{m}]^2} = 9.07\,\text{m/s}^2$$

$$W = (70\,\text{kg})\left(9.07\,\frac{\text{m}}{\text{s}^2}\right) = 635\,\text{N}$$

At an altitude of 965 km:

$$g = \frac{GM_{\text{earth}}}{R^2} = \frac{\left(6.673 \times 10^{-11} \frac{\text{m}^3}{\text{kg}\cdot\text{s}^2}\right)(5.97 \times 10^{24}\,\text{kg})}{[(6378 \times 10^3 + 965 \times 10^3)\,\text{m}]^2} = 7.38\,\text{m/s}^2$$

$$W = (70\,\text{kg})\left(7.38\,\frac{\text{m}}{\text{s}^2}\right) = 517\,\text{N}$$

Note that at these altitudes an astronaut still has a significant weight. The near weightless conditions that you see on TV are created by the orbital speed of the shuttle. For example, when the space shuttle circles the earth at an altitude of 935 km at a speed of 7744 m/s, it creates a normal acceleration of 8.2 m/s^2. It is the difference between g and normal acceleration that creates the condition of weightlessness.

Units of Force

One newton is defined as the force that will accelerate 1 kilogram of mass at a rate of 1 m/s^2. This relationship is based on Newton's law of motion.

$$1\ \text{newton} = (1\,\text{kg})(1\,\text{m/s}^2) \tag{7.6}$$

In comparison, 1 pound force will accelerate 1 slug at a rate of 1 ft/s^2, as given by

$$1\,\text{lb} = (1\,\text{slug})(1\,\text{ft/s}^2) \tag{7.7}$$

And 1 pound force is equal to 4.48 newtons ($1\,\text{lb}_\text{f} = 4.448\,\text{N}$).

7.3 PRESSURE AND STRESS—FORCE ACTING OVER AN AREA

Pressure provides a measure of intensity of a force acting over an area. It can be defined as the ratio of force over the contact surface area:

$$\text{pressure} = \frac{\text{force}}{\text{area}} \tag{7.8}$$

To better understand what the magnitude of a pressure represents, consider the situations shown in Figure 7.8. Let us first look at the situation depicted in Figure 7.8(a), in which we lay a solid brick, in the form of a rectangular prism with dimensions of 21.6 × 6.4 × 10.2 cm ($8\frac{1}{2} \times 2\frac{1}{2} \times 4$ in.) that weighs 28 N (6.4 lb), flat on its face. Using Equation (7.8) for this orientation, the pressure at the contact surface is

$$\text{pressure} = \frac{\text{force}}{\text{area}} = \frac{28\,\text{N}}{(0.216\,\text{m})(0.102\,\text{m})} = 1271\frac{\text{N}}{\text{m}^2} = 1271\,\text{Pa}$$

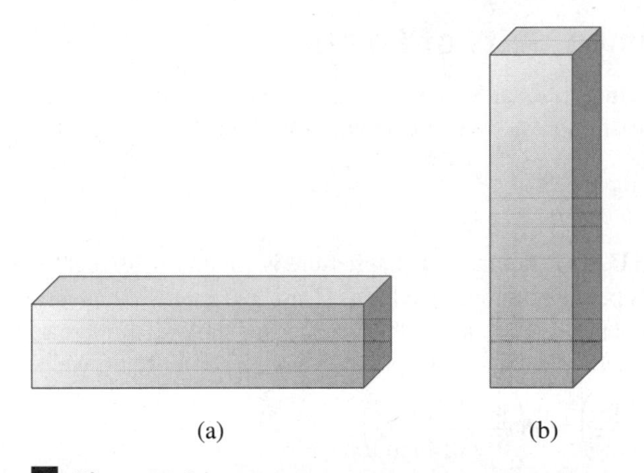

Figure 7.8 An experiment demonstrating the concept of pressure. (a) A solid brick resting on its face. (b) A solid brick resting on its end. In position (b), the block creates a higher pressure on the surface.

Note that one newton per squared meter is called one pascal ($1 \text{ N}/1 \text{ m}^2 = 1 \text{ Pa}$). Now if we were to lay the brick on its end as depicted in Figure 7.8(b), the pressure due to the weight of the brick becomes

$$\text{pressure} = \frac{\text{force}}{\text{area}} = \frac{28 \text{ N}}{(0.064 \text{ m})(0.102 \text{ m})} = 4289 \frac{\text{N}}{\text{m}^2} = 4289 \text{ Pa}$$

It is important to note here that the weight of the brick is 28 N, regardless of how it is laid. But the pressure that is created at the contact surface depends on the magnitude of the contact surface area. The smaller the contact area, the larger the pressure created by the same force. You already know this from your everyday experiences—which situation would create more pain, pushing on someone's arm with a finger or a thumbtack?

In Chapter 4, when we were discussing the importance of area and its role in engineering applications, we briefly mentioned the importance of understanding pressure in situations such as foundations of buildings, hydraulic systems, and cutting tools. We also said that in order to design a sharp knife, one must design the cutting tool such that it creates large pressures along the cutting edge. This is achieved by reducing the contact surface area along the cutting edge. Understanding fluid pressure distributions is very important in engineering problems in hydrostatics, hydrodynamics, and aerodynamics. *Hydrostatic* refers to water at rest and its study; however, the understanding of other fluids is also considered in hydrostatics. A good example of a hydrostatic problem is calculating the water pressure acting on the surface of a dam and how it varies along the height of the dam. Understanding pressure is also important in hydrodynamics studies, which deal with understanding the motion of water and other fluids such as those encountered in the flow of oil or water in pipelines or the flow of water around the hull of a ship or a submarine. Air pressure distributions play important roles in the analysis of air resistance to the movement of vehicles or in creating lift forces over the wings of an airplane. Aerodynamics deals with understanding the motion of air around and over surfaces.

Common Units of Pressure

In the International System of units, pressure units are expressed in pascal. One pascal is the pressure created by one newton force acting over a surface area of 1 m^2:

$$1\ \text{Pa} = \frac{1\ \text{N}}{\text{m}^2} \tag{7.9}$$

In U.S. Customary Units, pressure is commonly expressed in pounds per square inch or pounds per square foot; one lb/in^2 (1 psi) represents the pressure created by a 1-pound force over an area of 1 in^2, and 1 lb/ft^2 represents the pressure created by a 1-pound force over an area of 1 ft^2. To convert the pressure from lb/ft^2 to lb/in^2, we take the following step:

$$P\left(\frac{\text{lb}}{\text{in}^2}\right) = P\left(\frac{\text{lb}}{\text{ft}^2}\right)\left(\frac{1\ \text{ft}^2}{144\ \text{in}^2}\right) \tag{7.10}$$

Note that 1 lb/in^2 is usually referred to as 1 **psi**, which reads one **p**ound per **s**quare **i**nch. Many other units are also commonly used for pressure. For example, the atmospheric pressure is generally given in inches of mercury (in·Hg) or millimeters of mercury (mm·Hg). In air-conditioning applications, often inches of water are used to express the magnitude of air pressure. These units and their relationships with Pa and psi will be explained further in Example 7.6, after we explain how the pressure of a fluid at rest changes with its depth.

For fluids at rest, there are two basic laws: (1) Pascal's law, which explains that the pressure of a fluid at a point is the same in all directions, and (2) the pressure of fluid increases with its depth. Even though these laws are simple, they are very powerful tools for analyzing various engineering problems.

Pascal's Law

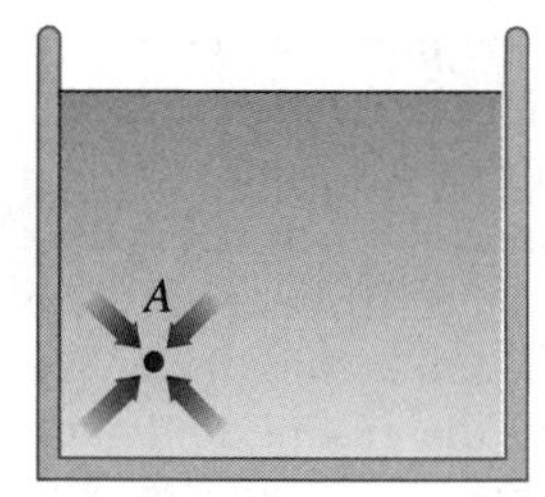

Figure 7.9 In a static fluid, pressure *at a point* is the same in all directions.

Pascal's law states that for a fluid at rest, *pressure at a point* is the same in all directions. Note carefully that we are discussing pressure at a point. To demonstrate this law, consider the vessel shown in Figure 7.9; the pressure at point A is the same in all directions.

Another simple, yet important concept for a fluid at rest states that pressure increases with the depth of fluid. So the hull of a submarine is subjected to more water pressure when cruising at 300 m than at 100 m. Some of you have directly experienced the variation of pressure with depth when you have gone scuba diving or swimming in a lake. You recall from your experience that as you go deeper, you feel higher pressure on your body. Referring to Figure 7.10, the relationship between the gauge pressure and the height of a fluid column above it is given by

$$P = \rho g h \tag{7.11}$$

where

P = is the fluid pressure at point B (see Figure 7.10) (in Pa or lb/ft^2)

ρ = is the density of the fluid (in kg/m^3 or slugs/ft^3)

g = is the acceleration due to gravity (g = 9.81 m/s^2 or g = 32.2 ft/s^2)

h = is the height of fluid column (in m or ft)

Also shown in Figure 7.10 is the force balance between the pressure acting on the bottom surface of a fluid column and its weight. Equation (7.11) is derived under the assumption of constant fluid density.

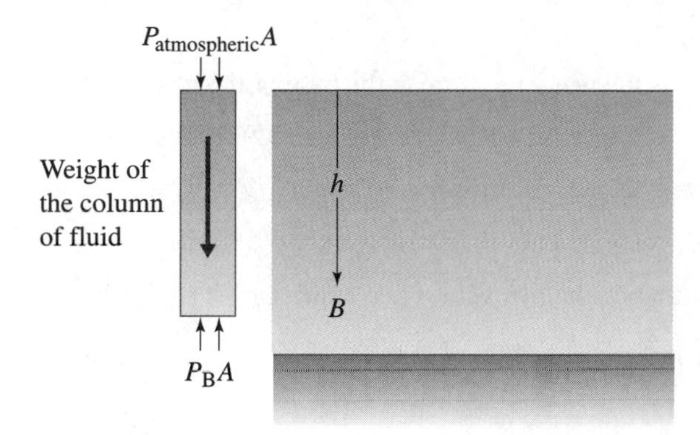

Figure 7.10 The variation of pressure with depth

Another concept, which is closely related to fluid pressure distribution is ***buoyancy***. As we discussed in Chapter 4, buoyancy is the force that a fluid exerts on a submerged object. The net upward buoyancy force arises from the fact that the fluid exerts a higher pressure at the bottom surfaces of the object than it does on the top surfaces of the object. Thus, the net effect of fluid pressure distribution acting over the submerged surface of an object is the buoyancy force. The magnitude of the buoyancy force is equal to the weight of the volume of the fluid displaced, which is given here again for convenience.

$$F_B = \rho V g$$

where F_B is the buoyancy force (N), ρ represents the density of the fluid (kg/m^3), and g is acceleration due to gravity (9.81 m/s^2).

EXAMPLE 7.5

Most of you may have seen water towers in and around small towns. The function of a water tower is to create a desirable municipal water pressure for household and other usage in a town. To achieve this purpose, water is stored in large quantities in elevated tanks. You may also have noticed that sometimes the water towers are placed on top of hills or at high points in a town. The municipal water pressure may vary from town to town, but it generally falls somewhere between 50 psi and 80 psi. We are interested in developing a table that shows the relationship between the height of water aboveground in the water tower and the water pressure in a pipeline located at the base of the water tower, as shown in Figure 7.11. We will calculate the water pressure, in psi, in a pipe located at the base of the water tower as we vary the height of water in increments of 5 ft. We will create a table showing our results up to 100 psi. And using the height–pressure table, we will answer the question, "What should be the water level in the water tower to create a 50-psi water pressure in a pipe at the base of the water tower?" The relationship between the height of water aboveground in the water tower and the water pressure in a pipeline located at the base of the water tower is given by Equation (7.11):

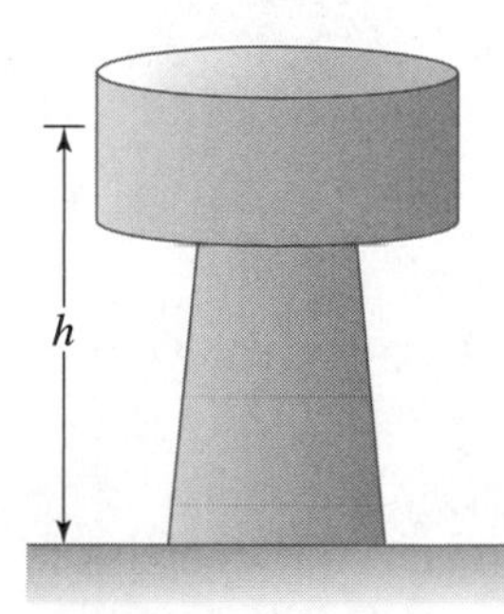

Figure 7.11 The water tower of Example 7.5

$$P = \rho g h$$

where

P = is the water pressure at the base of the water tower (lb/ft^2)

ρ = is the density of water (equal to 1.94 slugs/ft^3)

g = is the acceleration due to gravity (g = 32.2 ft/s^2)

h = is the height of water aboveground (in ft)

Substituting the known values into Equation (7.11) leads to

$$P\left(\frac{\text{lb}}{\text{ft}^2}\right) = 1.94\left(\frac{\text{slugs}}{\text{ft}^3}\right)\left(32.2\,\frac{\text{ft}}{\text{s}^2}\right)[h(\text{ft})]$$

Recall that to convert the pressure from lb/ft^2 to lb/in^2, we take the following step:

$$P\left(\frac{\text{lb}}{\text{in}^2}\right) = P\left(\frac{\text{lb}}{\text{ft}^2}\right)\left(\frac{1\ \text{ft}^2}{144\ \text{in}^2}\right)$$

TABLE 7.2 The Relationship Between the Height of Water in a Water Tower and the Pressure in a Pipeline at Its Base

Water Level in the Tower (ft)	Water Pressure (lb/in^2)
10	4.34
20	8.68
30	13.01
40	17.35
50	21.69
60	26.03
70	30.37
80	34.70
90	39.04
100	43.38
110	47.72
120	52.06
130	56.39
140	60.73
150	65.07
160	69.41
170	73.75
180	78.08
190	82.42
200	86.76
210	91.10
220	95.44
230	99.77
240	104.11

And performing the conversion,

$$P\left(\frac{\text{lb}}{\text{in}^2}\right) = 1.94\left(\frac{\text{slugs}}{\text{ft}^3}\right)\left(32.2\,\frac{\text{ft}}{\text{s}^2}\right)[h(\text{ft})]\left(\frac{1\ \text{ft}^2}{144\ \text{in}^2}\right) = (0.4338)[h(\text{ft})]$$

Note that the factor 0.4338 has the proper units that give the pressure in lb/in^2 when the value h is input in feet. Next, we generate Table 7.2 by substituting for h in increments of 5 ft in the preceding equation. From Table 7.2 we can see that in order to create a 50-psi water pressure in a pipeline at the base of the tower, the water level in the water tower should be approximately 120 ft.

Atmospheric Pressure

Earth's atmospheric pressure is due to the weight of the air in the atmosphere above the surface of the earth. It is the weight of the column of air (extending all the way to the outer edge of the atmosphere) divided by a unit area at the base of the air column. Standard atmosphere at sea level has a value of 101.325 kPa. From the definition of atmospheric pressure you should realize that it is a function of altitude. The variation of standard atmospheric pressure and air density with altitude is given in Table 7.3. For example,

TABLE 7.3 Variation of Standard Atmosphere with Altitude

Altitude (m)	Atmospheric Pressure (kPa)	Air Density (kg/m^3)
0 (sea level)	101.325	1.225
500	95.46	1.167
1000	89.87	1.112
1500	84.55	1.058
2000	79.50	1.006
2500	74.70	0.957
3000	70.11	0.909
3500	65.87	0.863
4000	61.66	0.819
4500	57.75	0.777
5000	54.05	0.736
6000	47.22	0.660
7000	41.11	0.590
8000	35.66	0.526
9000	30.80	0.467
10,000	26.50	0.413
11,000	22.70	0.365
12,000	19.40	0.312
13,000	16.58	0.266
14,000	14.17	0.228
15,000	12.11	0.195

Source: Data from U.S. Standard Atmosphere (1962).

the atmospheric pressure and air density at the top of Mount Everest is approximately 31 kPa and 0.467 kg/m^3, respectively. The elevation of Mount Everest is approximately 9000 m (exactly 8848 m), and thus less air lies on top of Mount Everest than there is at sea level. Moreover, the air pressure at the top of Mount Everest is 30% of the value of atmospheric pressure at sea level, and the air density at its top is only 38% of air density at sea level. New commercial planes have a cruising altitude capacity of approximately 11,000 m. Referring to Table 7.3, you see the atmospheric pressure at that altitude is approximately one-fifth of the sea-level value, so there is a need for pressurizing the cabin. Also, because of the lower air density at that altitude, the power required to overcome air resistance (to move the plane through atmosphere) is not as much as it would be at lower altitudes. Atmospheric pressure is commonly expressed in one of the following units: pascals, millimeters of mercury, and inches of mercury. Their values are

$$1 \text{ atm} = 101325 \text{ Pa} = 101.3 \text{ kPa}$$
$$1 \text{ atm} = 14.69 \text{ lb/in}^2$$
$$1 \text{ atm} = 760 \text{ mm·Hg}$$
$$1 \text{ atm} = 29.92 \text{ in·Hg}$$

See Example 7.6 to see how these and other units of atmospheric pressure are related.

EXAMPLE 7.6

Starting with an atmospheric pressure of 101,325 Pa, express the magnitude of the pressure in the following units: (a) millimeters of mercury (mm·Hg), (b) inches of mercury (in·Hg), (c) meters of water, and (d) feet of water. The densities of water and mercury are $\rho_{H_2O} = 1{,}000 \text{ kg/m}^3$ and $\rho_{Hg} = 13{,}550 \text{ kg/m}^3$ respectively.

(a) We start with the relationship between the height of a fluid column and the pressure at the base of the column, which is

$$P = \rho g h = 101{,}325\left(\frac{\text{N}}{\text{m}^2}\right) = 13{,}550\left(\frac{\text{kg}}{\text{m}^3}\right)\left[9.81\left(\frac{\text{m}}{\text{s}^2}\right)\right]h(\text{m})$$

And solving for h, we have

$$h = 0.762 \text{ m} = 762 \text{ mm}$$

Therefore, the pressure due to a standard atmosphere is equal to the pressure created at a base of a 762-mm-tall column of mercury. Stated another way, 1 atm = 762 mm·Hg.

(b) Next, we will convert the magnitude of pressure from millimeters of mercury to units that express it in inches of mercury.

$$762 \text{ (mm·Hg)}\left(\frac{0.03937 \text{ in.}}{1 \text{ mm}}\right) = 30 \text{ in·Hg}$$

(c) To express the magnitude of pressure in meters of water, as we did in part (a), we begin with the relationship between the height of a fluid column and the pressure at the base of the column, which is

$$P = \rho g h = 101{,}325\left(\frac{\text{N}}{\text{m}^2}\right) = 1{,}000\left(\frac{\text{kg}}{\text{m}^3}\right)\left[9.81\left(\frac{\text{m}}{\text{s}^2}\right)\right]h(\text{m})$$

And then solve for h, which results in

$$h = 10.328 \text{ m}$$

Therefore, the pressure due to a standard atmosphere is equal to the pressure created at a base of a 10.328-m-tall column of water, that is, 1 atm = 10.328 m·H_2O.

(d) Finally we convert the units of pressure from meters of water to feet of water.

$$10.328(\text{m}\cdot\text{H}_2\text{O})\left(\frac{3.280\ \text{ft}}{1\ \text{m}}\right) = 33.87\ \text{ft}\cdot\text{H}_2\text{O}$$

Absolute Pressure and Gauge Pressure

Most pressure gauges show the magnitude of the pressure of a gas or a liquid relative to the local atmospheric pressure. For example, when a tire pressure gauge shows 32 psi, this means the pressure of air inside the tire is 32 psi above the local atmospheric pressure. In general, we can express the relationship between the absolute and the gauge pressure by

$$P_{\text{absolute}} = P_{\text{gauge}} + P_{\text{atmospheric}} \tag{7.12}$$

Vacuum refers to pressures below atmospheric level. Thus, negative gauge pressure readings indicate vacuum. Some of you, in your high school physics class, may have seen demonstrations dealing with an enclosed container connected to a vacuum pump. As the vacuum pump draws air out of the container, thus creating a vacuum, the atmospheric pressure acting on the outside of the container made the container collapse. As more air is drawn, the container continued to collapse. An absolute zero pressure reading in a container indicates absolute vacuum, meaning there is no more air left in the container. In practice, achieving absolute zero pressure is not possible, meaning at least a little air will remain in the container.

EXAMPLE 7.7

We have used a tire gauge and measured the air pressure inside a car tire to be 35.0 psi. What is the absolute pressure of the air inside the tire, if the car is located in (a) a city located at sea level, (b) a city located in Colorado with an elevation of 1500 m? Express your results in both units of psi and pascals.

The absolute pressure is related to the gauge pressure according to

$$P_{\text{absolute}} = P_{\text{gauge}} + P_{\text{atmospheric}}$$

(a) For a city located at sea level, $P_{\text{atmospheric}} = 14.69\ \text{psi} \approx 14.7\ \text{psi}$,

$$P_{\text{absolute}} = 35.0\ \text{psi} + 14.7\ \text{psi} = 49.7\ \text{psi}$$

$$P_{\text{absolute}} = 49.7\ (\text{psi})\left(\frac{6895\ \text{Pa}}{1\ \text{psi}}\right) = 342{,}680\ \text{Pa} = 342.68\ \text{kPa}$$

(b) We can use Table 7.3 to look up the standard atmospheric pressure for a city located in Colorado with an elevation of 1500 m. The atmospheric pressure at an elevation of 1500 m is $P_{\text{atmospheric}} = 84.55$ kPa.

$$P_{\text{gauge}} = 35.0(\text{psi})\left(\frac{6895\ \text{Pa}}{1\ \text{psi}}\right) = 241{,}325\ \text{Pa}$$

$$P_{\text{absolute}} = 241{,}325 + 84{,}550 = 325{,}875\ \text{Pa} = 325.875\ \text{kPa}$$

$$P_{\text{absolute}} = 325{,}875(\text{Pa})\left(\frac{1\ \text{psi}}{6895\ \text{Pa}}\right) = 47.3\ \text{psi}$$

EXAMPLE 7.8

A vacuum pressure gauge monitoring the pressure inside a container reads 200 mm·Hg vacuum. What are the gauge and absolute pressures? Express results in pascals.

The stated vacuum pressure is the gauge pressure, and to convert its units from millimeters of mercury to pascal, we use the conversion factor 1 mm·Hg = 133.3 Pa. This results in

$$P_{\text{gauge}} = -200(\text{mm}\cdot\text{Hg})\left(\frac{133.3\ \text{Pa}}{1\ \text{mm}\cdot\text{Hg}}\right) = -26{,}660\ \text{Pa}$$

The absolute pressure is related to the gauge pressure according to

$$P_{\text{absolute}} = P_{\text{gauge}} + P_{\text{atmospheric}}$$

$$P_{\text{absolute}} = -26{,}660 + 101{,}325 = 74{,}665\ \text{Pa}$$

Note that the negative gauge pressure indicates vacuum, or a pressure below atmospheric level.

Vapor Pressure of a Liquid

You may already know from your everyday experience that under the same surrounding conditions some fluids evaporate faster than others do. For example, if you were to leave a pan of water and a pan of alcohol side by side in a room, the alcohol would evaporate and leave the pan long before you would notice any changes in the level of water. That is because the alcohol has a higher vapor pressure than water. Thus, under the same conditions, fluids with low vapor pressure, such as glycerin, will not evaporate as quickly as those with high values of vapor pressure. Stated another way, at a given temperature the fluids with low vapor pressure require relatively smaller surrounding pressure at their free surface to prevent them from evaporating.

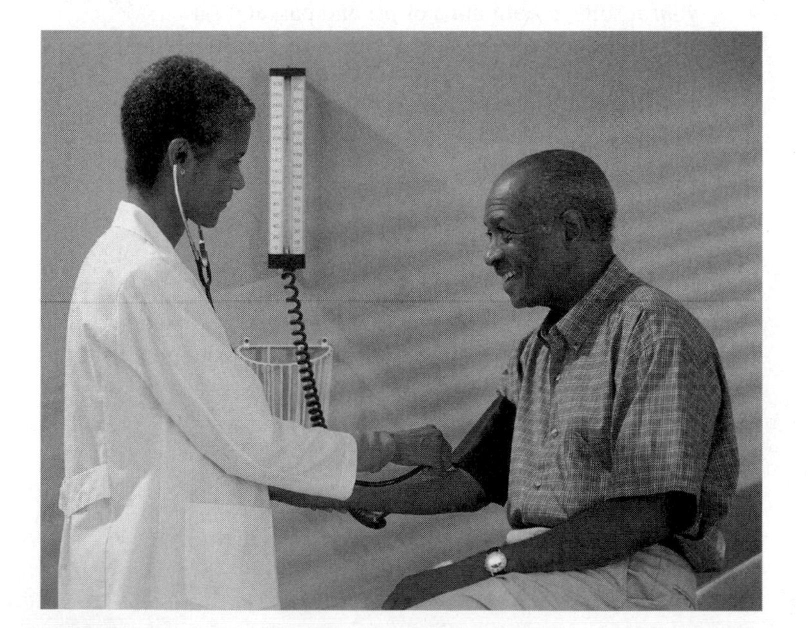

Measuring blood pressure

Blood Pressure We all have been to a doctor's office. One of the first things that a doctor or a nurse will measure is our blood pressure. Blood pressure readings are normally given by two numbers, for example 115/75 (mm·Hg). The first, or the higher, number corresponds to systolic pressure, which is the maximum pressure exerted when our heart contracts. The second, or the lower number, measures the diastolic blood pressure, which is the pressure in the arteries when the heart is at rest. The pressure unit commonly used to represent blood pressure is millimeters of mercury (mm·Hg). You may already know that blood pressure may change with level of activity, diet, temperature, emotional state, and so on.

Hydraulic Systems

Hydraulic digging machines and bulldozers perform tasks in hours that used to take days or weeks. Today, hydraulic systems are readily used in many applications, including the braking and steering systems in cars, the aileron or flap control system of an airplane, and as jacks to lift cars. To understand hydraulic systems, you must first fully understand the concept of pressure. In an enclosed fluid system, when a force is applied at one point, creating a pressure at that point, that force is transmitted through the fluid to other points, provided the fluid is incompressible. When you press your car's brake pedal, that action creates a force and subsequently a pressure in the brake master cylinder. The force is then transmitted through the hydraulic fluid from the master cylinder to the wheel cylinders where the pistons in the wheel cylinder push the brake shoes against the brake drums (rotator). The mechanical power is converted to hydraulic power and back to mechanical power. We will discuss the idea of what we mean by power later in this chapter and in Chapter 10. The typical hydraulic fluid pressure in a car's hydraulic brake lines is approximately 10 MPa (1500 psi).

Now let us formulate a general relationship among force, pressure, and area in a hydraulic system such as the simple hydraulic system shown in Figure 7.12. Because the pressure is nearly constant in the hydraulic system shown, the relationship between the forces F_1 and F_2 could be established in the following manner:

$$P_1 = \frac{F_1}{A_1} \tag{7.13}$$

$$P_2 = \frac{F_2}{A_2} \tag{7.14}$$

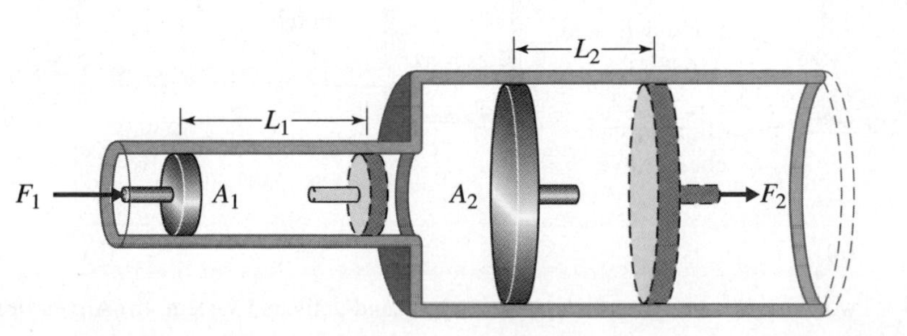

Figure 7.12
An example of a simple hydraulic system

$$P_1 = P_2 = \frac{F_1}{A_1} = \frac{F_2}{A_2} \tag{7.15}$$

$$F_2 = \frac{A_2}{A_1} F_1 \tag{7.16}$$

It is important to realize that when the force F_1 is applied, the drive piston, piston 1, moves by a distance L_1, while the driven piston, piston 2, moves by a lesser distance, L_2. However, the volume of the fluid displaced is constant, as shown in Figure 7.12. Knowing that in an enclosed system, the volume of the fluid displaced remains constant, we can then determine the distance L_2 by which piston 2 moves, provided the displacement of piston 1, L_1, is known.

$$\text{fluid volume displaced} = A_1 L_1 = A_2 L_2 \tag{7.17}$$

$$L_2 = \frac{A_1}{A_2} L_1 \tag{7.18}$$

Moreover, a relationship between the speed of the drive piston and the driven piston can be formulated by dividing both sides of Equation (7.18) by the time that it takes to move each piston. This leads to the following equation:

$$V_2 = \frac{A_1}{A_2} V_1 \tag{7.19}$$

where V_1 represents the speed of the drive piston, and V_2 is the speed at which the driven piston moves.

Next, we will explain the operation and components of manually activated and pump-driven hydraulic systems. Examples of these types of hydraulic systems are depicted in Figure 7.13(a) and (b). Figure 7.13(a) shows a schematic diagram of a hand-driven hydraulic jack. The system consists of a reservoir, a hand pump, the load piston, a relief valve, and a high-pressure check valve. To raise the load, the arm of the hand pump is

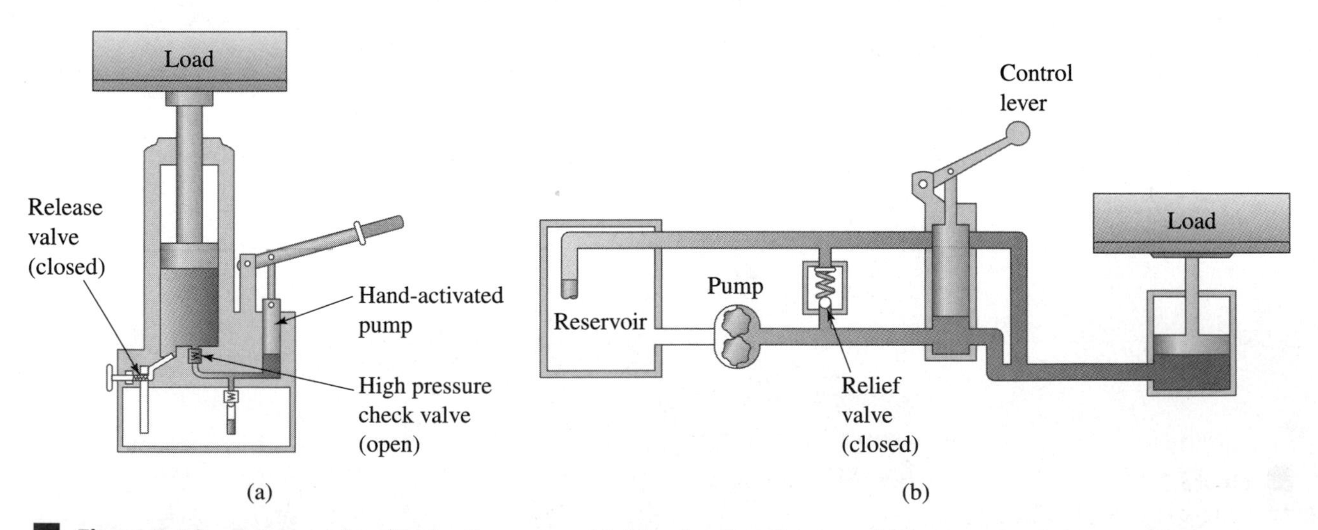

Figure 7.13 Two examples of hydraulic systems. (a) A hand-activated system. (b) A gear or rotary pump system.

pushed downward; this action pushes the fluid into the load cylinder, which in turn creates a pressure that is transmitted to the load piston, and consequently the load is raised. To lower the load, the release valve is opened. The amount by which the release valve is opened will determine the speed at which the load will be lowered. Of course, the viscosity of the hydraulic fluid and the magnitude of the load will also affect the lowering speed. In the system shown, the fluid reservoir is necessary to supply the line with as much fluid as needed to extend the driven piston to any desired level.

The hydraulic system shown in Figure 7.13(b) replaces the hand-activated pump by a gear or a rotary pump that creates the necessary pressure in the line. As the control handle is moved up, it opens the passage that allows the hydraulic fluid to be pushed into the load cylinder. When the control handle is pushed down, the hydraulic fluid is routed back to the reservoir, as shown. The relief valves shown in Figure 7.13 could be set at desired pressure levels to control the fluid pressure in the lines by allowing the fluid to return to the reservoir as shown.

EXAMPLE 7.9

Determine the load that can be lifted by the hydraulic system shown in Figure 7.14.

$$F_1 = m_1 g = (100\ \text{kg})(9.81\ \text{m/s}^2) = 981\ \text{N}$$

$$F_2 = \frac{A_2}{A_1}F_1 = \frac{\pi(0.15\ \text{m})^2}{\pi(0.05\ \text{m})^2}(981\ \text{N}) = 8829\ \text{N}$$

$$F_2 = 8829\ \text{N} = (m_2\ \text{kg})(9.81\ \text{m/s}^2) \quad \Rightarrow \quad m_2 = 900\ \text{kg}$$

Note that for this problem we could have started with the equation that relates F_2 to F_1, and then simplified the similar quantities such as π and g in the following manner:

$$F_2 = \frac{A_2}{A_1}F_1 = m_2 g = \frac{\pi(R_2)^2}{\pi(R_1)^2}(m_1 g)$$

$$m_2 = \frac{(R_2)^2}{(R_1)^2}m_1 = \frac{(15\ \text{cm})^2}{(5\ \text{cm})^2}(100\ \text{kg}) = 900\ \text{kg}$$

This approach is preferred over the direct substitution of values into the equation right away because it allows us to change a value of a variable, such as m_1 or the areas, and see what happens to the result. For example, using the second approach, we can see clearly that if m_1 is increased to a value of 200 kg, then m_2 changes to 1800 kg.

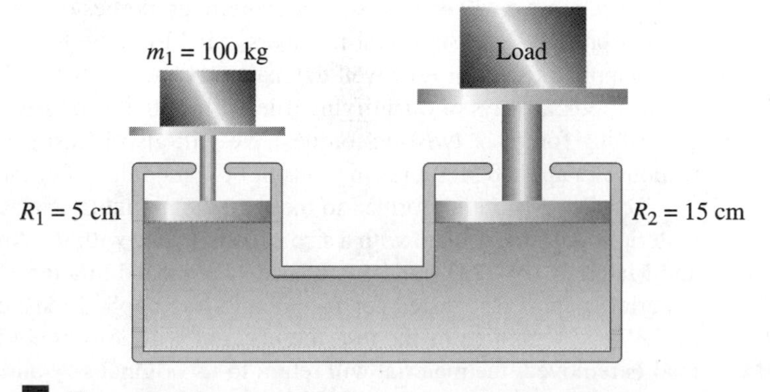

Figure 7.14 The hydraulic system of Example 7.9

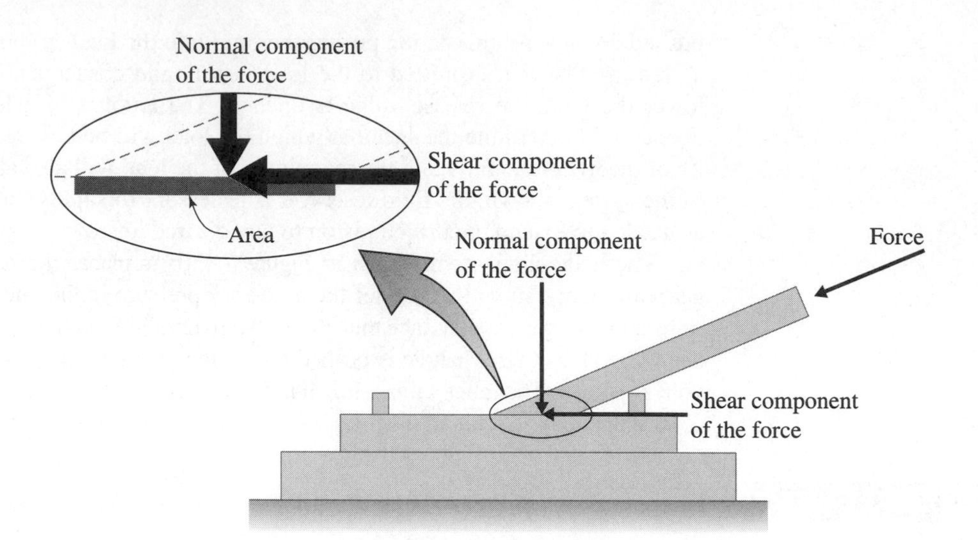

Figure 7.15 An anchoring plate subjected to a compressive and a shearing force

Stress

Stress provides a measure of the intensity of a force acting over an area. Consider the situation shown in Figure 7.15. The plate shown is subjected to the compressive force. Because the force is applied at an angle, it has two components: a horizontal and a vertical component. The tendency of the horizontal component of the force is to shear the plate, and the tendency of the vertical component is to compress the plate.

The ratio of the normal (vertical) component of the force to the area is called the ***normal stress***, and the ratio of the horizontal component of the force (the component of the force that is parallel to the plate surface) to the area is called ***shear stress***. The normal stress component is often called *pressure*.

7.4 MODULUS OF ELASTICITY, MODULUS OF RIGIDITY, AND BULK MODULUS OF COMPRESSIBILITY

In this section we will discuss some important properties of materials. Engineers, when designing products and structural members, need to know how a selected material behaves under applied forces or how well the material conducts thermal energy or electricity. Here we will look at ways of quantifying (measuring) the response of solid materials to pulling or pushing forces or twisting torques. We will also investigate the behavior of fluids in response to applied pressures in terms of bulk modulus of compressibility.

Tensile tests are performed to measure the modulus of elasticity and strength of solid materials. A test specimen with a size in compliance with the American Society for Testing and Materials (ASTM) standards is made and placed in a tensile testing machine. When a material specimen is tested for its strength, the applied tensile load is increased slowly. In the very beginning of the test, the material will deform elastically, meaning that if the load is removed, the material will return to its original size and shape without any permanent deformation. The point to which the material exhibits this elastic behavior is called

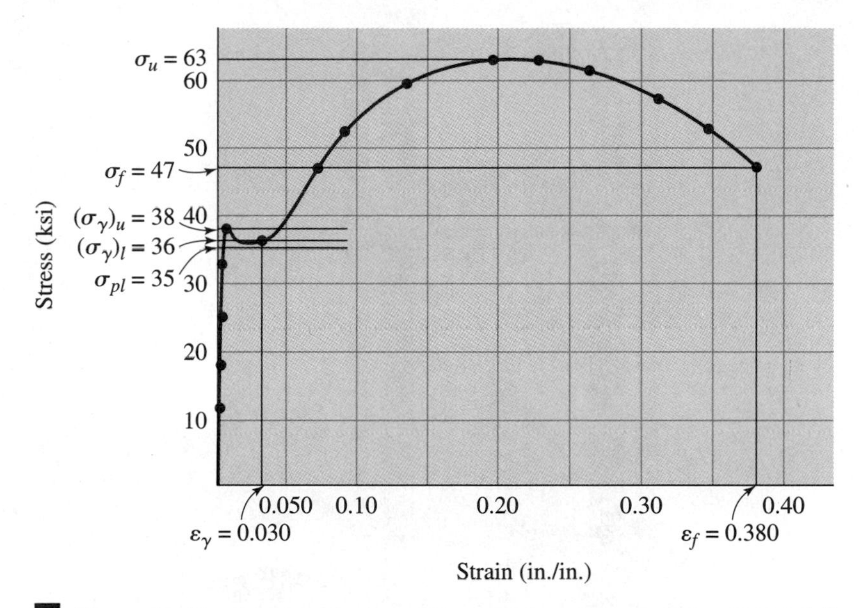

Figure 7.16 Stress–strain diagram for a mild steel sample (ksi = 1000 lb/in^2, σ_{pl} = elastic stress, σ_u = ultimate stress, σ_f = fracture stress).
Source: R. C. Hibbler, *Mechanics of Materials.*

elastic point. As the material stretches, the normal stress (the normal force divided by the cross-sectional area) is plotted versus the strain (deformation divided by the original length of the specimen). An example of such test results for a steel sample is shown in Figure 7.16.

The modulus of elasticity, or Young's modulus, is computed by calculating the slope of a stress–strain diagram over the elastic region. The ***modulus of elasticity*** is a measure of how easily a material will stretch when pulled (subject to a tensile force) or how well the material will shorten when pushed (subject to a compressive force). The larger the value of the modulus of elasticity, the larger the force required to stretch or shorten the material by a certain amount. In order to better understand what the value of the modulus of elasticity represents, consider the following example: Given a piece of rubber, a piece of aluminum, and a piece of steel, all having the same rectangular shape, cross-sectional area, and original length as shown in Figure 7.17, which piece will stretch more when subjected to the same force F?

As you already know from your experience, it will require much less effort to elongate the piece of rubber than it does to elongate the aluminum piece or the steel bar. In fact, the steel bar will require the greatest force to be elongated by the same amount when compared to the other samples. This is because steel has the highest modulus of elasticity value of the three samples. The results of our ob-

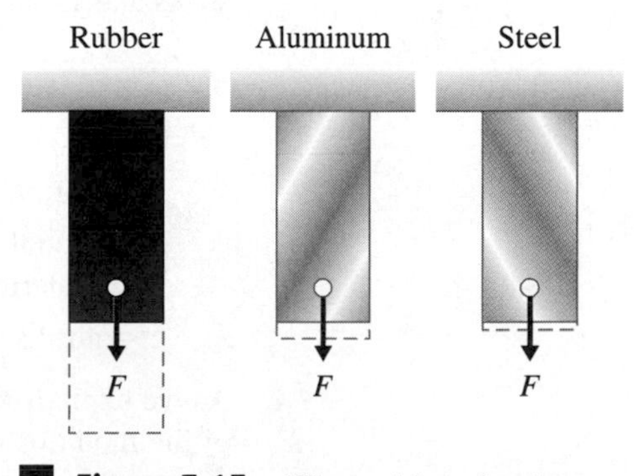

Figure 7.17 When subjected to the same force, which piece of material will stretch more?

An example of a tensile test machine

servation—in terms of the example cited—are expressed, in a general form that applies to all solid materials, by Hooke's law. We discussed Hooke's law earlier when we were explaining springs. Hooke's law also applies to stretching or compressing a solid piece of material. However, for solid pieces of material, Hooke's law is expressed in terms of stress and strain according to

$$\sigma = E\varepsilon \tag{7.20}$$

where

σ = normal stress (N/m^2 or lb/in^2)

E = modulus of elasticity (or sometimes called Young's modulus), a property of material (N/m^2 or lb/in^2)

ε = normal strain, the ratio of change in length to original length (dimensionless)

As we explained earlier, the slope of a stress–strain diagram is used to compute the value of the modulus of elasticity for a solid specimen. Equation (7.20) defines the equation of the line that relates the stress and strain on the stress–strain diagram.

As an alternative approach, we will next explain a simple procedure that can be used to measure E. Although the following procedure is not the formal way to obtain the modulus

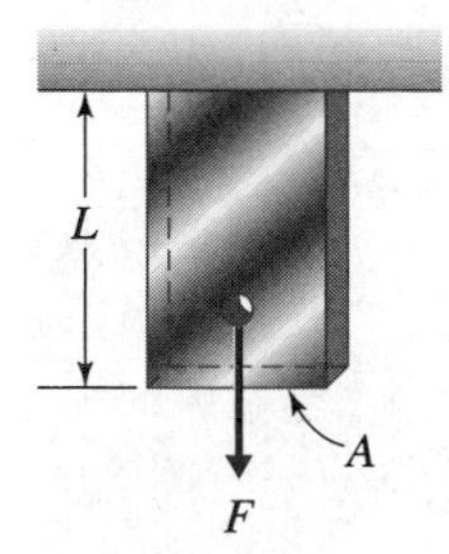

Figure 7.18
A rectangular bar subjected to a tensile load

of elasticity value for a material, it is a simple procedure that provides additional insight into the modulus of elasticity. Consider a given piece of rectangular-shaped material with an original length L, and cross-sectional area A, as shown in Figure 7.18. When subjected to a known force F, the bar will stretch; by measuring the amount the bar stretched (final length of the bar minus the original length), we can determine the modulus of elasticity of the material in the following manner. Starting with Hooke's law, $\sigma = E\varepsilon$, and substituting for σ and the strain ε, in terms of their elementary definitions, we have

$$\frac{F}{A} = E\frac{x}{L} \tag{7.21}$$

Rearranging Equation (7.21) to solve for modulus of elasticity E, we have

$$E = \frac{FL}{Ax} \tag{7.22}$$

where

E = modulus of elasticity (N/m^2)
F = the applied force (N)
L = original length of the bar (m)
A = cross-sectional area of the bar (m^2)
x = elongation, final length minus original length (m)

Note that the physical variables involved include the fundamental base dimensions length and area from Chapter 4 and the physical variable force from this chapter. Note also from Equation (7.22) that the modulus of elasticity, E, is inversely proportional to elongation x; the more the material stretches, the smaller the value of E. Values of the modulus of elasticity E for selected materials are shown in Table 7.4.

Another important mechanical property of material is the ***modulus of rigidity*** or ***shear modulus***. The modulus of rigidity is a measure of how easily a material can be twisted or sheared. The value of the modulus of rigidity shows the resistance of a given material to shear deformation. Engineers consider the value of shear modulus when selecting materials for shafts or rods that are subjected to twisting torques. For example, the modulus of rigidity or shear modulus for aluminum alloys is in the range of 26 to 36 GPa, whereas the shear modulus for steel is in the range of 75 to 80 GPa. Therefore, steel is approximately 3 times more rigid in shear when compared to aluminum. The shear modulus is measured using a torsional test machine. A cylindrical specimen of known dimensions is twisted with a known torque. The angle of twist is measured and is used to determine the value of the shear modulus (see Example 7.11). The values of shear modulus for various solid materials are given in Table 7.4.

Based on their mechanical behavior, solid materials are commonly classified as either *ductile* or *brittle*. A ductile material, when subjected to a tensile load, will go through significant permanent deformation before it breaks. Steel and aluminum are good examples of ductile materials. On the other hand, a brittle material shows little or no permanent deformation before it ruptures. Glass and concrete are examples of brittle materials.

The tensile and compressive strength are other important properties of materials. To predict failure, engineers perform stress calculations that are compared to the tensile and compressive strength of materials. The tensile strength of a piece of material is

TABLE 7.4 Modulus of Elasticity and Shear Modulus of Selected Materials

Material	Modulus of Elasticity (GPa)	Shear Modulus (GPa)
Aluminum alloys	70–79	26–30
Brass	96–110	36–41
Bronze	96–120	36–44
Cast iron	83–170	32–69
Concrete (compression)	17–31	
Copper alloys	110–120	40–47
Glass	48–83	19–35
Magnesium alloys	41–45	15–17
Nickel	210	80
Plastics		
Nylon	2.1–3.4	
Polyethylene	0.7–1.4	
Rock (compression)		
Granite, marble, quartz	40–100	
Limestone, sandstone	20–70	
Rubber	0.0007–0.004	0.0002–0.001
Steel	190–210	75–80
Titanium alloys	100–120	39–44
Tungsten	340–380	140–160
Wood (bending)		
Douglas fir	11–13	
Oak	11–12	
Southern pine	11–14	

Source: Adapted from J. M. Gere, *Mechanics of Materials.*

determined by measuring the maximum tensile load a material specimen in a shape of a rectangular bar or cylinder can carry without failure. The ***tensile strength*** or ultimate strength of a material is expressed as the maximum tensile force per unit of the original cross-sectional area of the specimen. As we mentioned earlier, when a material specimen is tested for its strength, the applied tensile load is increased slowly so that one can determine the modulus of elasticity and the elastic region. In the very beginning of the test, the material will deform elastically, meaning that if the load is removed, the material will return to its original size and shape without any permanent deformation. The point to which the material exhibits this elastic behavior is the elastic point. The elastic strength represents the maximum load that the material can carry without any permanent deformation. In many engineering design applications, the elastic strength or yield strength (the yield value is very close to the elastic strength value) is used as the tensile strength.

Some materials are stronger in compression than they are in tension; concrete is a good example. The compression strength of a piece of material is determined by measuring the maximum compressive load a material specimen in a shape of a cube can carry without failure. The ultimate ***compressive strength*** of a material is expressed as the maximum compressive force per unit of the cross-sectional area of the specimen. Concrete has a compressive strength in the range of 10 to 70 MPa. The strength of some selected materials is given in Table 7.5.

TABLE 7.5 The Strength of Selected Materials

Material	Yield Strength (MPa)	Ultimate Strength (MPa)
Aluminum alloys	35–500	100–550
Brass	70–550	200–620
Bronze	82–690	200–830
Cast iron (tension)	120–290	69–480
Cast iron (compression)		340–1,400
Concrete (compression)		10–70
Copper alloys	55–760	230–830
Glass		30–1,000
Plate glass		70
Glass fibers		7,000–20,000
Magnesium alloys	80–280	140–340
Nickel	100–620	310–760
Plastics		
Nylon		40–80
Polyethylene		7–28
Rock (compression)		
Granite, marble, quartz		50–280
Limestone, sandstone		20–200
Rubber	1–7	7–20
Steel		
High-strength	340–1,000	550–1,200
Machine	340–700	550–860
Spring	400–1,600	700–1,900
Stainless	280–700	400–1,000
Tool	520	900
Steel wire	280–1,000	550–1,400
Structural steel	200–700	340–830
Titanium alloys	760–1,000	900–1,200
Tungsten		1,400–4,000
Wood (bending)		
Douglas fir	30–50	50–80
Oak	40–60	50–100
Southern Pine	40–60	50–100
Wood (compression parallel to grain)		
Douglas fir	30–50	40–70
Oak	30–40	30–50
Southern pine	30–50	40–70

EXAMPLE 7.10

A structural member with a rectangular cross section, as shown in Figure 7.19, is used to support a load of 4000 N distributed uniformly over the cross-sectional area of the member. What type of material should be used to carry the load safely?

Material selection for structural members depends on a number of factors, including the density of the material, its strength, its toughness, its reaction to the surrounding environment, and its appearance. In this example, we will only consider the strength of the material as the design factor. The average normal stress in the member is given by

$$\sigma = \frac{4000\text{ N}}{(0.05\text{ m})(0.005\text{ m})} = 16\text{ MPa}$$

Aluminum alloy or structural steel material with the yield strength of 50 MPa and 200 MPa, respectively, could carry the load safely.

10 cm
5 mm
5 cm

Figure 7.19 The structural member of Example 7.10

One of the goals of most structural analysis is to check for failure. The prediction of failure is quite complex in nature; consequently, many investigators have been studying this topic. In engineering analysis, to compensate for what we do not know about the exact behavior of material and/or to account for future loading for which we may have not accounted but to which someone may subject the part or the structural member, we introduce a factor of safety (F.S.), which is defined as

$$\text{F.S.} = \frac{P_{\text{max}}}{P_{\text{allowable}}} \tag{7.23}$$

where P_{max} is the load that can cause failure. For certain situations, it is also customary to define the factor of safety in terms of the ratio of maximum stress that causes failure to the allowable stresses if the applied loads are linearly related to the stresses. The factor of safety for Example 7.10 using an alumnium alloy with a yield strength of 50 MPa is 3.1.

EXAMPLE 7.11

A setup similar to the one shown in Figure 7.20 is commonly used to measure the shear modulus, G. A specimen of known length L and diameter D is placed in the test machine. A known torque T is applied to the specimen and the angle of twist ϕ is measured. The shear modulus is then calculated from

$$G = \frac{32TL}{\pi D^4 \phi} \tag{7.24}$$

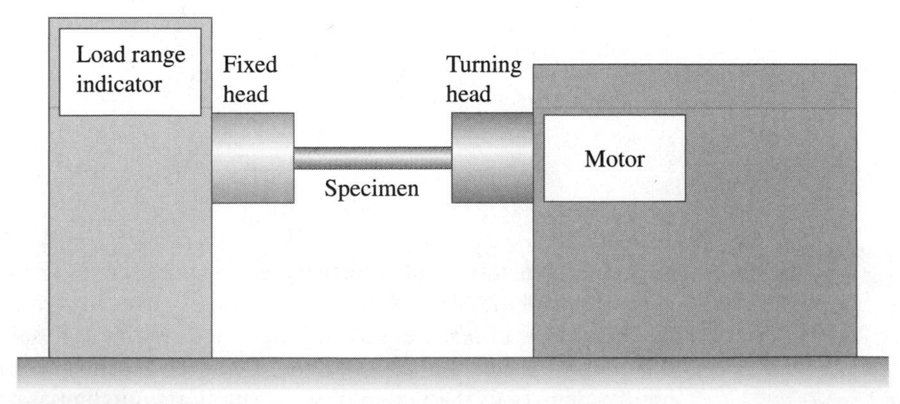

Figure 7.20 A setup to measure the shear modulus of a material

Using Equation (7.24), calculate the shear modulus for a given specimen and test results of $T = 3450$ N·m, $L = 20$ cm, $D = 5$ cm, $\phi = 0.015$ rad.

$$G = \frac{32TL}{\pi D^4 \phi} = \frac{32(3450\,N\cdot\text{m})(0.2\text{ m})}{\pi(0.05\text{ m})^4(0.015\text{ rad})} = 75\text{ GPa}$$

Bulk Modulus of Compressibility

Most of you have pumped air into a bicycle tire at one time or another. From this and other experiences you know that gases are more easily compressed than liquids. In engineering, to see how compressible a fluid is, we look up the value of a bulk modulus of compressibility of the fluid. The value of fluid bulk modulus shows how easily the volume of the fluid can be reduced when the pressure acting on it is increased.

The bulk modulus of compressibility is formally defined as

$$E_v = \frac{\text{increase in pressure}}{\dfrac{\text{decrease in volume}}{\text{original volume}}} = \frac{\text{increase in pressure}}{\dfrac{\text{increase in density}}{\text{original density}}} \tag{7.25}$$

Equation (7.25) expressed mathematically is

$$E_V = \frac{dP}{-\dfrac{dV}{V}} = \frac{dP}{\dfrac{d\rho}{\rho}} \tag{7.26}$$

where

E_V = modulus of compressibility (N/m^2)

dP = change in pressure (increase in pressure) (N/m^2)

dV = change in volume (decrease in volume, negative value) (m^3)

V = original volume (m^3)

$d\rho$ = change in density (increase in density) (kg/m^3)

ρ = original density (kg/m^3)

When the pressure is increased (a positive change in value), the volume of the fluid is decreased (a negative change in value), and thus the need for a minus sign in front of the dV term to make E_V a positive value. In Equation (7.26), also note that when the pressure is increased (a positive change in value), the density of the fluid is also increased (a positive change in value), and thus there is no need for a minus sign in front of the $d\rho$ term to make E_V a positive value. The compressibility values for some common fluids are given in Table 7.6.

EXAMPLE 7.12

When we were defining the bulk modulus of compressibility we said that as the pressure is increased, the volume of fluid is decreased, and consequently the density of the fluid is increased. Why does the density of a fluid increase when its volume is decreased?

Recall that density is defined as the ratio of mass to volume, as given by

$$\text{density} = \frac{\text{mass}}{\text{volume}}$$

From examining this equation, for a given mass you can see that as the volume is reduced, the density is increased.

TABLE 7.6 The Values of Compressibility Modulus for Some Common Fluids

Fluid	Compressibility Modulus (N/m^2)	Compressibility Modulus (lb_f/in^2)
Ethyl alcohol	1.06×10^9	1.54×10^5
Gasoline	1.3×10^9	1.9×10^5
Glycerin	4.52×10^9	6.56×10^5
Mercury	2.85×10^{10}	4.14×10^6
SAE 30 oil	1.5×10^9	2.2×10^5
Water	2.24×10^9	3.25×10^5

EXAMPLE 7.13

For water, starting with 1 m^3 of water in a container, what is the pressure required to decrease the volume of water by 1%?

From Table 7.6, the compressibility modulus for water is 2.24×10^9 N/m^2. It would require a pressure of 2.24×10^7 N/m^2 to decrease the original 1 m^3 volume of water to a final volume of 0.99 m^3, or, said another way, by 1%. It would require an equivalent pressure of 221 atm (2.24×10^7 N/m^2) to reduce the pressure of a unit volume of water by 1%.

7.5 MOMENT, TORQUE—FORCE ACTING AT A DISTANCE

As we mentioned earlier, the two tendencies of an unbalanced force acting on an object are to translate the object (i.e., to move it in the direction of the unbalanced force) and to rotate or bend or twist the object. In this section, we will focus our attention on understanding the tendency of an unbalanced force to rotate, bend, or twist objects. Being able to calculate moments created by forces about various points and axes is important in many engineering analyses. For example, all of you have noticed that the street light poles or the traffic light poles are thicker near the ground than they are on the top. One of the main reasons for this is that wind loading creates a bending moment, which has a maximum value about the base of the pole. Thus, a bigger section is needed at the base to prevent failure. Nature understands the concept of bending moments well; that is why trees have big trunks near the ground to support the bending moment created by the weight of the branches and the wind loading. Understanding moments, or torques, is also important when designing objects that rotate, such as a shaft, gear, or wheel.

To better grasp the concept of moment, let us consider a simple example. When you open a door you apply a pulling or a pushing force on the doorknob (or handle). The application of this force will make the door rotate about its hinges. In mechanics, this tendency of force is measured in terms of a ***moment of a force*** about an axis or a point. *Moment has both direction and magnitude.* In our example, the direction is defined by the sense of rotation of the door. As shown in Figure 7.21, looking at the door from the top, when you open the door by applying the force shown, the direction of moment is clockwise, and when you close the same door, the force creates a counterclockwise

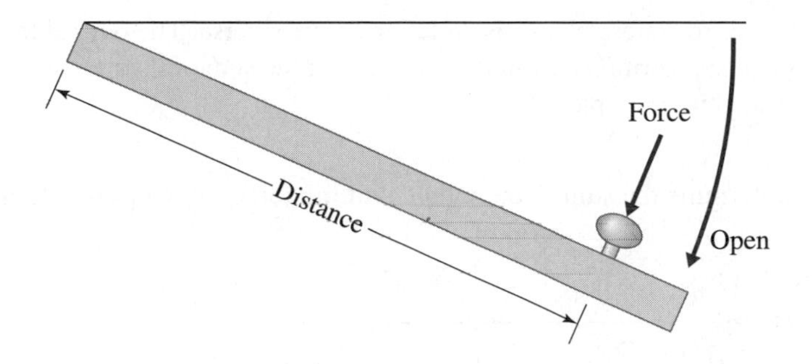

Figure 7.21 Moment of the force created by someone opening or closing a door

moment. The magnitude of the moment is obtained from the product of the moment arm times the force. Moment arm represents the perpendicular distance between the line of action of the force and the point about which the object could rotate.

In this book, we will only consider the moment of a force about a point. Most of you will take a mechanics class, where you will learn how to calculate the magnitude and the direction of moment of a force about an arbitrary axis. For an object that is subject to force $\boldsymbol{F}$, the relationship between the line of action of the force and moment arm is shown in Figure 7.22. The magnitude of moment about arbitrary points A, B, and C is given by

$$M_A = d_1 F \tag{7.27}$$

$$M_B = d_2 F \tag{7.28}$$

$$M_C = 0 \tag{7.29}$$

Note that the moment of the force F about point C is zero because the line of action of the force goes through point C, and thus the value of moment arm is zero. When calculating the sum of moments created by many forces about a certain point, make sure you use the correct moment arm for each force, and also make sure to ask yourself whether the tendency of a given force is to rotate the object clockwise or counterclockwise. As a bookkeeping procedure, you may assign a positive value to clockwise rotations and negative

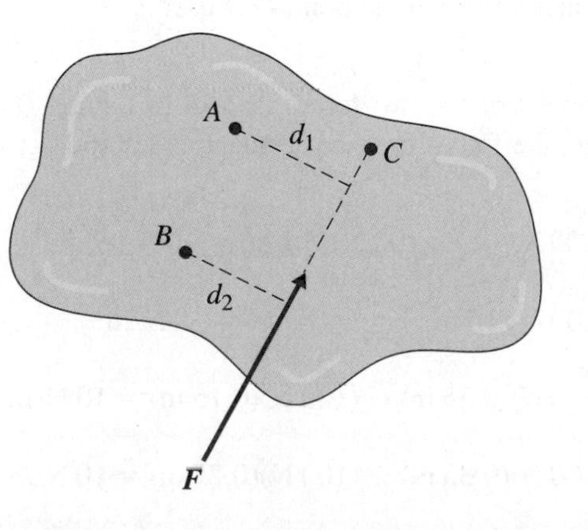

Figure 7.22
The moment of the force $\boldsymbol{F}$ about points A, B, and C

values to counterclockwise rotations or vice versa. The overall tendency of the forces could then be determined from the final sign of the sum of the moments. This approach is demonstrated in Example 7.14.

EXAMPLE 7.14

Determine the sum of the moment of the forces about point O shown in Figure 7.23.

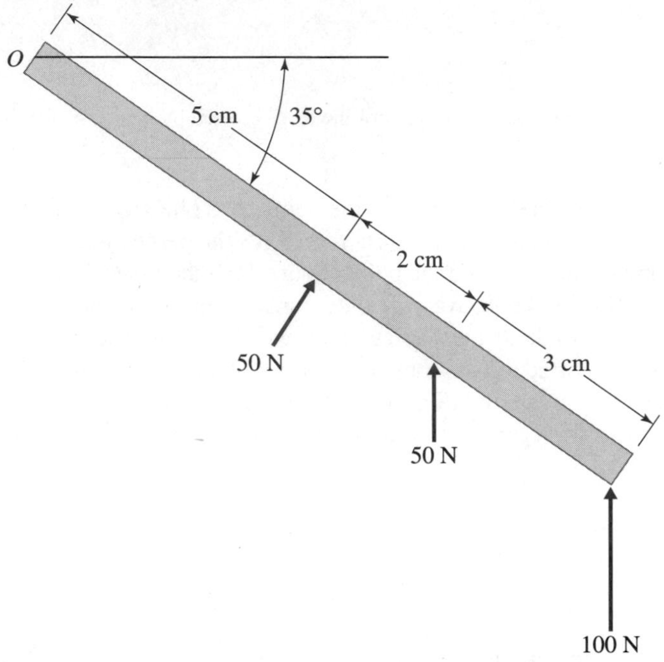

O
5 cm
35°
2 cm
3 cm
50 N
50 N
100 N

Figure 7.23 The diagram for Example 7.14

$$\circlearrowleft + \sum M_O = (50\text{ N})(0.05\text{ m}) + (50\text{ N})(0.07\ \cos 35) + (100\text{ N})(0.1\ \cos 35) = 13.55\text{ N}\cdot\text{m}$$

Note that in calculating the moment of each force, the perpendicular distance between the line of action of each force and the point O is used.

EXAMPLE 7.15

Determine the moment of the two forces, shown in Figure 7.24, about points A, B, C, and D. As you can see, the forces have equal magnitudes and act in opposite directions to one another.

$$\circlearrowright + \sum M_A = (100\text{ N})(0) + (100\text{ N})(0.1\text{ m}) = 10\text{ N}\cdot\text{m}$$

$$\circlearrowright + \sum M_B = (100\text{ N})(10\text{ m}) + (100\text{ N})(0) = 10\text{ N}\cdot\text{m}$$

$$\circlearrowright + \sum M_C = (100\text{ N})(0.25\text{ m}) - (100\text{ N})(0.15\text{ m}) = 10\text{ N}\cdot\text{m}$$

$$\circlearrowright + \sum M_D = (100\text{ N})(0.35\text{ m}) - (100\text{ N})(0.25\text{ m}) = 10\text{ N}\cdot\text{m}$$

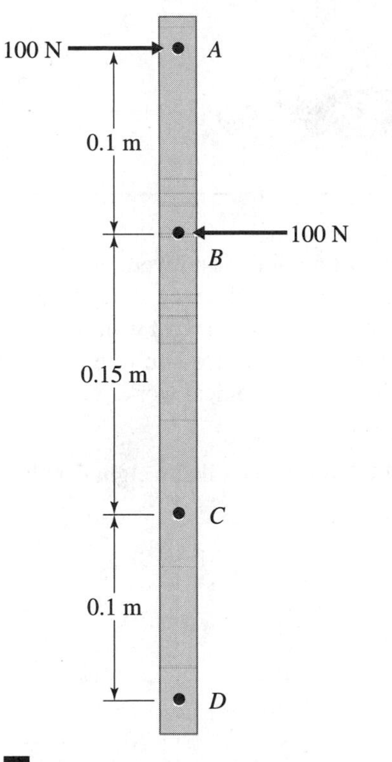

Figure 7.24 A schematic diagram for Example 7.15

Referring to Example 7.15, note that two forces that are equal in magnitude and opposite in direction (not having the same line of action) constitute *a couple*. As you can see from the results of this example, the moment created by a couple is equal to the magnitude of either force times the perpendicular distance between the line of action of the forces involved.

7.6 WORK—FORCE ACTING OVER A DISTANCE

When you push on a car that has run out of gas and move it through a distance, you perform mechanical work. When you push on a lawn mower and move it in a certain direction, you are doing work. Mechanical work is done when the applied force moves the object through a distance. Simply stated, ***mechanical work*** is defined as the component of the force that moves the object times the distance the object moves. Consider the car shown in Figure 7.25. The work done by the pushing force moving the car from position 1 to position 2 is given by

$$W_{1-2} = (F\cos\theta)(d) \tag{7.30}$$

Note from examining Equation (7.30) that the normal component of the force does not perform mechanical work. That is because the car is not moving in a direction normal to the ground. So next time you are mowing the lawn, ask yourself if you are pushing the lawn mower horizontally or at an angle? Another point you should remember is that if

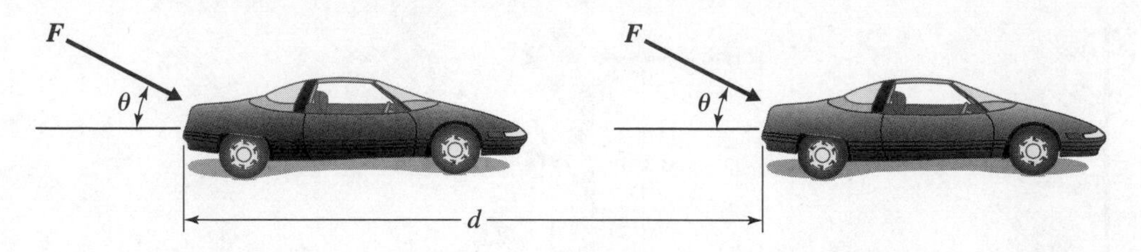

Figure 7.25 The work done on the car by the force F is equal to $W_{1-2} = (F \cos\theta)(d)$

you were to push hard against an object and were not able to move it, then by definition you are not doing any mechanical work, even though this action could make you tired. For example, if you were to push against a rigid wall in a building, regardless of how hard you were to push with your hands, you wouldn't be able to move the wall, and thus, by definition, you are not doing any mechanical work.*

EXAMPLE 7.16

Determine the work required to lift a box that weighs 100 N, 1.5 m above the ground.

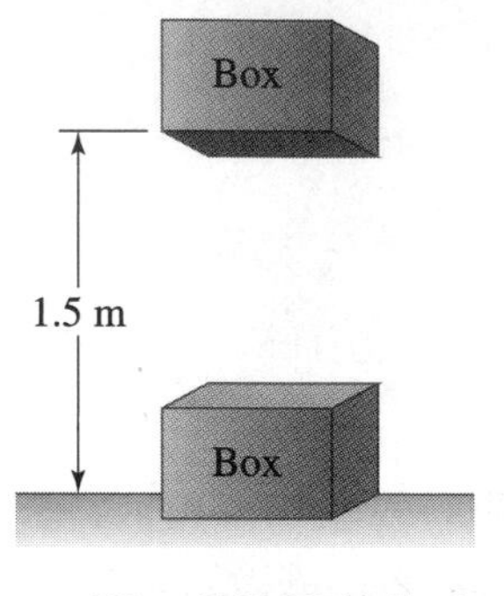

$$W = (100\text{ N})(1.5\text{ m}) = 150\text{ N·m}$$

The SI unit for work is N·m, which is called *joule*; thus the unit joule (J) is defined as the work done by a 1-N force through a distance of 1 m.

$$1\text{ joule} = (1\text{ N})(1\text{ m})$$

Students often confuse the concept of work with power. Power represents how fast you want to do the work. It is the time rate of doing work, or said another way, it is the work done divided by the time that it took to perform it. For instance, in Example 7.16, if we want to lift the box in 3 seconds, then the power required is

$$\text{power} = \frac{\text{work}}{\text{time}} = \frac{150\text{ J}}{3\text{ s}} = 50\text{ J/s} = 50\text{ watts}$$

Note that 1 J/s is called 1 watt (W). If you wanted to lift the box in 1.5 seconds, then the required power to do this task is: 150 J/1.5 s or 100 W. Thus, twice as much power is required. It is important to note that the work done in each case is the same; however, the power requirement is different. Remember if you want to do work in a shorter time, then you are going to need more power. The shorter the time, the more power required to do that work. We will discuss power and its units in much more detail in Chapter 10.

*Of course, your effort goes into deforming the wall on a very, very small scale that cannot be detected by the naked eye. The forces created by your push deform the wall material, and your effort is stored in the material in the form of what is called the *strain energy*.

7.7 LINEAR IMPULSE—FORCE ACTING OVER TIME

Up to this point, we have defined the effects of a force acting at a distance in terms of creating moment, and through a distance as doing work. Now we will consider force acting over a period of time. Understanding linear impulse and impulsive forces is important in the design of products such as air bags and sport helmets to prevent injuries. Understanding impulsive forces also helps with designing cushion materials to prevent damage to products when dropped or when subjected to impact. On TV or in the movies you have seen a stuntman jumping off a roof of a multistory building onto an air mat on the ground and not getting hurt. Had he jumped onto concrete pavement, the same stuntman would most likely be killed. Why is that? Well, by using an inflated air mat, the stuntman is increasing his time of contact with the ground through staying in touch with the air mat for a long time. This statement will make more sense after we show the relationship between the linear impulse and linear momentum.

Linear impulse represents the net effect of a force acting over a period of time. There is a relationship between the linear impulse and the linear momentum. We explained what we mean by linear momentum in Section 6.6. A force acting on an object over a period of time creates a linear impulse that brings about a change in the linear momentum of the object, according to

$$F_{\text{average}}\,\Delta t = mV_{\text{final}} - mV_{\text{inital}} \tag{7.31}$$

where

F_{average} = average magnitude of the force acting on the object (N)
Δt = time period over which the force acts on the object (s)
m = mass of the object (kg)
V_{f} = final velocity of the object (m/s)
V_{i} = initial velocity of the object (m/s)

It is important to note that Equation (7.31) is a vectoral relationship, meaning it has both magnitude and direction. Using Equation (7.31), let us now explain why the stuntman does not hurt himself when he jumps onto the air mat. For the sake of demonstration, let us assume that the stuntman is jumping off a ten-story building, where the average height of each floor is 15 ft. Moreover, let us assume that he jumps onto an inflated air mat that is 15 ft tall. Neglecting air resistance during his jump to make the calculation simpler, the stuntman's velocity right before he hits the air mat could be determined from

$$V_{\text{i}} = \sqrt{2gh} \tag{7.32}$$

where V_{i} represents the velocity of the stuntman right before he hits the air mat, g is the acceleration due to gravity ($g = 32.2$ ft/s^2), and h is the height of the building minus the height of the air mat ($h = 135$ ft). Substituting for g and h in Equation (7.32) leads to an initial velocity of $V_{\text{i}} = 93.2$ ft/s. Also, we realize that the air mat reduces the velocity of the stuntman to a final velocity of zero ($V_{\text{f}} = 0$). Now we can solve for F_{average} from Equation (7.31), assuming different values for Δt, and assuming a mass of 4.65 slugs (150 lb$_{\text{m}}$). The results of these calculations are summarized and given in Table 7.7. It is important to note that F_{average} represents the force that the stuntman exerts on the air mat and subsequently through the air to the ground. But you recall from Newton's third law that for

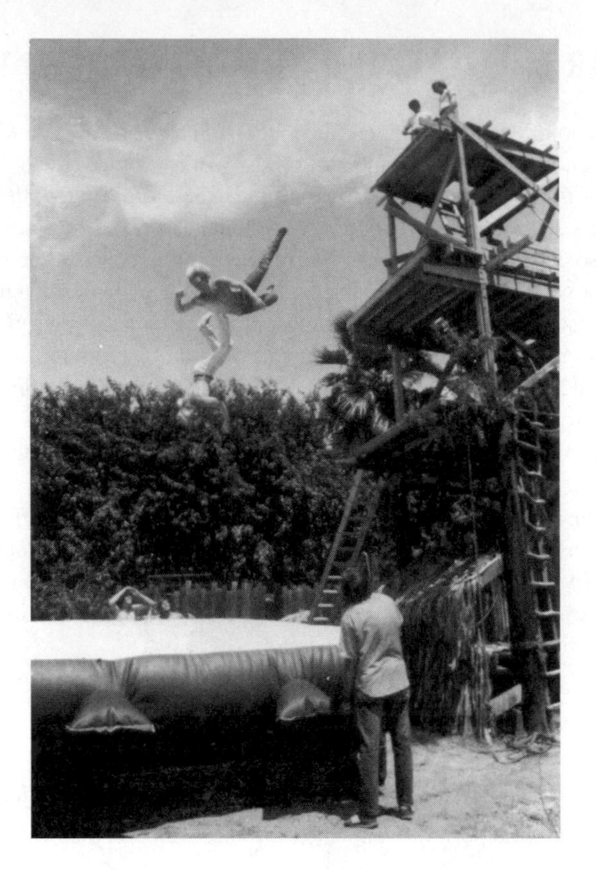

Two stuntmen practice a fall

every action there is a reaction equal in magnitude and opposite in direction. Therefore, the average force exerted by the ground on the stuntman is equal in magnitude to F_{average}. Also note that the reaction force will be distributed over the back surface of the stuntman's body and thus create a relatively small pressure distribution on his back. You can see from these calculations that the longer the time of contact, the lower the reaction force, and consequently the lower the pressure on the stuntman's back.

TABLE 7.7 The Average Reaction Force Acting on a Stuntman

Time of Contact (s)	Average Reaction Force (lb_f)
0.1	4334
0.5	867
1.0	433
2.0	217
5.0	87
10.0	43

SUMMARY

Now that you have reached this point in the text

- You should have a good understanding of what is meant by force and its common units. You should also know that there are different types of force. You should understand the tendency of an unbalanced force, which is to translate and/or rotate objects. The application of forces can also shorten, elongate, bend, and twist objects.
- You should be familiar with the Newton's laws used in mechanics. These laws are the basis of analysis in many engineering problems.
- You should understand what is meant by pressure. For static fluids, you should understand Pascal's law, which states that fluid pressure at a point is the same in all directions. You should also remember the relationship between the fluid pressure and the depth of the fluid. You should understand what the fluid properties such as viscosity and bulk modulus of compressibility mean. You should also have a good idea of how hydraulic systems work.
- You should realize that engineers, when designing products, need to know how a selected material behaves under applied forces. You should know what material properties mean, such as modulus of elasticity, shear modulus, and tensile and compressive strength.
- You should understand that when a force is applied to an object, it can affect the object's state in a number of ways. In engineering, these effects are measured in terms of stress, moment, work, and linear impulse. Remember that the intensity of a force acting within the material, in terms of pulling it apart, compressing, or shearing it, is measured in terms of stress; the effect of a force acting at a distance is measured in terms of creating a moment about a point or an axis; the effect of a force acting over a distance is represented by mechanical work; and the effect of a force acting over a period of time is measured in terms of a linear impulse.

PROBLEMS

1. Design a mass–spring system that can be taken to Mars to measure the acceleration due to gravity at the surface of Mars. Explain the basis of your design and how it should be calibrated and used.
2. In the past, scientists and engineers have used pendulums to measure the value of g at a location. Design a pendulum that can be used to measure the value of g at your location. The formula to use to measure the acceleration due to gravity is

$$T = 2\pi\sqrt{\frac{L}{g}}$$

where T is the period of oscillation of the pendulum, the time that it takes the pendulum to complete one cycle. The distance between the pivot point and the center of the mass of the suspended deadweight is represented by L.

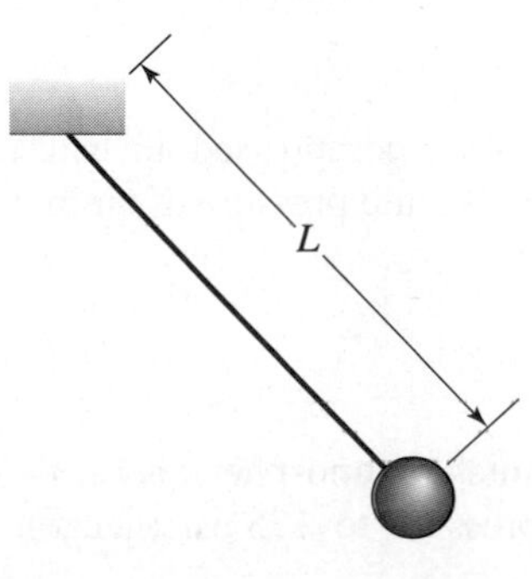

3. An astronaut has a mass of 70 kg. What is the weight of the astronaut on earth at sea level? What are the mass and the weight of the astronaut on the moon, and on Mars? What is the ratio of the pressure exerted by the astronaut's shoe on earth to Mars?
4. Basketball player Shaquille O'Neal weighs approximately 335 lb and wears a size 23 shoe. Estimate the pressure he exerts on a floor. What pressure would he exert if his shoe size were 13? State all your assumptions, and show how you arrived at a reasonable solution.
5. Investigate the relationship between pressure for the following situations:
   a. when you are standing up barefooted
   b. when you are lying on your back
   c. when you are sitting on a tall chair with your feet off the ground and the chair
6. Investigate the pressure created on a surface by the following objects:
   a. a bulldozer
   b. a car
   c. a bicycle
   d. a pair of cross-country skis.
7. Do you think there is relationship between a person's height, weight, and foot size? If so, verify your answer. Designers often learn from their natural surroundings; do the bigger animals have bigger feet?
8. Calculate the pressure exerted by water on the hull of a submarine that is cruising at a depth of 500 ft below ocean level. Assume the density of the ocean water is $\rho = 1025\ \text{kg/m}^3$.
9. Investigate the operation of pressure-measuring devices such as manometers, Bourden tubes, and diaphragm pressure sensors. Write a brief report discussing your findings.
10. Convert the following pressure readings from inches of mercury units to psi and Pascal units:
    a. 28.5 in·Hg
    b. 30.5 in·Hg
11. If a pressure gauge on a compressed air tank reads 120 psi, what is the absolute pressure of air in
    a. psi
    b. pascal
    c. feet of water
    d. feet of mercury
12. An air compressor intakes atmospheric air at 14.7 psi and increases its pressure to 175 psi (gauge). The pressurized air is stored in a horizontal tank. What is the absolute pressure of air inside the tank? Express your answers in
    a. psi
    b. kPa
    c. bar
    d. standard atm (recall that 1 atm = 14.7 psi = 101.325 kPa)
13. Calculate the pressure exerted by water on a scuba diver who is swimming at a depth of 50 ft below the water surface.
14. Investigate the typical operating pressure range in the hydraulic systems used in the following applications. Write a brief report to discuss your findings.
    a. controlling wing flaps in airplanes
    b. your car's steering system
    c. jacks used to lift cars in a service station
15. Using the information given in Table 7.3, determine the ratio of local pressure and density to sea-level values. Estimate the value of air density at the cruising altitude of most commercial airliners.

Altitude (m)	$\frac{\rho}{\rho_{\text{sea level}}}$	$\frac{P}{P_{\text{sea level}}}$
0 (sea level)		
1000		
3000		
5000		
8000		
10,000		
12,000		
14,000		
15,000		

16. Bourdon-type pressure gauges are used in thousands of applications. A deadweight tester is a device that is used to calibrate pressure gauges. Investigate the operation of a deadweight pressure tester. Write a brief report to discuss your findings.
17. Investigate what the typical range of pressure is in the following applications: a bicycle tire, home water line, natural gas line, and refrigerant in your refrigerator's evaporator and condenser lines. Present your findings in a brief report.
18. Convert the 130/85 systolic/diastolic pressures given in millimeters of Hg to Pa, psi, and in·H_2O.

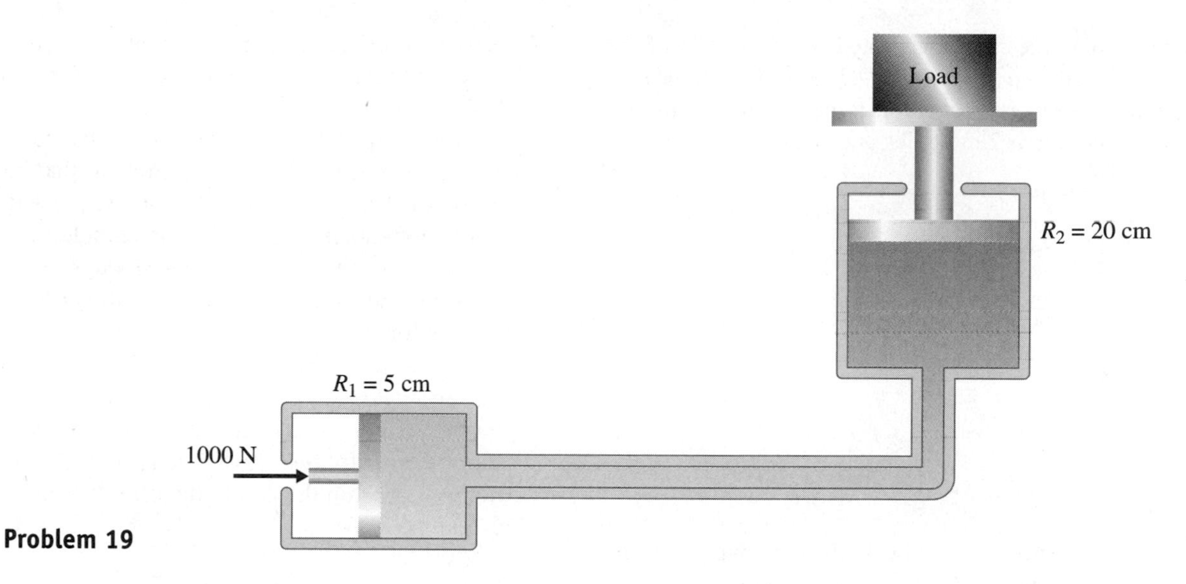

Problem 19

19. Determine the load that can be lifted by the hydraulic system shown in the figure above. All of the necessary information is shown in the figure.
20. Determine the pressure required to decrease the volume of the following fluids by 2%:
    a. water
    b. glycerin
    c. SAE motor oil
21. Compute the deflection of a structural member made of aluminum if a load of 500 lb is applied to the bar. The member has a uniform rectangular cross section with the dimensions of $\frac{1}{4}$ in. × 4 in., as shown in the accompanying figure.

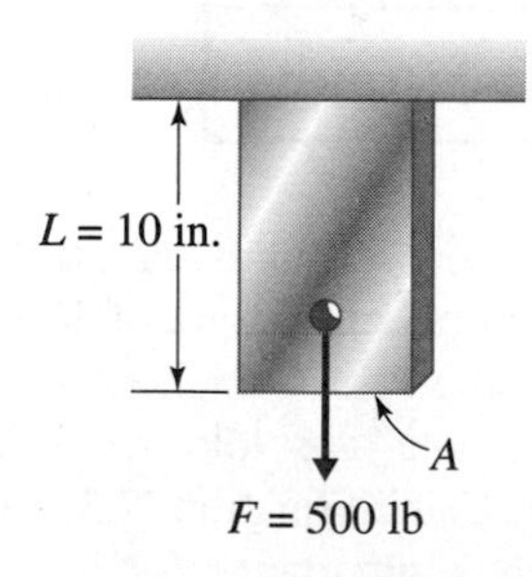

22. A structural member with a rectangular cross section as shown in the accompanying figure is used to support a load of 2500 N. What type of material do you recommend be used to carry the load safely? Base your calculations on the yield strength and a factor of safety of 2.0.

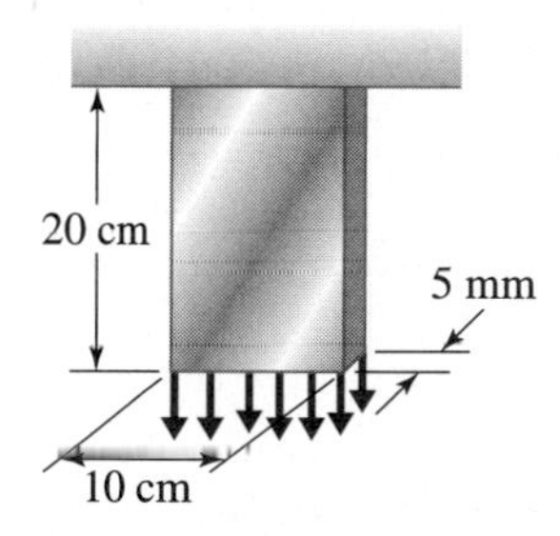

23. The tire wrench shown in the accompanying figure is used to tighten the bolt on a wheel. Given the information on the diagram, determine the moment about point O for the two loading situations shown:
    a. pushing perpendicular to the wrench arm
    b. pushing at a 75° angle, as shown

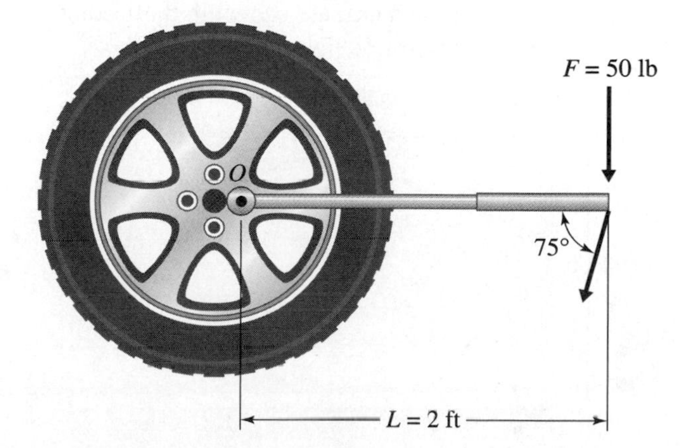

24. Determine the moment created by the weight of the suspended sign about point O. Dimensions of the sign and the support are shown in the accompanying figure. The sign is 2 mm thick and is made of aluminum.

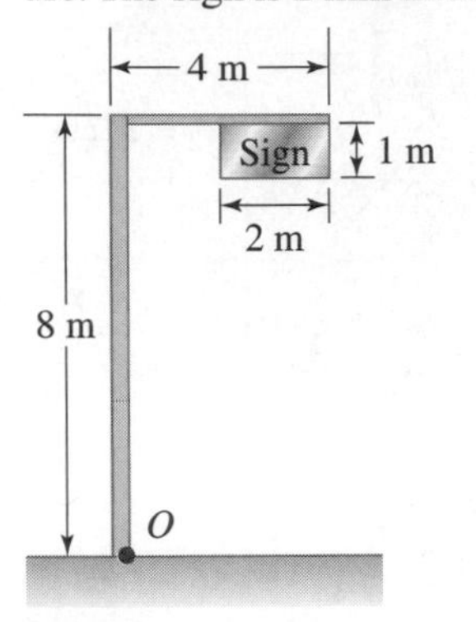

25. Determine the moment created by the weight of the lamp about point O. The dimensions of the street light post and arm are shown in the accompanying figure. The lamp weighs 20 lb.

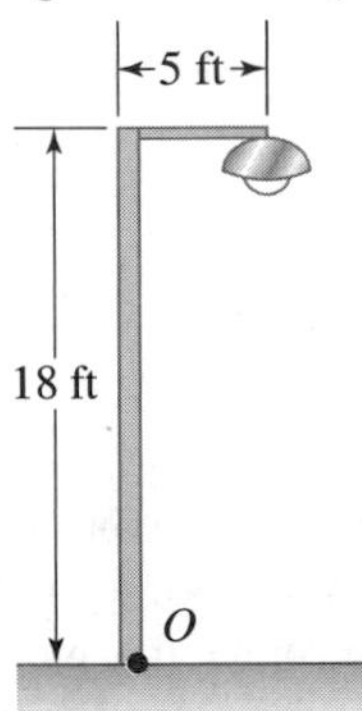

26. Determine the moment created by the weight of the traffic light about point O. The dimensions of the traffic light post and arm are shown in the accompanying figure. The traffic light weighs 40 lb.

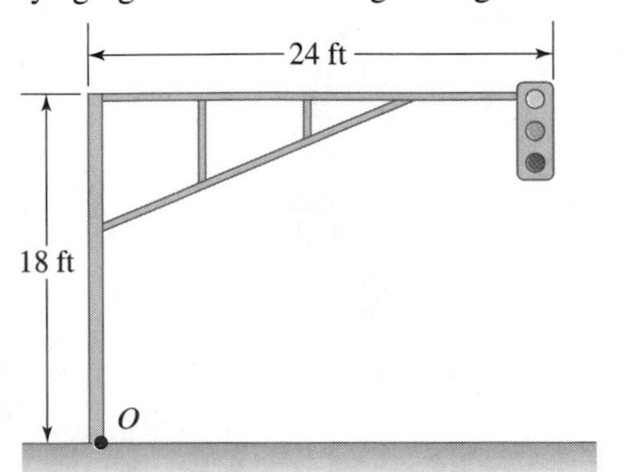

27. Determine the sum of the moments created by the forces shown in the figure on page 175 about points A, B, C, and D.
28. Determine the amount of work done by you and a friend when each of you push on a car that has run out of gas. Assume you and your friend each push with a horizontal force with a magnitude of 200 N, a distance of 100 m. Also, suggest ways of actually measuring the magnitude and the direction of the pushing force.
29. Determine the work done by an electric motor lifting an elevator weighing 1000 lb through five floors; assume a distance of 15 ft between each floor. Compute the power requirements for the motor to go from the first to the fifth floor in
   a. 5 s
   b. 8 s

   Express your results in horsepower (1 hp = 550 ft·lb/s).
30. If a laptop computer weighing 22 N is dropped from a distance of 1 m onto a floor, determine the average reaction force from the floor if the time of contact is changed by using different cushion materials, as shown in the accompanying table.

Time of Contact (s)	Average Reaction Force (N)
0.01	
0.05	
0.1	
1.0	
2.0	

31. Obtain the values of vapor pressures of alcohol, water, and glycerin at a room temperature of 20°C.
32. When learning to play some sports, such as tennis, golf, or baseball, often you are told to follow through with your swing. Using Equation (7.31), explain why follow-through is important.
33. In many applications, calibrated springs are commonly used to measure the magnitude of a force. Investigate how typical force-measuring devices work. Write a brief report to discuss your findings.
34. Calculate the moment created by the forces shown in the accompanying figure (on p. 175) about point O.

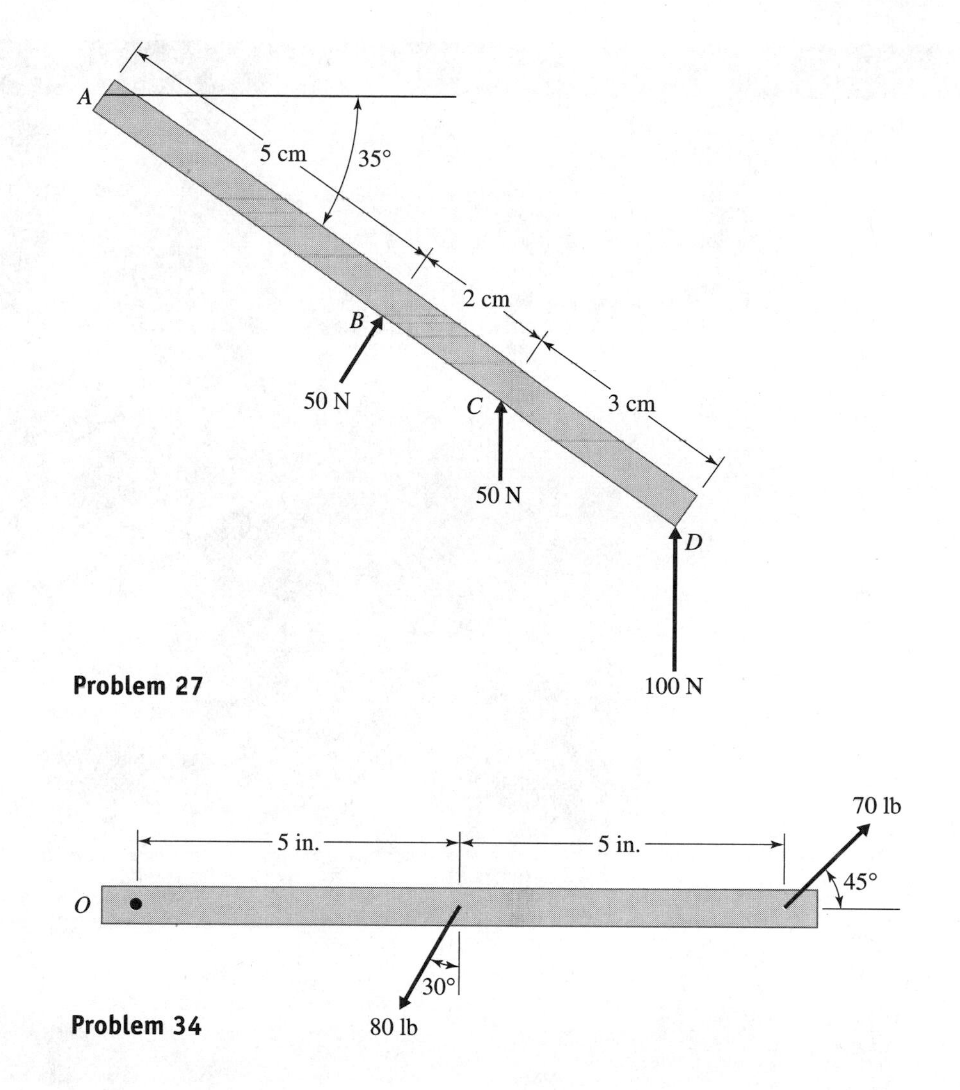

Problem 27

Problem 34

35. We have used an experimental setup similar to Example 7.1 to determine the value of a spring constant. The deflection caused by the corresponding weights are given in the accompanying table. What is the value of the spring constant?

Weight (lb)	The Deflection of the Spring (in.)
5.0	0.48
10.0	1.00
15.0	1.90
20.0	2.95

36. If an astronaut and her space suit weight 250 lb on Earth, what should be the volume of her suit if she is to practice for weightless conditions in an underwater neutral buoyant simulator, similar to the one used by NASA as shown in Chapter 4?

Caterpillar 797 Mining Truck*

The Caterpillar® 797, which features many patented innovations, was developed using a clean-sheet approach. But the design also draws on the experience gained from hundreds of thousands of operating hours accumulated by more than 35,000 Caterpillar construction and mining trucks working worldwide. The 797 builds on such field-proven designs as the mechanical power train, electronics that manage and monitor all systems, and structures that provide durability and long life. Engineers from various fields, including mechanical, electrical, structural, hydraulic, and mining, collaborated to design the Caterpillar 797 mining truck.

*Materials were adapted with permission from Caterpillar® documents.

Caterpillar developed the 797 in response to mining companies that were seeking a means to reduce cost per ton in large-scale operations. Therefore, the 797 is sized to work efficiently with loading shovels used in large mining operations, and Caterpillar matched the body design to the material being hauled to optimize payloads.

The new 797 mining truck is the largest of the extensive Caterpillar off-highway truck line, and it is the largest mining truck ever constructed. The Caterpillar 797 mining truck has a payload capacity of 360 tons (326 metric tons) and design operating mass of 1,230,000 lb (557,820 kg). The Caterpillar 797 Mining Truck specifications include:

Engine model: 3524B high displacement
Gross power at 1750 rpm: 3400 hp (2537 kW)
Transmission: 7-speed planetary power shift
Top speed: 40 mph (64 km/h)
Operating mass: 610,082 lb (1,345,000 kg)
Nominal payload capacity: 360 tons (326 metric tons)
Body capacity (SAE 2:1): 290 yd^3 (220 m^3)
Tire size: 58/80R63

A summary of some information about various components of the 797 truck follows.

Comfortable and Efficient Operator Station The 797 cab is designed to reduce operator fatigue, enhance operator performance, and promote safe operation. The controls and layout provide greater operator comfort along with an automotive feel, while enhancing functionality and durability. The cab and frame design meets SAE standards for rollover and falling-object protection.

The truck and cab design provides exceptional all-around visibility. For example, the clean deck to the right of the cab improves sight lines. The spacious cab includes two full-sized air suspension seats, which allow a trainer to work with the operator.

The steering column tilts and telescopes so the operator can adjust it for comfort and optimum control. The Vital Information Management System (VIMS) display and keypad provide precise machine status information. The new hoist control is fingertip actuated, allowing the operator to easily and precisely control hoist functions—raise, hold, float, and lower. Electrically operated windows and standard air conditioning add to operator comfort. The cab is resiliently mounted to dampen noise and vibration, and sound-absorbing material in the doors and side panels cuts noise further.

Cast Frame Mild steel castings comprise the entire load-bearing frame for durability and resistance to impact loads. The nine major castings are machined for precise fit before being joined using a robotic welding technology that ensures full-penetration welds. The frame design reduces the number of weld joints and ensures a durable foundation for the 797.

The suspension system, which uses oil-over-nitrogen struts similar to other Caterpillar mining trucks, is designed to dissipate haul road and loading impacts.

Electronically Controlled Engine The new Cat 3524B high-displacement diesel engine produces 3400 gross hp (2537 kW). It is turbocharged and aftercooled, and features electronic unit injection (EUI) technology, which helps the engine meet year 2000 emissions regulations.

The engine uses two in-line blocks linked by an innovative coupler and precisely controlled by electronic controllers. These electronic controllers integrate engine information with mechanical power train information to optimize truck performance, extend component life, and improve operator comfort.

The engine uses a hydraulically driven fan for efficient cooling. The fan design and operation also reduce fuel consumption and noise levels. The engine and its components are designed to minimize service time, which helps keep availability of the truck high.

Efficient Mechanical Power Train The 797 power train includes a new torque converter with lockup clutch that delivers high mechanical efficiency. The new automatic-shift transmission features seven speeds forward and one speed reverse. Electronic clutch pressure control (ECPC) technology smoothes shifts, reduces wear, and increases reliability. Large clutch discs give the transmission high torque capacity and extend transmission life. The transmission and torque converter enable the truck to maintain good speed up grades and to reach a top speed of 40 mph (64 km/h).

The differential is rear mounted, which improves access for maintenance. The differential is pressure lubricated, thus promoting greater efficiency and long life. Wide wheel-bearing spread reduces bearing loads and helps ensure durability. A hydraulically driven lube and cooling system operates independently of ground speed and pumps a continuous supply of filtered oil to each final drive.

Oil-cooled, multiple disc brakes provide fade-resistant braking and retarding. The electronically managed automatic retarder control (ARC) is an integral part of the intelligent power train. ARC controls the brakes on grade to maintain optimum engine rpm and oil cooling. Automatic electronic traction aid (AETA) uses the rear brakes to optimize traction. A combination of constant-displacement and variable-displacement pumps delivers a regulated flow of brake-cooling oil for constant retarding capability and peak truck performance on downhill grades.

Electrohydraulic Hoist Control New hoist hydraulics include an electronic hoist control, independent metering valve (IMV), and a large hoist pump. These features allow automatic body snubbing for reduced impact on the frame, hoist cylinders, and operator. This design also allows the operator to modulate flow and control overcentering when dumping.

Serviceability Caterpillar designed the 797 to reduce service time thus ensuring maximum availability. Routine maintenance points, such as fluid fill and check points, are close to ground level. Easily accessed connectors allow technicians to download data and to calibrate machine functions, thus extending maintenance intervals which further increases availability. The 797 is also designed to make major components accessible, which effectively reduces removal, installation, and in-truck maintenance time for all drivetrain components.

To allow technicians to raise the truck body inside a standard-height maintenance building, the canopy portion of the body is hinged so that it can be folded back. In the folded and raised position, the 797 body height is no higher than the 793, Caterpillar's 240-ton-capacity (218-metric ton) mining truck.

Testing Several 797 mining trucks were tested at the Caterpillar proving grounds. Mine evaluation performances were also done during 1999–2000. The test results are used to correct unexpected problems and fine-tune various components of the truck. The 797 has been commercially available worldwide since January 2001.

PROBLEMS

1. Calculate the braking force required to bring the truck from its top speed to full stop in a distance of 250 ft.
2. Calculate the linear momentum of the truck when fully loaded and cruising near its top speed. Calculate the force required to bring the truck from its top speed to a full stop in 5 s.
3. Estimate the rotational speed of the tires when the truck is moving along near its top speed.
4. Estimate the net torque output of the engine from the following equation:

$$\text{power} = T\omega$$

where

power is in $lb_f{\cdot}ft/s$
T = torque ($lb_f{\cdot}ft/s$)
ω = angular speed (rad/s)

CHAPTER

8 Temperature and Temperature-Related Parameters

Molten steel is being poured by a ladle into ingot molds for transport to another processing plant to make steel products. Depending on the carbon content of steel, the temperature of molten steel can exceed 1300°C.

In this chapter, we will explain what temperature, thermal energy, and heat transfer mean. We will take a close look at the role of temperature and heat transfer in engineering design and examine how they play hidden roles in our everyday lives. We will discuss temperature and its various scales, including Celsius, Fahrenheit, Rankine, and Kelvin. We will also explain a number of temperature-related properties of materials, such as specific heat, thermal expansion, and thermal conductivity. After studying this chapter, you will have learned that thermal energy transfer occurs whenever there exists a temperature difference within an object, or whenever there is a temperature difference between two bodies or a body and its surroundings. We will briefly discuss the various modes of heat transfer, what the R-value of insulation means, and what the term metabolic rate *means. We will also explain the heating values of fuels.*

In order to see how this chapter fits into what you have been studying so far, recall from our discussion in Chapter 3 that based on what we know about our physical world today, we need seven fundamental *or* base dimensions *to correctly express our natural world. The seven fundamental dimensions are* length, mass, time, temperature, electric current, amount of substance, *and* luminous intensity. *Recall that with the help of these base dimensions we can explain all other necessary physical quantities that describe how nature works. In the previous chapters, you studied the roles of length, mass, and time in engineering applications and in your everyday lives. In this chapter, we will discuss the role of temperature, another fundamental, or base, dimension, in engineering analysis.* ■

8.1 TEMPERATURE AS A FUNDAMENTAL DIMENSION

Understanding what temperature means and what its magnitude or value represents is very important in understanding our surroundings. Recall from our discussion in Chapter 3 that in order to describe how cold or hot something is, humans needed a physical quantity, or physical dimension, which we now refer to as temperature. Think about the important role of temperature in your everyday lives in describing various states of things. Do you know the answer to some of these questions: What is your body temperature? What is the room air temperature? What is the temperature of the water that you used this morning to take a shower? What is the temperature of the air inside your refrigerator that kept the milk cold overnight? What is the temperature inside the freezer section of your refrigerator? What is the temperature of the air coming out of your hair dryer? What is the surface temperature of your stove's heating element when set on high? What is the surface temperature of a 100-W light bulb? What is the average operating temperature of the electronic chips inside your TV or your computer? What is the temperature of combustion

products coming out of your car's engine? Once you start thinking about the role of temperature in quantifying what goes on in our surroundings, you realize that you could ask hundreds of similar questions.

Regardless of which engineering discipline you are planning to pursue, you need to develop a good understanding of what is meant by temperature and how it is quantified. Figure 8.1 illustrates some of the systems for which this understanding is important. Electronic and computer engineers, when designing computers, televisions, or any electronic equipment, are concerned with keeping the temperature of various electronic components at a reasonable operating level so that the electronic components will function properly. In

Figure 8.1
Examples of engineering systems for which understanding of temperature and heat transfer is important

fact, they use heat sinks (fins) and fans to cool the electronic chips. Civil engineers need to have a good understanding of temperature when they design bridges and other structures. They must design the structures in such a way as to allow for expansion and contraction of materials, such as steel and concrete, that occur due to changes in the surrounding temperatures. Mechanical engineers design heating, ventilating, and air-conditioning (HVAC) equipment to create the comfortable environment in which we rest, work, and play. They need to understand heat transfer processes and the properties of air, including its temperature and moisture content, when designing this equipment. Automotive engineers need to have a good understanding of temperature and heat transfer rates when designing the cooling system of an engine.

Material properties are a function of temperature. Physical and thermal properties of solids, liquids, and gases vary with temperature. For example, as you know, cold air is denser than warm air. The air resistance to your car's motion is greater in winter than it is in summer, provided the car is moving at the same speed. Most of you who live in a cold climate know it is harder to start your car in the morning in the winter. As you may know, starting difficulty is the result of the viscosity of oil increasing as the temperature decreases. The variation in density of air with temperature is shown in Figure 8.2. The temperature dependence of SAE 10W, SAE 10W-30, and SAE 30 oil is shown in Figure 8.3. These are but a few examples of why as engineers you need to have a good understanding of temperature and its role in design.

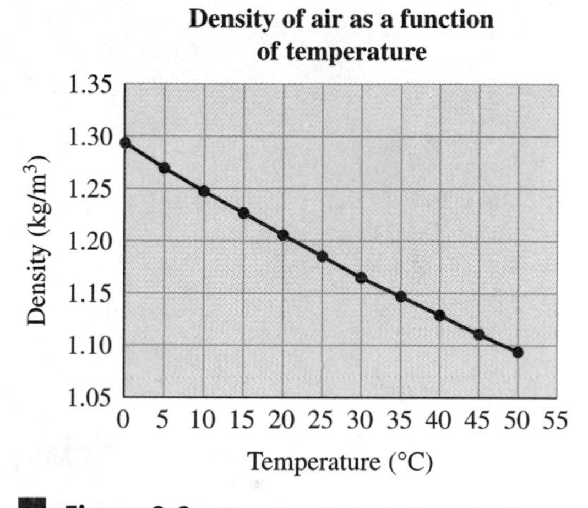

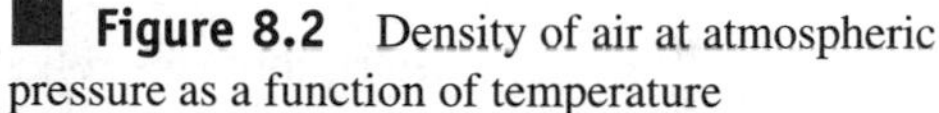
Figure 8.2 Density of air at atmospheric pressure as a function of temperature

Let us now examine more closely what we mean by temperature. Temperature provides a measure of molecular activity and the internal energy of an object. Recall that all objects and living things are made of matter, and matter itself is made up of atoms or chemical elements. Moreover, atoms are combined naturally, or in a laboratory setting, to create molecules. For example, as you already know, water molecules are made of two atoms of hydrogen and one atom of oxygen. Temperature represents the level of molecular activity of a substance. The molecules of a substance at a high temperature are more active than at a lower temperature. Perhaps a simple way to visualize this is to imagine the molecules of a gas as being the popcorn in a popcorn popper; the molecules that are at a higher temperature move, rotate, and bounce around faster than the colder ones in the popper. Therefore, temperature quantifies or provides a measure of how active these molecules are on a microscopic level. For example, air molecules are more active at, say, 50°C than they are at 25°C. You may want to think of temperature this way: We have bundled all the microscopic molecular movement into a single, macroscopic, measurable value that we call *temperature*.

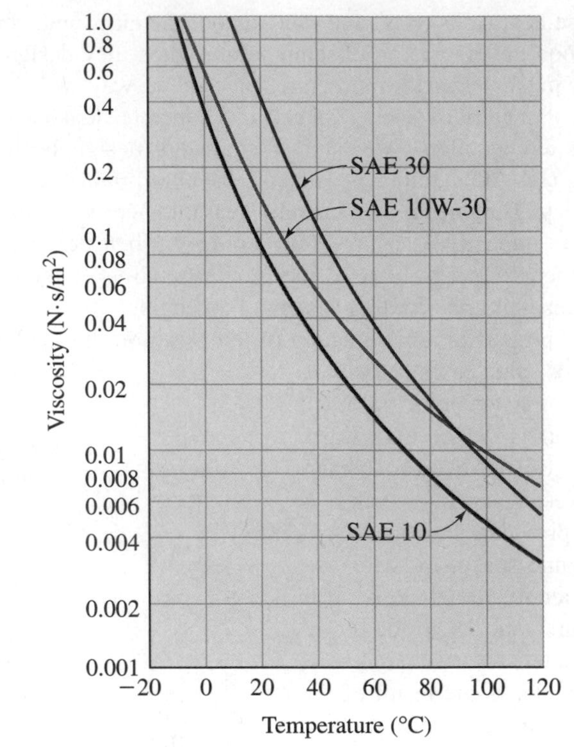

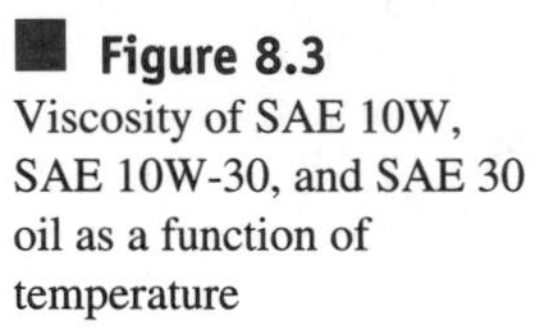

Figure 8.3
Viscosity of SAE 10W, SAE 10W-30, and SAE 30 oil as a function of temperature

8.2 MEASUREMENT OF TEMPERATURE AND ITS SI UNITS

Early humans relied on the sense of touch or vision to measure how cold or how warm something was. In fact, we still rely on touch today. When you are planning to take a bath, you first turn the hot and cold water on and let the bath fill with water. Before you enter the tub, however, you first touch the water to feel how warm it is. Basically, you are using your sense of touch to get an indication of the temperature. Of course, using touch alone, you can't quantify the temperature of water accurately. You cannot say, for example, that the water is at 21.5°C. So note the need for a more precise way of quantifying what the temperature of something is. Moreover, when we express the temperature of water, we need to use a number that is understood by all. In other words, we need to establish and use the same units and scales that are understood by everyone.

Another example of how people relied on their senses to quantify temperature is the way blacksmiths used to use their eyes to estimate how hot a fire was. They judged the temperature by the color of the burning fuel before they placed the horseshoe or an iron piece in the fire. In fact, this relationship between the color of heated iron and its actual temperature has been measured and established. Table 8.1 shows this relationship.

From these examples, you see that our senses are useful in judging how cold or how hot something is, but they are limited in accuracy and cannot quantify a value for a temperature. Thus, we need a measuring device that can provide information about the temperature of something more accurately and effectively.

This need led to the development of thermometers, which are based on thermal expansion or contraction of a fluid, such as alcohol, or a liquid metal, such as mercury. All of you know that almost everything will expand and its length increase when you increase its temperature, and it will contract and its length decrease when you decrease its temperature. We will discuss thermal expansion of material in more detail later in this chapter. But for now, remember that a mercury thermometer is a temperature sensor that works on the principle of expansion or contraction of mercury when its temperature is changed. Most of you have seen a thermometer, a graduated glass rod that is filled with mercury. On the Celsius scale, under standard atmospheric conditions, the value zero was *arbitrarily* assigned to the temperature at which water freezes, and the value of 100 was assigned to the temperature at which water

TABLE 8.1 The Relation of Color to Temperature of Iron

Color	Temperature (°F)
Dark blood red, black red	990
Dark red, blood red, low red	1050
Dark cherry red	1175
Medium cherry red	1250
Cherry, full red	1375
Light cherry, light red	1550
Orange	1650
Light orange	1725
Yellow	1825
Light yellow	1975
White	2200

Source: T. Baumeister et al., *Mark's Handbook*.

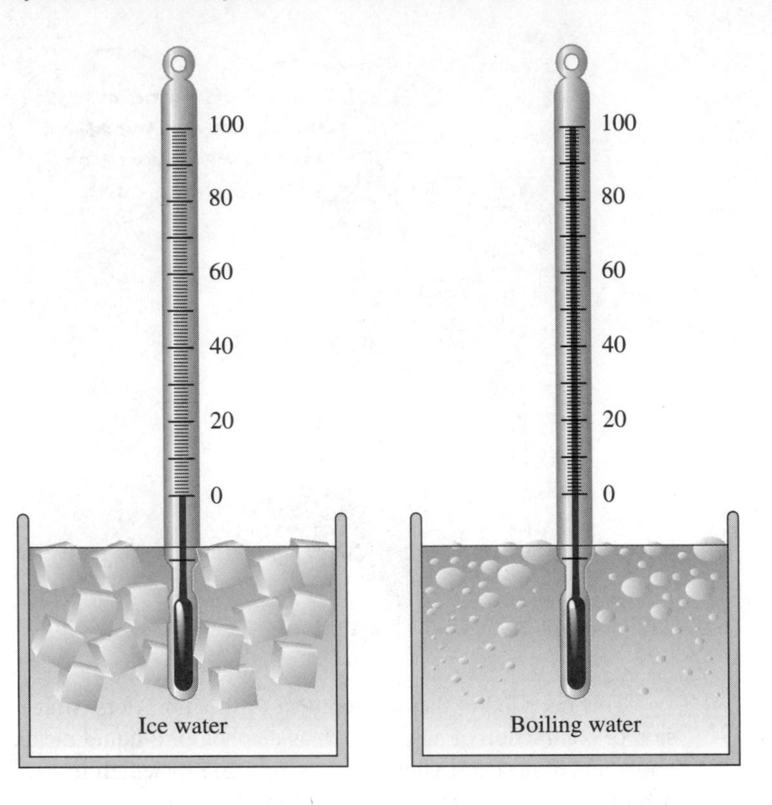

Figure 8.4
Calibration of a mercury thermometer

boils. This procedure is called *calibration* of an instrument and is depicted in Figure 8.4. It is important for you to understand that the numbers were assigned arbitrarily; had someone decided to assign a value of 100 to the ice water temperature and a value of 1000 to boiling water, we would have had a very different type of temperature scale today! In fact, on a Fahrenheit temperature scale, under standard atmospheric conditions, the temperature at which water freezes is assigned a value of 32°, and the temperature at which the water boils is assigned a value of 212°. The relationship between the two temperature scales is given by

$$T(°C) = \frac{5}{9}[T(°F) - 32] \tag{8.1}$$

$$T(°F) = \frac{9}{5}T(°C) + 32 \tag{8.2}$$

As with other instruments, thermometers evolved over time into today's accurate instruments that can measure temperature to 1/100°C increments.

Today we also use other changes in properties of matter, such as electrical resistance or optical or emf (electromotive force) changes to measure temperature. These property changes occur within matter when we change its temperature. Thermocouples and thermistors, shown in Figure 8.5, are examples of temperature-measuring devices that use these properties. A thermocouple consists of two dissimilar metals. A relatively small voltage output is created when a difference in temperature between the two junctions of a thermocouple exists. The small voltage output is proportional to the difference in temperature between the two junctions. Two of the common combinations of dissimilar metals used in thermocouple wires include iron/constantan (J-type) and copper/constantan (T-type). (The J and the T are the American National

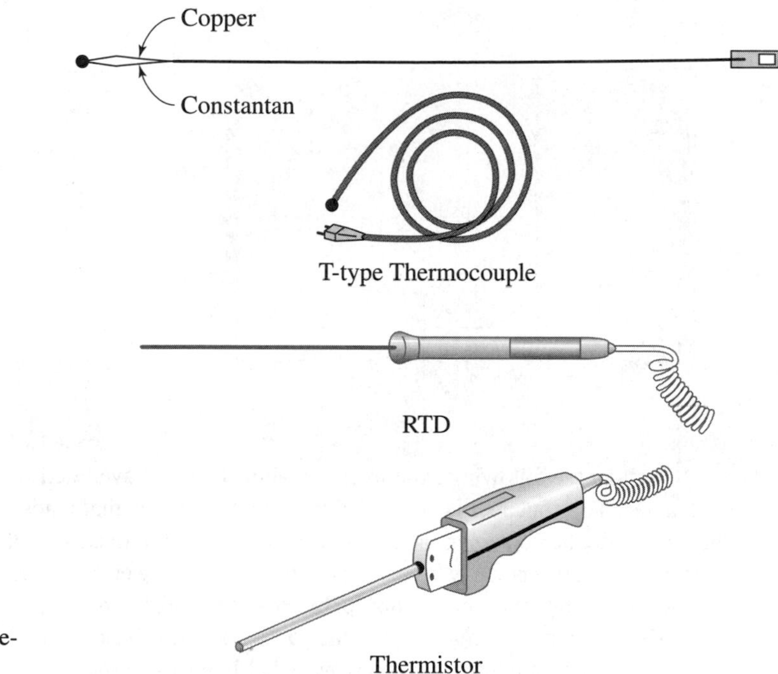

Figure 8.5
Examples of temperature-measuring devices

Standards Institute (ANSI) symbols used to refer to these thermocouple wires.) A thermistor is a temperature-sensing device composed of a semiconductor material with such properties that a small temperature change creates large changes in the electrical resistance of the material. Therefore, the electrical resistance of a thermistor is correlated to a temperature value.

Absolute Zero Temperature

Because both the Celsius and the Fahrenheit scales are arbitrarily defined, as we have explained, scientists recognized a need for a better temperature scale. This need led to the definition of an absolute scale, the Kelvin and Rankine scales, which are based on the behavior of an ideal gas. You have observed what happens to the pressure inside your car's tires during a cold winter day, or what happens to the air pressure inside a basketball if it is left outside during a cold night. At a given pressure, as the temperature of an ideal gas is decreased, its volume will also decrease. Well, for gases under certain conditions, there is a relationship between the pressure of the gas, its volume, and its temperature as given by what is commonly called the ***ideal gas law***. The ideal gas law is given by

$$PV = mRT \tag{8.3}$$

where

P = absolute pressure of the gas (Pa or lb/ft^2)

V = volume of the gas (m^3 or ft^3)

m = mass (kg or lb$_m$)

R = gas constant $\left(\frac{\text{J}}{\text{kg}\cdot\text{K}} \text{ or } \frac{\text{ft}\cdot\text{lb}}{\text{lb}_m\cdot\text{R}}\right)$

T = absolute temperature (Kelvin or Rankine, which we will explain soon)

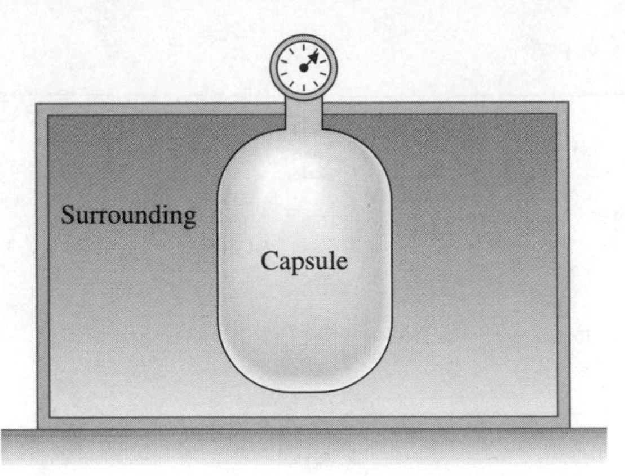

Figure 8.6
The capsule used in the absolute zero temperature example

Consider the following experiment. Imagine that we have filled a rigid container (capsule) with a gas. The container is connected to a pressure gauge that reads the absolute pressure of the gas inside the container, as shown in Figure 8.6. Moreover, imagine that we immerse the capsule in a surrounding whose temperature we can lower. Of course, if we allow enough time, the temperature of the gas inside the container will reach the temperature of its surroundings. Also keep in mind that because the gas is contained inside a rigid container, it has a constant volume and a constant mass. Now, what happens to the pressure of the gas inside the container as indicated by the pressure gauge as we decrease the surrounding temperature? The pressure will decrease as the temperature of the gas is decreased. We can determine the relationship between the pressure and the temperature using the ideal gas law, Equation (8.3).

Now, let us proceed to run a series of experiments. Starting with the surrounding temperature equal to some reference temperature—say, T_r—and allowing enough time for the container to reach equilibrium with the surroundings, we then record the corresponding pressure of the gas, P_r. The pressure and temperature of the gas are related according to the ideal gas law:

$$T_r = \frac{P_r V}{mR} \tag{8.4}$$

Now imagine that we lower the surrounding temperature to T_1, once equilibrium is reached, recording the pressure of the gas and denoting the reading by P_1. Because the temperature of the gas was lowered, the pressure would be lowered as well.

$$T_1 = \frac{P_1 V}{mR} \tag{8.5}$$

Dividing Equation (8.5) by Equation (8.4), we have

$$\frac{T_1}{T_r} = \frac{\frac{P_1 V}{mR}}{\frac{P_r V}{mR}} \tag{8.6}$$

And after canceling out the *m*, the *R*, and the *V*, we have

$$\frac{T_1}{T_r} = \frac{P_1}{P_r} \tag{8.7}$$

$$T_1 = T_r\left(\frac{P_1}{P_r}\right) \tag{8.8}$$

Equation (8.8) establishes a relationship between the temperature of the gas, its pressure, and the reference pressure and temperature. If we were to proceed by lowering the surrounding temperature, a lower gas pressure would result, and if we were to extrapolate the results of our experiments, we would find that we eventually reach zero pressure at zero temperature. This temperature is called the *absolute thermodynamic temperature* and is related to the Celsius and Fahrenheit scales. The relationship between the Kelvin (K) and degree Celsius (°C) in SI units is

$$T(\text{K}) = T(°\text{C}) + 273.15 \tag{8.9}$$

The relationship between degree Rankine (°R) and degree Fahrenheit (°F) in U.S. Customary Units is

$$T(°\text{R}) = T(°\text{F}) + 459.67 \tag{8.10}$$

Note that the experiment cannot be completely carried out, because as the temperature of the gas decreases, it reaches a point where it will liquify and thus the ideal gas law will not be valid. This is the reason for extrapolating the result to obtain a theoretical absolute zero temperature. It is also important to note that while there is a limit as to how cold something can be, there is no theoretical upper limit as to how hot something can be. Finally, we can also establish a relationship between the degree Rankine and the Kelvin by

$$T(\text{K}) = \frac{5}{9}T(°\text{R}) \tag{8.11}$$

In Equations (8.9) and (8.10), unless you are dealing with very precise experiments, when converting from degree Celsius to Kelvin you can round down the 273.15 to 273. The same is true when converting from degree Fahrenheit to Rankine; round up the 459.67 to 460.

EXAMPLE 8.1

What is the equivalent value of $T = 50°\text{C}$ in degrees Fahrenheit, Rankine, and Kelvin?

We can use Equations (8.2), (8.9), and (8.10):

$$T(°\text{F}) = \frac{9}{5}T(°\text{C}) + 32 = \left(\frac{9}{5}\right)(50) + 32 = 122°\text{F}$$

$$T(°\text{R}) = T(°\text{F}) + 460 = 122 + 460 = 582°\text{R}$$

$$T(\text{K}) = T(°\text{C}) + 273 = 50 + 273 = 323\ \text{K}$$

Note that we also could have converted the 582°R to Kelvin directly using Equation (8.11):

$$T(\text{K}) = \frac{5}{9}T(°\text{R}) = \left(\frac{5}{9}\right)(582) = 323\ \text{K}$$

EXAMPLE 8.2

On a summer day, in Phoenix, Arizona, the inside room temperature is maintained at 68°F while the outdoor air temperature is a sizzling 110°F. What is the outdoor-indoor temperature difference in (a) degree Fahrenheit, (b) degree Rankine, (c) degree Celsius, and (d) Kelvin? Is a 1° temperature difference in Celsius equal to a 1° temperature difference in Kelvin, and is a 1° temperature difference in Fahrenheit equal to a 1° temperature difference in Rankine? If so, why?

We will first answer these questions the long way, and then we will discuss the short way.

(a) $T_{outdoor} - T_{indoor} = 110°F - 68°F = 42°F$

(b) $T_{outdoor}(°R) = T_{outdoor}(°F) + 460 = 110 + 460 = 570°R$

$T_{indoor}(°R) = T_{indoor}(°F) + 460 = 68 + 460 = 528°R$

$T_{outdoor} - T_{indoor} = 570°R - 528°R = 42°R$

Note that the temperature difference expressed in degrees Fahrenheit is equal to the temperature difference expressed in degrees Rankine.

(c) $T_{outdoor}(°C) = \frac{5}{9}(T_{outdoor}(°F) - 32) = \frac{5}{9}(110 - 32) = 43.3°C$

$T_{indoor}(°C) = \frac{5}{9}(T_{indoor}(°F) - 32) = \frac{5}{9}(68 - 32) = 20°C$

$T_{outdoor} - T_{indoor} = 43.3°C - 20°C = 23.3°C$

(d) $T_{outdoor}(K) = T_{outdoor}(°C) + 273 = 43.3 + 273 = 316.3\text{ K}$

$T_{indoor}(K) = T_{indoor}(°C) + 273 = 20 + 273 = 293\text{ K}$

$T_{outdoor} - T_{indoor} = 316.3\text{ K} - 293\text{ K} = 23.3\text{ K}$

Note that the temperature difference expressed in degrees Celsius is equal to the temperature difference expressed in Kelvin.

It should be clear by now that a 1° temperature difference in Celsius is equal to a 1° temperature difference in Kelvin, and a 1° temperature difference in Fahrenheit is equal to a 1° temperature difference in Rankine. Of course, this relationship is true because when you are computing the difference between two temperatures and converting to the absolute temperature scale, you are adding the same base value to each temperature. For example, to compute the temperature difference in degrees Rankine between two temperatures T_1 and T_2 given in degrees Fahrenheit, you first add 460 to each temperature to convert T_1 and T_2 from degrees Fahrenheit to degrees Rankine. This step is shown next.

$$T_1(°R) - T_2(°R) = \overbrace{[T_1(°F) + 460]}^{T_1(°R)} - \overbrace{[T_2(°F) + 460]}^{T_2(°R)} = T_1(°F) + 460 - T_2(°F) - 460$$

And simplifying this relationship leads to

$$T_1(°R) - T_2(°R) = T_1(°F) - T_2(°F)$$

Also note that you could have converted the temperature difference in degrees Rankine to Kelvin directly in the following manner:

$$\Delta T(K) = \frac{5}{9}\Delta T(°R) = \left(\frac{5}{9}\right)(42) = 23.3\text{ K}$$

8.3 TEMPERATURE DIFFERENCE AND HEAT TRANSFER

Thermal energy transfer occurs whenever there exists a temperature difference within an object, or whenever there is a temperature difference between two bodies, or a temperature difference between a body and its surroundings. This form of energy transfer that

TABLE 8.2 Conversion Factors for Thermal Energy and Thermal Energy per Unit Time (Power)

Relationship Between the Units of Thermal Energy	Relationship Between the Units of Thermal Energy per Unit Time (Power)
1 Btu = 1055 J	1 W = 1 J/s
1 Btu = 252 cal	1 W = 3.4123 Btu/h
1 cal = 4.186 J	1 cal/s = 4.186 W

occurs between bodies of different temperatures is called ***heat transfer***. Additionally, heat always flows from a high-temperature region to a low-temperature region. This statement can be confirmed by observation of our surroundings. When hot coffee in a cup is left in a surrounding such as a room with a lower temperature, the coffee cools down. Thermal energy transfer takes place from the hot coffee through the cup and from its open surface to the surrounding room air. The thermal energy transfer occurs as long as there is a temperature difference between the coffee and its surroundings. At this point, make sure you understand the difference between *temperature* and *heat*. Heat is a form of energy that is transferred from one region to the next region as a result of a temperature difference between the regions, whereas temperature represents on a macroscopic level, by a single number, the level of microscopic molecular movement in a region.

There are three units that are commonly used to quantify thermal energy: (1) the British thermal unit (Btu), (2) the calorie, and (3) the joule. The ***British thermal unit (Btu)*** is defined as the amount of heat required to raise the temperature of 1 pound mass (1 lb_m) of water by 1 degree Fahrenheit (1°F). The ***calorie*** is defined as the amount of heat required to raise the temperature of 1 gram of water by 1°C. Note, however, that the energy content of food is typically expressed in *Calories*, which is equal to 1000 calories. In SI units, no distinction is made between the units of thermal energy and mechanical energy and therefore energy is defined in terms of the fundamental dimensions of mass, length, and time. We will discuss this in more detail in Chapter 10. In the SI System of Units, the ***joule*** is the unit of energy and is defined as

$$1 \text{ joule} = 1 \text{ N}\cdot\text{m} = 1 \text{ kg}\cdot\text{m}^2/\text{s}^2$$

The conversion factors among various units of heat are given in Table 8.2.

EXAMPLE 8.3

Using the units of energy and time, show that 1 watt (W) is equal to 3.4123 Btu/h, as shown in Table 8.2.

$$1 \text{ W} = 1\left(\frac{\text{J}}{\text{s}}\right)\left(\frac{1 \text{ Btu}}{1055 \text{ J}}\right)\left(\frac{3600 \text{ s}}{1 \text{ h}}\right) = 3.412\frac{\text{Btu}}{\text{h}}$$

EXAMPLE 8.4

In many parts of the United States, in order to keep a house warm in the winter months, a gas furnace is used. If a gas furnace puts out 60,000 Btu/h to compensate for heat loss from a house, what is the equivalent value of the thermal power (energy per unit time) output of the furnace in watts?

From Table 8.2, you know that: 1 Btu = 1055 J; you also know that 1 h = 3600 s. Substituting for these values,

$$q' = 60{,}000\left(\frac{\text{Btu}}{\text{h}}\right)\left(\frac{1{,}055\ \text{J}}{1\ \text{Btu}}\right)\left(\frac{1\ \text{h}}{3{,}600\ \text{s}}\right) = 17{,}583\left(\frac{\text{J}}{\text{s}}\right) = 17{,}583\ \text{W} = 17.583\ \text{kW}$$

Or you could have used the direct conversion factor between Btu/h and W, as shown:

$$q' = 60{,}000\left(\frac{\text{Btu}}{\text{h}}\right)\left(\frac{1\ \text{W}}{3.4123\left(\frac{\text{Btu}}{\text{h}}\right)}\right) = 17583\ \text{W} = 17.583\ \text{kW}$$

Now that you know heat transfer or thermal energy transfer occurs as a result of temperature difference in an object or between objects, let us look at different modes of heat transfer. There are three different mechanisms by which energy is transferred from a high-temperature region to a low-temperature region. These are referred to as the *modes* of heat transfer. The three modes of heat transfer are conduction, convection, and radiation.

Conduction

Conduction refers to that mode of heat transfer that occurs when a temperature difference (gradient) exists in a medium. The energy is transported within the medium from the region with more-energetic molecules to the region with less-energetic molecules. Of course, it is the interaction of the molecules with their neighbors that makes the transfer of energy possible. To better demonstrate the idea of molecular interactions, consider the following example of conduction heat transfer. All of you have experienced what happens when you heat up some soup in an aluminum container on a stove. Why do the handles or the lid of the soup container get hot, even though the handles and the lid are not in direct contact with the heating element? Well, let us examine what is happening.

Because of the energy transfer from the heating element, the molecules of the container in the region near the heating element are more energetic than those molecules further away.

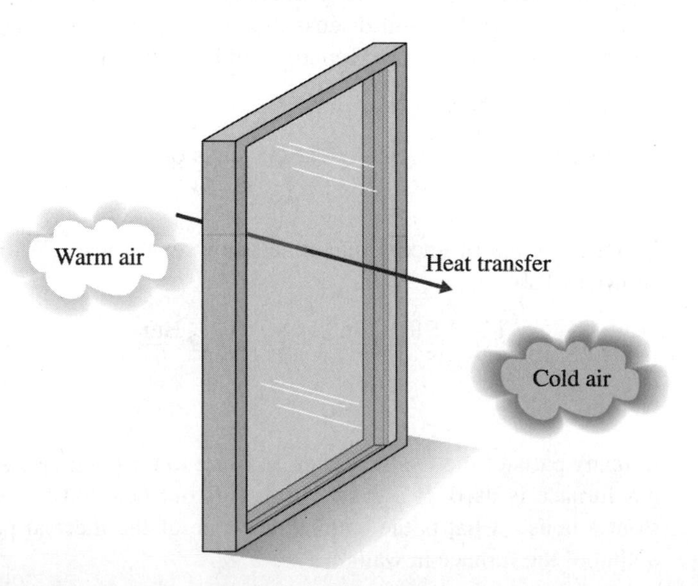

Figure 8.7
Conduction heat transfer through a glass window

The more energetic molecules share or transfer some of their energy to the neighboring regions, and the neighboring regions do the same thing, until the energy transfer eventually reaches the handles and the lid of the container. The energy is transported from the high-temperature region to the low-temperature region by molecular activity. The rate of heat transfer by conduction is given by ***Fourier's law***, which states that the rate of heat transfer through a material is proportional to the temperature difference, normal area A, through which heat transfer occurs, and the type of material involved. The law also states that the heat transfer rate is inversely proportional to the material thickness over which the temperature difference exists. For example, referring to Figure 8.7, we can write the Fourier's law for a single-glass window as

$$q = kA\frac{T_1 - T_2}{L} \quad (8.12)$$

where

q = heat transfer rate (in W or Btu/h)

k = thermal conductivity $\left(\frac{\text{W}}{\text{m}\cdot°\text{C}} \text{ or } \frac{\text{Btu}}{\text{h}\cdot\text{ft}\cdot°\text{F}}\right)$

A = cross-sectional area normal to heat flow (m^2 or ft^2)

$T_1 - T_2$ = temperature difference across the material of L thickness

L = material thickness (m or ft)

The temperature difference $T_1 - T_2$ is commonly referred to as the *temperature gradient*. Again, keep in mind that a temperature gradient must exist in order for heat transfer to occur. Thermal conductivity is a property of materials that shows how good the material is in transferring thermal energy (heat) from a high-temperature region to a low-temperature region within the material. In general, solids have a higher thermal condutivity than liquids, and liquids have a higher thermal conductivity than gases. The thermal conductivity of some materials is given in Table 8. 3 on page 194.

EXAMPLE 8.5

Calculate the heat transfer rate from a single-pane glass window with an inside surface temperature of approximately 20°C and an outside surface temperature of 5°C. The glass is 1 m tall, 1.8 m wide, and 8 mm thick. The thermal conductivity of the glass is approximately $k = 1.4$ W/m·K.

$$L = 8\,(\text{mm})\left(\frac{1\text{ m}}{1000\text{ mm}}\right) = 0.008\text{ m}$$

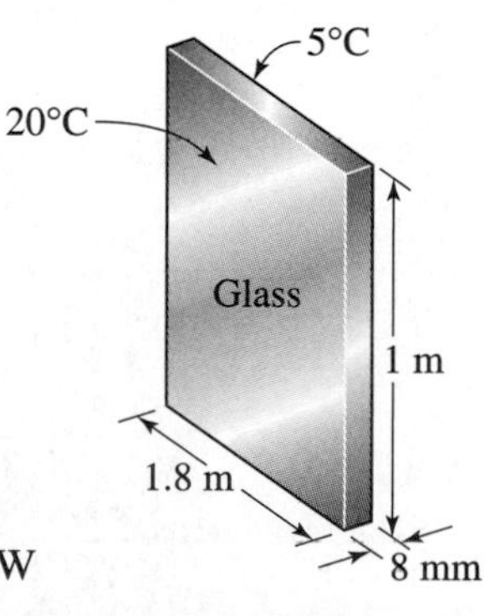

$$A = (1\text{ m})(1.8\text{ m}) = 1.8\text{ m}^2$$

$$T_1 - T_2 = 20°\text{C} - 5°\text{C} = 15°\text{C} = 15\text{ K}$$

Note, as we explained before, a 15°C temperature difference is equal to a 15 K temperature difference. Substituting for the values of k, A, $(T_1 - T_2)$, and L in Equation (8.12), we have

$$q = kA\frac{T_1 - T_2}{L} = (1.4)\left(\frac{\text{W}}{\text{m}\cdot\text{K}}\right)(1.8\text{ m}^2)\left(\frac{15\text{ K}}{0.008\text{ m}}\right) = 4725\text{ W}$$

(Be careful with lowercase k, denoting thermal conductivity, and the capital K, representing Kelvin, an absolute temperature scale.)

TABLE 8.3 Thermal Conductivity of Some Materials at 300 K

Material	Thermal Conductivity (W/m·K)
Air (at atmospheric pressure)	0.0263
Aluminum (pure)	237
Aluminum alloy-2024-T6 (4.5% copper, 1.5% magnesium, 0.6% manganese)	177
Asphalt	0.062
Bronze (90% copper, 10% aluminum)	52
Brass (70% copper, 30% zinc)	110
Brick (fire clay)	1.0
Concrete	1.4
Copper (pure)	401
Glass	1.4
Gold	317
Human fat layer	0.2
Human muscle	0.41
Human skin	0.37
Iron (pure)	80.2
Stainless steels (AISI 302, 304, 316, 347)	15.1, 14.9, 13.4, 14.2
Lead	35.3
Paper	0.18
Platinum (pure)	71.6
Sand	0.27
Silicon	148
Silver	429
Zinc	116
Water (liquid)	0.61

Thermal Resistance

In this section, we will explain what the *R*-values of insulating materials mean. Most of you understand the importance of having a well-insulated house because the better insulated a house is, the less the heating or cooling cost of the house. For example, you may have heard that in order to reduce heat loss through the attic, some people add enough insulation to their attic so that the *R*-value of insulation is 40. But what does the *R*-value of 40 mean, and what does the *R*-value of an insulating material mean in general? Let us start by rearranging Equation (8.12) in the following manner. Starting with Equation (8.12),

$$q = kA\frac{T_1 - T_2}{L} \tag{8.12}$$

and rearranging it, we have

$$q = \frac{T_1 - T_2}{\dfrac{L}{kA}} = \frac{\text{temperature difference}}{\text{thermal resistance}} \tag{8.13}$$

and thermal resistance $= L/kA$.

Figure 8.8 depicts the idea of thermal resistance and how it is related to the material's thickness, area, and thermal conductivity. When examining Equation (8.13), you should note the following: (1) The heat transfer (flow) rate is directly proportional to the temperature difference; (2) the heat flow rate is inversely proportional to the thermal resistance—the higher the value of thermal resistance, the lower the heat transfer rate will be.

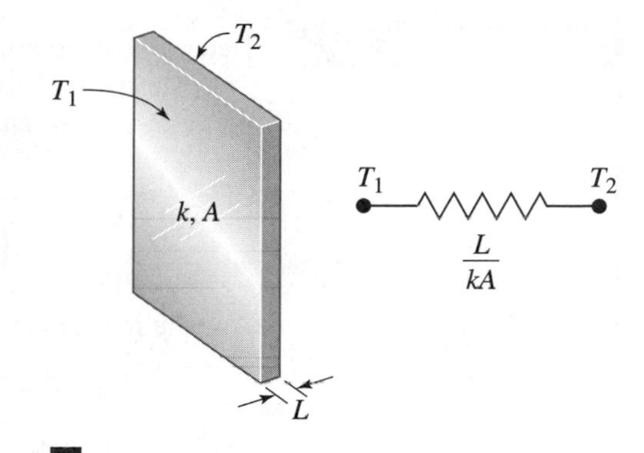

Figure 8.8 A slab of material and its thermal resistance

When expressing Fourier's law in the form of Equation (8.13), we are making an analogy between the flow of heat and the flow of electricity in a wire. Ohm's law, which relates the voltage V to current I and the electrical resistance R_e, is analogous to heat flow. Ohm's law is expressed as

$$V = R_e I \tag{8.14}$$

or

$$I = \frac{V}{R_e} \tag{8.15}$$

We will discuss Ohm's law in more detail in Chapter 9. Comparing Equation (8.13) to Equation (8.15), note that the heat flow is analogous to electric current, the temperature difference to voltage, and the thermal resistance to electrical resistance.

Now turning our attention back to thermal resistance and Equation (8.13), we realize that the thermal resistance for a unit area of a material is defined as

$$R' = \frac{L}{kA} \tag{8.16}$$

R' has the units of °C/W (K/W) or °F·h/Btu (°R·h/Btu). When Equation (8.16) is expressed per unit area of the material, it is referred to as the R-value or the R-factor.

$$R = \frac{L}{k} \tag{8.17}$$

where R has the units of

$$\frac{\text{m}^2\cdot{}^\circ\text{C}}{\text{W}}\left(\frac{\text{m}^2\cdot\text{K}}{\text{W}}\right)$$

or

$$\frac{\text{ft}^2\cdot{}^\circ\text{F}}{\frac{\text{Btu}}{\text{h}}}\left(\frac{\text{ft}^2\cdot{}^\circ\text{R}}{\frac{\text{Btu}}{\text{h}}}\right)$$

Note that neither R' nor R is dimensionless and sometimes the R-values are expressed per unit thickness. The R-value or R-factor of a material provides a measure of resistance to heat flow: The higher the value, the more resistance to heat flow the material offers. Finally, when the materials used for insulation purposes consist of various components, the total R-value of the composite material is the sum of resistance offered by the various components.

EXAMPLE 8.6

Determine the thermal resistance R' and the R-value for the glass window of Example 8.5.

The thermal resistance R' and the R-value of the window can be determined from Equations (8.16) and (8.17), respectively.

$$R' = \frac{L}{kA} = \frac{0.008 \text{ m}}{1.4\left(\frac{\text{W}}{\text{m}\cdot\text{K}}\right)(1.8 \text{ m}^2)} = 0.00317 \frac{\text{K}}{\text{W}}$$

And the R-value or the R-factor for the given glass pane is

$$R = \frac{L}{k} = \frac{0.008 \text{ m}}{1.4\left(\frac{\text{W}}{\text{m}\cdot\text{K}}\right)} = 0.0057 \frac{\text{m}^2\cdot\text{K}}{\text{W}}$$

EXAMPLE 8.7

For Example 8.6, convert the thermal resistance R' and R-factor results from SI units to U.S. Customary Units.

$$R' = 0.00317\left(\frac{\text{K}}{\text{W}}\right)\left(\frac{1 \text{ W}}{3.4123 \frac{\text{Btu}}{\text{h}}}\right)\left(\frac{\frac{5}{9} \text{ °R}}{\text{K}}\right) = 5.161 \times 10^{-4} \frac{\text{°R}}{\frac{\text{Btu}}{\text{h}}}$$

$$R = \frac{L}{k} = 0.0057\left(\frac{\text{m}^2\cdot\text{K}}{\text{W}}\right)\left(\frac{1 \text{ W}}{3.4123 \frac{\text{Btu}}{\text{h}}}\right)\left(\frac{\frac{5}{9} \text{ °R}}{\text{K}}\right)\left(\frac{3.28 \text{ ft}}{1 \text{ m}}\right)^2 = 0.01 \frac{\text{°R}\cdot\text{ft}^2}{\frac{\text{Btu}}{\text{h}}}$$

Or, if the R-value is expressed in terms of in^2,

$$R = \frac{L}{k} = 0.0057\left(\frac{\text{m}^2\cdot\text{K}}{\text{W}}\right)\left(\frac{1 \text{ W}}{3.4123 \frac{\text{Btu}}{\text{h}}}\right)\left(\frac{\frac{5}{9} \text{ °R}}{\text{K}}\right)\left(\frac{39.37 \text{ in.}}{1 \text{ m}}\right)^2 = 1.43 \frac{\text{°R}\cdot\text{in}^2}{\frac{\text{Btu}}{\text{h}}}$$

EXAMPLE 8.8

A double-pane glass window consists of two pieces of glass, each having a thickness of 8 mm, with a thermal conductivity of $k = 1.4$ W/m·K. The two glass panes are separated by an air gap of 10 mm, as shown in Figure 8.9. Assuming the thermal conductivity of air to be $k = 0.025$ W/m·K, determine the total R-value for this window.

The total thermal resistance of the window is obtained by adding the resistance offered by each pane of glass and the air gap in the following manner:

$$R_{total} = R_{glass} + R_{air} + R_{glass} = \frac{L_{glass}}{k_{glass}} + \frac{L_{air}}{k_{air}} + \frac{L_{glass}}{k_{glass}}$$

substituting for L_{glass}, k_{glass}, L_{air}, k_{air}, we have

$$R_{total} = \frac{0.008(\text{m})}{1.4\left(\frac{\text{W}}{\text{m}\cdot\text{K}}\right)} + \frac{0.01(\text{m})}{0.025\left(\frac{\text{W}}{\text{m}\cdot\text{K}}\right)} + \frac{0.008(\text{m})}{1.4\left(\frac{\text{W}}{\text{m}\cdot\text{K}}\right)} = 0.4\left(\frac{\text{m}^2\cdot\text{K}}{\text{W}}\right)$$

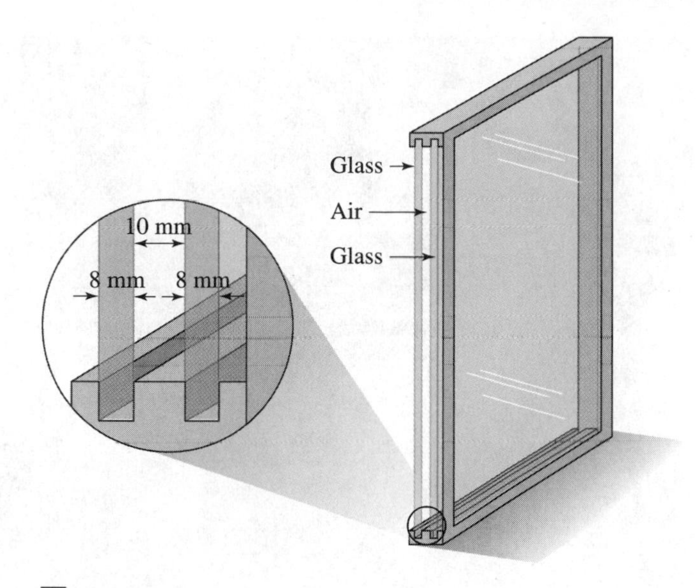

Figure 8.9 The double-pane glass window of Example 8.8

Note the units of R-value. The R-value for the double-pane glass window in U.S. Customary Units is

$$R_{\text{total}} = 0.4\left(\frac{\text{m}^2\cdot\text{K}}{\text{W}}\right)\left(\frac{1\ \text{W}}{3.4123\,\dfrac{\text{Btu}}{\text{h}}}\right)\left(\frac{\frac{5}{9}\,^\circ\text{R}}{\text{K}}\right)\left(\frac{3.28\ \text{ft}}{1\ \text{m}}\right)^2 = 0.7\,\frac{^\circ\text{R}\cdot\text{ft}^2}{\dfrac{\text{Btu}}{\text{h}}}$$

As you can see from the results of Example 8.6 and Example 8.8, ordinary glass windows do not offer much resistance to heat flow. To increase the R-value of windows, some manufacturers make windows that use triple glass panes and fill the spacing between the glass panes with Argon gas. See Example 8.11 for a sample calculation for total thermal resistance of a typical exterior frame wall of a house consisting of siding, sheathing, insulation material, and gypsum wallboard (drywall).

Convection

Convection heat transfer occurs when a fluid (a gas or a liquid) in motion comes into contact with a surface whose temperature differs from the moving fluid. For example, on a hot summer day when you sit in front of a fan to cool down, the heat transfer rate that occurs from your warm body to the cooler moving air is by convection. Or when you are cooling a hot food, such as freshly baked cookies, by blowing on it, you are using the principles of convection heat transfer. The cooling of computer chips by blowing air across them is another example of cooling something by convection heat transfer. There are two broad areas of convection heat transfer: *forced convection* and *free (natural) convection*. Forced convection refers to situations where the flow of fluid is

CPU being cooled by a fan

caused or forced by a fan or a pump. Free convection, on the other hand, refers to situations where the flow of fluid occurs naturally due to density variation in the fluid. Of course, the density variation is caused by the temperature distribution within the fluid. When you leave a hot pie to cool on the kitchen counter, the heat transfer is by natural convection. The heat loss from the exterior surfaces of the hot oven is also by natural convection. To cite another example of natural convection, the large electrical transformers that sit outdoors at power substations are cooled by natural convection as well.

For both the forced and the free convection situations, the overall heat transfer rate between the fluid and the surface is governed by Newton's law of cooling, which is given by

$$q = hA(T_s - T_f) \tag{8.18}$$

where h is the heat transfer coefficient in W/m^2·K (or Btu/hr·ft^2·R), A is the area of the exposed surface in m^2 (ft^2), T_s is the surface temperature, and T_f represents the temperature of moving fluid. The value of the heat transfer coefficient for a particular situation is determined from experimental correlation; these values are available in many books about heat transfer. At this stage of your education, you need not be concerned about how to obtain the numerical values of heat transfer coefficients. However, it important for you to know that the value of h is higher for forced convection than it is for free convection. Of course, you already know this! When you are trying to cool down rapidly, do you sit in front of a fan or do you sit in an area of the room where the air is still? Moreover, the heat transfer coefficient h is higher for liquids than it is for gases. Have you noticed that you can walk around comfortably in a T-shirt when the outdoor air temperature is 70°F, but if you went into a swimming pool whose water temperature was 70°F, you would feel cold? That is because the liquid water has a higher heat transfer coefficient than does air, and therefore, according to Newton's law of cooling, Equation (8.18), water removes more heat from your body. The typical range of heat transfer coefficient values is given in Table 8.4.

TABLE 8.4 Typical Values of Heat Transfer Coefficients

Convection Type	Heat Transfer Coefficient, *h* (W/m²·K)	Heat Transfer Coefficient, *h* (Btu/h·ft²·°R)
Free Convection		
Gases	2 to 25	0.35 to 4.4
Liquids	50 to 1000	8.8 to 175
Forced Convection		
Gases	25 to 250	4.4 to 44
Liquids	100 to 20,000	17.6 to 3500

In the field of heat transfer, it is also common to define a resistance term for the convection process, similar to the *R*-value in conduction. The thermal convection resistance is defined as:

$$R' = \frac{1}{hA} \tag{8.19}$$

Again, R' has the units of °C/W (K/W) or °F·h/Btu (°R·h/Btu). Equation (8.19) is commonly expressed per unit area of solid surface exposure and is called *film resistance* or *film coefficient*.

$$R = \frac{1}{h} \tag{8.20}$$

where R has the units of

$$\frac{\text{m}^2\cdot°\text{C}}{\text{W}}\left(\frac{\text{m}^2\cdot\text{K}}{\text{W}}\right)$$

or

$$\frac{\text{ft}^2\cdot\text{F}}{\frac{\text{Btu}}{\text{h}}}\left(\frac{\text{ft}^2\cdot\text{R}}{\frac{\text{Btu}}{\text{h}}}\right)$$

It is important to realize once again that neither R' nor R is dimensionless, and they provide a measure of resistance to heat flow; the higher the values of R, the more resistance to heat flow to or from the surrounding fluid.

EXAMPLE 8.9

Determine the heat transfer rate by convection from an electronic chip whose surface temperature is 35°C and has an exposed surface area of 9 cm². The temperature of the surrounding air is 20°C. The heat transfer coefficient for this situation is $h = 40$ W/m²·K.

$$A = 9(\text{cm}^2)\left(\frac{1\ \text{m}^2}{10{,}000\ \text{cm}^2}\right) = 0.0009\ \text{m}^2$$

$$T_s - T_f = 35°\text{C} - 20°\text{C} = 15°\text{C} = 15\ \text{K}$$

Note again that the 15°C temperature difference is equal to a 15 K temperature difference. We can determine the heat transfer rate from the chip by substituting for the values of h, A, and $(T_s - T_f)$ in Equation (8.18), which results in

$$q_{\text{convection}} = hA(T_s - T_f) = 40\left(\frac{\text{W}}{\text{m}^2\cdot\text{K}}\right)(0.0009\ \text{m}^2)(15\ \text{K}) = 0.54\ \text{W}$$

EXAMPLE 8.10

Calculate the R-factor (film resistance) for the following situations: (a) wind blowing over a wall, $h = 5.88$ Btu/h·°F·ft^2, and (b) still air inside a room near a wall, $h = 1.47$ Btu/h·°F·ft^2.

For the situation where wind is blowing over a wall:

$$R = \frac{1}{h} = \frac{1}{5.88\dfrac{\text{Btu}}{\text{h}\cdot\text{ft}^2\ {}^\circ\text{F}}} = 0.17\frac{\text{h}\cdot\text{ft}^2\ {}^\circ\text{F}}{\text{Btu}}$$

And for still air inside a room, near a wall:

$$R = \frac{1}{h} = \frac{1}{1.47\dfrac{\text{Btu}}{\text{h}\cdot\text{ft}^2\ {}^\circ\text{F}}} = 0.68\frac{\text{h}\cdot\text{ft}^2\ {}^\circ\text{F}}{\text{Btu}}$$

EXAMPLE 8.11

A typical exterior frame wall (made up of 2 × 4 studs) of a house contains the materials shown in the following table. For most residential buildings, the inside room temperature is kept around 70°F. Assuming an outside temperature of 20°F and an exposed area of 150 ft^2, we are interested in determining the heat loss through the wall.

Items	Thermal Resistance (h·ft^2·°F/Btu)
1. Outside film resistance (winter, 15 mph wind)	0.17
2. Siding, wood (1/2 × 8 lapped)	0.81
3. Sheathing (1/2 in. regular)	1.32
4. Insulation batt (3–3 $\frac{1}{2}$ in.)	11.0
5. Gypsum wallboard (1/2 in.)	0.45
6. Inside film resistance (winter)	0.68

In general, the heat loss through the walls, windows, doors, or roof of a building occurs due to conduction heat losses through the building materials—including siding, insulation material, gypsum wallboard (drywall), glass, and so on—and convection losses through the wall surfaces exposed to the indoor warm air and the outdoor cold air. The total resistance to heat flow is the sum of resistances offered by each component in the path of heat flow. For a plane wall we can write:

$$q = \frac{T_{\text{inside}} - T_{\text{outside}}}{\sum R'} = \frac{(T_{\text{inside}} - T_{\text{outside}})A}{\sum R}$$

The total resistance to heat flow is given by

$$\sum R = R_1 + R_2 + R_3 + R_4 + R_5 + R_6 = 0.17 + 0.81 + 1.32 + 11.0 + 0.45 + 0.68 = 14.43$$

$$q = \frac{(T_{\text{inside}} - T_{\text{outside}})A}{\sum R} = \frac{(70 - 20)(150)}{14.43} = 520\,\frac{\text{Btu}}{\text{h}}$$

The equivalent thermal resistance circuit for this problem is shown in the accompanying diagram.

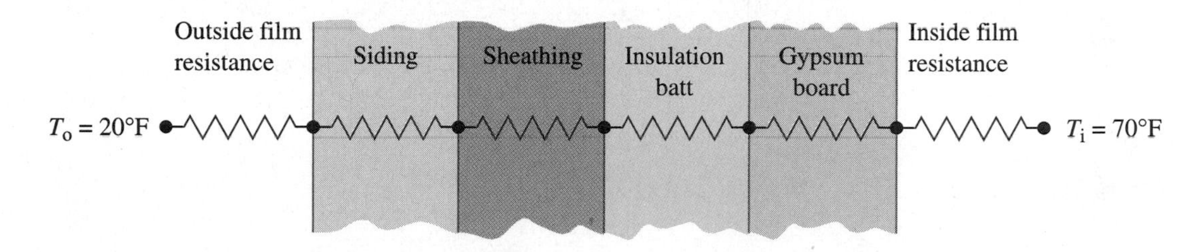

Windchill Factor

As all of you know, the rate of heat transfer from your body to the surrounding air increases on a cold, windy day. Simply stated, you lose more body heat on the cold, windy day than you do on a calm, cold day. By now you understand the difference between natural and forced convection heat transfer. The windchill index is intended to account for the combined effect of wind speed and the air temperature. Thus, it is supposed to account for the additional body heat loss that occurs on a cold, windy day. However, keep in mind that the common windchill correlations are based on a series of experiments performed in a very cold climate, where the time that it took to freeze water in a plastic container under various surrounding air temperatures and wind conditions was studied. A common correlation used to determine the windchill index is

$$\text{WCI} = (10.45 - V + 10\sqrt{V})(33 - T_a) \tag{8.21}$$

where

WCI = windchill index (kcal/m²·h)

V = wind speed (m/s)

T_a = ambient air temperature (°C)

and the value 33 is the body surface temperature in degrees Celsius. The more common equivalent windchill temperature $T_{\text{equivalent}}$ (°C) is given by

$$T_{\text{equivalent}} = 0.045(5.27V^{0.5} + 10.45 - 0.28V)(T_a - 33) + 33 \tag{8.22}$$

Note that in Equation (8.22) V is expressed in kilometers per hour (km/h). Table 8.5 shows the windchill temperatures for the range of ambient air temperature $-30°\text{C} < T_a < 10°\text{C}$ and wind speed of $20\text{ km/h} < V < 80\text{ km/h}$.

Radiation

All matter emits thermal ***radiation***. This rule is true as long as the body in question is at a nonzero absolute temperature. The higher the temperature of the surface of the object, the

TABLE 8.5 Windchill Temperatures

Wind Speed (km/h)	Ambient Temperature (°C)								
	10	5	0	−5	−10	−15	−20	−25	−30
20	3.6	−2.8	−9.2	−15.6	−22.0	−28.4	−34.8	−41.2	−47.6
30	1.0	−6.0	−12.9	−19.9	−26.8	−33.8	−40.7	−47.7	−54.6
40	−0.7	−8.1	−15.4	−22.7	−30.0	−37.4	−44.7	−52.0	−59.4
50	−1.9	−9.5	−17.1	−24.7	−32.2	−39.8	−47.4	−55.0	−62.6
60	−2.7	−10.4	−18.2	−25.9	−33.7	−41.5	−49.2	−57.0	−64.7
70	−3.2	−11.0	−18.9	−26.8	−34.6	−42.5	−50.3	−58.2	−66.1
80	−3.4	−11.3	−19.3	−27.2	−35.1	−43.0	−50.9	−58.8	−66.8

more thermal energy is emitted by the object. A good example of thermal radiation is the heat you can literally feel radiated by a fire in a fireplace. Simply stated, the amount of radiant energy emitted by a surface is given by the equation

$$q = \varepsilon \sigma A T_s^4 \qquad (8.23)$$

where q represents the rate of thermal energy, per unit time, emitted by the surface; ε is the emissivity of the surface, $0 < \varepsilon < 1$, and σ is the Stefan–Boltzmann constant ($\sigma = 5.67 \times 10^{-8}$ W/m^2·K^4); A represents the area of the surface in m^2, and T_s is the surface temperature of the object expressed in Kelvin. Emissivity, ε, is a property of the surface of the object, and its value indicates how well the object emits thermal radiation compared to a black body (an ideal perfect emitter). It is important to note here that unlike the conduction and convection modes, heat transfer by radiation can occur in a vacuum. A daily example of this is the radiation of the sun reaching the earth's atmosphere as it travels through a vacuum in space. Because all objects emit thermal radiation, it is the net energy exchange among the bodies that is of interest to us. Because of this fact, thermal radiation calculations are generally complicated in nature and require an in-depth understanding of the underlying concepts and geometry of the problem.

EXAMPLE 8.12

On a hot summer day, the flat roof of a tall building reaches 45°C in temperature. The area of the roof is 400 m^2. Estimate the heat radiated from this roof to the sky in the evening when the temperature of the surrounding air or sky is at 20°C. The temperature of the roof decreases as it cools down. Estimate the rate of energy radiated from the roof, assuming roof temperatures of 50, 40, 30, and 25°C. Assume $\varepsilon = 0.9$ for the roof.

We can determine the amount of thermal energy radiated by the surface from Equation (8.23):

$$q = \varepsilon \sigma A T_s^4 = (0.9)\left(5.67 \times 10^{-8}\left(\frac{\text{W}}{\text{m}^2\cdot\text{K}^4}\right)\right)(400\ \text{m}^2)(323\ \text{K})^4 = 222{,}000\ \text{W}$$

The rest of the solution is shown in Table 8.6.

TABLE 8.6 The results of Example 8.13

Surface Temperature (°C)	Surface Temperature (K) $T(K) = T(°C) + 273$	Energy Emitted by the Surface (W) $q = \varepsilon\sigma A T_s^4$
50	323	$(0.9)(5.67 \times 10^{-8})(400)(323)^4 = 222{,}000$ W
40	313	$(0.9)(5.67 \times 10^{-8})(400)(313)^4 = 196{,}000$ W
30	303	$(0.9)(5.67 \times 10^{-8})(400)(303)^4 = 172{,}000$ W
25	298	$(0.9)(5.67 \times 10^{-8})(400)(298)^4 = 161{,}000$ W

Most of you will take a heat transfer or a transport phenomenon class during your third year where you will learn in more detail about various modes of heat transfer. You will also learn how to estimate heat transfer rates for various situations, including the cooling of electronic devices and the design of fins for transformers or motorcycle and lawn mower engine heads and other heat exchangers, like the radiator in your car or the heat exchangers in furnaces and boilers. The intent of this section was to briefly introduce you to the concept of heat transfer and its various modes.

8.4 THERMAL COMFORT, METABOLIC RATE, AND CLOTHING INSULATION

Human thermal comfort is of special importance to bioengineers and mechanical engineers. For example, mechanical engineers design the heating, ventilating, and air-conditioning (HVAC) systems for homes, public buildings, hospitals, and manufacturing facilities. When sizing the HVAC systems, the engineer must design not only for the building's heat losses or gains but also for an environment that occupants feel comfortable within. What makes us thermally comfortable in an environment? As you know, the temperature of the environment and the humidity of the air are among the important factors that define thermal comfort. For example, most of us feel comfortable in a room that has a temperature of 70°F and a relative humidity of 40 to 50%. Of course, if you are exercising on a treadmill and watching TV, then perhaps you feel more comfortable in a room with a temperature lower than 70°F, say 50° to 60°F. Thus, the level of activity is also an important factor. In general, the amount of energy that a person generates depends on the person's age, gender, size, and activity level. A person's body temperature is controlled by (1) convective and radiative heat transfer to the surroundings, (2) sweating, (3) respiration by breathing surrounding air and exhaling it at near the body's temperature, (4) blood circulation near the surface of the skin, and (5) metabolic rate. Metabolic rate determines the rate of conversion of chemical to thermal energy within a person's body. The metabolic rate depends on the person's activity level. A unit commonly used to express the metabolic rate for an average person under sedentary conditions, per unit surface area, is called *met*; 1 met is equal to 58.2 W/m^2 or, in U.S. Customary Units, 1 met = 18.4 Btu/h·ft^2. For an average person, a heat transfer surface area of 1.82 m^2 or 19.6 ft^2 was assumed when defining the unit of met. Table 8.7 shows the metabolic rate for various activities. As you expect, clothing also affects thermal comfort. A unit that is generally used to express the insulating

TABLE 8.7 Typical Metabolic Heat Generation for Various Human Activities

Activity	Heat Generation (Btu/h·ft^2)	Heat Generation (met)
Resting		
Sleeping	13	0.7
Seated, quiet	18	1.0
Standing, relaxed	22	1.2
Walking on Level Surface		
2 miles/h	37	2.0
3 miles/h	48	2.6
4 miles/h	70	3.8
Office Work		
Reading, seated	18	1.0
Writing	18	1.0
Typing	20	1.1
Filing, seated	22	1.2
Filing, standing	26	1.4
Driving/Flying		
Car	18 to 37	1.0 to 2.0
Aircraft, routine	22	1.2
Aircraft, instrument landing	33	1.8
Aircraft, combat	44	2.4
Heavy vehicle	59	3.2
Miscellaneous Housework		
Cooking	29 to 37	1.6 to 2.0
House cleaning	37 to 63	2.0 to 3.4
Miscellaneous Leisure Activities		
Dancing, social	44 to 81	2.4 to 4.4
Calisthenics/exercise	55 to 74	3.0 to 4.0
Tennis, singles	66 to 74	3.6 to 4.0
Basketball	90 to 140	5.0 to 7.6
Wrestling, competitive	130 to 160	7.0 to 8.7

Source: American Society of Heating, Refrigerating, and Air-Conditioning Engineers.

value of clothing is called *clo*. 1 clo is equal to 0.155 m^2·°C/W, or, in U.S. Customary Units, 1 clo = 0.88°F·ft^2·h/Btu. Table 8.8 shows the insulating values of various clothing.

EXAMPLE 8.13

Use Table 8.7 to calculate the amount of energy dissipated by an average adult person doing the following things: (a) driving a car for 3 h, (b) sleeping for 8 h, (c) walking at the speed of 3 mph on a level surface for 2 h, (d) dancing for 2 h.

(a) Using average values, the amount of energy dissipated by an average adult person driving a car for 3 h is

$$\left(\frac{18+37}{2}\right)\left(\frac{\text{Btu}}{\text{h}\cdot\text{ft}^2}\right)(19.6\ \text{ft}^2)(3\ \text{h}) = 1617\ \text{Btu}$$

(b) Using average values, the amount of energy dissipated by an average adult person sleeping for 8 h is

$$13\left(\frac{\text{Btu}}{\text{h}\cdot\text{ft}^2}\right)(19.6\ \text{ft}^2)(8\ \text{h}) = 2038\ \text{Btu}$$

(c) Walking at the speed of 3 mph on a level surface for 2 h:

$$48\left(\frac{\text{Btu}}{\text{h}\cdot\text{ft}^2}\right)(19.6\ \text{ft}^2)(2\ \text{h}) = 1882\ \text{Btu}$$

(d) Dancing for 2 h:

$$\left(\frac{44+81}{2}\right)\left(\frac{\text{Btu}}{\text{h}\cdot\text{ft}^2}\right)(19.6\ \text{ft}^2)(2\ \text{h}) = 2450\ \text{Btu}$$

In Problem 8.23 you are asked to convert these results from Btu to calories.

TABLE 8.8 Typical Insulating Values for Clothing

Clothing	**Insulation Values (clo)** **1 clo = 0.155 m²·°C/W or** **1 clo = 0.88°F·ft²·h/Btu**
Walking shorts, short-sleeve shirt	0.36
Trousers, short-sleeve shirt	0.57
Trousers, long-sleeve shirt	0.61
Trousers, long-sleeve shirt, plus suit jacket	0.96
Same as above, plus vest and T-shirt	1.14
Sweatpants, sweatshirt	0.74
Knee-length skirt, short-sleeve shirt, panty hose, sandals	0.54
Knee-length skirt, short-sleeve shirt, full slip, panty hose	0.67
Knee-length skirt, long-sleeve shirt, half slip, panty hose, long-sleeve sweater	1.10
Same as above, replace sweater with suit jacket	1.04

Source: American Society of Heating, Refrigerating, and Air-Conditioning Engineers.

8.5 SOME TEMPERATURE-RELATED MATERIAL PROPERTIES

Thermal Expansion

As we mentioned earlier in this chapter, accounting for thermal expansion and contraction of materials due to temperature fluctuations is important in engineering problems, including the design of bridges, roads, piping systems (hot water or steam pipes), engine blocks, gas turbine blades, electronic devices and circuits, cookware, tires, and in many manufacturing processes. In general, as the temperature of a material is increased, the material will expand—increase in length—and if the temperature of the material is decreased, it will contract—decrease in length. The magnitude of this elongation or contraction due to

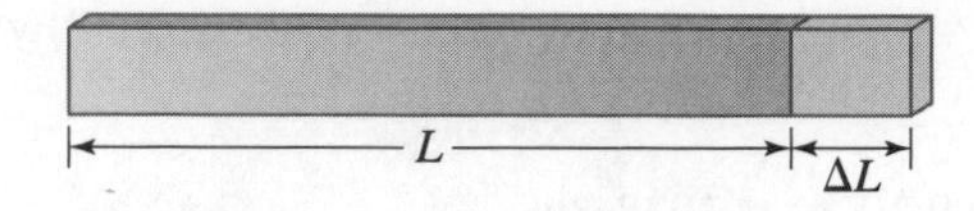
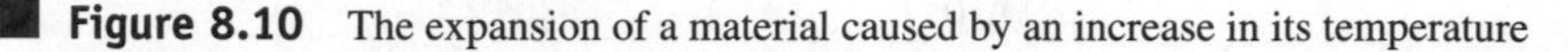

Figure 8.10 The expansion of a material caused by an increase in its temperature

temperature rise or temperature drop depends on the composition of the material. The coefficient of linear thermal expansion provides a measure of the change in length that occurs due to any temperature fluctuations. This effect is depicted in Figure 8.10.

The coefficient of linear expansion, α_L, is defined as the change in the length, ΔL, per original length L per degree rise in temperature, ΔT, and is given by the following relationship:

$$\alpha_L = \frac{\Delta L}{L\,\Delta T} \qquad (8.24)$$

Note that the coefficient of linear expansion is a property of a material and has the units of 1/°F or 1/°C. Because a 1° Fahrenheit temperature difference is equal to a 1° Rankine temperature difference, and a 1° Celsius temperature difference is equal to a 1° Kelvin temperature difference, the units of α_L can also be expressed using 1/°R and 1/K. The values of the coefficient of linear expansion itself depend on temperature; however, average values may be used for a specific temperature range. The values of the coefficient of thermal expansion for various solid materials for a temperature range of 0°C to 100°C are given in Table 8.9. Equation (8.24) could be expressed in such a way as to allow for direct calculation of the change in the length that occurs due to temperature change in the following manner:

$$\Delta L = \alpha_L L\,\Delta T \qquad (8.25)$$

For liquids and gases, in place of the coefficient of linear expansion, it is customary to define the coefficient of volumetric thermal expansion, α_V. The coefficient of volumetric expansion is defined as the change in the volume, ΔV, per original volume V per degree rise in temperature, ΔT, and is given by the following relationship:

$$\alpha_V = \frac{\Delta V}{V\,\Delta T} \qquad (8.26)$$

Note that the coefficient of volumetric expansion also has the units of 1/°F or 1/°C. Equation (8.26) could also be used for solids. Moreover, for homogeneous solid materials, the relationship between the coefficient of linear expansion and the coefficient of volumetric expansion is given by

$$\alpha_V = 3\alpha_L \qquad (8.27)$$

EXAMPLE 8.14

Calculate the change in length of a 1000-ft-long stainless steel cable when its temperature changes by 100°F.

We will use Equation (8.25) and Table 8.9 to solve this problem. From Table 8.9 the coefficient of thermal expansion for stainless steel is $\alpha = 2.9 \times 10^{-6}$ 1/°F. Using Equation (8.25), we have

$$\Delta L = \alpha_L L\,\Delta T = (2.9 \times 10^{-6} 1/°\text{F})(1000\text{ ft})(100°\text{F}) = 0.29\text{ ft} = 3.48\text{ in.}$$

TABLE 8.9 The Coefficients of Linear Thermal Expansion for Various Solid Materials (mean value over 0°C to 100°C or 32°F to 212°F)

Solid Material	Mean Value of α_L (1/°F)	Mean Value of α_L (1/°C)
Brick	5.3×10^{-6}	2.9×10^{-6}
Bronze	10.0×10^{-6}	5.5×10^{-6}
Cast iron	5.9×10^{-6}	3.3×10^{-6}
Concrete	8.0×10^{-6}	4.4×10^{-6}
Glass (plate)	5.0×10^{-6}	2.8×10^{-6}
Glass (Pyrex)	1.8×10^{-6}	1.0×10^{-6}
Glass (thermometer)	4.5×10^{-6}	2.5×10^{-6}
Masonry	$2.5 \times 10^{-6} - 5.0 \times 10^{-6}$	$1.4 \times 10^{-6} - 2.8 \times 10^{-6}$
Solder	13.4×10^{-6}	7.4×10^{-6}
Stainless steel (AISI 316)	2.9×10^{-6}	1.6×10^{-6}
Steel (hard-rolled)	5.6×10^{-6}	3.1×10^{-6}
Steel (soft-rolled)	6.3×10^{-6}	3.5×10^{-6}
Wood (oak) normal to fiber	3.0×10^{-6}	1.7×10^{-6}
Wood (oak) parallel to fiber	2.7×10^{-6}	1.5×10^{-6}
Wood (pine) parallel to fiber	3.0×10^{-6}	1.7×10^{-6}

Source: T. Baumeister et al., *Mark's Handbook.*

Specific Heat

Have you noticed that some materials get hotter than others when exposed to the same amount of thermal energy? For example, if we were to expose 1 kg of water and 1 kg of concrete to a heat source that puts out 100 J every second, you would see that the concrete would experience a higher temperature rise. The reason for this material behavior is that when compared to water, concrete has a lower heat capacity. More explanation regarding our observation will be given in Example 8.16.

Specific heat provides a quantitative way to show how much thermal energy is required to raise the temperature of a 1 kg mass of a material by 1° Celsius. Or, using U.S. Customary Units, the specific heat is defined as the amount of thermal energy required to raise the temperature of a 1-lb mass of a material by 1° Fahrenheit. The values of specific heat for various materials at constant pressure are given in Table 8.10. For solids and liquids in the absence of any phase change, the relationship among the required thermal energy ($E_{thermal}$), mass of the given material (m), its specific heat (c), and the temperature rise ($T_{final} - T_{initial}$) that will occur is given by

$$E_{thermal} = mc(T_{final} - T_{initial}) \tag{8.28}$$

where

$E_{thermal}$ = thermal energy (J or Btu)

m = mass (kg or lb_m)

c = specific heat (J/kg·K or J/kg·°C or Btu/lb_m·°R or Btu/lb_m·°F)

$T_{final} - T_{initial}$ = temperature rise (°C or K, °F or °R)

TABLE 8.10 Specific Heat (at Constant Pressure) of Some Materials at 300 K

Material	Specific Heat (J/kg·K)
Air (at atmospheric pressure)	1007
Aluminum (pure)	903
Aluminum alloy-2024-T6 (4.5% copper, 1.5 % Mg, 0.6% Mn)	875
Asphalt	920
Bronze (90% copper, 10% aluminum)	420
Brass (70% copper, 30% zinc)	380
Brick (fire clay)	960
Concrete	880
Copper (pure)	385
Glass	750
Gold	129
Iron (pure)	447
Stainless steels (AISI 302, 304, 316, 347)	480, 477, 468, 480
Lead	129
Paper	1340
Platinum (pure)	133
Sand	800
Silicon	712
Silver	235
Zinc	389
Water (liquid)	4180

Next we will look at two examples that demonstrate the use of Equation (8.28).

EXAMPLE 8.15

An aluminum circular disk with a diameter, d, of 15 cm and a thickness of 4 mm is exposed to a heat source that puts out 200 J every second. The density of the aluminum is 2700 kg/m^3. Assuming no heat loss to the surrounding, estimate the temperature rise of the disk after 15 s.

We will make use of Equation (8.28) and Table 8.10 to solve this problem, but first we need to calculate the mass of the disk using the information given.

$$\text{mass} = m = (\text{density})\,(\text{volume})$$

$$\text{Volume} = V = \frac{\pi}{4}d^2\,(\text{thickness}) = \frac{\pi}{4}(0.15\text{ m})^2(0.004\text{ m}) = 7.06858 \times 10^{-5}\text{ m}^3$$

$$m = (7.06858 \times 10^{-5}\text{ m}^3)(2700\text{ kg/m}^3) = 0.191\text{ kg}$$

$$E_{\text{thermal}} = mc(T_{\text{final}} - T_{\text{inital}})$$

$$200\text{ J} = (0.191\text{ kg})(875\text{ J/kg}\cdot\text{K})(T_{\text{final}} - T_{\text{initial}})$$

$$(T_{\text{final}} - T_{\text{initial}}) = 1.2\text{ K}$$

And after 15 s the temperature rise will be 18°C or 18 K.

EXAMPLE 8.16

We have exposed 1 kg of water, 1 kg of brick, and 1 kg of concrete, each to a heat source that puts out 100 J every second. Assuming that all of the supplied energy goes to each material and they were all initially at the same temperature, which one of these materials will have a greater temperature rise after 10 s?

We can answer this question using Equation (8.28) and Table 8.10. We will first look up the values of the specific heat for water, brick, and concrete, which are $c_{water} = 4180$ J/kg·K, $c_{brick} = 960$ J/kg·K, and $c_{concrete} = 880$ J/kg·K. Now applying Equation (8.28), $E_{thermal} = mc(T_{fiina} - T_{initial})$ to each situation, it should be clear that although each material has the same amount of mass and is exposed to the same amount of thermal energy, the concrete will experience a higher temperature rise because it has the lowest heat capacity value among the three given materials.

8.6 HEATING VALUES OF FUELS

As engineers you need to know what the heating value of a fuel means. Why? Where does the energy that drives your car come from? Where does the energy that makes your home warm and cozy during the cold winter months come from? How is electricity generated in a conventional power plant that supplies power to manufacturing companies, homes, and offices? The answer to all of these questions is that the initial energy comes from fuels. Most conventional fuels that we use today to generate power come from coal, natural gas, oil, or gasoline. All these fuels consist of carbon and hydrogen.

When a fuel is burned, whether it is gas, oil, etc., thermal energy is released. The heating value of a fuel quantifies the amount of energy that is released when a unit mass (kilogram or pound) or a unit volume (cubic meter or cubic foot) of a fuel is burned. Different fuels have different heating values. Moreover, based on the phase of water in the combustion product, whether it is in the liquid form or in vapor form, two different heating values are reported. The higher heating value of a fuel, as the name implies, is the higher end of energy released by the fuel when the combustion by-products include water in liquid form. The lower heating value refers to the amount of energy that is released during combustion when the combustion products include water vapor. The typical heating values of liquid fuels, coal, and natural gas are given in Tables 8.11 through 8.13.

TABLE 8.11 Typical Heating Values of Standard Grades of Fuel Oil and Gasoline

Grade No.	Density (lb/gal)	Heating Value (Btu/gal)
1	6.950 to 6.675	137,000 to 132, 900
2	7.296 to 6.960	141,800 to 137,000
4	7.787 to 7.396	148,100 to 143,100
5L	7.940 to 7.686	150,000 to 146,800
5H	8.080 to 7.890	152,000 to 149,400
6	8.448 to 8.053	155,900 to 151,300
Gasoline	6.0 to 6.2	108,500 to 117,000

Source: American Society of Heating, Refrigerating, and Air-Conditioning Engineers.

TABLE 8.12 The Typical Heating Value of Some Coal

Coal from County and State of:	Higher Heating Value (Btu/lb_m)
Musselshell, Montana	12,075
Emroy, Utah	13,560
Pike, Kentucky	15,040
Cambria, Pennsylvania	15,595
Williamson, Illinois	13,710
McDowell, West Virginia	15,600

Source: Babcock and Wilcox, *Steam: Its Generation and Use.*

TABLE 8.13 The Typical Higher Heating Value of Natural Gas

Source of Gas	Heating Value (Btu/lb_m)	Heating Value (Btu/ft^3) @ 60°F and 30 in·Hg
Pennsylvania	23,170	1129
Southern California	22,904	1116
Ohio	22,077	964
Louisiana	21,824	1002
Oklahoma	20,160	974

Source: Babcock and Wilcox, *Steam: Its Generation and Use.*

EXAMPLE 8.17

How much thermal energy is released when 5 lb of a coal sample from Emroy, Utah is burned?

The total amount of thermal energy, $E_{thermal}$, released when some fuel is burned is determined by multiplying the mass of the fuel, m, by the heating value of fuel, H_V.

$$E_{thermal} = mH_V = (5\ lb_m)\ 13{,}560\left(\frac{Btu}{lb_m}\right) = 67{,}800\ Btu$$

EXAMPLE 8.18

Calculate the total amount of thermal energy released when 60 ft^3 of natural gas from Louisiana is burned inside a gas furnace in 1 h.

$$E_{thermal} = (60\ ft^3)\ 1002\left(\frac{Btu}{ft^3}\right) = 60{,}120\ Btu$$

SUMMARY

Now that you have reached this point in the text

- You should have a good understanding of what temperature means, how it is measured, and how heat transfer occurs.

- You should know the difference among temperature scales and understand how they are related.

$$T(°C) = \frac{5}{9}[T(°F) - 32] \qquad T(°F) = \frac{9}{5}T(°C) + 32$$

$$T(K) = T(°C) + 273.15 \qquad T(°R) = T(°F) + 459.67$$

$$T(K) = \frac{5}{9}T(°R) \qquad T(°R) = \frac{9}{5}T(K)$$

- You should understand what absolute zero temperature means.
- You should know what we mean by the term *heat*, know its common units, and be familiar with different modes of heat transfer.
- You should know what the *R*-value or *R*-factor for insulation means.
- You should be able to perform some simple heat transfer calculations.
- You should be familiar with some thermophysical properties of materials such as thermal conductivity, specific heat, and thermal expansion.
- You should understand what the heating value of a fuel means.

PROBLEMS

1. Investigate the value of temperature for the following items. Write a brief report discussing how these values are used in their respective areas.
   a. What is a normal body temperature?
   b. What is the temperature range that clinically is referred to as fever?
   c. What is a normal surface temperature of your body? Is this value constant?
   d. What is a comfortable room temperature range? What is its significance in terms of human thermal comfort? What is the role of humidity?
   e. What is the operating temperature range of the condenser in a household refrigerator? The condenser section of a household refrigerator is the black tubing in the back of the refrigerator where heat is being rejected to the surrounding air.
2. Using Excel, or a spreadsheet of your choice, create a degrees Fahrenheit to degrees Celsius conversion table for the following temperature range: from −50°F to 130°F in increments of 5°F.
3. Alcohol thermometers can measure temperatures in the range of −100°F to 200°F. Determine the temperature at which an alcohol thermometer with a Fahrenheit scale will read the same number as a thermometer with a Celsius scale.
4. Obtain information about K-, E-, and R-type thermocouple wires. Write a brief report discussing their accuracy, temperature range of application, and in what application they are commonly employed.
5. A manufacturer of loose-fill cellulose insulating material provides a table showing the relationship between the thickness of the material and its *R*-value. The manufacturer's data is shown in the accompanying table.

R-value (units?)	Thickness (in.)
R-40	11
R-32	9
R-24	6.5
R-19	5.25
R-13	3.5

   Calculate the thermal conductivity of the insulating material. Also, determine how thick the insulation should be to provide *R*-values of
   a. R-30
   b. R-20
6. Calculate the *R*-value for the following materials:
   a. 4-in.-thick brick
   b. 10-cm-thick brick
   c. 12-in.-thick concrete slab

d. 20-cm-thick concrete slab
e. 1-cm-thick human fat layer

7. Calculate the thermal resistance due to convection for the following situations:
a. warm water with $h = 200$ W/m^2·K
b. warm air with $h = 10$ W/m^2·K
c. warm moving air (windy situation) $h = 30$ W/m^2·K

8. A typical exterior masonry wall of a house, shown in the accompanying figure, consists of the items in the accompanying table. Assume an inside room temperature of 68°F and an outside air temperature of 10°F, with an exposed area of 150 ft^2. Calculate the heat loss through the wall.

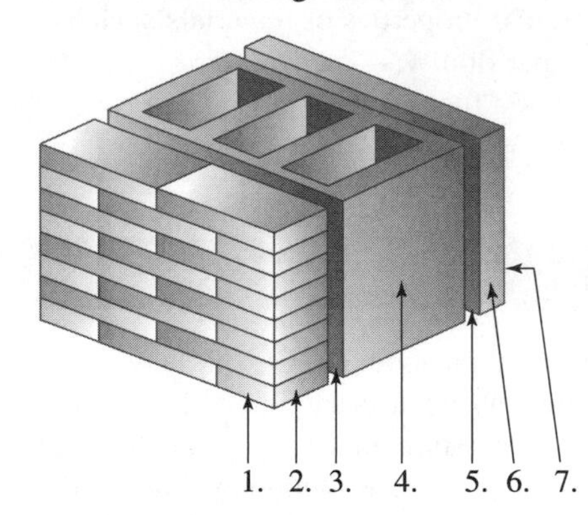

Items	Resistance (h·ft^2·F/Btu)
1. Outside film resistance (winter, 15 mph wind)	0.17
2. Face brick (4 in.)	0.44
3. Cement mortar (1/2 in.)	0.1
4. Cinder block (8 in.)	1.72
5. Air space (3/4 in.)	1.28
6. Gypsum wallboard (1/2 in.)	0.45
7. Inside film resistance (winter)	0.68

9. In order to increase the thermal resistance of a typical exterior frame wall, such as the one shown in Example 8.11, it is customary to use 2 × 6 studs instead of 2 × 4 studs to allow for placement of more insulation within the wall cavity. A typical exterior (2 × 6) frame wall of a house consists of the materials shown in the accompanying figure. Assume an inside room temperature of 68°F and an outside air temperature of 20°F with an exposed area of 150 ft^2. Determine the heat loss through this wall.

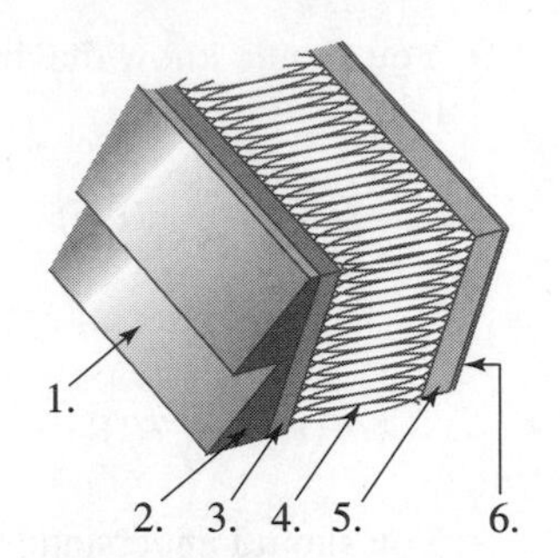

Items	Resistance (h·ft^2·°F/Btu)
1. Outside film resistance (winter, 15 mph wind)	0.17
2. Siding, wood (1/2 × 8 lapped)	0.81
3. Sheathing (1/2 in. regular)	1.32
4. Insulation batt ($5\frac{1}{2}$ in.)	19.0
5. Gypsum wallboard (1/2 in.)	0.45
6. Inside film resistance (winter)	0.68

10. A typical ceiling of a house consists of items shown in the accompanying table. Assume an inside room temperature of 70°F and an attic air temperature of 15°F, with an exposed area of 1000 ft^2. Calculate the heat loss through the ceiling.

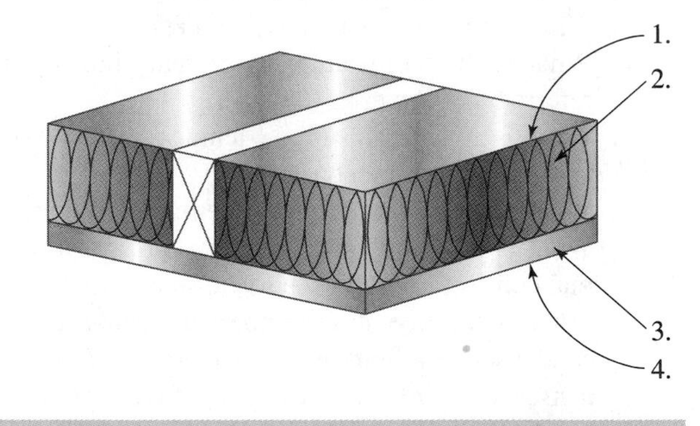

Items	Resistance (h·ft^2·°F/Btu)
1. Inside attic film resistance	0.68
2. Insulation batt (6 in.)	19.0
3. Gypsum wallboard (1/2 in.)	0.45
4. Inside film resistance (winter)	0.68

11. Estimate the change in the length of a power transmission line in your state when the temperature changes by 50°F. Write a brief memo to your instructor discussing your findings.

12. Calculate the change in 5-m-long copper wire when its temperature changes by 130°F.
13. Determine the temperature rise that would occur when 2 kg of the following materials are exposed to a heating element putting out 500 J. Discuss your assumptions.
    a. The material is copper.
    b. The material is aluminum.
    c. The material is concrete.
14. The thermal conductivity of a solid material can be determined using a setup similar to the one shown in the accompanying figure. The thermocouples are placed at 2.5-cm intervals in the known material (copper) and the unknown sample, as shown. The known material is heated on the top by a heating element, and the bottom surface of the sample is cooled by running water through the heat sink shown. Determine the thermal conductivity of the unknown sample for the set of data given in the accompanying table. Assume no heat loss to the surroundings and perfect thermal contact at the common interface of the sample and the copper.

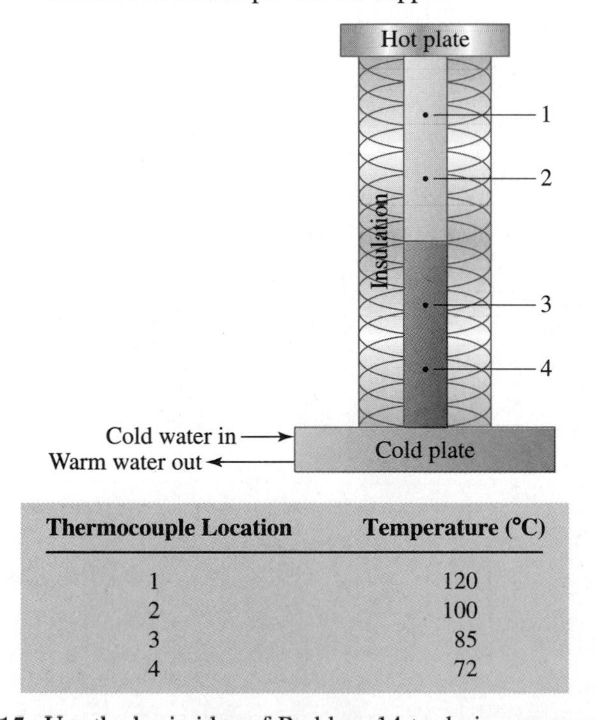

Thermocouple Location	Temperature (°C)
1	120
2	100
3	85
4	72

15. Use the basic idea of Problem 14 to design an apparatus that could be constructed to measure the thermal conductivity of solid samples in the range of 50 to 300 W/m·K. Write a brief report discussing in detail the overall size of the apparatus; components of the apparatus, including the heating source, the cooling source, insulation materials; and the measurement elements. In the report include drawings and a sample experiment. Estimate the cost of such a setup and briefly discuss it in your report. Also, write a brief experimental procedure that could be used by someone who is unfamiliar with the apparatus.
16. A copper plate, with dimensions of 3 cm × 3 cm × 5 cm (length, width, and thickness, respectively), is exposed to a thermal energy source that puts out 150 J every second, as shown in the accompanying figure. The density of copper is 8900 kg/m^3. Assuming no heat loss to the surrounding block, determine the temperature rise in the plate after 10 seconds.

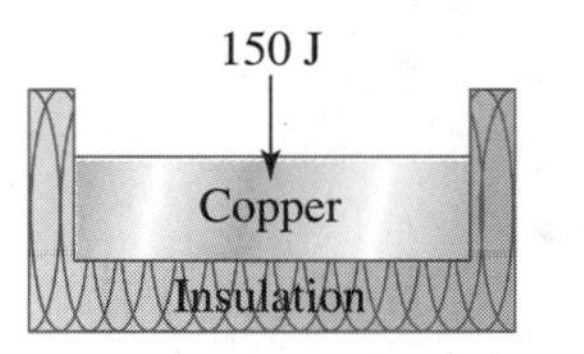

17. An aluminum plate, with dimensions of 3 cm × 3 cm × 5 cm (length, width, and thickness, respectively), is exposed to a thermal energy source that puts out 150 J every second as shown in the accompanying figure. The density of aluminum is 2700 kg/m^3. Assuming no heat loss to the surrounding block, determine the temperature rise in the plate after 10 seconds.
18. Use the basic idea of Problem 16 to design an apparatus that could be constructed to measure the heat capacity of solid samples in the range of 500 to 800 W/kg·K. Write a brief report discussing in detail the overall size of the apparatus; components of the apparatus, including the heating source, the supporting block, insulation materials; and the measurement elements. In the report include drawings and a sample experiment. Estimate the cost of such a setup, and briefly discuss it in your report. Also, write a brief experimental procedure that could be used by someone who is unfamiliar with the apparatus.
19. How would you use the principle given in Problem 16 to measure the heat output of something? For example, use the basic idea behind Problem 16 to design a setup that can be used to measure the heat output of an ironing press.

20. Calorimeters are devices that are commonly used to measure the heating value of fuels. For example, a Junkers flow calorimeter is used to measure the heating value of gaseous fuels. A bomb calorimeter, on the other hand, is used to measure the heating value of liquid or solid fuels, such as kerosene, heating oil, or coal. Perform a search to obtain information about these two types of calorimeters. Write a brief report discussing the principles behind the operation of these calorimeters.

21. Refer to Tables 8.12 and 8.13 to answer this question. What is the maximum amount of energy released when a 10-lb_m sample of coal from McDowell, West Virginia is burned? Also calculate the amount of energy released when 15 ft^3 of natural gas from Oklahoma is burned.

22. Contact the natural gas provider in your city and find out how much you are being charged for each ft^3 of natural gas. Also, contact your electric company and determine how much they are charging on average per kWh usage of electricity. If a hot-air gas furnace has an efficiency of 94% and an electric heater has an efficiency of 100%, what is the more economical way of heating your home: a gas furnace or an electric heater?

23. Convert the results of Example 8.13 from Btu to calories.

CHAPTER

9 Electricity

Engineers understand the importance of electricity and electrical power and the role they play in our everyday lives. As future engineers you should know what is meant by *voltage, electric current,* and know the difference between *direct* and *alternating current*. You should also know the various sources of electricity and understand how electricity is generated.

Every engineer needs to understand the fundamentals of electricity and magnetism. Look around and you will see all the devices, appliances, and machines that are driven by electrical power. The objective of this chapter is to introduce the fundamentals of electricity. We will briefly discuss what is meant by electric charge, electric current (both alternating current, ac, and direct current, dc), electrical resistance, and voltage. We will also define what is meant by an electrical circuit and its components. We will then look at the role of electric motors in our everyday lives and identify the factors that engineers consider when selecting a motor for a specific application. ■

9.1 ELECTRIC CURRENT AS A FUNDAMENTAL DIMENSION

As explained in Chapter 3, based on our understanding of our physical world today, we need *seven fundamental* or *base dimensions* to correctly express the physical laws that govern our world. They are *length, mass, time, temperature, electric current, amount of substance*, and *luminous intensity*. Moreover, with the help of these base dimensions we can derive all other necessary physical quantities that describe how nature works.

In the previous chapters, we discussed the role of the fundamental dimension length and length-related parameters such as area and volume, the fundamental dimension mass and mass-related parameters, the fundamental dimension time and time-related parameters, and the fundamental dimension temperature and temperature-related parameters in engineering analysis and design. We now turn our attention to the fundamental dimension *current*. As explained in Chapter 3, it was not until 1946 that the proposal for *ampere* as a base unit for electric current was approved by the General Conference on Weights and Measures (CGPM). In 1954, CGPM included *ampere* among the base units. The ***ampere*** is defined formally as that constant current which, if maintained in two straight parallel conductors of infinite length, of negligible circular cross section, and placed 1 meter apart in a vacuum, would produce between these conductors a force equal to 2×10^{-7} newton per meter of length.

To better understand what the ampere represents, we need to take a closer look at the behavior of material at the subatomic level. In Chapter 6 we explained what is meant by atoms and molecules. An atom has three major subatomic particles, namely, electrons, protons, and neutrons. Neutrons and protons form the nucleus of an atom. How a material conducts electricity is influenced by the number and the arrangement of electrons. Electrons have negative charge, whereas protons have a positive charge, and neutrons have no charge.

Simply stated, the basic law of electric charges states that *unlike charges attract each other while like charges repel.* In SI units, the unit of charge is the coulomb (C). One coulomb is defined as the amount of charge that passes a point in a wire in 1 second when a current of 1 ampere is flowing through the wire. In Chapter 7, we explained the universal law of gravitational attraction between two masses. Similarly, there exists a law that describes the attractive electric force between two opposite-charge particles. The electric force exerted by one point charge on another is proportional to the magnitude of each charge and is inversely proportional to the square of the distance between the point charges. Moreover, the electric force is attractive if the charges have opposite signs, and it is repulsive if the charges have the same sign. The electric force between two point charges is given by Coulomb's law:

$$F_{12} = \frac{kq_1q_2}{r^2} \tag{9.1}$$

where $k = 8.99 \times 10^9\ \text{N·m}^2/\text{C}^2$, q_1 and q_2 are the point charges, and r is the distance between them. Another important fact that one must keep in mind is that the electric charge is conserved, meaning the electric charge is not created nor destroyed, it can only be transferred from one object to another.

You may already know that in order for water to flow through a pipe, a pressure difference must exist. Moreover, the water flows from the high-pressure region to the lower-pressure region. In Chapter 8 we also explained that whenever there is a temperature difference in a medium, or between bodies, thermal energy flows from the high-temperature region to the low-temperature region. In a similar way, whenever there exists a difference in electric potential between two bodies, electric charge will flow from the higher electric potential to the lower potential region. The flow of charge will occur when the two bodies are connected by an electrical conductor such as a copper wire. The flow of electric charge is called ***electric current*** or simply, ***current***. The electric current, or the flow of charge, is measured in amperes. One ampere or "amp" (A) is defined as the flow of 1 unit of charge per second. For example, a toaster that draws 6 amps has 6 units of charge flowing through the heating element each second. The amount of current that flows through an electrical element depends on the electrical potential, or voltage, available across the element and the resistance the element offers to the flow of charge.

9.2 VOLTAGE

Voltage represents the amount of work required to move charge between two points, and the amount of charge that is moving between the two points per unit time is called *current*. ***Electromotive force*** (***emf***) represents the electric potential difference between an area with an excess of free electrons (negative charge) and an area with an electron deficit (positive charge). The voltage, or the electromotive force, induces current to flow in a circuit. The most common sources of electricity are chemical reaction, light, and magnetism.

Batteries

All of you have used batteries for different purposes at one time or another. In all batteries, electricity is produced by the chemical reaction that takes place within the battery. When a device that uses batteries is on, its circuits create paths for the electrons to flow through. When the device is turned off, there is no path for the electrons to flow, thus the chemical reaction stops.

A battery cell consists of chemical compounds, internal conductors, positive and negative connections, and the casing. Examples of cells include sizes N, AA, AAA, C, and D. A cell that cannot be recharged is called a *primary cell*. An alkaline battery is an example of a primary cell. On the other hand, a *secondary cell* is a cell that can be recharged. The recharging is accomplished by reversing the current flow from the positive to the negative areas. Lead acid cells in your car battery and nickel-cadmium (NiCd), and nickel-metal

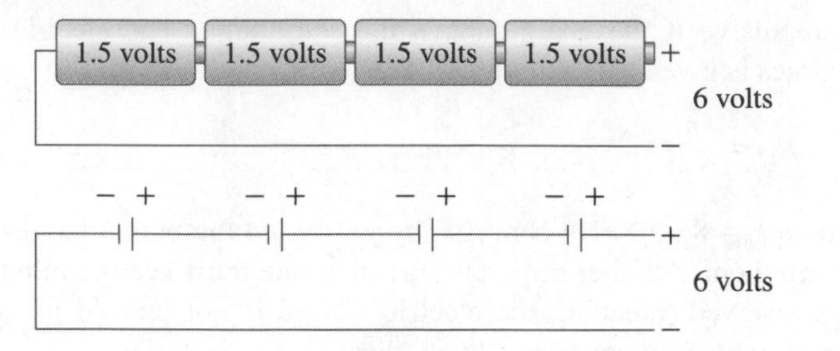

Figure 9.1
Batteries connected in a series arrangement

hydride (NiMH) cells are examples of secondary cells. The NiCd batteries are some of the most common rechargeable batteries used in cordless phones, toys, and some cellular phones. The NiMH batteries, which are smaller, are used in many smaller cellular phones because of their size and capacity.

To increase the voltage output, batteries are often placed in a series arrangement. If we connect batteries in a series arrangement, the batteries will produce a net voltage, which will be the sum of the individual batteries placed in series. For example, if we were to connect four 1.5-volt batteries in series, the resulting potential will be 6 volts, as shown in Figure 9.1. Batteries connected in a parallel arrangement, as shown in Figure 9.2, produce the same voltage, but more current.

Photoemission

Photoemisssion is another principle used to generate electricity. When light strikes a surface that has certain properties, electrons can be freed; thus electric power is generated. You may have seen examples of photovoltaic devices such as light meters used in photography, photovoltaic cells in hand-held calculators, and solar cells used in remote areas to generate electricity.

Photovoltaic devices are becoming increasingly common in many applications because they do not pollute the environment. In general, there are two ways in which the sun's radiation is converted to electricity: photothermal and photovoltaic. In photothermal plants, solar radiation (the sun's radiation) is used to make steam by heating water flowing through a pipe. The pipe runs through a parabolic solar collector, and then the steam is used to drive a generator. In a photovoltaic cell, the light is converted directly to electricity. Photovoltaic cells, which are made of gallium arsenide, have conversion efficiencies of approximately 20%. The cells are grouped together in many satellites to create an array

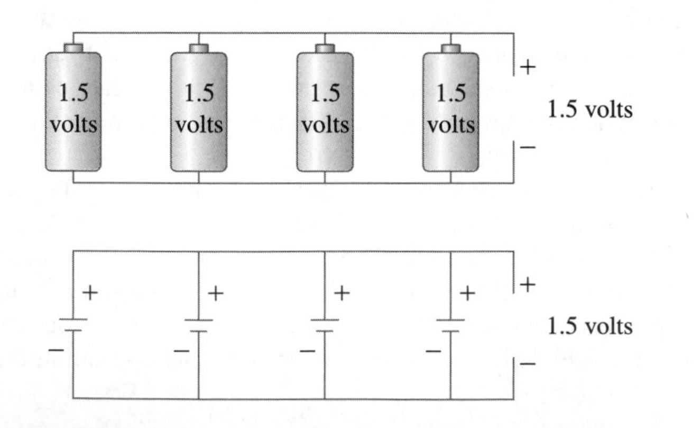

Figure 9.2
Batteries connected in a parallel arrangement

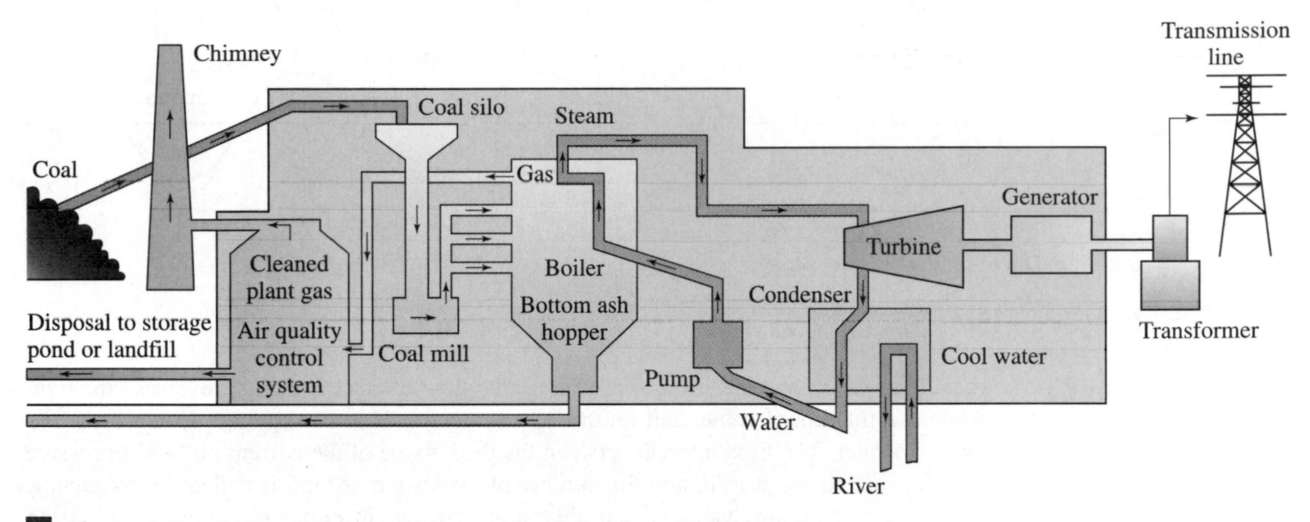

Figure 9.3 A schematic of a steam power plant (*Courtesy of Xcel Energy*).

that is used to generate electric power. Photovoltaic solar farms are also becoming common. A solar farm consists of a vast area where a great number of solar arrays are put together to convert the sun's radiation into electricity.

Power Plants

Electricity that is consumed at homes, schools, malls, and by various industries is generated in a power plant. Water is used in all steam power-generating plants to produce electricity. A simple schematic of a power plant is shown in Figure 9.3. Fuel is burned in a boiler to generate heat, which in turn is added to liquid water to change its phase to steam; steam passes through turbine blades, turning the blades, which in effect runs the generator connected to the turbine, creating electricity. The electricity is generated by turning a coil of wire inside a magnetic field. A conductor placed in a changing magnetic field will have a current induced in it. Magnetism is the most common method for generating electricity. The low-pressure steam leaving the turbine liquefies in a condenser and is pumped through the boiler again, closing a cycle, as shown in Figure 9.3.

9.3 DIRECT CURRENT AND ALTERNATING CURRENT

Direct current (***dc***) is the flow of electric charge that occurs in one direction, as shown in Figure 9.4(a). Direct current is typically produced by batteries and direct current generators. In the late 19th century, given the limited understanding of fundamentals and technology and for economic reasons, direct current could not be transmitted over long distances. Therefore, it was succeeded by alternating current (ac). Direct current was not economically feasible to transform because of the high voltages needed for long-distance transmission. However, developments in the 1960s have led to techniques that now allow the transmission of direct current over long distances.

Alternating current (***ac***) is the flow of electric charge that periodically reverses. As shown in Figure 9.4 (b), the magnitude of the current starts from zero, increases to a

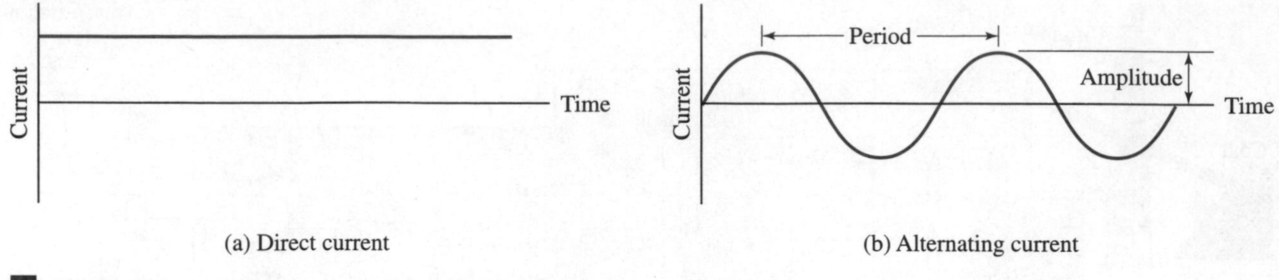

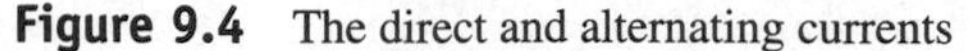

Figure 9.4 The direct and alternating currents

maximum value, and then decreases to zero; the flow of electric charge reverses direction, reaches a maximum value, and returns to zero again. This flow pattern is repeated in a cyclic manner. The time interval between the peak value of the current on two successive cycles is called the *period*, and the number of cycles per second is called the *frequency*. The peak (maximum) value of the alternating current in either direction is called the *amplitude*. Alternating current is created by generators at power plants. The current drawn by various electrical devices at your home is alternating current. The alternating current in domestic and commercial power use is 60 cycles per second (hertz) in the United States.

Kirchhoff's Current Law

Kirchhoff's current law is one of the basic laws in electricity that allows for the analysis of currents in electrical circuits. The law states that at any given time, the sum of the currents entering a node must be equal to the sum of the current leaving the node. This statement is demonstrated in Figure 9.5. As explained earlier, physical laws are based on observations, and the Kirchhoff's current law is no exception. This law represents the physical fact that charge is always conserved in an electrical circuit. The charge cannot accumulate or deplete at an electrical node; consequently, the sum of the currents entering a node must equal the sum of the currents leaving the node. For example, Kirchhoff's current law applied to the circuit shown in Figure 9.5 leads to: $i_1 + i_2 = i_3 + i_4$.

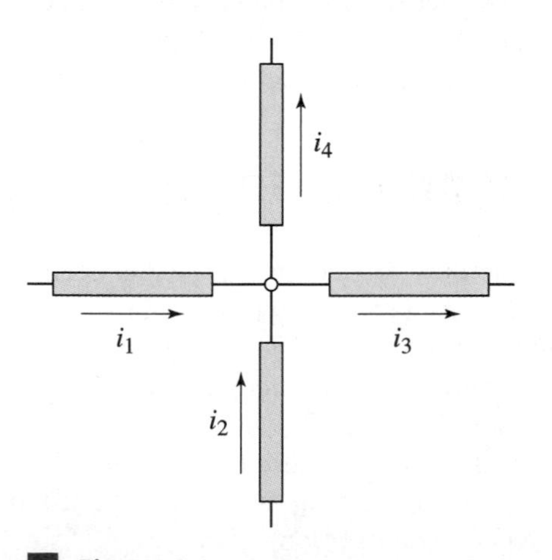

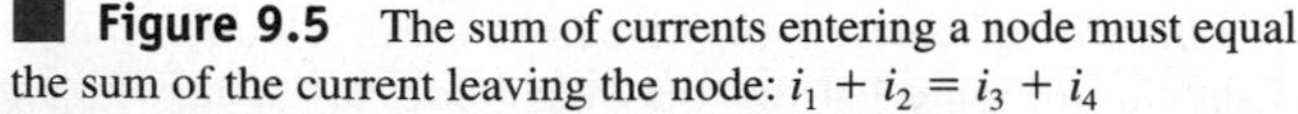

Figure 9.5 The sum of currents entering a node must equal the sum of the current leaving the node: $i_1 + i_2 = i_3 + i_4$

EXAMPLE 9.1

Determine the value of current i_3 in the circuit shown in Figure 9.6.

N

$i_1 = 3$ A $\quad i_3 = ?$

$i_2 = 1$ A

 **Figure 9.6** The circuit for Example 9.1

We can solve this simple problem by applying Kirchhoff's current law to node N. This leads to:

$$i_1 = i_2 + i_3$$

$$i_3 = 2A$$

Residential Power Distribution

An example of a typical residential power distribution system is shown in Figure 9.7, which gives examples of amperage requirements for outlets, lights, kitchen appliances,

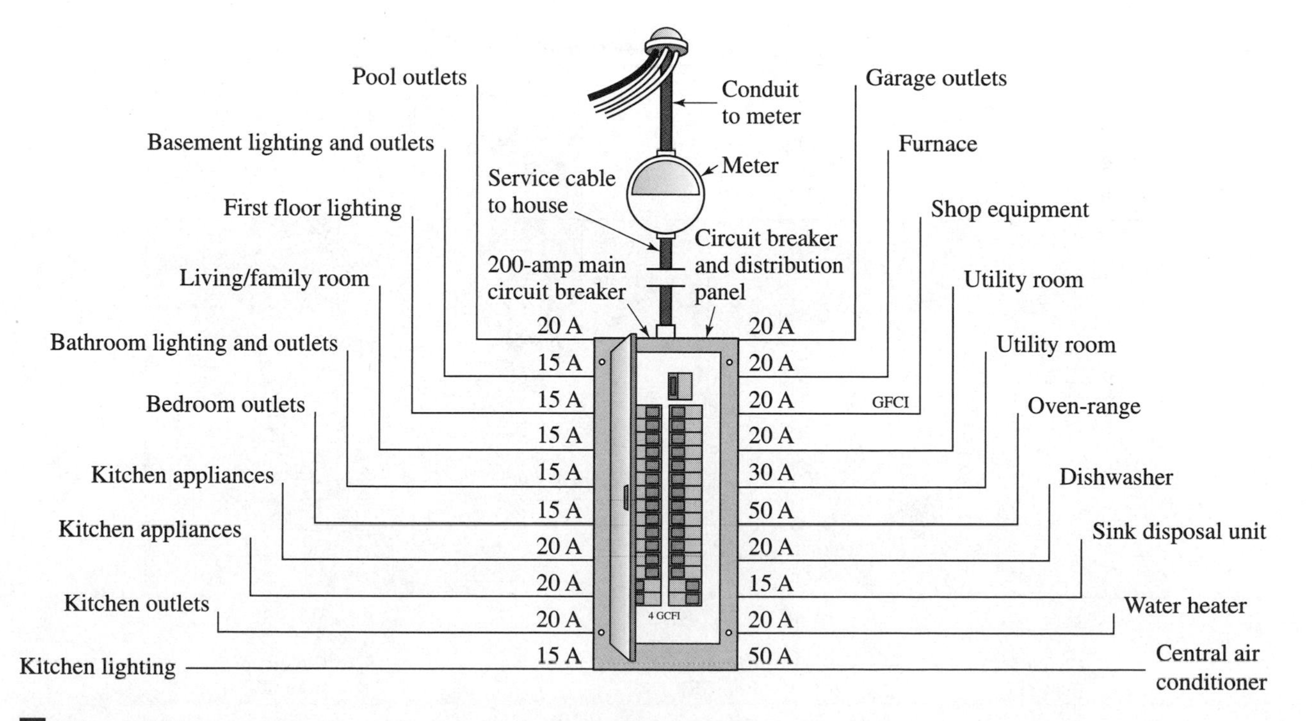

Figure 9.7 An example of an electrical distribution system for a residential building
Source: AAVIM, *Electrical Wiring*.

Common Electrical Symbols

Symbol	Description	Symbol	Description
	Ceiling outlet		Service entrance panel
	Wall outlet	S	Single-pole switch
PS	Ceiling outlet with pull switch	$S_{\|2}$	Double-pole switch
PS	Wall outlet with pull switch	S_3	3-way switch
	Duplex convenience outlet	S_4	4-way switch
WP	Weatherproof convenience outlet	S_P	Switch with pilot light
1,|3	Convenience outlet 1 = single 3 = triple		Push button
R	Range outlet	CH	Bell or chimes
S	Convenience outlet with switch		Telephone
D	Dryer outlet	TV	Television outlet
	Split-wired duplex outlet	S	Switch wiring
	Special-purpose outlet		Fluorescent ceiling fixture
D	Electric door opener		Fluorescent wall fixture

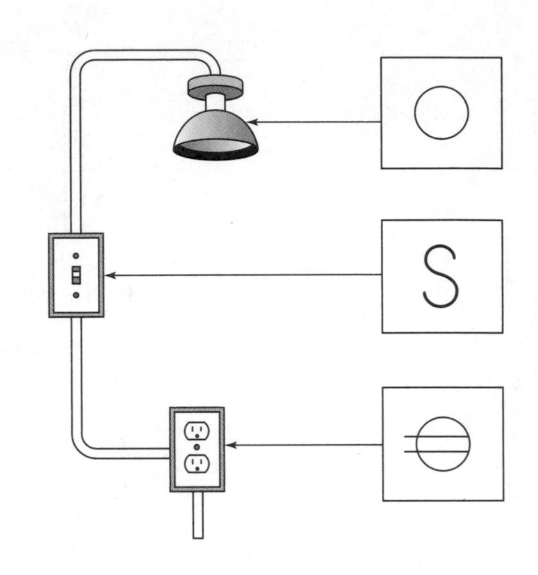

Figure 9.8 Examples of electrical symbols in a house plan
Source: AAVIM, *Electrical Wiring.*

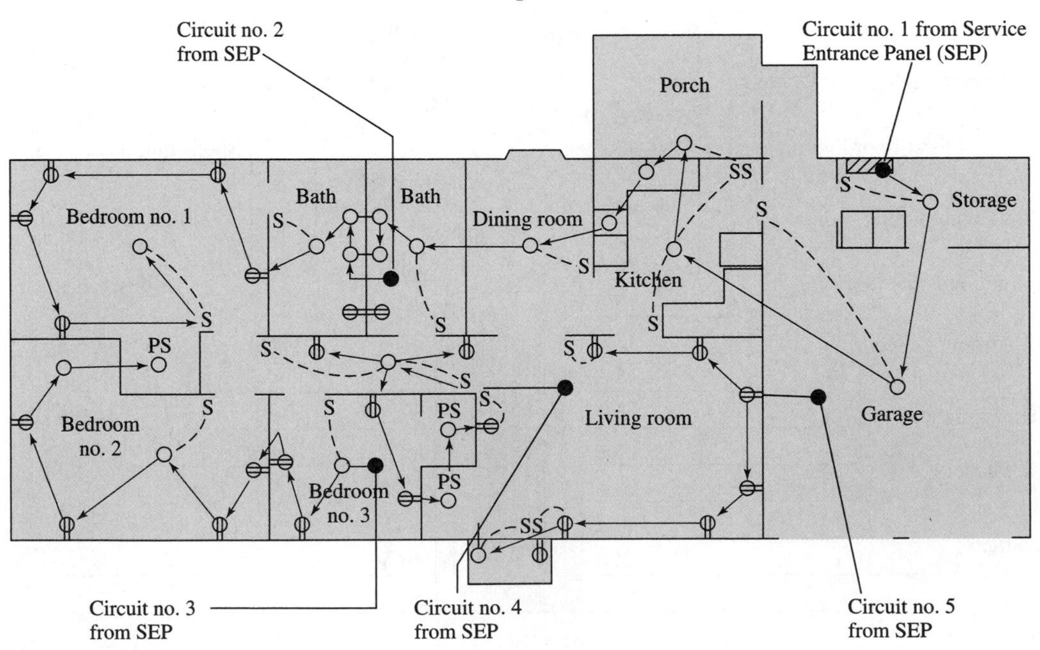

Figure 9.9 An example of an electrical plan for a house
Source: AAVIM, *Electrical Wiring.*

and central air conditioning. To wire a building, an electrical plan for the building must first be developed. In the plan the location and the types of switches and outlets, including outlets for the range and dryer, must be specified. We will discuss engineering symbols in more detail in Chapter 13. For now, examples of electrical symbols used in a house plan are shown in Figure 9.8. Moreover, an example of an electrical plan for a residential building is shown in Figure 9.9.

9.4 ELECTRICAL CIRCUITS AND COMPONENTS

An ***electrical circuit*** refers to the combination of various electrical components that are connected together. Examples of electrical components include wires (conductors), switches, outlets, resistors, and capacitors. First, let us take a closer look at electrical wires. In a wire the resistance to electrical current depends on the material from which the wire is made and its length, diameter, and temperature. Materials show varying amount of resistance to the flow of electric current. ***Resistivity*** is a measure of resistance of a piece of material to electric current. The resistivity values are usually measured using samples made of a 1-cm cube or a cylinder having a diameter of 1 mil and length of 1 ft. The resistance of the sample is then given by

$$R = \frac{\rho \ell}{a} \tag{9.2}$$

where ρ is the resistivity, ℓ is the length of the sample, and a is the cross-sectional area of the sample. The resistance for 1-ft-long wire having a diameter of 1 mil made of various materials is given in Table 9.1. The electrical resistance of a material varies with temperature. In general, the resistance of conductors (except for carbon) increases with increasing temperature. Some materials exhibit near-zero resistance at very low temperatures (temperatures approaching absolute zero). This behavior is commonly referred to as *superconductivity*.

TABLE 9.1 Electrical Resistance for 1-ft-long Wire Made of Various Metals Having a Diameter of 1 mil at 20°C

Metal	Resistance (Ω)
Aluminum	17.01
Brass	66.80
Copper	10.37
Gold	14.7
Iron	59.9
Lead	132.0
Nickel	50.8
Platinum	63.8
Silver	9.8
Tin	70.0
Tungsten	33.2
Zinc	35.58

Ohm's Law

Ohm's law describes the relationship among voltage, V, resistance, R, and current, I, according to

$$V = RI \tag{9.3}$$

Note from Ohm's law, Equation (9.3), that current is directly proportional to voltage and inversely proportional to resistance. As electric potential is increased, so is the current; and if the resistance is increased, the current will decrease. The electric resistance is measured in units of ohms. An element with 1 ohm resistance allows a current flow of 1 amp when there exists a potential of 1 volt across the element. Stated another way, when there exists an electrical potential of 1 volt across a conductor with a resistance of 1 ohm, then 1 ampere of electric current will flow through the conductor.

The American Wire Gage (AWG)

Electrical wires are typically made of copper or aluminum. The actual size of wires is commonly expressed in terms of gage number. The American Wire Gage (AWG) is based on successive gage numbers having a constant ratio of approximately 1.12 between their

TABLE 9.2 Examples of American Wire Gage (AWG) for Solid Copper Wire

American Wire Gauge (AWG) Number	Diameter (mils)	Resistance per 1000 ft (ohms) @ 77°F
0000	460.0	0.0500
000	410.0	0.0630
00	365.0	0.0795
1	289.0	0.126
2	258.0	0.159
5	182.0	0.319
6	162.0	0.403
7	144.0	0.508
10	91.0	1.28
12	81.0	1.62
14	64.0	2.58
16	51.0	4.09
18	40.0	6.51
20	32.0	10.4
22	25.3	16.5
24	20.1	26.2
26	15.9	41.6
28	12.6	66.2
30	10.0	133
32	8.0	167
34	6.3	266
36	5.0	423
38	4.0	673
40	3.1	1070

diameters. For example, the ratio of the diameter of No. 1 AWG wire to No. 2 is 1.12 (289 mils/258 mils = 1.12), or, as another example, the ratio of diameter of No. 0000 to No. 000 is 1.12 (460 mils/410 mils = 1.12). Moreover, the ratio of cross sections of successive gage numbers is approximately $(1.12)^2 = 1.25$. The ratio of the diameter of wires differing by 6 gage numbers is approximately 2.0. For example, the ratio of diameters of wire No. 1 to No. 7 is 2 (289 mils/144 mils ≈ 2.0), or the ratio of diameters of No. 30 to No. 36 is 2 (10.0 mils/5.0 mils = 2.0). Table 9.2 shows the gage number, the diameter, and the resistance for copper wires. When examining Table 9.2, note that the smaller the gage number, the bigger the wire diameter. Also note from data given in Table 9.2 that the electric resistance of a wire is increased as its diameter is decreased.

The National Electrical Code, published by the Fire Protection Association, contains specific information on the type of wires used for general wiring. The code describes the wire types, maximum operating temperatures, insulating materials, outer cover sheaths, the type usage, and the specific location where a wire should be used.

Electric Power

The electric power consumption of various electrical components can be determined using the following power formula:

$$P = VI \tag{9.4}$$

where P is power in watts, V is the voltage, and I is the current in amps. The kilowatt hour units are used in measuring the rate of consumption of electricity by homes and business. One kilowatt hour represents the amount of power consumed during 1 hour by a device that uses one kilowatt (kW), or 1000 joules per second.

EXAMPLE 9.2

The electric resistance of a light bulb is 145 Ω. Determine the value of current flowing through the lamp when it is connected to a 120-volt source.

Using Ohm's law, Equation (9.3), we have

$$V = RI$$

$$I = \frac{V}{R} = \frac{120}{145} = 0.83 \text{ A}$$

EXAMPLE 9.3

Assuming that your electric power company is charging you 10 cents for each kWh usage, estimate the cost of leaving five 100-W light bulbs on from 6 P.M. until 11 P.M. every night for 30 nights.

$$(5 \text{ light bulbs})\left(100 \frac{\text{W}}{\text{light bulb}}\right)\left(\frac{1 \text{ kW}}{1000 \text{ W}}\right)\left(5 \frac{\text{h}}{\text{nights}}\right)(30 \text{ nights})\left(10 \frac{\text{cents}}{\text{kWh}}\right) = 750 \text{ cents}$$

$$= \$7.50$$

Series Circuit

Electrical components can be connected in either a series or a parallel arrangement. As we mentioned earlier in this chapter, there are many different types of electrical components. Here in this section we will only look at resistors and capacitors.

A ***resistor*** is an electrical component that resists the flow of either direct or alternating current. Resistors are commonly used to protect sensitive components or to control the flow of current in a circuit. Moreover, a resistor can be used to divide or control voltages in a circuit. There are two broad groups of resistors: fixed-value resistors and variable resistors. As the name implies, the fixed-value resistor has a fixed value. On the other hand, the variable resistors, sometimes called *potentiometers*, can be adjusted to a desired value. The variable resistors are used in various circuits to adjust the current in the circuit. For example, variable resistors are used in light switches that allow you to dim the light. They are also used to adjust the volume in a radio. A resistor dissipates heat as electric current flows through it. The amount of the heat dissipated depends on the magnitude of the resistor and the amount of the current that is passing through the resistor.

Similar to a constant flow of water in a series of pipes of varying size, the electric current flowing through a series of elements in an electric circuit is the same (constant). Moreover, for a circuit that has elements in a series arrangement, the following is true:

- The voltage drop across each element can be determined using Ohm's law.
- The sum of the voltage drop across each element is equal to the total voltage supplied to the circuit.
- The total resistance is the sum of resistance in the circuit.

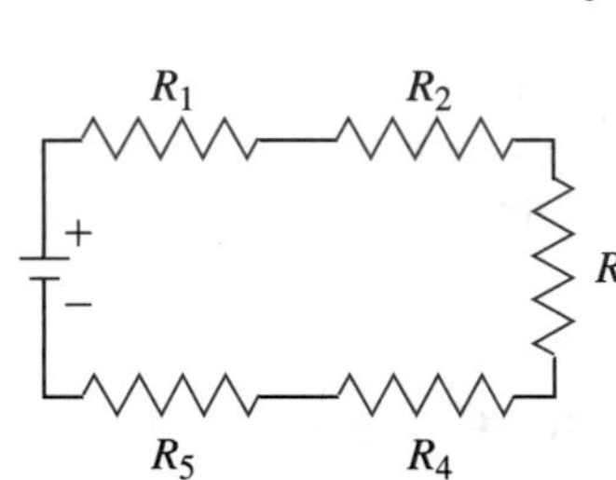

Figure 9.10 Resistors in series, $R_{total} = R_1 + R_2 + R_3 + R_4 + R_5$

For resistors in series, the total equivalent resistance is equal to the sum of the individual resistors, as shown in Figure 9.10.

$$R_{total} = R_1 + R_2 + R_3 + R_4 + R_5 \tag{9.5}$$

One of the problems with a circuit that has elements in a series arrangement is that if one of the elements fails, that failure prevents the current from flowing through other elements in the circuit; thus the entire circuit fails. You may have experienced this problem with a series of light bulbs in a string of Christmas tree lights. In a series arrangement, when one light fails, the current to other lights stops, resulting in all of the lights being off.

EXAMPLE 9.4

Determine the total resistance and the current flowing in the circuit shown in Figure 9.11.

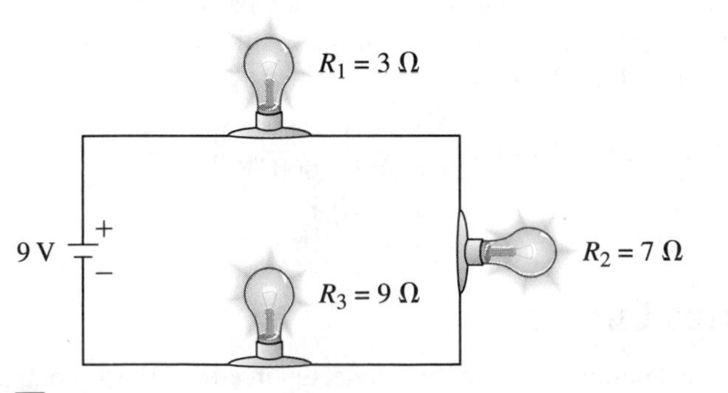

Figure 9.11 The circuit for Example 9.4

The light bulbs are connected in a series arrangement; therefore, the total resistance is given by

$$R_{\text{total}} = R_1 + R_2 + R_3 = 3\ \Omega + 7\ \Omega + 9\ \Omega = 19\ \Omega$$

We can now use Ohm's law to determine the current flowing through the circuit.

$$I = \frac{V}{R_{\text{total}}} = \frac{9}{19} = 0.47\ \text{A} \approx 0.5\ \text{A}$$

We can also obtain the voltage drop across each lamp using Ohm's law:

$$V_{1-2} = R_1 I = (3)(0.47) = 1.41\ \text{V}$$

$$V_{2-3} = R_2 I = (7)(0.47) = 3.29\ \text{V}$$

$$V_{3-4} = R_3 I = (9)(0.47) = 4.23\ \text{V}$$

Note that, neglecting rounding-off errors, the sum of the voltage drops across each light bulb should be 9 volts.

Parallel Circuit

Consider the circuit shown in Figure 9.12. The resistive elements in the given circuit are connected in a parallel arrangement. For this situation, the electric current is divided among each branch. For the parallel arrangement shown in Figure 9.12, the electric potential, or the voltage, across each branch is the same. Moreover, the sum of the current in each branch is equal to the total current flowing in the circuit. The current flow in each branch can be determined using Ohm's law. Note that unlike a series circuit, if one branch fails in a parallel circuit, depending on the failure mode, the other branches could still remain operational. This is the reason parallel circuit arrangement is used when wiring different zones of a building. However, it is worth noting that if one branch fails in a parallel circuit, it could result in an increased current flow in other branches, which could be undesirable.

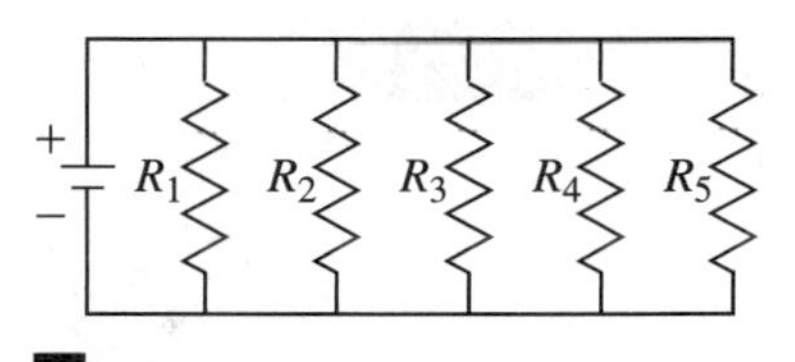

Figure 9.12 Resistors in parallel,

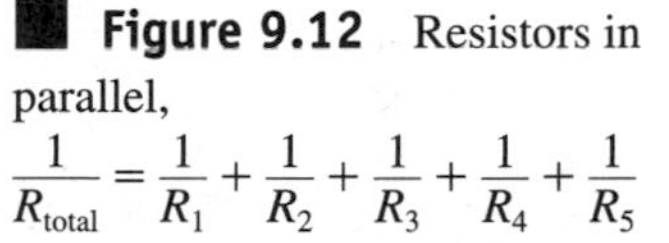

$$\frac{1}{R_{\text{total}}} = \frac{1}{R_1} + \frac{1}{R_2} + \frac{1}{R_3} + \frac{1}{R_4} + \frac{1}{R_5}$$

EXAMPLE 9.5

The light bulbs in the circuit shown in Example 9.4 are placed in a parallel arrangement, as shown in Figure 9.13. Determine the current flow through each branch. Also compute the total resistance offered by all light bulbs to current flow.

Because the light bulbs are connected in a parallel arrangement, the voltage drop across each light bulb is equal to 9 volts. We use Ohm's law to determine the current in each branch in the following manner:

$$V = R_1 I_1 \quad \Rightarrow \quad 9 = 3I_1 \quad \Rightarrow \quad I_1 = 3.0\ \text{A}$$

$$V = R_2 I_2 \quad \Rightarrow \quad 9 = 7I_2 \quad \Rightarrow \quad I_2 = 1.3\ \text{A}$$

$$V = R_3 I_3 \quad \Rightarrow \quad 9 = 9I_3 \quad \Rightarrow \quad I_3 = 1.0\ \text{A}$$

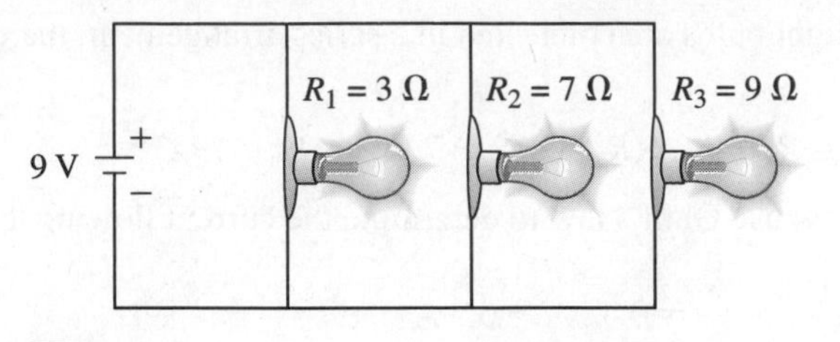

Figure 9.13 The circuit for Example 9.5

The total current drawn by the circuit is

$$I_{\text{total}} = I_1 + I_2 + I_3 = 3.0 + 1.3 + 1.0 = 5.3\ \text{A}$$

The total resistance is given by

$$\frac{1}{R_{\text{total}}} = \frac{1}{R_1} + \frac{1}{R_2} + \frac{1}{R_3} = \frac{1}{3} + \frac{1}{7} + \frac{1}{9} \quad \Rightarrow \quad R_{\text{total}} = 1.7\ \Omega$$

Note that we could have obtained the total current drawn by the circuit using the total resistance and the Ohm's law in the following manner:

$$V = R_{\text{total}} I_{\text{total}} \quad \Rightarrow \quad 9\ \text{V} = (1.7\ \Omega)(I_{\text{total}}) \quad \Rightarrow \quad I_{\text{total}} = 5.3\ \text{A}$$

EXAMPLE 9.6

Determine the total current drawn by the circuit shown in Figure 9.14.

The circuit given in this example has components that are in both series and parallel arrangements. We will first combine the light bulbs in the parallel branches into one equivalent resistance.

$$\frac{1}{R_{\text{equivalent}}} = \frac{1}{R_2} + \frac{1}{R_3} = \frac{1}{7} + \frac{1}{9} \quad \Rightarrow \quad R_{\text{equivalent}} = 3.9\ \Omega$$

Next, we add the equivalent resistance to R_1, noting that the two resistors are in series now.

$$R_{\text{total}} = R_1 + R_{\text{equivalent}} = 3 + 3.9 = 6.9\ \Omega$$

$$V = R_{\text{total}} I_{\text{total}} \quad \Rightarrow \quad 9\ \text{V} = (6.9\ \Omega)(I_{\text{total}}) \quad \Rightarrow \quad I_{\text{total}} = 1.3\ \text{A}$$

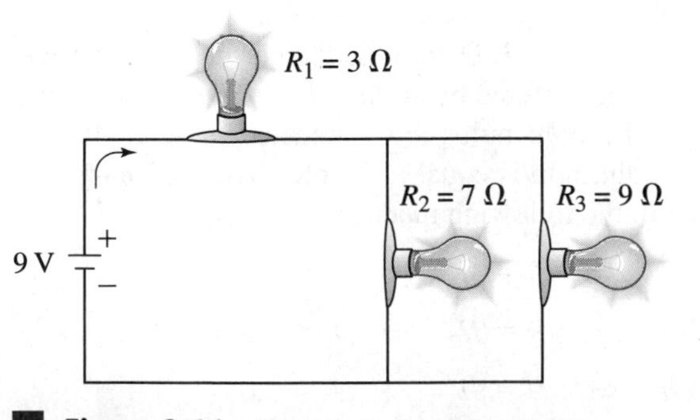

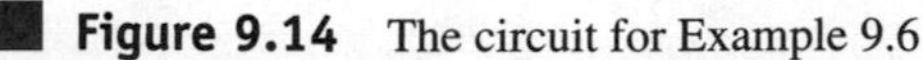

Figure 9.14 The circuit for Example 9.6

Capacitors

Capacitors are electrical components that store electrical energy. A capacitor has two oppositely charged electrodes with a dielectric material inserted between the electrodes. Dielectric material is a poor conductor of electricity. Capacitors are used in many applications to serve as filters to protect sensitive components in electrical circuits against power surges. They are also used in large computer memories to store information during a temporary loss of electric power to the computer. You will also find capacitors in tuned circuits for radio receivers, audio filtering applications (e.g., bass and treble controls), timing elements for strobe lights, and intermittent wipers in automobiles.

The ***farad*** **(F)** is the basic unit used to designate the size of a capacitor. One farad is equal to 1 coulomb per volt. Because the farad is a relatively large unit, many capacitor sizes are expressed in microfarad ($\mu F = 10^{-6}$ F) or picofarad ($pF = 10^{-12}$ F).

9.5 ELECTRIC MOTORS

As engineering students, most of you will take at least one class in basic electrical circuits wherein you will learn about circuit theory and various elements, such as resistors, capacitors, inductors, and transformers. You will also be introduced to different types of motors and their operations. Even if you are studying to be a civil engineer, you should pay special attention to this class in electrical circuits, especially the part about motors, because motors drive many devices and equipment that make our lives comfortable and less laborious. To understand the significant role motors play in our everyday lives, look around you. You will find motors running all types of equipment in homes, commercial buildings, hospitals, recreational equipment, automobiles, computers, printers, copiers, and so on. For example, in a typical home you can identify a large number of motors operating quietly all around you that you normally don't think about. Here, we have identified a few household appliances with motors:

- Refrigerator: compressor motor, fan motor
- Garbage disposer
- Microwave with a turning tray
- Stove hood with a fan
- Exhaust fan in the bathroom
- Room ceiling fan
- Tape player in a VCR
- Hand-held power screwdriver or hand-held drill
- Heating, ventilating, or cooling system fan
- Vacuum cleaner
- Hair dryer
- Electric shaver
- Computer: cooling fan, hard drive

Some of the factors that engineers consider when selecting a motor for an application are: (a) motor type, (b) motor speed (rpm), (c) motor performance in terms of torque output, (d) efficiency, (e) duty cycles, (f) cost, (g) life expectancy, (h) noise level, (i) maintenance and service requirements. Most of these are self-explanatory. We discussed speed and torque in previous chapters; next we will discuss motor types and duty cycles.

TABLE 9.3 Examples of Motors Used in Different Applications

Schematic Diagrams for AC Motors

Split-phase	Repulsion-start induction	Capacitor-start	Repulsion-induction	Capacitor-motor
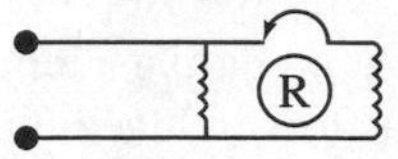 R	 Centrifugal switch	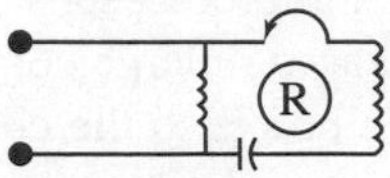 R	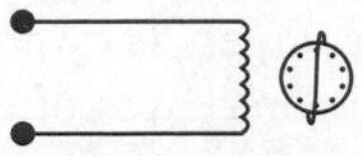	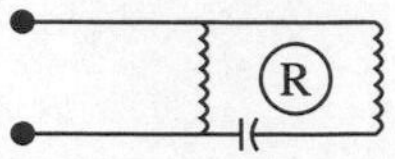 R

Load Type	Motor Type (1)	Starting Ability (Torque) (2)	Starting Current (3)	Size hp (4)	Electrical Power Requirements: Phase (5)	Electrical Power Requirements: Voltage (6)
Easy Starting Loads	Shaded-pole induction	Very low 1/2 to 1 times running torque	Low	.035–.2 kW (1/20–1/4 hp)	Single	Usually120
	Split-phase	Low 1 to $1\frac{1}{2}$ times running torque	High 6 to 8 times running current	.035–.56 kW (1/20–3/4 hp)	Single	Usually 120
	Permanent-split, capacitor-induction	Very low 1/2 to 1 times running torque	Low	.035–.8 kW (1/20–1 hp)	Single	Single voltage 120 or 240
	Soft-start	Very low 1/2 to 1 times running torque	Low 2 to $2\frac{1}{2}$ times running current	5.6–25 kW ($7\frac{1}{2}$–50 hp)	Single	240
Difficult Starting Loads	Capacitor-start, induction-run	High 3 to 4 times running torque	Medium 3 to 6 times running current	.14–8 kW (1/6–10 hp)	Single	120–240
	Repulsion-start, induction-run	High 4 times running torque	Low $2\frac{1}{2}$ to 3 times running current	.14–16 kW (1/6–20 hp)	Single	120–240
	Capacitor-start, capacitor-run	High $3\frac{1}{2}$ to $4\frac{1}{2}$ times running torque	Medium 3 to 5 times running current	.4–20 kW (1/2–25 hp)	Single	120–240
	Repulsion-start, capacitor-run	High 4 times running torque	Low $2\frac{1}{2}$ to 3 times running current	.8–12 kW (1–15 hp)	Single	Usually 240
	Three-phase, general-purpose	Medium 2 to 3 times running torque	High 3 to 6 times running current	.4–300 kW (1/2–400 hp)	Three	120–240; 240–480 or higher

Source: AAVIM, *Electric Motors.*

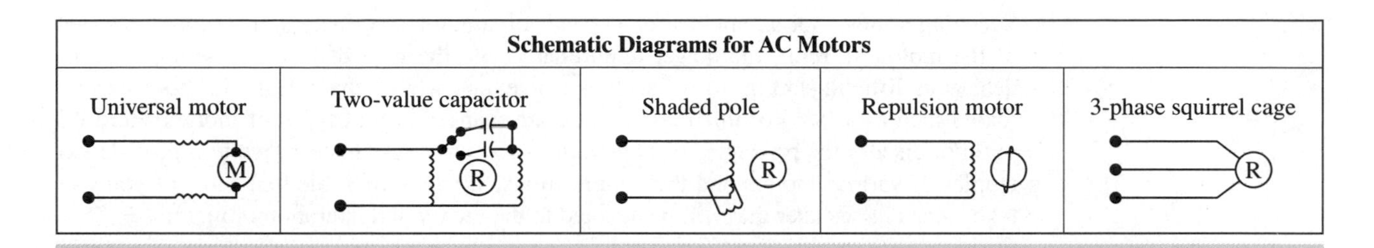

Load Type	Speed Range (7)	Reversible (8)	Relative Cost (9)	Other Characteristics (10)	Typical Uses (11)
Easy Starting Loads	900, 1200, 1800, 3600	No	Very low	Light duty, low in efficiency	Small fans, freezer blowers, arc welder blower, hair dryers
	900, 1200, 1800, 3600,	Yes	Low	Simple construction	Fans, furnace blowers, lathes, small shop tools, jet pumps
	Variable 900–1800	Yes	Low	Usually custom-designed for special application	Small compressors, fans
	1800, 3600	Yes	High	Used in motor sizes normally served by3-phase power when 3-phase power not available	Centrifugal pumps, crop dryer fans, feed grinder
Difficult Starting Loads	900, 1200, 1800, 3600	Yes	Moderate	Long service, low maintenance, very popular	Water systems, air compressors, ventillating fans, grinders, blowers
	1200, 1800, 3600	Yes	Moderate to high	Handles large load variations with little variation in current demand	Grinders, deep-well pumps, silo unloaders, grains conveyors, barn cleaners
	900, 1200, 1800, 3600	Yes	Moderate	Good starting ability and full-load efficiency	Pumps, air compressors, drying fans, large conveyors, feed mills
	1200, 1800, 3600	Yes	Moderate to high	High efficiency, requires more service than most motors	Conveyors, deep-well pumps, feed mills, silo unloaders
	900, 1200, 1800, 3600	Yes	Very low	Very simple construction, dependable, service-free	Conveyors, dryers, elevators, hoists, irrigation pumps

Motor Type

Selecting a motor for an application depends on a number of factors, including the speed of the motor (in rpm), the power requirement, and the type of load. Some applications deal with difficult starting loads, such as conveyors, while others deal with easy starting loads, such as a fan. For this reason, there are many different types of motors. Here we will not discuss the principles of how various motors work. Instead, we will provide examples of various motors and their applications, as given in Table 9.3. Most of you will take some classes later that will be devoted to the theory and operation of motors.

Duty Cycle

Manufacturers of motors classify them according to the amount of time the motor needs to be operated. The motors are generally classified as *continuous duty* or *intermittent duty*. The continuous duty motors are used in applications where the motor is expected to operate over a period of an hour or longer. In some applications, continuous operation of the motor may be required. In applications where the motor is expected to operate for short periods of time and then rest, the intermittent duty motors are found. The continuous duty motors are more expensive than the intermittent duty motors.

In this chapter we introduced you to electricity and some basic electrical components. Most of you will take a basic electrical circuit class where you will learn in more detail about electricity, electrical components, and motors, so you now see the importance of paying attention and studying carefully.

SUMMARY

Now that you have reached this point in the text

- You should understand the importance of electricity and electrical power in our everyday lives.
- You should know what is meant by electric current and the difference between direct current and alternating current.
- You should have an idea about various sources of electricity.
- You should understand the concept of electrical resistance and its unit.
- You should be familiar with the AWG wire designations.
- You should be familiar with the role of electric motors in our everyday lives.

PROBLEMS

1. How are batteries connected in the following products: a hand-held calculator such as a TI 85, a flashlight, and a portable radio? Are the batteries connected in a series arrangement or in a parallel arrangement?
2. Identify the types of batteries used in the following products:
   a. a laptop computer
   b. an electric shaver
   c. a cordless drill
   d. a camcorder
   e. a cellular phone
   f. a flashlight
   g. a wristwatch
   f. a camera.
3. Investigate the size and the material used for heating elements in the following products: a toaster,

hair dryer, pressing iron, coffeemaker, and an electric stovetop.

4. As explained in this chapter potentiometers (rheostats) are used in applications to adjust the electric current in a circuit. Investigate the different types of resistance element used for various applications. The resistance element of the rheostat is typically made from metal wires or ribbons, carbon, or a conducting liquid. For example, in applications where the current in the circuit is relatively small, the carbon type is used. Write a brief report discussing your findings.
5. Electrical furnaces are used in the production of steel that is consumed in the structural, automotive, tool, and aircraft industries in the United States. Electrical furnaces are typically classified into resistance furnaces, arc furnaces, and induction furnaces. Investigate the operation of these three types of furnaces. Write a brief report describing your findings.
6. Identify examples of motors used in a new automobile. For example, a motor is used to run the fan that delivers warm or cold air into the car.
7. Obtain information about the electric current and voltage ratings of your own residential building or a building belonging to someone you know. If possible draw a diagram showing the power distribution, similar to Figure 9.7.
8. What is the current that flows through each of the following light bulbs: 40 W, 60 W, 75 W, 100 W? Each light is connected to a 120-volt line.
9. If a 1500-W hair dryer is connected to a 120-V line, what is the maximum current drawn?
10. Refering to Table 9.1, create a table that shows the relative resistance of 1-ft-long wires having diameters of 1 mil made of the metals given in Table 9.1 to 1-ft-long copper with a diameter of 1 mil. For example, using the data given in Table 9.1, the relative resistance of 1-foot-long, 1-mil-diameter aluminum wire to 1-ft-long, 1-mil copper wire is 1.64 (17.01 Ω/10.37 Ω = 1.64).
11. Investigate how an alkaline battery and an automobile maintenance-free battery (lead acid, gel cell) work. Write a brief report discussing your findings.
12. When subjected to pressure, certain materials create a relatively small voltage. Materials that behave in this manner are called piezoelectrics. Investigate the applications in which piezoelectrics are used. Write a brief report discussing your findings.
13. Prepare an electrical circuit plan similar to the one shown in Figure 9.9 for your apartment, your home, or a section of your dormitory.
14. The National Electrical Code (NEC) covers the safe and proper installation of wiring, electrical devices, and equipment in private and public buildings. NEC is published by the National Fire Protection Association (NFPA) every three years. As an example of an NEC provision, the receptacle outlets in a room in a dwelling should be placed such that no point on the wall space is more than 6 ft away from the outlet in order to minimize the use of extension cords. After performing a Web search or obtaining a copy of the NEC handbook, give at least three other examples of National Electric Codes for a family dwelling.
15. Obtain a multimeter (voltmeter, ohm, and current meter) and measure the resistance and the voltage of the heating element in the back window of a car. Determine the power output of the heater.
16. Visit a hardware store and obtain information on the sizes of heating elements used in home hot water heaters. If a hot water heater is connected to a line with 240 V, determine the current drawn by the hot water heater and its power consumption.
17. As explained in this chapter, the National Electrical Code gives specific information on the type of wires used for general wiring. Perform a search and obtain information on wire types, maximum operating temperatures, insulating materials, outer cover sheaths, and usage types. Prepare a table that shows examples of these codes; show wire size, temperature rating, application provisions, insulation, and outer covering.
18. Determine the total resistance and the current flow for the circuit shown in the accompanying figure.

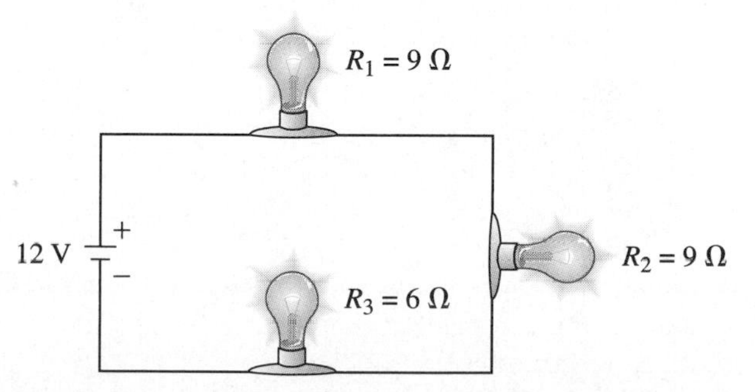

19. Determine the total resistance and the current flow in each branch for the circuit shown in the accompanying figure.

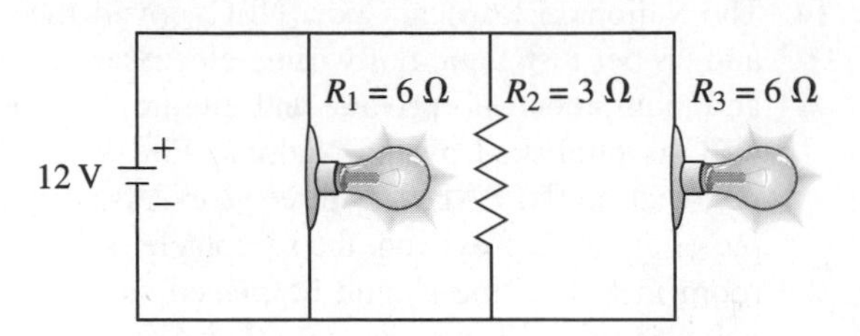

20. Use Kirchhoff's current law to determine the missing current in the circuit shown in the accompanying figure.

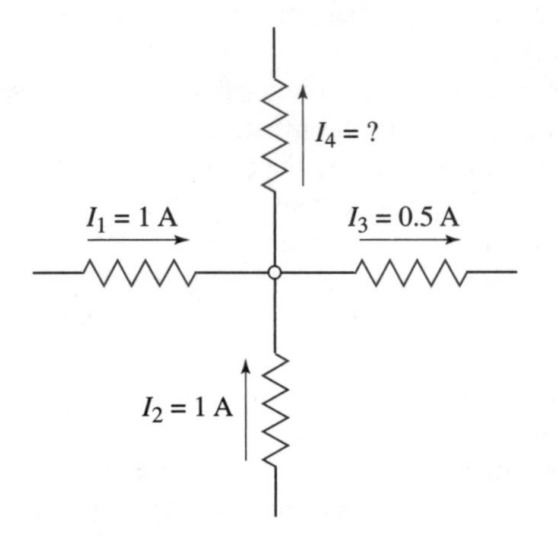

21. Obtain a flashlight and a voltmeter. Measure the resistance of the light bulb used in the flashlight. Draw the electrical circuit for the flashlight. Estimate the current drawn by the light when the flashlight is on.
22. Contact your electric utility company and obtain information on what the company charges for each kilowatt hour usage. Estimate the cost of your electric power consumption for a typical day. Make a list of your activities for the day and estimate the power consumption for the devices that you used. Write a brief report discussing your findings.
23. For Example 9.6, we computed the total current drawn by the circuit. Determine the current in each branch of the circuit given in Example 9.6.
24. For an automobile battery, investigate what is meant by the following terms: *ampere-hour rating*, *cold-cranking rating*, and *reserve capacity rating*. Write a brief report discussing your findings.
25. For a primary cell such as an alkaline battery, what does the term *amp-hour* represent? Gather information on the amp-hour rating for some alkaline batteries. Write a one-page summary of your findings.
26. As you know, a fuse is a safety device that is commonly placed in an electrical circuit to protect the circuit against excessive current. Investigate the various types of fuses, their shapes, materials, and sizes. Write a brief report discussing your findings.
27. There are many different types of capacitors, including ceramic, air, mica, paper, and electrolytic. Investigate their usage in electrical and electronic applications. Write a brief report discussing your findings.

CHAPTER

10 Energy and Power

A powerful crane lifts and lowers a structural member in place. We need energy to build shelter, to cultivate and process food, to make goods, and to maintain our living places at comfortable settings. To quantify the requirements to move objects, to lift objects, or to heat or cool something, energy is defined and classified into different categories. Power is the time rate of doing work. The value of power required to perform a task represents how fast you want the task done. If you want a task done in a shorter period of time, then more power is required.

The objectives of this chapter are to introduce the concept of energy, the various types of energy, and what is meant by the term power. *We will explain various mechanical forms of energy including kinetic energy, potential energy, and elastic energy. We will also revisit the definition of thermal energy forms from Chapter 8, including heat and internal energy. Next, we will present conservation of energy and its applications. We will define power as the rate of doing work and explain in detail the difference between work/energy and power. After our discussion, these differences should be clear to you. The common units of power, watts and horsepower, are also explained. Once you have a good grasp of the concepts of work, energy, and power, then you can better understand the manufacturer's power ratings of engines and motors. Moreover, in this chapter we will also explain what is meant by the term* efficiency *and look at the efficiencies of power plants, internal combustion engines (car engines), electric motors, pumps, and heating, air-conditioning, and refrigeration systems.* ■

10.1 WORK, MECHANICAL ENERGY, THERMAL ENERGY

As we explained in Chapter 7, mechanical work is performed when a force moves an object through a distance. But what is energy? Energy is one of those abstract terms that you already have a good feel for. For instance, you already know that we need energy to create goods, to build shelter, to cultivate and process food, and to maintain our living places at comfortable temperature and humidity settings. But what you may not know is that energy can have different forms. Recall that scientists and engineers define terms and concepts to explain various physical phenomena that govern nature. To better explain quantitatively the requirements to move objects, to lift things, to heat or cool objects, and to stretch materials, energy is defined and classified into different categories. Let us begin with the definition of ***kinetic energy***. When work is done on or against an object, it changes the kinetic energy of the object (see Figure 10.1). In fact, as you will learn in more detail in your physics and dynamics class, mechanical work performed on an object brings about a change in the kinetic energy of the object according to

$$\text{work}_{1-2} = \frac{1}{2}mV_f^2 - mV_i^2 \tag{10.1}$$

where m is the mass of the object and V_i (position 1) and V_f (position 2) are the initial and the final velocity of the object, respectively. To better demonstrate Equation (10.1),

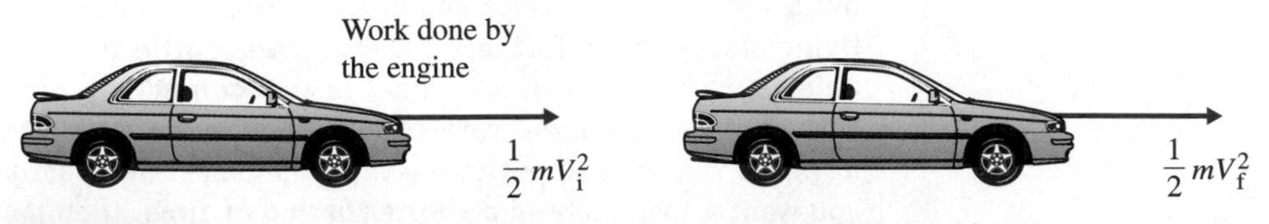

■ **Figure 10.1** The relationship between work and change in kinetic energy

consider the following example. When you push on a lawn mower, which is initially at rest, you perform mechanical work on the lawn mower and move it, consequently changing its kinetic energy from a zero value to some nonzero value.

Kinetic Energy

An object having a mass m and moving with a speed V has a kinetic energy, which is equal to

$$\text{kinetic energy} = \frac{1}{2}mV^2 \tag{10.2}$$

The SI unit for kinetic energy is the joule, which is a derived unit. The unit of joule is obtained by substituting kg for the units of mass, and m/s for the units of velocity, as shown here.

$$\text{kinetic energy} = \frac{1}{2}mV^2 = (\text{kg})\left(\frac{\text{m}}{\text{s}}\right)^2 = (\text{kg})\left(\frac{\text{m}}{\text{s}^2}\right)(\text{m}) = \text{N}\cdot\text{m} = \text{joule} = \text{J}$$

Note that the $\frac{1}{2}$ factor in the kinetic energy equation is unitless. As we discussed in the previous section, it is the change in kinetic energy, ΔKE, that is used in engineering analysis as given by Equation (10.1). The change in the kinetic energy is given by

$$\Delta\text{KE} = \frac{1}{2}mV_2^2 - \frac{1}{2}mV_1^2 \tag{10.3}$$

EXAMPLE 10.1

Determine the net force needed to bring a car that is traveling at 90 km/h to a full stop in a distance of 100 m. The mass of the car is 1400 kg.

We begin the analysis by first changing the units of speed to m/s and then use Equation (10.1) to analyze the problem.

$$V_i = 90\left(\frac{\text{km}}{\text{h}}\right)\left(\frac{1\text{ hr}}{3600\text{ s}}\right)\left(\frac{1000\text{ m}}{1\text{ km}}\right) = 25\,\frac{\text{m}}{\text{s}}$$

$$V_f = 0$$

$$\text{work}_{1-2} = (\text{force})(\text{distance}) = \frac{1}{2}mV_f^2 - mV_i^2$$

$$(\text{force})(100\text{ m}) = 0 - \frac{1}{2}(1400\text{ kg})\left(25\,\frac{\text{m}}{\text{s}}\right)^2$$

$$\text{force} = 4375\text{ N}$$

Note that in Example 10.1, change in the kinetic energy was used in the analysis.

Potential Energy

The work required to lift an object with a mass m by a vertical distance Δh is called ***gravitational potential energy***. It is the mechanical work that must be performed to overcome the gravitational pull of the earth on the object (see Figure 10.2). The change in the potential energy of the object when its elevation is changed is given by

$$\text{change in potential energy} = \Delta\text{PE} = mg\,\Delta h \tag{10.4}$$

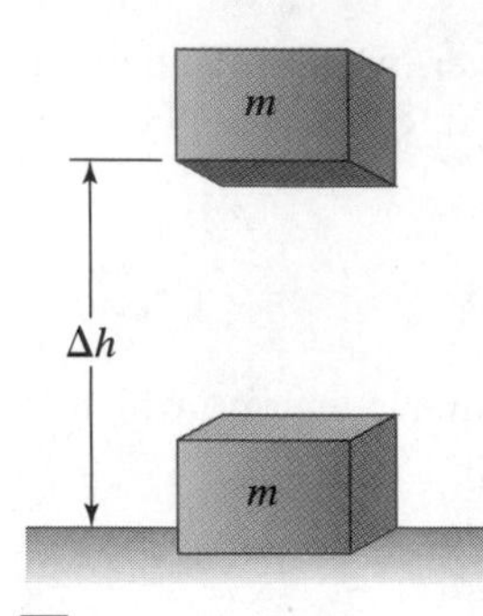

Figure 10.2
Change in potential energy of an object

where

m = mass of the object (kg)
g = acceleration due to gravity (9.81 m/s^2)
Δh = change in the elevation (m)

The SI unit for potential energy is also joule, a derived unit, and is obtained by substituting kg for the units of mass, and m/s^2 for the units of acceleration due to gravity, and m for the elevation change:

$$\text{potential energy} = mgh = \overbrace{(\text{kg})\left(\frac{\text{m}}{\text{s}^2}\right)}^{\text{N}}(\text{m}) = \text{N}\cdot\text{m} = \text{J}$$

As in the case with kinetic energy, keep in mind that it is the change in the potential energy that is of significance in engineering calculations. For example, the energy required to lift an elevator from the first floor to the second floor is the same as lifting the elevator from the third floor to the fourth floor, provided that the distance between each floor is the same.

EXAMPLE 10.2

Calculate the energy required to lift an elevator and its occupant with a mass of 2000 kg for the following situations: (a) between the first and the second floor, (b) between the third and the fourth floor, (c) between the first and the fourth floor. (See Figure 10.3.) The vertical distance between each floor is 4.5 m.

We can use Equation (10.4) to analyze this problem; the energy required to lift the elevator is equal to the change in its potential energy, starting with

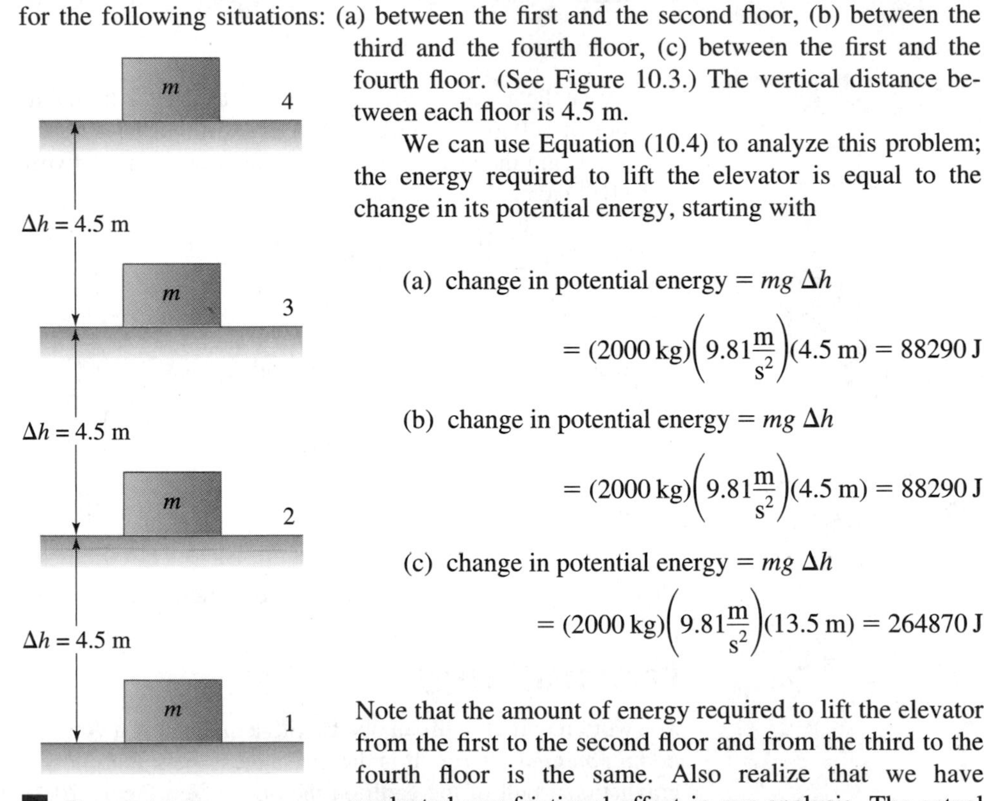

Figure 10.3
A schematic diagram for Example 10.2

(a) change in potential energy $= mg\,\Delta h$

$$= (2000\ \text{kg})\left(9.81\frac{\text{m}}{\text{s}^2}\right)(4.5\ \text{m}) = 88290\ \text{J}$$

(b) change in potential energy $= mg\,\Delta h$

$$= (2000\ \text{kg})\left(9.81\frac{\text{m}}{\text{s}^2}\right)(4.5\ \text{m}) = 88290\ \text{J}$$

(c) change in potential energy $= mg\,\Delta h$

$$= (2000\ \text{kg})\left(9.81\frac{\text{m}}{\text{s}^2}\right)(13.5\ \text{m}) = 264870\ \text{J}$$

Note that the amount of energy required to lift the elevator from the first to the second floor and from the third to the fourth floor is the same. Also realize that we have neglected any frictional effect in our analysis. The actual energy requirement would be greater in the presence of frictional effect.

Elastic Energy

When a spring is stretched or compressed from its unstretched position, ***elastic energy*** is stored in the spring, energy that will be released when the spring is allowed to return to its unstretched position (see Figure 10.4). The elastic energy stored in a spring when stretched or compressed is given by

$$\text{elastic energy} = \frac{1}{2}kx^2 \tag{10.5}$$

where

k = spring constant (N/m)

x = deflection of spring from its unstretched position (m)

The SI unit for elastic energy is also the joule. It is obtained by substituting N/m for the units of spring constant and m for the units of deflection, as shown:

$$\text{elastic energy} = \frac{1}{2}kx^2 = \left(\frac{\text{N}}{\text{m}}\right)(\text{m})^2 = \text{N·m} = \text{J}$$

Note once again that the $\frac{1}{2}$ factor in the elastic energy equation is unitless. Let us now consider the spring shown in Figure 10.5; the spring is stretched by x_1 to position 1 and then stretched by x_2 to position 2. The elastic energy stored in the spring in position 1 is given by

$$\text{elastic energy} = \frac{1}{2}kx_1^2$$

$$\text{elastic energy} = \frac{1}{2}kx_2^2$$

$$\text{change in elastic energy} = \Delta\text{EE} = \frac{1}{2}kx_2^2 - \frac{1}{2}kx_1^2 \tag{10.6}$$

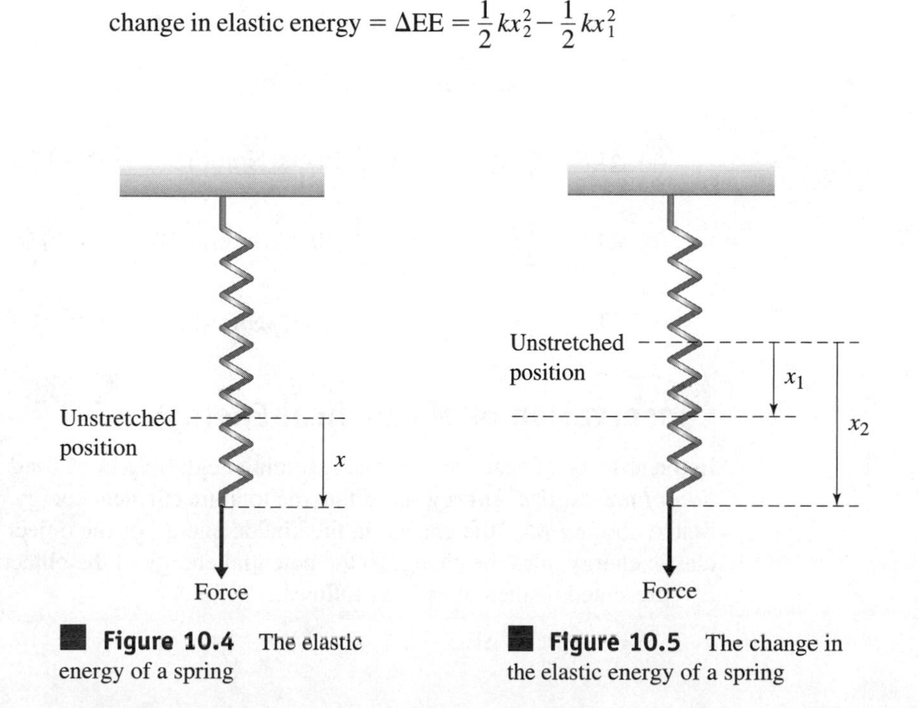

Figure 10.4 The elastic energy of a spring

Figure 10.5 The change in the elastic energy of a spring

EXAMPLE 10.3

Determine the change in the elastic energy of the spring shown in Figure 10.6 when it is stretched from: (a) position **1** to position **2**, (b) position **2** to position **3**, and position **1** to position **3**. The spring constant is $k = 100$ N/cm. Additional information is shown in Figure 10.6.

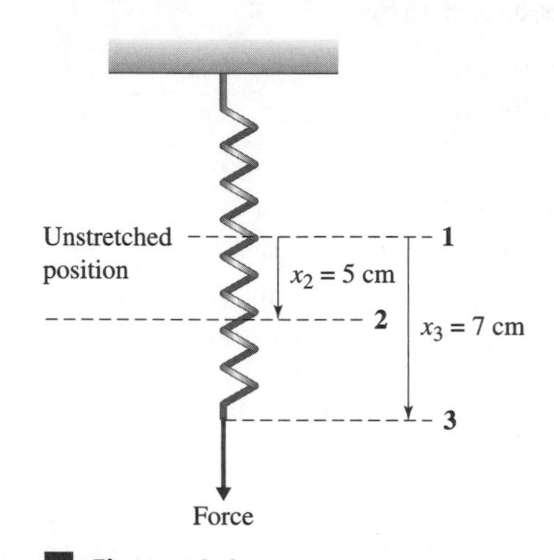

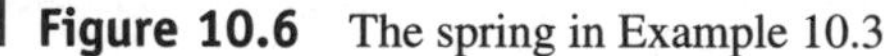

Figure 10.6 The spring in Example 10.3

We begin by converting the units of the spring constant from N/cm to N/m in the following manner:

$$(100 \text{ N/cm})(100 \text{ cm/m}) = 10{,}000 \text{ N/m}$$

Using Equation (10.6), we can now answer the questions:

$$\text{change in elastic energy} = \Delta\text{EE} = \frac{1}{2}kx_2^2 - kx_1^2$$

$$\text{(a)}\ \Delta\text{EE} = \frac{1}{2}kx_2^2 - \frac{1}{2}kx_1^2 = \frac{1}{2}(10{,}000 \text{ N/m})(0.05)^2 - 0 = 12.5 \text{ J}$$

$$\text{(b)}\ \Delta\text{EE} = \frac{1}{2}kx_3^2 - \frac{1}{2}kx_2^2 = \frac{1}{2}(10{,}000 \text{ N/m})(0.07)^2 - \frac{1}{2}(10{,}000 \text{ N/m})(0.05)^2 = 12 \text{ J}$$

$$\text{(c)}\ \Delta\text{EE} = \frac{1}{2}kx_3^2 - \frac{1}{2}kx_1^2 = \frac{1}{2}(10{,}000 \text{ N/m})(0.07)^2 - 0 = 24.5 \text{ J}$$

Conservation of Mechanical Energy

In the absence of heat transfer, and assuming negligible losses and no work, the ***conservation of mechanical energy*** states that the total mechanical energy of a system is constant. Stated another way, the change in the kinetic energy of the object, plus the change in the elastic energy, plus the change in the potential energy of the object is zero. This statement is represented mathematically as follows.

$$\Delta\text{KE} + \Delta\text{PE} + \Delta\text{EE} = 0 \qquad (10.7)$$

Equation (10.7) states that energy content of a system can change form, but the total energy content of the system is constant.

EXAMPLE 10.4

In a manufacturing process carts are rolling down an inclined surface as shown in Figure 10.7. Estimate the height from which the cart must be released so that as it reaches point A, the cart has a velocity of 2.5 m/s. Neglect rolling friction.

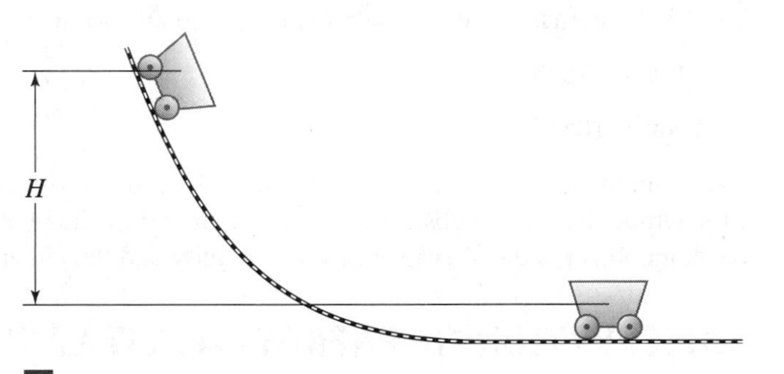

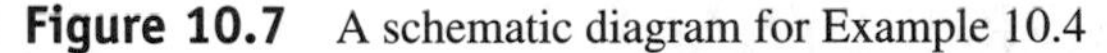
Figure 10.7 A schematic diagram for Example 10.4

We can solve this problem using Equation (10.7)

$$\Delta\text{KE} + \Delta\text{PE} + \Delta\text{EE} = 0$$

$$\Delta\text{KE} = \frac{1}{2}mV_{\text{f}}^2 - \frac{1}{2}mV_{\text{i}}^2 = \frac{1}{2}m(2.5\ \text{m/s})^2 - 0$$

$$\Delta\text{EE} = 0$$

$$\Delta\text{PE} = mg\,\Delta h = -m\left(9.81\frac{\text{m}}{\text{s}^2}\right)H$$

$$\frac{1}{2}m\,(2.5\ \text{m/s})^2 + -m\left(9.81\frac{\text{m}}{\text{s}^2}\right)H = 0$$

And solving for H we have

$$H = 0.318\ \text{m}$$

Thermal Energy Units

In Chapter 8, we explained that thermal energy transfer occurs whenever there exists a temperature difference within an object, or whenever there is a temperature difference between two bodies, or a temperature difference between a body and its surroundings. This form of energy transfer is called *heat*. Remember the fact that heat always flows from a high-temperature region to the low-temperature region. Moreover, we discussed the three different modes of heat transfer: conduction, convection, and radiation. We also discussed three units that are commonly used to quantify thermal energy (1) the British thermal unit, (2) the calorie, and (3) the joule.

As we explained Btu (British thermal unit) in Chapter 8, one Btu is formally defined as the amount of thermal energy needed to raise the temperature of 1 lb_m of water by 1°F.

The calorie is defined as the amount of heat required to raise the temperature of 1 g of water by 1°C. And as you may also recall from our discussion in Chapter 8, in SI units no distinction is made between the units of thermal energy and mechanical energy, and therefore the units of thermal energy are defined in terms of fundamental dimensions of mass, length, and time. In the SI system of units, the joule is the unit of energy and is defined as

$$1 \text{ joule} = 1 \text{ N} \cdot \text{m} = 1 \text{ kg} \cdot \text{m}^2/\text{s}^2$$

The U.S. Customary Unit of thermal energy is related to mechanical energy through

$$1 \text{ Btu} = 778 \text{ lb} \cdot \text{ft}$$

$$1 \text{ Btu} = 1055 \text{ J}$$

Finally, internal energy is a measure of the molecular activity of a substance and is related to the temperature of a substance. As we explained in Chapter 8, the higher the temperature of an object, the higher its molecular activity and thus its internal energy.

10.2 CONSERVATION OF ENERGY—FIRST LAW OF THERMODYNAMICS

Earlier in this chapter, we discussed the conservation of mechanical energy. We stated that in the absence of heat transfer, and assuming negligible losses and no work, the conservation of mechanical energy states that the total mechanical energy of the system is constant. In this section, we will discuss the effects of heat and work in conservation of energy. There are a number of different ways that we can describe the general form of the conservation of energy, or the first law of thermodynamics. Expressed simply, the ***first law of thermodynamics*** states that energy is conserved. It cannot be created or destroyed; energy can only change forms. Another more elaborate statement of the first law says that for a system having a fixed mass, the net heat transfer to the system minus the work done by the system is equal to the change in total energy of the system (see Figure 10.8) according to

$$Q - W = \Delta E \tag{10.8}$$

where

Q = net heat transfer into the system $(\Sigma Q_{in} - \Sigma Q_{out})$ in joules

W = net work done by the system $(\Sigma W_{out} - \Sigma W_{in})$ in joules

ΔE = net change in total energy of the system in joules, where E represents the sum of the internal energy, kinetic energy, potential energy, elastic energy, and other forms of energy of the system

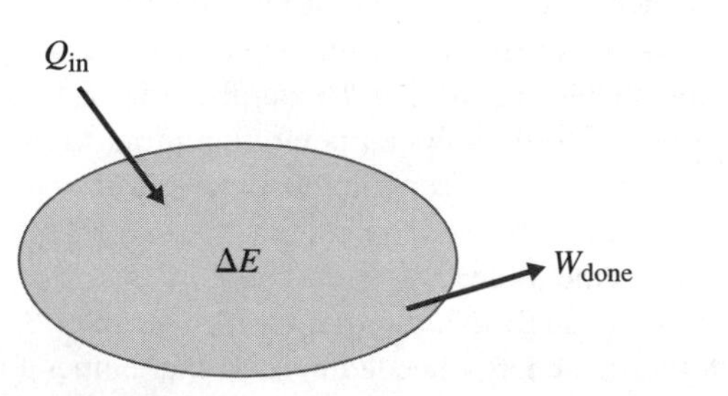

Figure 10.8 The first law of thermodynamics for a system having a fixed mass

There is a sign convention associated with Equation (10.8) that must be followed carefully. Heat transfer into the system, or the work done by the system, is considered a positive quantity, whereas the heat transfer out of the system, or the work done on the system, is a negative quantity. The reason for this type of sign convention is to show that work done on the system increases the total energy of the system, while work done by the system decreases the total energy of the system. Also note that using this sign convention, heat transfer into the system increases the total energy of the system, while heat transfer out of the system decreases its total energy.

You may also think of the first law of thermodynamics in the following manner: When it comes to energy, the best you can do is to break even. You cannot get more energy out of a system than the amount you put into it. For example, if you put 100 joules into a system as work, you can get 100 joules out of the system in the form of change in internal, kinetic, or potential energy of the system. As you learn more about energy you will also learn that according to the second law of thermodynamics, unfortunately you cannot even break even, because there are always losses associated with processes. We will discuss the effect of losses in terms of performance and efficiency of various systems later in this chapter.

EXAMPLE 10.5

Determine the change in the total energy of the system shown in Figure 10.9. The heater puts 150 watts (J/s) into the water pot. The heat loss from the water pot to the atmosphere is 60 watts. Calculate the change in total energy of water in the pot after 5 minutes.

We can use Equation (10.8) to solve this problem.

$$Q - W = \Delta E$$

$$W = 0$$

$$(150\ \text{J/s})(300\ \text{s}) - (60\ \text{J/s})(300\ \text{s}) = \Delta E$$

$$\Delta E = 27\ \text{kJ}$$

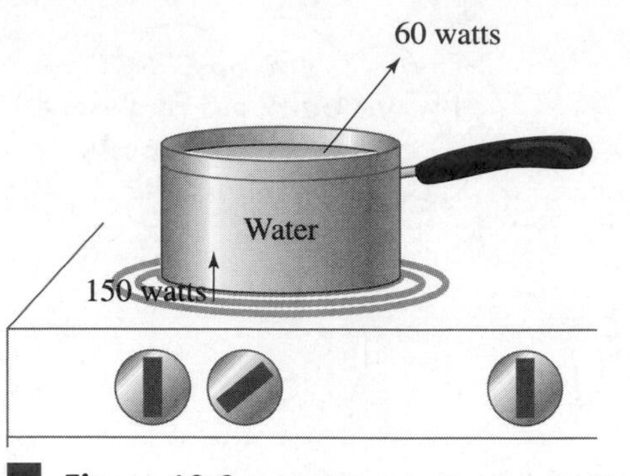

Figure 10.9 A schematic diagram for Example 10.9

10.3 UNDERSTANDING WHAT WE MEAN BY POWER

In Section 10.1, we reviewed the concept of work as presented in Chapter 7 and explained the different forms of energy. We now consider what is meant by the term *power*. ***Power*** is formally defined as the time rate of doing work, or stated simply, the required work, or energy, divided by the time required to perform the task.

$$\text{power} = \frac{\text{work}}{\text{time}} = \frac{(\text{force})(\text{distance})}{\text{time}} \tag{10.9}$$

or

$$\text{power} = \frac{\text{energy}}{\text{time}}$$

From Equation (10.9), the definition of power, it should be clear that the value of power required to perform a task represents how fast you want the task done. If you want a task done in a shorter time, then more power is required. For the sake of demonstrating this point better, imagine that in order to perform a task 3600 joules are required. The next question then becomes, how fast do we want this task done? If we want the task done in 1 second, 3600 J/s power is required; if we want the task done in 1 minute, then 60 J/s power is needed; and if we want the task done in 1 hour, then the required power is 1 J/s. From this simple example, you should see clearly that to perform the same task in a shorter period of time, more power is required. More power means more energy expenditure per second. Another example that you have a direct experience with is the following situation: Which requires more power, to walk up a flight of stairs or to run up the stairs? Of course, as you already know, it requires more power to run up the stairs, because as compared to walking, when running up the stairs, you perform the same amount of work in a shorter time period. Many engineering managers understand the concept of power well, for they understand the benefit of teamwork. In order to finish a project in a shorter period of time, instead of assigning a task to an individual, the task is divided among several team members. More useful energy expenditure per day is expected from a team than from a single person, thus the project or the task can be done in less time.

10.4 WATTS AND HORSEPOWER

As we explained in the previous section, power is defined as the time rate of doing work, or stated another way, work or energy divided by time. The units of power in SI units are defined in the following manner:

$$\text{power} = \frac{\text{work}}{\text{time}} = \frac{(\text{force})(\text{distance})}{\text{time}} = \frac{\text{N}\cdot\text{m}}{\text{s}} = \frac{\text{J}}{\text{s}} = \text{W} \tag{10.10}$$

Note the following: 1 N·m is called 1 joule (J), and 1 J/s is called 1 watt (W). In U.S. Customary Units, the units of power are expressed in $\text{lb}_f\cdot\text{ft/s}$ and horsepower (hp), in the following manner:

$$\text{power} = \frac{\text{work}}{\text{time}} = \frac{(\text{force})(\text{distance})}{\text{time}} = \frac{\text{lb}_f\cdot\text{ft}}{\text{s}} \tag{10.11}$$

and

$$1\ \text{hp} = 550\frac{\text{lb}_f\cdot\text{ft}}{\text{s}} \tag{10.12}$$

The U.S. Customary Units of power are related to the SI unit of power, watt, in the following manner:

$$1\frac{\text{lb}_f\cdot\text{ft}}{\text{s}} = 1.3558\ \text{W} \cong 1.36\ \text{W} \tag{10.13}$$

$$1\ \text{hp} = 745.69\ \text{W} \cong 746\ \text{W} \tag{10.14}$$

Remember that $1(\text{lb}_f\cdot\text{ft/s})$ is slightly greater in magnitude than 1 watt. Also keep in mind that 1 hp is slightly smaller than 1 kW. Another unit that is sometimes confused for the unit of power is kilowatt hour, used in measuring the consumption of electricity by homes and the manufacturing sector. First, kilowatt hour (kWh) is a unit of energy—not power. One kilowatt hour represents the amount of energy consumed during 1 hour by a device that uses one kilowatt or 1000 joules per second. Therefore,

$$1\ \text{kW} = 1000\ \text{W} = 1000\ \text{J/s}$$

$$1\ \text{kWh} = (1000\ \text{J/s})(3600\ \text{s}) = 3{,}600{,}000\ \text{J} = 3.6\ \text{MJ}$$

$$1\ \text{kWh} = 3.6\ \text{MJ}$$

In heating, ventilating, and air-conditioning (HVAC) applications, Btu per hour (Btu/h) is used to represent the heat loss from a building during cold months and the heat gained by the building during summer months. The units of Btu/h are related to the unit of watt in the following manner:

$$1\ \text{Btu/h} = 1055\ \text{W} = 1.055\ \text{kW}$$

Another common unit used in the United States in air-conditioning and refrigeration systems is *ton of refrigeration or cooling*. One ton of refrigeration represents the capacity of a refrigeration system to freeze 2000 lb_m or 1 ton of liquid water at 32°F into 32°F ice in 24 hours. It is

$$1\ \text{ton of refrigeration} = 12{,}000\ \text{Btu/h}$$

In the case of an air-conditioning unit, one ton of cooling represents the capacity of the air-conditioning system to remove 12,000 Btu thermal energy from a building in 1 hour. Clearly, the capacity of a residential air-conditioning system depends on the size of the building, its construction, shading, the orientation of its windows, and its climatic location. Residential air-conditioning units generally have a 1- to 5-ton capacity. The sizes of home gas furnaces in the United States are also expressed in units of Btu/h. The size of a typical single-family-home gas furnace used in moderate winter conditions is 60,000 Btu/h.

To get a feel for the relative magnitudes that watt and horsepower physically represent, consider the following examples.

EXAMPLE 10.6

Determine the power required to move 30 people, with an average mass of 61 kg (135 lb_m) per person, between two floors of a building, a vertical distance of 5 m (16 ft) in 2 s.

The required power is determined by

$$\text{power} = \frac{\text{work}}{\text{time}} = \frac{(30\text{ persons})\left(61\,\dfrac{\text{kg}}{\text{person}}\right)\left(9.81\,\dfrac{\text{m}}{\text{s}^2}\right)(5\text{ m})}{2\text{ s}} \cong 45{,}000\text{ W}$$

The minimum energy requirement for this task is equivalent to providing electricity to fifteen 100-W light bulbs for 1 minute. Next time you feel lazy and are thinking about taking the elevator to go up one floor, reconsider and think about the total amount of energy that could be saved if people would take the stairs instead of taking the elevator to go up one floor. As an example, if 1 million people decided to take the stairs on a daily basis, the minimum amount of energy saved during a year, based on an estimate of 220 working days in a year, would be

$$\text{energy savings} = \left(\frac{90000\text{ J}}{30\text{ persons}}\right)\left(\frac{1}{\text{day}}\right)(1{,}000{,}000\text{ persons})(220\text{ days})$$

$$= 66 \times 10^9\text{ J} = 660\text{ GJ}$$

We will revisit this problem, after we discuss efficiency, to determine the amount of fuel needed in a power plant to provide the amount of energy that we just calculated in Example 10.6.

EXAMPLE 10.7

Determine the power required to move a person who weighs 220 lb a vertical distance of 2.5 ft in 1 s.

$$\text{power} = \frac{\text{work}}{\text{time}} = \frac{(220\text{ lb}_f)(2.5\text{ ft})}{1\text{ s}} = 550\,\frac{\text{lb}_f\cdot\text{ft}}{\text{s}} = 1\text{ hp}$$

Therefore, 1 horsepower represents the power required to lift a person weighing 220 lb_f, a distance of 2.5 ft in 1 second. There are a number of other ways to think about what 1 horsepower physically represents. It could also be interpreted as the power required to lift an object weighting 100 lb_f a distance of 5.5 ft in 1 second. How powerful are you?

EXAMPLE 10.8

Determine the power required to move an object that weighs 800 N (179.85 lb_f), a vertical distance of 4 m (13.12 ft) in 2 s.

The power requirement expressed in SI units is given by

$$\text{power} = \frac{\text{work}}{\text{time}} = \frac{(800\ \text{N})(4\ \text{m})}{2\text{s}} = 1600\ \text{W}$$

And using U.S. Customary Units, the power requirement is

$$\text{power} = \frac{\text{work}}{\text{time}} = \frac{(179.85\ \text{lb}_f)(13.12\ \text{ft})}{2\text{s}} = 1180\,\frac{\text{lb}_f\cdot\text{ft}}{\text{s}}$$

The power expressed in horsepower is

$$\text{power} = 1180\left(\frac{\text{lb}_f\cdot\text{ft}}{\text{s}}\right)\left(\frac{\text{hp}}{500\ \frac{\text{lb}_f\cdot\text{ft}}{\text{s}}}\right) = 2.14\ \text{hp}$$

In this example, think about the relationships among various units of power needed to perform the same amount of work in the same amount of time.

EXAMPLE 10.9

Most of you have seen car advertisements where the car manufacturer brags about how fast one of its cars can go from 0 to 60 mph. According to the manufacturer, the BMW 750iL 2001 model can go from 0 to 60 mph in 6.7 seconds. This performance is usually measured on a test track. The engine in the car is rated at 326 hp at 5000 rpm. The car has a reported weight of 4597 lb_f. Is the claim by the manufacturer justifiable?

Well, to answer this question you may agree it would be more fun to drive to a BMW dealership and take the car for a test run on a racing track. But let us answer this question with the knowledge you have gained so far in this course. We need to make some assumptions first; we can assume that the driver weighs 180 lb_f and the reported weight of the car includes enough gasoline for this test. Next, we convert the speed and the mass values into appropriate units.

$$V_i = 0\frac{\text{ft}}{\text{s}}$$

$$V_f = \left(60\frac{\text{mi}}{\text{h}}\right)\left(\frac{1\text{ h}}{3600\text{ s}}\right)\left(\frac{5280\text{ ft}}{1\text{ mi}}\right) = 88\frac{\text{ft}}{\text{s}}$$

$$m = \frac{\text{weight}}{g} = \frac{(4597 + 180)\text{ lb}_f}{32.2\dfrac{\text{ft}}{\text{s}^2}} = 148\text{ slugs}$$

Using Equation (10.1), we can determine the required work to go from 0 to 60 mph.

$$\text{work}_{1-2} = \frac{1}{2}mV_f^2 - \frac{1}{2}mV_i^2$$

$$\text{work}_{1-2} = \frac{1}{2}(148\text{ slugs})\left(88\frac{\text{ft}}{\text{s}}\right)^2 - 0 = 573{,}056\text{ lb}_f\text{·ft}$$

The power requirement to perform this work in 6.7 seconds is

$$\text{power} = \frac{\text{work}}{\text{time}} = \frac{573{,}056\text{ lb}_f\text{·ft}}{6.7\text{ s}} = 85{,}530\frac{\text{lb}_f\text{·ft}}{\text{s}}$$

The power expressed in horsepower is

$$\text{power} = 85{,}530\left(\frac{\text{lb}_f\text{·ft}}{\text{s}}\right)\left(\frac{\text{hp}}{550\dfrac{\text{lb}_f\text{·ft}}{\text{s}}}\right) = 155.5\text{ hp}$$

Keeping in mind that power is needed to overcome air resistance and that there are always additional mechanical losses in the car, it is still safe to say the claim is good.

10.5 EFFICIENCY

As we mentioned earlier, there is always some loss associated with a dynamic system. In engineering, when we wish to show how well a machine or a system is functioning, we express its efficiency. In general, the overall efficiency of a system is defined as:

$$\text{efficiency} = \frac{\text{actual output}}{\text{required input}} \tag{10.15}$$

All machines and engineering systems require more input than what they put out. In the next few sections, we will look at the efficiencies of common engineering components and systems.

Power Plant Efficiency

Water is used in all steam power-generating plants to produce electricity. A simple schematic of a power plant is shown in Figure 10.10. Fuel is burned in a boiler to generate heat, which in turn is added to liquid water to change its phase to steam; steam passes

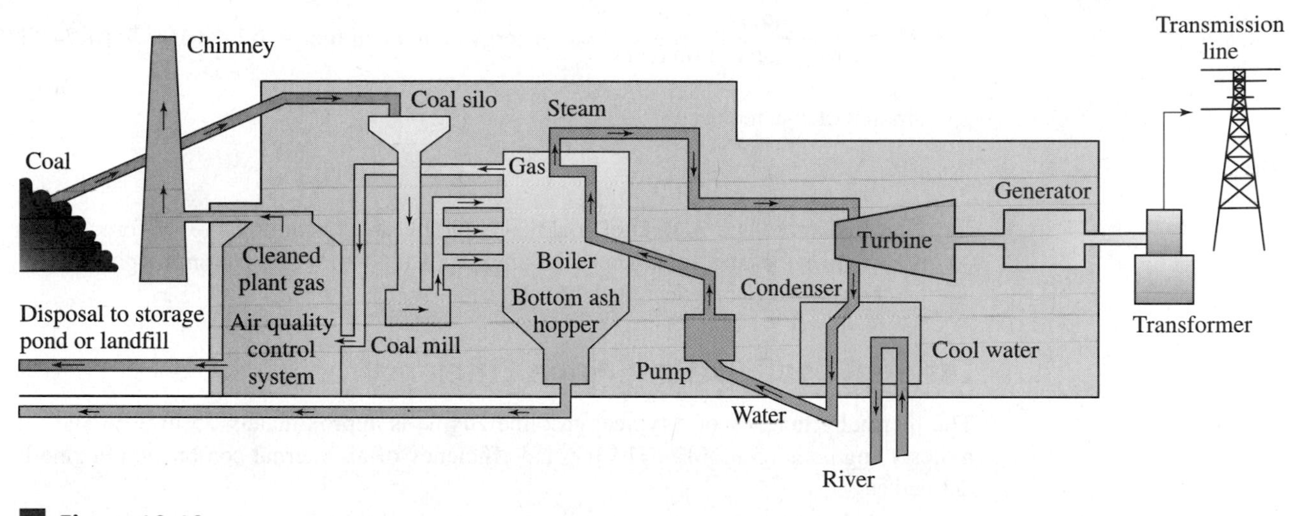

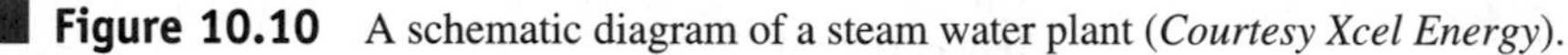

Figure 10.10 A schematic diagram of a steam water plant (*Courtesy Xcel Energy*).

through turbine blades, turning the blades, which in effect runs the generator connected to the turbine, creating electricity. The low-pressure steam liquefies in a condenser and is pumped through the boiler again, completing a cycle, as shown in Figure 10.10. The overall efficiency of a steam power plant is defined as

$$\text{power plant efficiency} = \frac{\text{energy generated}}{\text{energy input from fuel}} \tag{10.16}$$

The efficiency of today's power plants where a fossil fuel (oil, gas, coal) is burned in the boiler is near 40%, and for the nuclear power plants the overall efficiency is nearly 34%.

Electricity is also generated by liquid water stored behind dams. The water is guided into water turbines located in hydroelectric power plants housed within the dam to generate electricity. The potential energy of the water stored behind the dam is converted to kinetic energy as the water flows through the turbine and consequently spins the turbine, which turns the generator.

EXAMPLE 10.10

In Example 10.6 we determined the power required to move 30 people, with an average mass of 61 kg (135 lb_m) per person, between two floors of a building, a vertical distance of 5 m (16 ft) in 2 s. The energy and the power requirements were 90,000 J and 45,000 W, respectively. Moreover, we estimated savings in energy if 1 million people decided to walk up a floor instead of taking the elevator on a daily basis. The minimum amount of energy saved during a year, based on an estimate of 220 working days in a year, would be 660 GJ. Let us now estimate the amount of fuel, such as coal, that can be saved in a power plant, assuming a 38% overall efficiency for the power plant and a heating value of approximately 7.5 MJ/kg for coal.

$$\text{power plant efficiency} = \frac{\text{energy generated}}{\text{energy input from fuel}}$$

$$0.38 = \frac{660\ \text{GJ}}{\text{energy input from fuel}} \quad \Rightarrow \quad \text{energy input from fuel} = 1.74 \times 10^{12}\ \text{J} = 1.74\ \text{TJ}$$

$$\text{amount of coal required} = \frac{1.74 \times 10^{12}\ \text{J}}{7.5 \times 10^{6}\ \frac{\text{J}}{\text{kg}}} = 232{,}000\ \text{kg}\ (511{,}472\ \text{lb}_\text{m})$$

As you can see, the amount of coal that could be saved is quite large! Before you get on an elevator next time, think about the amount of fuel—not to mention the pollution—that can be saved if people just walk up a floor!

Internal Combustion Engine Efficiency

The thermal efficiency of a typical gasoline engine is approximately 25 to 30% and for a diesel engine is 35 to 40%. The thermal efficiency of an internal combustion engine is defined as

$$\text{thermal efficiency} = \frac{\text{power output}}{\text{heat power input as fuel is burned}} \tag{10.17}$$

Keep in mind that when expressing the overall efficiency of a car, one must account for the mechanical losses as well.

Motor and Pump Efficiency

As we explained in Chapter 9, motors run many devices and equipment that make our lives comfortable and less laborious. As an example, we identified a large number of motors in various devices at home, including motors that run the compressor of your refrigerator, garbage disposer, exhaust fans, tape player in a VCR, vacuum cleaner, turntable of a microwave, hair dryer, electric shaver, computer fan, and computer hard drive. When selecting

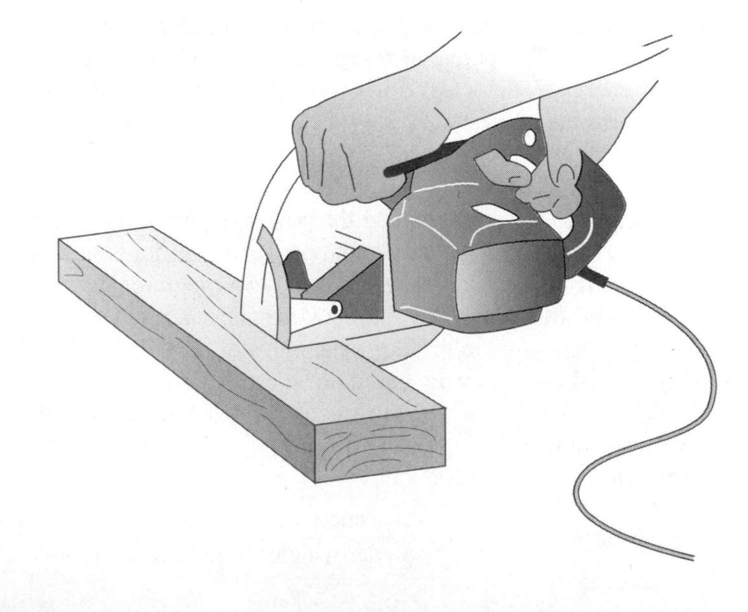

motors for these products, engineers consider the efficiency of the motor as one of the design criteria. The efficiency of an electric motor can be simply defined as

$$\text{efficency} = \frac{\text{power input to the device being driven}}{\text{electric power input to the motor}} \tag{10.18}$$

The efficiency of motors is a function of load and speed. The electric motor manufacturers provide performance curves and tables for their products that show, among other information, the efficiency of the motor.

You find pumps in hydraulic systems, the fuel system of your car, and systems that deliver water to a city piping network. Pumps are also used in food processing and in petrochemical plants. The function of a pump is to increase the pressure of a liquid entering the pump. The pressure rise in the fluid is used to overcome pipe friction and losses in fittings and valves and to transport the liquid to a higher elevation. Pumps themselves are driven by motors and engines. The efficiency of a pump is defined by

$$\text{efficiency} = \frac{\text{power input to the fluid by the pump}}{\text{power input to the pump by the motor}} \tag{10.19}$$

The efficiency of a pump, at a given operating speed, is a function of flow rate and the pressure rise (head) of the pump. Manufacturers of pumps provide performance curves or tables that show, among other performance data, the efficiency of the pump.

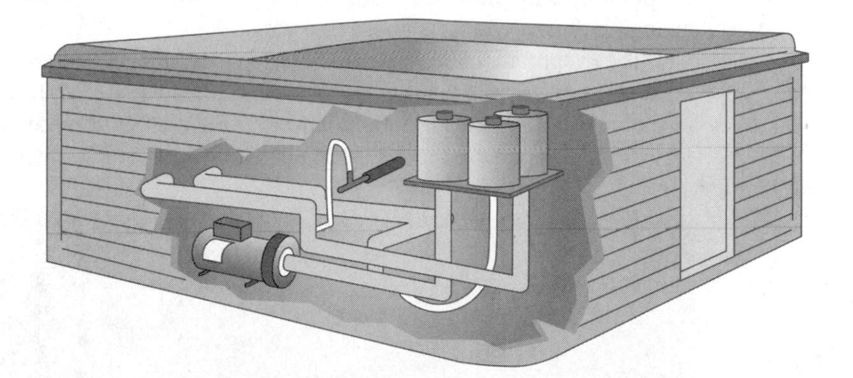

Efficiency of Heating, Cooling, and Refrigeration Systems

Before we discuss the efficiency of heating, cooling, and refrigeration systems, let us briefly explain the main components of these systems and how they operate. We begin with the cooling and refrigeration systems because their designs and operations are similar. Most of today's air-conditioning and refrigeration systems are designed according to a vapor–compression cycle. A schematic diagram of a simple vapor–compression cycle is shown in Figure 10.11. The main components of the vapor–compression cycle include a condenser, an evaporator, a compressor, and a throttling device, such as an expansion valve or a capillary tube, as shown in Figure 10.11.

Refrigerant is the fluid that transports thermal energy from the evaporator, where thermal energy or heat is absorbed, to the condenser, where the thermal energy is ejected to the surroundings. Referring to Figure 10.11, at state 1, the refrigerant exlsts as a mixture of

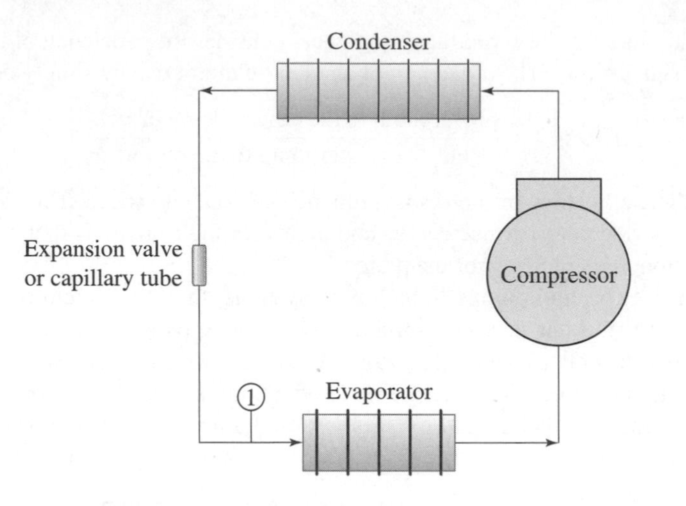

Figure 10.11
A schematic diagram of a vapor–compression cycle

liquid and gas. As the refrigerant flows through the evaporator, its phase completely changes to vapor. The phase change occurs because of the heat transfer from the surroundings to the evaporator and consequently to the refrigerant. The refrigerant enters the evaporator tube in a liquid/vapor mixture phase at very low temperature and pressure. The temperature of the air surrounding the evaporator is higher than the temperature of the evaporator, and thus heat transfer occurs from the surrounding air to the evaporator, changing the refrigerant's phase into vapor.

The evaporator in a refrigerator is a coil that is made from series of tubes that are located inside the freezer section of the refrigerator (see Figure 10.12). In an air-conditioning unit, the evaporator coil is located in the ductwork, near the fan–furnace unit inside the house. After leaving the evaporator, the refrigerant enters the compressor, where the temperature and the

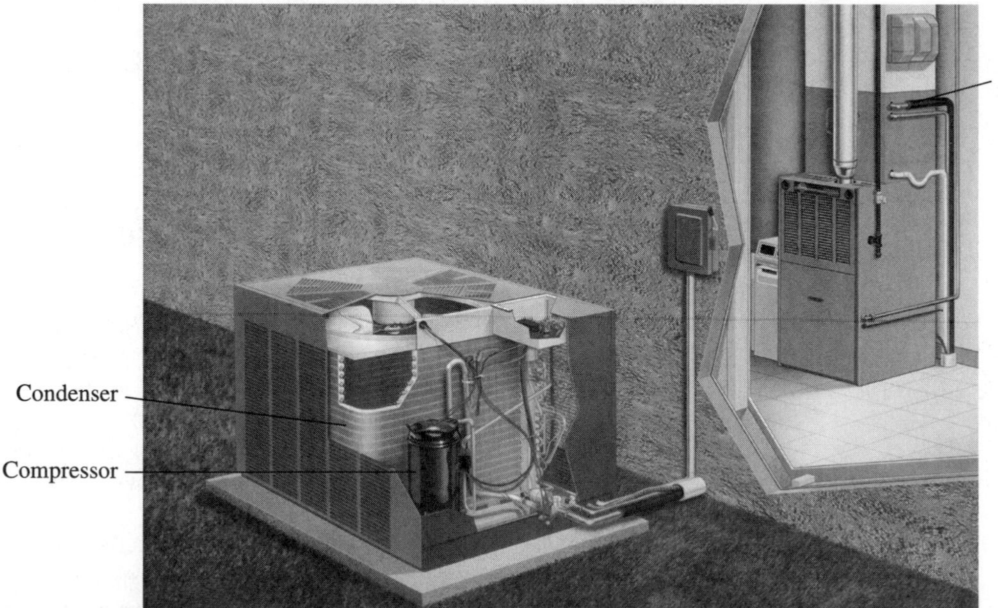

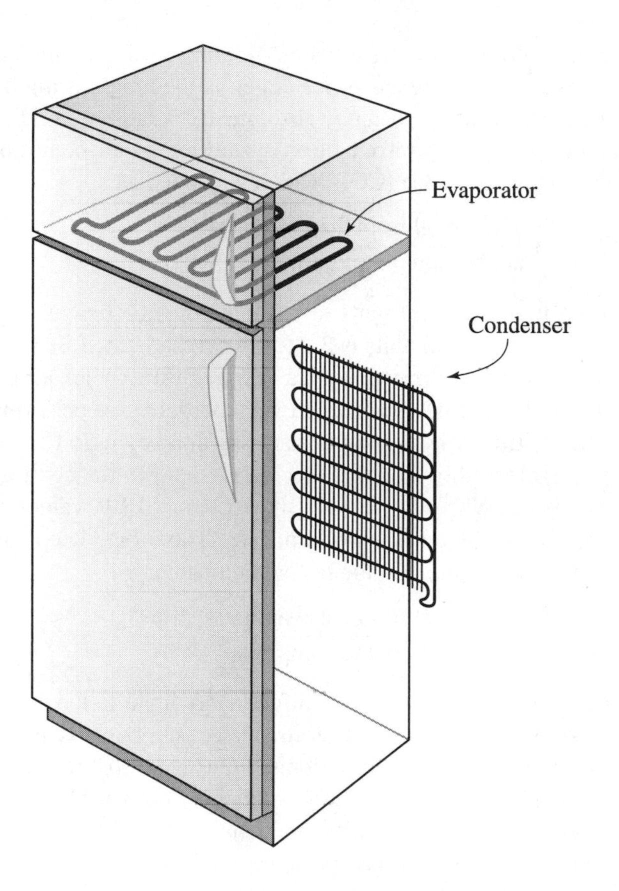

Figure 10.12
The typical locations of the evaporator and the condenser in a household refrigerator

pressure of the refrigerant is raised. The discharge side of the compressor is connected to the inlet side of the condenser, where the refrigerant enters the condenser in gaseous phase at a high temperature and pressure. Because the refrigerant in the condenser has a higher temperature than the surrounding air, heat transfer to the surrounding air occurs, and consequently thermal energy is ejected to the surroundings. Like an evaporator, a condenser is also made of a series of tubes with good thermal conductivity. As the refrigerant flows through the condenser, more and more heat is removed (or transferred to the surroundings); consequently, the refrigerant changes phase from gas to liquid and it leaves the condenser coil in liquid phase. In a household refrigerator, the condenser is the series of black tubes located on the back of the refrigerator. In an air-conditioning unit, the condenser is located outside the house in a housing that also contains the compressor and a fan that forces air over the condenser. After leaving the condenser, the liquid refrigerant flows through an expansion valve or a long capillary tube, which makes the refrigerant expand. The expansion is followed by a drop in the refrigerant's temperature and pressure. The refrigerant leaves the expansion valve or the capillary tube and flows into the evaporator to complete the cycle shown in Figure 10.11.

Another point worth mentioning is that as the warm air flows over the evaporator section of an air-conditioning unit, and as the air cools, the moisture (the water vapor) in the air condenses on the outside of the evaporator coil. The condensation forming on the outside of the evaporator is eventually drained. Thus, the evaporator acts as a dehumidification device as well. This process is similar to what happens when hot, humid air comes into contact with a glass of ice water. You have all seen the condensation that forms on the outside surface of a glass of ice water as the neighboring hot, humid air cools down. We will discuss what we mean by absolute and relative humidity in Chapter 11.

The efficiency of a refrigeration system or an air-conditioning unit is given by the coefficient of performance (COP) which is defined as

$$\text{COP} = \frac{\text{heat removal from the evaporator}}{\text{energy input to the compressor}} \tag{10.20}$$

You should use consistent units to calculate the coefficient of performance. The COP of most vapor compression units is 2.9 to 4.9. In the United States, it is customary to express the coefficient of performance of a refrigeration or an air-conditioning system using mixed SI and U.S. units. Quite often, the coefficient of performance is called energy efficiency ratio (EER) or the seasonal energy efficiency ratio (SEER). In such cases, the heat removal is expressed in Btu, and the energy input to the compressor is expressed in watt-hours, and because 1 Wh = 3.412 Btu, EER or SEER values of greater than 10 are obtained for the coefficient of performance. Therefore, keep in mind that in the United States, the EER is defined in the following manner:

$$\text{EER} = \frac{\text{heat removal from the evaporator (Btu)}}{\text{energy input to the compressor (Wh)}} \tag{10.21}$$

The reason for using the units of Wh for energy input to the compressor is that compressors are powered by electricity, and electricity consumption is measured (even in the United States) in kWh. Today's air-conditioning units have SEER values that range from approximately 10 to 17. In fact, all new air-conditioning units sold in the United States must have a SEER value of at least 10. In 1992, the United States government established the minimum standard efficiencies for various appliances, including air-conditioning units and gas furnaces.

In a gas furnace, natural gas is burned, and as a result the hot combustion products go through the inside of a heat exchanger where thermal energy is transported to cold indoor air that is passing over the heat exchanger. The warm air is distributed to various parts of the house through conduits. As mentioned, in 1992 the United States government established a minimum annual fuel utilization efficiency (AFUE) rating of 78% for furnaces installed in new homes, so manufacturers must design their gas furnaces to adhere to this standard. Today, most high-efficiency furnaces offer AFUE ratings in the range of 80 to 96%.

EXAMPLE 10.11

An air-conditioning unit has a cooling capacity of 24,000 Btu/h. If the unit has a rated energy efficiency ratio (EER) of 10, how much electrical energy is consumed by the unit in 1 h? If a power company charges 12 cents per kWh usage, how much would it cost to run the air conditioning unit for a month (30 days), assuming the unit runs 10 hours a day? What is the coefficient of performance (COP) for the given air-conditioning unit?

We can compute the energy consumption of the given air-conditioning unit using Equation (10.21).

$$\text{EER} = \frac{\text{heat removal from the evaporator (Btu)}}{\text{energy input to the compressor (Wh)}}$$

$$10 = \frac{24{,}000 \text{ (Btu)}}{\text{energy input to the compressor (Wh)}}$$

Energy input to the unit for 1 h of operation = 2400 Wh = 2.4 kWh.

The cost to run the unit for a month for a period of 10 hours a day is calculated in the following manner:

$$\text{cost to operate the unit} = \left(\frac{2.4\ \text{kWh}}{1\ \text{h}}\right)\left(\frac{10\ \text{h}}{\text{day}}\right)\left(\frac{\$0.12}{\text{kWh}}\right)(30\ \text{days}) = \$86.40$$

The coefficient of performance (COP) is calculated from Equation (10.21):

$$\text{COP} = \frac{\text{heat removal from the evaporator}}{\text{energy input to the compressor}} = \frac{24{,}000\ \text{Btu}}{(2400\ \text{Wh})\left(\dfrac{1\ \text{Wh}}{3.412\ \text{Btu}}\right)} = 2.9$$

Note the relationship between the EER and COP:

$$\text{COP} = \frac{\text{EER}}{3.412} = \frac{10}{3.412} = 2.9$$

SUMMARY

Now that you have reached this point in the text

- You should understand how different forms of mechanical energy are defined. You should know how and when to use kinetic energy, potential energy, and elastic energy in engineering analyses.
- You should know what is meant by internal energy and heat.
- You should know how and when to use conservation of energy to solve engineering problems.
- You should clearly understand the definition of power, its common units, and how it is related to work and energy.
- You should know the basic definition of efficiency and be familiar with its various forms, including the definitions of thermal efficiency, SEER, and AFUE, which are commonly used to express the efficiencies of components such as pumps, compressors, and motors, as well as complete systems including heating, ventilating, and air-conditioning (HVAC) equipment and household appliances.

PROBLEMS

1. Identify ways that you can save energy. For example, walking up a floor instead of taking the elevator, or walking or riding your bike an hour a day instead of taking the car. Estimate the amount of energy that you could save every year with your proposal. Also, estimate the amount of fuel that can be saved in the same manner. State your assumptions, and present your detailed analysis in a report.

2. Look up the manufacturer's data for the most recent year of the following cars:
   a. Toyota Camry
   b. Honda Accord
   c. Buick LeSabre

   You can visit the cars.com Web site to gather information. For each car, perform calculations similar to Example 10.9 to determine the power required to accelerate the car from 0 to 60 mph. State all of your assumptions.

3. An elevator has a rated capacity of 2000 lb. It can transport people at the rated capacity between the first and the fifth floors, with a vertical distance of 15 ft between each floor, in 7 s. Estimate the power requirement for such an elevator.
4. Determine the gross force needed to bring a car that is traveling at 110 km/h to a full stop in a distance of 100 m. The mass of the car is 2100 kg. What happens to the initial kinetic energy? Where does it go or to what form of energy does the kinetic energy convert?
5. A centrifugal pump is driven by a motor. The performance of the pump reveals the following information:

   Power input to the pump by the motor (kW): 0.5, 0.7, 0.9, 1.0, 1.2
   Power input to the fluid by the pump (kW): 0.3, 0.55, 0.7, 0.9, 1.0

   Plot the efficiency curve.
6. A power plant has an overall efficiency of 30%. The plant generates 20 MW of electricity, and uses coal from Montana (see Table 8.12) as fuel. Determine how much coal must be burned to sustain the generation of 20 MW of electricity.
7. Estimate the amount of gasoline that could be saved if all of the passenger cars in the United States were driven 1000 miles less each year. State your assumptions and write a brief report discussing your findings.
8. Investigate the typical power consumption range of the following products:
   a. home refrigerator
   b. 25-inch television set
   c. clothes washer
   d. electric clothes dryer
   e. vacuum cleaner
   f. hair dryer

   Discuss your findings in a brief report.
9. Investigate the typical power consumption range of the following products:
   a. personal computer with a 19-inch monitor
   b. laser printer
   c. cellular phone
   d. palm calculator

   Discuss your findings in a brief report.
10. Look up the furnace size and the size of the air-conditioning unit in your own home or apartment. Investigate the SEER and the AFUE of the units.
11. Investigate the size of a gas furnace used in a typical single-family dwelling in upstate New York, and compare that size to the furnaces used in Minnesota and in Kansas.
12. An air-conditioning unit has a cooling capacity of 18,000 Btu/h. If the unit has a rated energy efficiency ratio (EER) of 11, how much electrical energy is consumed by the unit in 1 h? If a power company charges 14 cents per kWh usage, how much would it cost to run the air-conditioning unit for a month (31 days), assuming the unit runs 8 h a day? What is the coefficient of performance (COP) for the given air-conditioning unit?
13. Visit a store that sells window-mount air-conditioning units. Obtain information on their rated cooling capacities and EER values. Contact your local power company and determine the cost of electricity in your area. Estimate how much it will cost to run the air-conditioning unit during the summer. Write a brief report to your instructor discussing your findings and assumptions.

Hoover Dam*

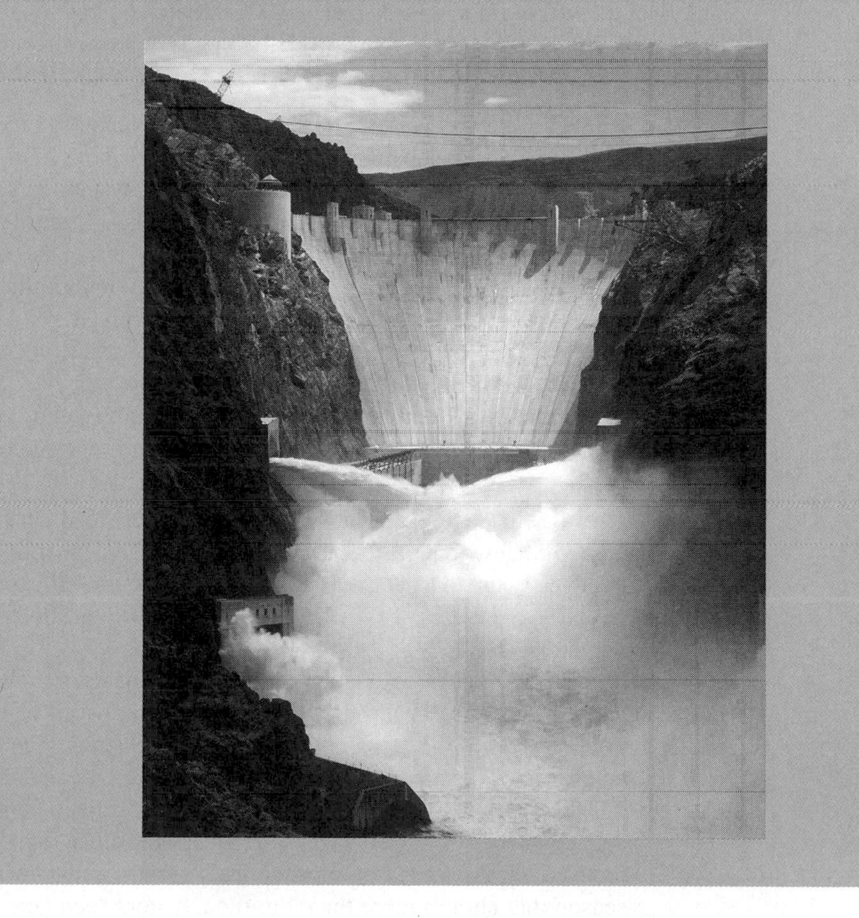

Hoover Dam is one of the Bureau of Reclamations' multipurpose projects on the Colorado River. These projects control floods; they store water for irrigation, municipal, and industrial use; and they provide hydroelectric power, recreation, and fish and wildlife habitat.

The Hoover Dam is a concrete arch–gravity type of dam, in which the water load is carried by both gravity action and horizontal arch action. The first concrete for the dam was placed on June 6, 1933, and the last concrete was placed in the dam on May 29, 1935. The following is a summary of some facts about the Hoover Dam.

*Materials were adapted from U.S. Bureau of Reclamation.

Dam dimensions: Height: 726.4 ft; length at crest: 1244 ft; width at top: 45 ft; width at base: 660 ft
Weight: 6.6 million tons
Reservoir statistics: Capacity: 28,537,000 acre-feet; length: 110 mi; shoreline: 550 mi; max depth: 500 ft; surface area:157,000 acres
Quantities of materials used in project: Concrete: 4,440,000 yd^3; explosives: 6,500,000 lb; plate steel and outlet pipes: 88,000,000 lb; pipe and fittings: 6,700,000 lb (840 mi); reinforcement steel: 45,000,000 lb; concrete mix proportions: cement: 1.00 part, sand: 2.45 parts, fine gravel: 1.75 parts, intermediate gravel: 1.46 parts, coarse gravel: 1.66 parts, cobbles (3 to 9 in.): 2.18 parts; water: 0.54 parts

The dam was built in blocks, or vertical columns, varying in size from approximately 60 ft square at the upstream face of the dam to about 25 ft square at the downstream face. Adjacent columns were locked together by a system of vertical keys on the radial joints and horizontal keys on the circumferential joints. After the concrete was cooled, a cement and water mixture called grout was forced into the spaces created between the columns by the contraction of the cooled concrete to form a monolithic (one-piece) structure.

Hoover Dam itself contains 3.25 million yd^3 of concrete. Altogether, there are 4,360,000 yd^3 of concrete in the dam, power plant, and appurtenant works. This much concrete would build a monument 100 ft square and 2-1/2 mi high; would rise higher than the Empire State Building (which is 1250 ft) if placed on an ordinary city block; or would pave a standard highway, 16 ft wide, from San Francisco to New York City.

The Reservoir At elevation 1221.4, Lake Mead, the largest man-made lake in the United States, contains 28,537,000 acre-feet (an acre-foot is the amount of water required to cover 1 acre to a depth of 1 foot). This reservoir will store the entire average flow of the river for two years. That much water would cover the entire state of Pennsylvania to a depth of 1 ft.

Lake Mead extends approximately 110 mi upstream toward the Grand Canyon and approximately 35 mi up the Virgin River. The width of Lake Mead varies from several hundred feet in the canyons to a maximum of 8 mi. The reservoir covers about 157,900 acres, or 247 square miles.

Recreation, although a by-product of this project, constitutes a major use of the lakes and controlled flows created by Hoover and other dams on the lower Colorado River today. Lake Mead is one of America's most popular recreation areas, with a 12-month season that attracts more than 9 million visitors each year for swimming, boating, skiing, and fishing. The lake and surrounding area are administered by the National Park Service as part of the Lake Mead National Recreation Area, which also includes Lake Mohave downstream from Hoover Dam.

The Power Plant There are 17 main turbines in Hoover Power plant. The original turbines were all replaced through an upgrading program between 1986 and 1993. With a rated capacity of 2,991,000 hp, and two station-service units rated at 3500 hp each, for a plant total of 2,998,000 hp, the plant has a nameplate capacity of 2,074,000 kW. This includes the two station-service units, which are rated at 2400 kW each.

Hoover Dam generates low-cost hydroelectric power for use in Nevada, Arizona, and California. Hoover Dam alone generates more than 4 billion kWh a year—enough to serve 1.3 million people. From 1939 to 1949, Hoover Power plant was the world's largest hydroelectric installation; with an installed capacity of 2.08 million kW, it is still one of this country's largest.

The $165 million cost of Hoover Dam has been repaid, with interest, to the federal treasury through the sale of its power. Hoover Dam energy is marketed by the Western Area Power Administration to 15 entities in Arizona, California, and Nevada under contracts that expire in 2017. More than half, 56%, goes to southern California users; Arizona contractors receive 19%, and Nevada users get 25%. The revenues from the sale of this power now pay for the dam's operation and maintenance. The power contractors also paid for the uprating of the power plant's nameplate capacity from 1.3 million to over 2.0 million kW.

PROBLEMS

1. Calculate the water pressure at the bottom of the dam when the water level is two-thirds of the height of the dam. Express your result in lb/in^2 and Pa. Also, calculate the magnitude of the force due to water pressure acting on a narrow strip (1 ft by 100 ft wide) located at the base of the dam.
2. As mentioned in the article describing Hoover Dam, Lake Mead contains 28,537,000 acre-feet of water (an acre-foot is the amount of water required to cover 1 acre to a depth of 1 ft). Express this water volume in gallons and m^3.
3. Hoover Dam generates more than 4 billion kWh a year. How many 100-watt light bulbs could be powered every hour by the Hoover Dam's power plant?
4. How much coal must be burned in a steam power plant with a thermal efficiency of 34% to generate enough power to equal the 4 billion KWh a year generated by Hoover Dam? Assume the coal comes from Montana (see Chapter 8 for heating value data).

11 Some Common Engineering Materials

Because pure aluminum is soft and has a relatively small tensile strength, it is alloyed with other metals to make it stronger, easier to weld, and to increase its resistance to corrosive environments. Lightweight metals, such as alloyed aluminum, are used in many structural and aerospace applications because of their small densities (relative to steel) and high strength-to-weight ratios.

As we discussed in Chapter 2, engineers design *millions of products and services that we use in our everyday lives: products such as cars, computers, aircraft, clothing, toys, home appliances, surgical equipment, heating and cooling equipment, health care devices, tools and machines that make various products, and so on. Engineers also* design *and supervise the construction of buildings, dams, highways, power plants, and mass transit systems.*

As design engineers, whether you are designing a machine part, a toy, a frame for a car, or a structure, the selection of material is an important design decision. There are a number of factors that engineers consider when selecting a material for a specific application. For example, they consider properties of material such as density, ultimate strength, flexibility, machinability, durability, thermal expansion, electrical and thermal conductivity, and resistance to corrosion. They also consider the cost of the material and how easily it can be repaired. Engineers are always searching for ways to use advanced materials to make products lighter and stronger for different applications.

In this chapter, we will look more closely at materials that are commonly used in various engineering applications. We will also discuss some of the basic physical characteristics of materials that are considered in design. We will examine solid materials such as metals and their alloys, plastics, glass, wood, and those that solidify over time, such as concrete. We will also investigate in more detail basic fluids, such as air and water, that not only are needed to sustain life but also play important roles in engineering. Did you ever stop to think about the important role that air plays in food processing, driving power tools, or in your car's tire to provide a cushiony ride? You may not think of water as an engineering material either, but we not only need water to live, we also need water to generate electricity in steam and hydroelectric power plants, and we use high-pressurized water, which functions like a saw, to cut materials. ■

11.1 MATERIAL SELECTION

Design engineers, when faced with selecting materials for their products, often ask questions such as: How strong will the material be when subjected to an expected load? Would it fail, and if not, how safely would the material carry the load? How would the material behave if its temperature were changed? Would the material remain as strong as it would under normal conditions if its temperature is increased? How much would it expand when its temperature is increased? How heavy and flexible is the material? What are its energy-absorbing properties? Would the material corrode? How would it react in the presence of some chemicals? How expensive is the material? Would it dissipate heat effectively? Would the material act as a conductor or as an insulator to the flow of electricity?

It is important to note here that we have only posed a few generic questions; we could have asked additional questions had we considered the specifics of the application. For example, when selecting materials for implants in bioengineering applications, one must consider many additional factors, including: Is the material toxic to the body? Can the material be sterilized? When the material comes into contact with body fluid will it corrode or deteriorate? Because the human body is a dynamic system, we should also ask: How would the material react to mechanical shock and fatigue? Are the mechanical properties of the implant material compatible with those of bone to ensure appropriate stress distributions at contact surfaces? These are examples of additional specific questions that one could ask to find suitable material for a specific application.

By now it should be clear that material properties and material cost are important design factors. However, in order to better understand material properties, we must first understand the phases of a substance. We discussed the phases of matter in Chapter 6; as a review and for the sake of continuity and convenience, we will briefly present the phases of matter again here.

The Phases of Matter: Solids, Liquids, Gases, and Plasma

As we discussed in Chapter 6, when you look around, you will find that matter exists in various forms and shapes. You will also notice that matter can change shape when its condition or its surroundings are changed. We also explained that all solid objects, liquids, gases, and living things are made of matter, and matter itself is made up of atoms or chemical elements. There are 106 known chemical elements to date. Atoms of similar characteristics are grouped together and shown in a table—the periodic table of chemical elements. Atoms are made up of even smaller particles we call *electrons*, *protons*, and *neutrons*. In your first chemistry class, you will study these ideas in more detail, if you have not yet done so. Some of you may decide to study chemical engineering, in which case you will spend a great deal of time studying chemistry. But for now, remember that atoms are the basic building blocks of all matter. Atoms are combined naturally, or in a laboratory setting, to create molecules. For example, as you already know, water molecules are made of two atoms of hydrogen and one atom of oxygen. A glass of water is made of billions and billions of homogeneous water molecules. A molecule is the smallest portion of a given matter that still possesses its microscopic characteristic properties.

Matter can exist in four states, depending on its own and the surrounding conditions: solid, liquid, gaseous, or plasma. Let us consider the water that we drink every day. As you already know, under certain conditions, water exists in a solid form that we call *ice*. At a standard atmospheric pressure, water exists in a solid form as long its temperature is kept under 0°C. Under standard atmospheric pressure, if you were to heat the ice and consequently change its temperature, the ice would melt and change into a liquid form. Under standard pressure at sea level, the water remains liquid up to a temperature of 100°C as you continue heating the water. If you were to carry out this experiment further by adding more heat to the liquid water, eventually the liquid water changes its phase from liquid into a gas. This phase of water we commonly refer to as *steam*. If you had the means to heat the water to even higher temperatures, temperatures exceeding 2000°C, you would find that you can break up the water molecules into their atoms, and eventually the atoms break up into free electrons and nuclei that we call *plasma*.

In general, the mechanical and thermophysical properties of a material depend on its phase. For example, as you know from your everyday experience, the density of ice is different from liquid water (ice cubes float in liquid water), and the density of liquid water is different from that of steam. Moreover, the properties of a material in a single phase could depend on its temperature and the surrounding pressure. For example, if you were to look up the density of liquid water in the temperature range of, say, 4° to 100°C, under standard atmospheric pressure, you would find that its density decreases with increasing temperature in that range. Therefore, properties of materials depend not only on their phase but also on their temperature and pressure. This is another important fact to keep in mind when selecting materials.

11.2 ELECTRICAL, MECHANICAL, AND THERMOPHYSICAL PROPERTIES OF MATERIALS

As we have been explaining up to this point, when selecting a material for an application, as an engineer you need to consider a number of material properties. In general, the properties of a material may be divided into three groups: electrical, mechanical, and thermal. In electrical and electronic applications, the electrical resistivity of materials is important. How much resistance to the flow of electricity does the material offer? In many mechanical, civil, and aerospace engineering applications, the mechanical properties of materials are important. These properties include modulus of elasticity, modulus of rigidity, tensile strength, compression strength, the strength-to-weight ratio, modulus of resilience, and modulus of toughness. In applications dealing with fluids (liquids and gases), thermophysical properties such as thermal conductivity, heat capacity, viscosity, vapor pressure, and compressibility are important properties. Thermal expansion of a material, whether solid or fluid, is also an important design factor. Resistance to corrosion is another important factor that must be considered when selecting materials.

Material properties depend on many factors, including how the material was processed, its age, its exact chemical composition, and any nonhomogeneity or defect within the material. Material properties also change with temperature and time as the material ages. Most companies that sell materials will provide upon request information on the important properties of their manufactured materials. Keep in mind that when practicing as an engineer, you should use the manufacturer's material property values in your design calculations. The property values given in this and other textbooks should be used as typical values—not as exact values.

In the previous chapters, we have explained what some properties of materials mean. The meaning of those properties and other properties that we have not explained already are summarized next.

Electrical Resistivity The value of electrical resistivity is a measure of the resistance of material to the flow of electricity. For example, plastics and ceramics typically have high resistivity, whereas metals typically have low resistivity, and among the best conductors of electricity are silver and copper.

Density Density is defined as mass per unit volume; it is a measure of how compact the material is for a given volume. For example, the average density of aluminum alloys is 2700 kg/m^3 and compared to steel density of 7850 kg/m^3, aluminum has a density that is approximately one-third the density of steel.

Modulus of Elasticity (Young's Modulus) Modulus of elasticity is a measure of how easily a material will stretch when pulled (subject to a tensile force) or how well the material will shorten when pushed (subject to a compressive force). The larger the value of the modulus of elasticity is, the larger the required force would be to stretch or shorten the material. For example, the modulus of elasticity of aluminum alloy is in the range of 70 to 79 GPa, whereas steel has a modulus of elasticity in the range of 190 to 210 GPa; therefore, steel is approximately 3 times stiffer than aluminum alloys.

Modulus of Rigidity (Shear Modulus) Modulus of rigidity is a measure of how easily a material can be twisted or sheared. The value of modulus of rigidity, also called *shear modulus*, shows the resistance of a given material to shear deformation. Engineers consider the value of shear modulus when selecting materials for shafts and rods that are subjected to twisting torques. For example, the modulus of rigidity or shear modulus for aluminum alloys is in the range of 26 to 36 GPa, whereas the shear modulus for steel is in the range of 75 to 80 GPa. Therefore, steel is approximately 3 times more rigid in shear than aluminum.

Tensile Strength The tensile strength of a piece of material is determined by measuring the maximum tensile load a material specimen in the shape of a rectangular bar or cylinder can carry without failure. The tensile strength or ultimate strength of a material is expressed as the maximum tensile force per unit cross-sectional area of the specimen. When a material specimen is tested for its strength, the applied tensile load is increased slowly. In the very beginning of the test, the material will deform elastically, meaning that if the load is removed, the material will return to its original size and shape without any permanent deformation. The point to which the material exhibits this elastic behavior is called *yield point*. The yield strength represents the maximum load that the material can carry without any permanent deformation. In certain engineering design applications (especially involving brittle materials), the yield strength is used as the tensile strength.

Compression Strength Some materials are stronger in compression than they are in tension; concrete is a good example. The compression strength of a piece of material is determined by measuring the maximum compressive load a material specimen in the shape of cylinder or cube can carry without failure. The ultimate compressive strength of a material is expressed as the maximum compressive force per unit cross-sectional area of the specimen. Concrete has a compressive strength in the range of 10 to 70 MPa.

Modulus of Resilience Modulus of resilience is a mechanical property of a material that shows how effective the material is in absorbing mechanical energy without sustaining any permanent damage.

Modulus of Toughness Modulus of toughness is a mechanical property of a material that indicates the ability of the material to handle overloading before it fractures.

Strength-to-Weight Ratio As the term implies, this is the ratio of the strength of the material to its specific weight (weight of the material per unit volume). Based on the application, engineers use either the yield or the ultimate strength of the material when determining the strength-to-weight ratio of a material.

Thermal Expansion The coefficient of linear expansion can be used to determine the change in the length (per original length) of a material that would occur if the temperature of the material were changed. This is an important material property to consider when designing products and structures that are expected to experience a relatively large temperature swing during their service lives.

Thermal Conductivity Thermal conductivity is a property of material that shows how good the material is in transferring thermal energy (heat) from a high-temperature region to a low-temperature region within the material.

Heat Capacity Some materials are better than others in storing thermal energy. The value of heat capacity represents the amount of thermal energy required to raise the temperature of 1 kilogram mass of a material by 1°C, or, using U.S. Customary Units, the amount of thermal energy required to raise one pound mass of a material by 1°F. Materials with large heat capacity values are good at storing thermal energy.

Viscosity, vapor pressure, and bulk modulus of compressibility are additional fluid properties that engineers consider in design.

Viscosity The value of viscosity of a fluid represents a measure of how easily the given fluid can flow. The higher the viscosity value is, the more resistance the fluid offers to flow. For example, it would require less energy to transport water in a pipe than it would to transport motor oil or glycerin.

Vapor Pressure Under the same conditions, fluids with low vapor pressure values will not evaporate as quickly as those with high values of vapor pressure. For example, if you were to leave a pan of water and a pan of glycerin side by side in a room, the water will evaporate and leave the pan long before you would notice any changes in the level of glycerin.

Bulk Modulus of Compressibility A fluid bulk modulus represents how compressible the fluid is. How easily can one reduce the volume of the fluid when the fluid pressure is increased? For example, as we discussed in Chapter 7, it would take a pressure of 2.24×10^7 N/m^2 to reduce 1 m^3 volume of water by 1% or, said another way, to a final volume of 0.99 m^3.

In this section, we explained the meaning and significance of some of the physical properties of materials. Tables 11.1 through 11.4 show some properties of the solid materials. In the following sections, we will examine the application and chemical composition of some common engineering materials.

TABLE 11.1 Modulus of Elasticity and Shear Modulus of Selected Materials

Material	Modulus of Elasticity (GPa)	Shear Modulus (GPa)
Aluminum alloys	70–79	26–30
Brass	96–110	36–41
Bronze	96–120	36–44
Cast iron	83–170	32–69
Concrete (compression)	17–31	
Copper alloys	110–120	40–47
Glass	48–83	19–35
Magnesium alloys	41–45	15–17
Nickel	210	80
Plastics		
Nylon	2.1–3.4	
Polyethylene	0.7–1.4	
Rock (compression)		
Granite, marble, quartz	40–100	
Limestone, sandstone	20–70	
Rubber	0.0007–0.004	0.0002–0.001
Steel	190–210	75–80
Titanium alloys	100–120	39–44
Tungsten	340–380	140–160
Wood (bending)		
Douglas fir	11–13	
Oak	11–12	
Southern pine	11–14	

Source: Adapted from J. M. Gere, *Mechanics of Materials*.

TABLE 11.2 Densities of Selected Materials

Material	Mass Density (kg/m^3)	Specific Weight (kN/m^3)
Aluminum alloys	2600–2800	25.5–27.5
Brass	8400–8600	82.4–84.4
Bronze	8200–8800	80.4–86.3
Cast iron	7000–7400	68.7–72.5
Concrete		
Plain	2300	22.5
Reinforced	2400	23.5
Lightweight	1100–1800	10.8–17.7
Copper	8900	87.3
Glass	2400–2800	23.5–27.5
Magnesium alloys	1760–1830	17.3–18.0
Nickel	8800	86.3
Plastics		
Nylon	880–1100	8.6–10.8
Polyethylene	960–1400	9.4–13.7

Continued

TABLE 11.2 Densities of Selected Materials (*continued*)

Material	Mass Density (kg/m3)	Specific Weight (kN/m3)
Rock		
Granite, marble, quartz	2600–2900	25.5–28.4
Limestone, sandstone	2000–2900	19.6–28.4
Rubber	960–1300	9.4–12.7
Steel	7850	77.0
Titanium alloys	4500	44.1
Tungsten	1900	18.6
Wood (air dry)		
Douglas fir	480–560	4.7–5.5
Oak	640–720	6.3–7.1
Southern pine	560–640	5.5–6.3

Source: Adapted from J. M. Gere, *Mechanics of Materials*.

TABLE 11.3 The Strength of Selected Materials

Material	Yield Strength (MPa)	Ultimate Strength (MPa)
Aluminum alloys	35–500	100–550
Brass	70–550	200–620
Bronze	82–690	200–830
Cast iron (tension)	120–290	69–480
Cast iron (compression)		340–1400
Concrete (compression)		10–70
Copper alloys	55–760	230–830
Glass		30–1000
Plate glass		70
Glass fibers		7000–20,000
Magnesium alloys	80–280	140–340
Nickel	100–620	310–760
Plastics		
Nylon		40–80
Polyethylene		7–28
Rock (compression)		
Granite, marble, quartz		50–280
Limestone, sandstone		20–200
Rubber	1–7	7–20
Steel		
High-strength	340–1000	550–1200
Machine	340–700	550–860
Spring	400–1600	700–1900
Stainless	280–700	400–1000
Tool	520	900
Steel wire	280–1000	550–1400
Structural steel	200–700	340–830

Continued

TABLE 11.3 The Strength of Selected Materials (*continued*)

Material	Yield Strength (MPa)	Ultimate Strength (MPa)
Titanium alloys	760–1000	900–1200
Tungsten		1400–4000
Wood (bending)		
Douglas fir	30–50	50–80
Oak	40–60	50–100
Southern pine	40–60	50–100
Wood (compression parallel to grain)		
Douglas fir	30–50	40–70
Oak	30–40	30–50
Southern pine	30–50	40–70

Source: Adapted from J. M. Gere, *Mechanics of Materials*.

TABLE 11.4 Coefficients of Thermal Expansion for Selected Materials*

Material	Coefficient of Thermal Expansion (1/°C) $\times 10^6$	Coefficient of Thermal Expansion (1/°F) $\times 10^6$
Aluminum alloys	23	13
Brass	19.1–21.2	10.6–11.8
Bronze	18–21	9.9–11.6
Cast iron	9.9–12	5.5–6.6
Concrete	7–14	4–8
Copper alloys	16.6–17.6	9.2–9.8
Glass	5–11	3–6
Magnesium alloys	26.1–28.8	14.5–16.0
Nickel	13	7.2
Plastics		
Nylon	70–140	40–80
Polyethylene	140–290	80–160
Rock	5–9	3–5
Rubber	130–200	70–110
Steel	10–18	5.5–9.9
High–strength	14	8.0
Stainless	17	9.6
Structural	12	6.5
Titanium alloys	8.1–11	4.5–6.0
Tungsten	4.3	2.4

*Note that you must multiply the coefficients given in this table by 10^{-6} to obtain the actual values of coefficients of thermal expansion.

Source: Adapted from J. M. Gere, *Mechanics of Materials*.

11.3 SOME COMMON SOLID ENGINEERING MATERIALS

In this section, we will briefly examine the chemical composition and common application of some solid materials. We will discuss light metals, copper and its alloys, iron and steel, concrete, wood, plastics, silicon, glass, and composite materials. Most of you will take a semester-long materials class during your sophomore or junior year and will learn more in depth about the atomic structure of various materials. Here our intent is to introduce you to materials and their applications.

Lightweight Metals

Aluminum, titanium, and magnesium, because of their small densities (relative to steel), are commonly referred to as *lightweight metals*. Because of their relatively high strength-to-weight ratios, lightweight metals are used in many structural and aerospace applications.

Aluminum and its alloys have densities that are approximately one-third the density of steel. Pure aluminum is very soft, thus it is generally used in electronics applications and in making reflectors and foils. Because pure aluminum is soft and has a relatively small tensile strength, it is alloyed with other metals to make it stronger, easier to weld, and to increase its resistance to corrosive environments. Aluminum is commonly alloyed with copper (Cu), zinc (Zn), magnesium (Mg), manganese (Mn), silicon (Si), and lithium (Li). The American National Standards Institute (ANSI) assigns designation numbers to specify aluminum alloys. Generally speaking, aluminum and its alloys resist corrosion; they are easy to mill and cut and can be brazed or welded. Aluminum parts can also be joined using adhesives. They are good conductors of electricity and heat and thus have relatively high thermal conductivity and low electrical resistance values. Aluminum is fabricated in sheets, plates, foil, rods, and wire and is extruded to make window frames or automotive parts. You are already familiar with everyday examples of common aluminum products, including beverage cans, household aluminum foil, nonrust staples in tea bags, building insulation, and so on.

Titanium has an excellent strength-to-weight-ratio. Titanium is used in applications where relatively high temperatures, exceeding 400° up to 600°C, are expected. Titanium alloys are used in the fan blades and the compressor blades of the gas turbine engines of commercial and military airplanes. In fact, without the use of titanium alloys the engines

Lightweight and durable, aluminum alloys are used to produce a wide range of products—from high-performance engines to soda cans.

Because of their excellent strength-to-weight ratio and resistance to corrosion, titanium alloys are used to produce a range of products—from bike frames to frames for glasses.

on commercial airplanes would not have been possible. Like aluminum, titanium is alloyed with other metals to improve its properties. Titanium alloys show excellent resistance to corrosion. Titanium is quite expensive compared to aluminum and it is heavier than aluminum, having a density which is roughly one-half that of steel. Because of their relatively high strength-to-weight ratios, titanium alloys are used in both commercial and military airplane airframes (fuselage and wings) and landing gear components. Titanium alloys are becoming a metal of choice in many products; you can find them in golf clubs, bicycle frames, tennis racquets, and spectacle frames. Because of their excellent corrosion resistance, titanium alloys have been used in the tubing in desalination plants as well. Replacement hips and other joints are examples of other applications where titanium is currently being used.

With its silvery white appearance, **magnesium** is another lightweight metal that looks like aluminum, but it is lighter, having a density of approximately 1700 kg/m^3. Pure magnesium does not provide good strength for structural applications and because of this fact, it is alloyed with other elements such as aluminum, manganese, and zinc to improve its mechanical characteristics. Magnesium and its alloys are used in nuclear applications, in drycell batteries, and in aerospace applications and some automobile parts as sacrificial anodes to protect other metals from corrosion. The mechanical properties of the lightweight metals are shown in Tables 11.1 through 11.4.

Copper and Its Alloys

Copper is a good conductor of electricity and because of this property is commonly used in many electrical applications, including home wiring. Copper and many of its alloys are

Examples of products made of copper alloys

also good conductors of heat, and this thermal property makes copper a good choice for heat exchanger applications in air conditioning and refrigeration systems. Copper alloys are also used as tubes, pipes, and fittings in plumbing and heating applications. Copper is alloyed with zinc, tin, aluminum, nickel, and other elements to modify its properties. When copper is alloyed with zinc it is commonly called *brass*. The mechanical properties of brass depend on the exact composition of percent copper and percent zinc. *Bronze* is an alloy of copper and tin. Copper is also alloyed with aluminum and is referred to as *aluminum bronze*. Copper and its alloys are also used in water tubes, heat exchangers, hydraulic brake lines, pumps, and screws.

Iron and Steel

Steel is a common material that is used in the framework of buildings, bridges, the body of appliances such as refrigerators, ovens, dishwashers, washers and dryers, and cooking utensils. Steel is an alloy of iron with approximately 2% or less carbon. Pure iron is soft and thus not good for structural applications, but the addition of even a small amount of carbon to iron hardens it and gives steel better mechanical properties, such as greater strength. The properties of steel can be modified by adding other elements, such as chromium, nickel, manganese, silicon, and tungsten. For example, chromium is used to increase the resistance of steel to corrosion. In general, steel can be classified into three broad groups: (1) the carbon steels containing approximately 0.015 to 2% carbon, (2) low-alloy steels having a maximum of 8% alloying elements, and (3) high-alloy steels containing more than 8% of alloying elements. Carbon steels constitute most of the world's steel consumption, thus you will commonly find them in the body of appliances

Alloy steels, because of mechanical properties such as superior strength, are used in a wide range of applications—from structural components to machine parts.

and cars. The low-alloy steels have good strength and are commonly used as machine or tool parts and as structural members. The high-alloy steels, such as stainless steels, could contain approximately 10 to 30% chromium and could contain up to 35% nickel. The 18/8 stainless steels, which contain 18% chromium and 8% nickel, are commonly used for tableware and kitchenware products. Finally, cast iron is also an alloy of iron that has 2 to 4% carbon. Note that the addition of extra carbon to the iron changes its properties completely. In fact, cast iron is a brittle material, whereas most iron alloys containing less than 2% carbon are ductile.

Concrete

Today, concrete is commonly used in construction of roads, bridges, buildings, tunnels, and dams. What is normally called concrete consists of three main ingredients: aggregate, cement, and water. Aggregate refers to materials such as gravel and sand, and cement refers to the bonding material that holds the aggregate together. The types and size (fine to coarse) of aggregate used in making concrete varies depending on the application. The amount of water used in making concrete could also influence its strength. Of course, the mixture must have enough water so that the concrete can be poured and have a consistent cement paste that completely wraps around all aggregates. The ratio of amount of cement to aggregate used in making concrete also affects the strength and durability of concrete.

Examples of concrete used in construction

Another factor that could influence the cured strength of concrete is the temperature of its surroundings when it is poured. Calcium chloride is added to cement when the concrete is poured in cold climates. The addition of calcium chloride will accelerate the curing process to counteract the effect of the low temperature of the surroundings. You may have also noticed as you walk by newly poured concrete for a driveway or sidewalk that water is sprayed onto the concrete for some time after it is poured. This is to control the rate of contraction of the concrete as it sets.

Concrete is a brittle material that can support compressive loads much better than it does tensile loads. Because of this fact, concrete is commonly *reinforced* with steel bars or steel mesh that consists of thin metal rods to increase its load-bearing capacity, especially in the sections where tensile stress is expected. Concrete is poured into forms that contain the metal mesh or steel bars. Reinforced concrete is used in foundations, floors, walls, and columns. Another common construction practice is the use of *precast concrete*. Precast concrete slabs, blocks, and structural members are fabricated in less time with less cost in factory settings where surrounding conditions are controlled. The precast concrete parts are then moved to the construction site where they are erected. This practice saves time and money. As we mentioned, concrete has a higher compressive strength than tensile strength. Because of this fact, concrete is also *prestressed* in the following manner. Before concrete is poured into forms that have the steel rods or wires, the steel rods or wires are stretched; after the concrete has been poured and after enough time has elapsed, the tension in the rods or wires is released. This process, in turn, compresses the concrete. The prestressed concrete then acts as a compressed spring, which will become uncompressed under the action of tensile loading. Therefore, the prestressed concrete section will not experience any tensile stress until the section has been completely uncompressed. It is important to note once again the reason for this practice is that concrete is weak under tension.

Wood

Throughout history, wood, because of its abundance in many parts of the world, has been a material of choice for many applications. Wood is a renewable source, and because of its ease of workability and its strength, it has been used to make many products. Wood also has been used as fuel in stoves and fireplaces. Today, wood is used in a variety of products ranging from telephone poles to toothpicks. Common examples of wood products include hardwood flooring, roof trusses, furniture frames, wall supports, doors, decorative items, window frames, trimming in luxury cars, tongue depressors, clothespins, baseball bats, bowling pins, fishing rods, and wine barrels (see Figure 11.1). Wood is also the main ingredient that is used to make various paper products. Whereas a steel structural member is susceptible to rust, wood, on the other hand, is prone to fire, termites, and rotting. Wood is anisotropic material, meaning that its properties are direction-dependent. For example, as you may already know, under axial loading (when pulled), wood is stronger in a direction parallel to a grain than it is in a direction across the grain. However, wood is stronger in a direction normal to the grain when it is bent. The properties of wood also depend on its moisture content; the lower the moisture content, the stronger the wood is. Density of wood is generally a good indication of its strength. As a rule of

Figure 11.1 Examples of wood products

thumb, the higher the density of wood, the higher its strength. Moreover, any defects, such as knots, would affect the load-carrying capacity of wood. Of course, the location of the knot and the extent of the defect will directly affect its strength.

Timber is commonly classified as *softwood* and *hardwood*. Softwood timber is made from trees that have cones (coniferous), such as pine, spruce, and Douglas fir. On the other hand, hardwood timber is made from trees that have broad leaves or have flowers. Examples of hardwoods include walnut, maple, oak, and beech. This classification of wood into softwood and hardwood should be used with caution, because there are some hardwood timbers that are softer than softwoods.

Plastics

In the latter part of the 20th century, plastics increasingly became the material of choice for many applications. They are very lightweight, strong, inexpensive, and easily made into various shapes. Over 100 million metric tons of plastic are produced annually worldwide. Of course, this number increases as the demand for inexpensive, durable, disposable material grows. Most of you are already familiar with examples of plastic products, including grocery and trash bags, plastic soft drink containers, home cleaning containers, vinyl siding, polyvinyl chloride (PVC) piping, valves, and fittings that are readily available in home improvement centers. Styrofoam™ plates and cups, plastic forks, knives, spoons, and sandwich bags are other examples of plastic products that are consumed every day.

Polymers are the backbone of what we call plastics. They are chemical compounds that have very large, molecular, chainlike structures. Plastics are often classified into two

Because of their low cost of production, polymer-based products are commonly used in our everyday lives.

categories: *thermoplastics* and *thermosets*. When heated to certain temperatures, the thermoplastics can be molded and remolded. For example, when you recycle Styrofoam dishes, they can be heated and reshaped into cups or bowls or other shapes. By contrast, thermosets can not be remolded into other shapes by heating. The application of heat to thermosets does not soften the material for remolding; instead, the material will simply break down. There are many other ways of classifying plastics; for instance, they may be classified on the basis of their chemical composition, or molecular structure, or the way molecules are arranged, or their densities. For example, based on their chemical composition, polyethylene, polypropylene, polyvinyl chloride, and polystyrene are the most commonly produced plastics. A grocery bag is an example of a product made from high-density polyethylene (HDPE). However, note that in a broader sense polyethylene and polystyrene, for example, are thermoplastics. In general, the way molecules of a plastic are arranged will influence its mechanical and thermal properties.

Plastics have relatively small thermal and electrical conductivity values. Some plastic materials such as Styrofoam cups are designed to have air trapped in them to reduce the heat conduction even more. Plastics are easily colored by using various metal oxides. For example, titanium oxide and zinc oxide are used to give a plastic sheet its white color. Carbon is used to give plastic sheets their black color, as is the case in black trash bags. Depending on the application, other additives are also added to the polymers to obtain specific characteristics such as rigidity, flexibility, enhanced strength, or a longer life span that excludes any change in the appearance or mechanical properties of the plastic over time. As with other materials, research is being performed every day to make plastics stronger and more durable and to control the aging process, to make plastics less susceptible to sun damage, and to control water and gas diffusion through them. The latter is especially important when the goal is to add shelf life to food that is wrapped in plastics. Those of you who are planning to study chemical engineering will take semester-long classes that will explore polymers in much more detail.

Silicon

Silicon is a nonmetallic chemical element that is used quite extensively in the manufacturing of transistors and various electronic and computer chips. Pure silicon is not found in nature; it is found in the form of silicon dioxide in sands and rocks or found combined

A computer chip

with other elements such as aluminum or calcium or sodium or magnesium in the form that is commonly referred to as *silicates*. Silicon, because of its atomic structure, is an excellent semiconductor, a material whose electrical conductivity properties can be changed to act either as a conductor of electricity or as an insulator (preventor of electricity flow). Silicon is also used as an alloying element with other elements such as iron and copper to give steel and brass certain desired characteristics.

Be sure not to confuse silicon with *silicones*, which are synthetic compounds consisting of silicon, oxygen, carbon, and hydrogen. You find silicones in lubricants, varnishes, and waterproofing products.

Glass

Glass is commonly used in products such as windows, light bulbs, TV CRT tubes, housewares such as drinking glasses, chemical containers, beverage and beer containers, and decorative items (see Figure 11.2). The composition of the glass depends on its application. The most widely used form of glass is soda–lime–silica glass. The materials used in making soda–lime–silica glass include sand (silicon dioxide), limestone (calcium carbonate), and soda ash (sodium carbonate). Other materials are added to create desired characteristics for specific applications. For example, bottle glass contains approximately 2% aluminum oxide, and glass sheets contain about 4% magnesium oxide. Metallic oxides are also added to give glass various colors. For example, silver oxide gives glass a yellowish stain, and copper oxide gives glass its blueish, greenish color, the degree depending on the amount added to the composition of the glass. Optical glasses have very specific chemical compositions and are quite expensive. The composition of optical glass will influence its refractive index and its light-dispersion properties. Glass that is made completely from silica (silicon dioxide) has properties that are sought after by many industries, such as fiber optics, but it is quite expensive to manufacture because the sand has to be heated to temperatures exceeding 1700°C. Silica glass has a low coefficient of

Figure 11.2
Examples of glass products

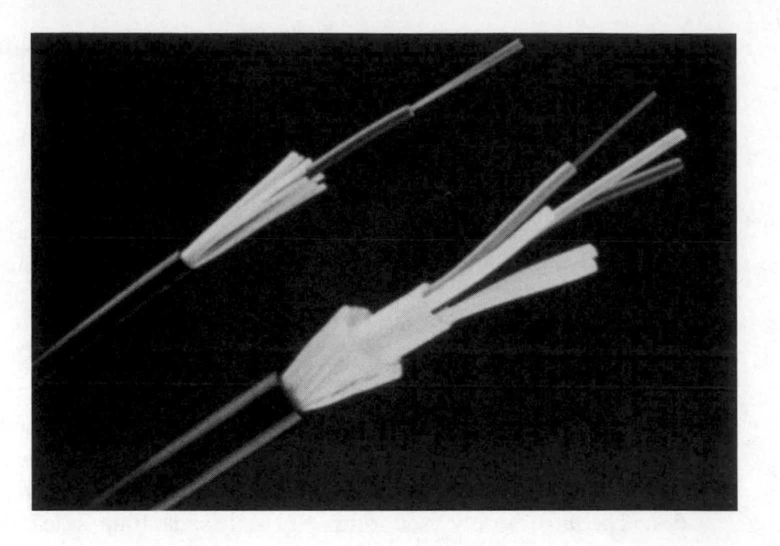

Examples of fiber optic cables made by Corning

thermal expansion, high electrical resistivity, and high transparency to ultraviolet light. Because silica glass has a low coefficient of thermal expansion, it can be used in high-temperature applications. Ordinary glass has a relatively high coefficient of thermal expansion; therefore, when its temperature is changed suddenly, it could break easily due to thermal stresses developed by the temperature rise. Cookware glass contains boric oxide and aluminum oxide to reduce its coefficient of thermal expansion.

Fiber Glass Silica glass fibers are commonly used today in fiber optics, that branch of science that deals with transmitting data, voice, and images through thin glass or plastic fibers. Every day, copper wires are being replaced by transparent glass fibers in telecommunication to connect computers together in networks. The glass fibers typically have an outer diameter of 0.125 mm (12 micron) with an inner transmitting core diameter of 0.01 mm (10 micron). Infrared light signals in the wavelength ranges of 0.8 to 0.9 m or 1.3 to 1.6 m are generated by light-emitting diodes or semiconductor lasers and travel through the inner core of glass fiber. The optical signals generated in this manner can travel to distances as far as 100 km without any need to amplify them again. Plastic fibers made of polymethylmethacrylate, polystyrene, or polycarbonate are also used in fiber optics. These plastic fibers are, in general, cheaper and more flexible than glass fibers. But when compared to glass fibers, plastic fibers require more amplification of signals due to their greater optical losses. They are generally used in networking computers in a building.

Composites

Because of their light weight and good strength, composite materials are becoming increasingly the materials of choice for a number of products and aerospace applications. Today you will find composite materials in military planes, helicopters, satellites, commercial planes, fast-food restaurant tables and chairs, and many sporting goods. They are also commonly used to repair bodies of automobiles. In comparison to conventional materials, such as metals, composite materials can be lighter and stronger. For this reason, composite materials are used extensively in aerospace applications.

Composites are created by combining two or more solid materials to make a new material that has properties that are superior to those of the individual components. Composite materials consist of two main ingredients: matrix material and fibers. Fibers are embedded in matrix materials, such as aluminum or other metals, plastics, or ceramics. Glass, graphite, and silicon carbide fibers are examples of fibers used in the construction of composite materials. The strength of the fibers is increased when embedded in the matrix material, and the composite material created in this manner is lighter and stronger. Moreover, once a crack starts in a single material, due to either excessive loading or imperfections in the material, the crack will propagate to the point of failure. In a composite material, on the other hand, if one or a few fibers fail, it does not necessarily lead to failure of other fibers or the material as a whole. Furthermore, the fibers in a composite material can be oriented either in a certain direction or many directions to offer more strength in the direction of expected loads. Therefore, composite materials are designed for specific load applications. For instance, if the expected load is uniaxial, meaning that it is applied in a single direction, then all the fibers are aligned in the direction of the expected load. For applications expecting multidirection loads, the fibers are aligned in different directions to make the material equally strong in various directions.

Depending upon what type of host matrix material is used in creating the composite material, the composites may be classified into three classes: (1) polymer-matrix composites, (2) metal-matrix composites, and (3) ceramic-matrix composites. We discussed the characteristics of matrix materials earlier when we covered metals and plastics.

11.4 SOME COMMON FLUID MATERIALS

Fluid refers to both liquids and gases. Air and water are among the most abundant fluids on earth. They are important in sustaining life and are used in many engineering applications. We will briefly discuss them next.

Air

We all need air and water to sustain life. Because air is readily available to us, it is also used in engineering as a cooling and heating medium in food processing, in controlling thermal comfort in buildings, as a controlling medium to turn equipment on and off, and to drive power tools. Compressed air in the tires of a car provides a cushioned medium to transfer the weight of the car to the road. Understanding the properties of air and how it behaves is important in many engineering applications, including understanding the lift and drag forces. Better understanding of how air behaves under certain conditions leads to the design of better planes and automobiles. The earth's atmosphere, which we refer to as air, is a mixture of approximately 78% nitrogen, 21% oxygen, and less than 1% argon. Small amounts of other gases are present in earth's atmosphere, as shown in Table 11.5.

There are other gases present in the atmosphere, including carbon dioxide, sulfur dioxide, and nitrogen oxide. The atmosphere also contains water vapor. The concentration level of these gases depends on the altitude and geographical location. At higher altitudes (10 to 50 km), the earth's atmosphere also contains ozone. Even though these gases make up a small percentage of earth's atmosphere, they play a significant role in maintaining a thermally comfortable environment for us and other living species. For example, the

TABLE 11.5 The Composition of Dry Air

Gases	Volume by Percent
Nitrogen (N_2)	78.084
Oxygen (O_2)	20.946
Argon (Ar)	0.934
Small amounts of other gases are present in atmosphere including:	
Neon (Ne)	0.0018
Helium (He)	0.000524
Methane (CH_4)	0.0002
Krypton (Kr)	0.000114
Hydrogen (H_2)	0.00005
Nitrous oxide (N_2O)	0.00005
Xenon (Xe)	0.0000087

ozone absorbs most of the ultraviolet radiation arriving from the sun that could harm us. The carbon dioxide plays an important role in sustaining plant life; however, if the atmosphere contains too much carbon dioxide, it will not allow the earth to cool down effectively by radiation. Water vapor in the atmosphere in the form of clouds allows for transport of water from the ocean to land in the form of rain and snow.

Humidity There are two common ways of expressing the amount of water vapor in air: absolute humidity or humidity ratio, and relative humidity. The absolute humidity is defined as the ratio of mass of water vapor in a unit mass of dry air, according to

$$\text{absolute humidity} = \frac{\text{mass of water vapor (kg)}}{\text{mass of dry air (kg)}} \tag{11.1}$$

For humans, the level of a comfortable environment is better expressed by relative humidity, which is defined as the ratio of the amount of water vapor or moisture in the air to the maximum amount of moisture that the air can hold at a given temperature. Therefore, relative humidity is defined as

$$\text{relative humidity} = \frac{\text{the amount of moisture in the air (kg)}}{\text{the maximum amount of moisture that air can hold (kg)}} \tag{11.2}$$

Most people feel comfortable when the relative humidity is around 30 to 50%. The higher the temperature of air, the more water vapor the air can hold before it is fully saturated. Because of its abundance, air is commonly used in food processing, especially in food drying processes to make dried fruits, spaghetti, cereals, and soup mixes. Hot air is transported over the food to absorb water vapors and thus remove them from the source.

Understanding how air behaves at given pressures and temperatures is also important when designing cars to overcome air resistance or designing buildings to withstand wind loading.

Water

You already know that every living thing needs water to sustain life. In addition to drinking water, we also need water for washing, laundry, grooming, cooking, and fire protection. You may also know that two-thirds of the earth's surface is covered with water, but most of this water cannot be consumed directly; it contains salt and other minerals that must be removed first. Radiation from the sun evaporates water; water vapors form into clouds and eventually, under favorable conditions, water vapors turn into liquid water or snow and fall back on the land and the ocean. On land, depending on the amount of precipitation, part of the water infiltrates the soil, part of it may be absorbed by vegetation, and part runs as streams or rivers and collects into natural reservoirs called lakes. Surface water refers to water in reservoirs, lakes, rivers, and streams. Groundwater, on the other hand, refers to the water that has infiltrated the ground; surface and groundwaters eventually return to the ocean, and the water cycle is completed. As we said earlier, everyone knows that we need water to sustain life, but what you may not realize is that water could be thought of as a common engineering material! Water is used in all steam power-generating plants to produce electricity. As we explained in Chapter 10, fuel is burned in a boiler to generate heat, which in turn is added to liquid water to change its phase to steam; steam passes through turbine blades, turning the blades, which in effect runs the generator connected to the turbine, creating electricity. The low-pressure steam liquefies in a condenser and is pumped through the boiler again, closing a cycle. Liquid water stored behind dams is also guided through water turbines located in hydroelectric power plants to generate electricity. Mechanical engineers need to understand the thermophysical properties of liquid water and steam when designing power plants.

We also need water to grow fruits, vegetables, nuts, cotton, trees, and so on. Irrigation channels are designed by civil engineers to provide water to farms and agricultural fields. Water is also used as a cutting tool. High-pressure water containing abrasive particles is used to cut marble or metals. Water is commonly used as a cooling or cleaning agent in a number of food processing plants and industrial applications. Thus, water is not only transported to our homes for our domestic use but it is also used in many engineering applications. So you see, understanding the properties of water and how it can be used to transport thermal energy, or what it takes to transport water from one location to the next, is important to mechanical engineers, civil engineers, manufacturing engineers, agricultural engineers, and so on. We will discuss the Environmental Protection Agency (EPA) standards for drinking water in Chapter 12.

SUMMARY

Now that you have reached this point in the text

- You should understand that engineers select materials for an application based on characteristics of materials, such as strength, density, corrosion resistance, durability, toughness, the ease of machining, and manufacturability. Moreover, you should understand that material cost is also an important selection criterion.
- You should be familiar with common applications of basic materials, such as light metals and their alloys, steel and its alloys, composite materials, and building materials such as concrete, wood, and plastics.

- You should be familiar with the application of fluids, such as air and water, in engineering. You should also be familiar with the composition of air and what the term *humidity* means.

PROBLEMS

1. Identify and list at least ten different materials that are used in your car.
2. Name at least five different materials that are used in a refrigerator.
3. Identify and list at least five different materials that are used in your TV set or computer.
4. List at least ten different materials that are used in a building envelope.
5. List at least five different materials used to fabricate window and door frames.
6. List the materials used in the fabrication of incandescent light bulbs.
7. Identify at least ten products around your home that make use of plastics.
8. In a brief report discuss the advantages and disadvantages of using Styrofoam, paper, glass, stainless steel, and ceramic materials for coffee or tea cups.
9. Every day you use a wide range of paper products at home or school. These paper products are made from different paper grades. Wood pulp is the main ingredient used in making a paper product. It is common practice to grind the wood first and cook it with some chemicals. Investigate the composition, processing methods, and the annual consumption rate of the following grades of paper products in the United States, and write a brief report discussing your findings. The paper products to investigate should include:
   a. printing papers
   b. sanitary papers
   c. glassine and waxing papers
   d. bag paper
   e. boxboard
   f. paper towels
10. As you already know, roofing materials keep the water from penetrating into the roof structure. There is a wide range of roofing products available on the market today. For example, asphalt shingles, which are made by impregnating a dry felt with hot asphalt, are used in some houses. Other houses use, for example, wood shingles, such as red-cedar or redwood shingles. A large number of houses in California use interlocking clay tiles as roofing materials. Investigate the properties and the characteristics of various roofing materials. Write a brief report discussing your findings.
11. Adhesives are used extensively to bond together parts made of the same or different materials. Discuss the characteristics that should be considered when selecting an adhesive for an application. For example, when selecting an adhesive engineers consider the adhesive's ability to bond dissimilar materials, cure time, service temperature, strength, and so on. Investigate the advantages and disadvantages of various adhesives, such as natural adhesives and thermosetting, thermoplastic, and synthetic elastomers.
12. Visit a home improvement center (hardware/lumber store) in your town, and try to gather information about various types of insulating materials that can be used in a house. Write a brief report discussing advantages, disadvantages, and the characteristics of various insulating materials, including their thermal characteristics in terms of R-value.
13. Investigate the characteristics of titanium alloys used in sporting equipment, such as bicycle frames, tennis racquets, and golf shafts. Write a brief summary report discussing your findings.
14. Investigate the characteristics of titanium alloys used in medical implants for hips and other joint replacements. Write a brief summary report discussing your findings.
15. Cobalt-chromium alloys, stainless steel, and titanium alloys are three common biomaterials that have been used as surgical implants. Investigate the use of these biomaterials, and write a brief report discussing the advantages and disadvantages of each.

16. According to the Aluminum Association, in 1998, 102 billion aluminum cans were produced and of these, 62 billion cans were recycled. Measure the mass of ten aluminum cans, and use an average mass for an aluminum can to estimate the total mass of aluminum cans that was recycled.
17. As we discussed in this chapter, when selecting materials for mechanical applications, the value of modulus of resilience for a material shows how good the material is at absorbing mechanical energy without sustaining any permanent damage. Another important characteristic of a material is its ability to handle overloading before it fractures. The value of modulus of toughness provides such information. Look up the values of modulus of resilience and modulus of toughness for the following materials:
    a. titanium
    b. steel
18. Investigate and discuss some of the characteristics of the materials that are used in bridge construction.
19. As we discussed in this chapter, the strength-to-weight ratio of material is an important criterion when selecting material for aerospace applications. Calculate the average strength-to-weight ratio for the following materials: aluminum alloy, titanium alloy, and steel. Use Tables 11.2 and 11.3 to look up appropriate values.
20. How much heavier, on average, will an aluminum-alloy tennis racquet be if it is made from titanium alloy? Obtain a tennis racquet, and take appropriate measurements to perform your analysis.
21. Tensile test machines are used to measure the mechanical properties of materials, such as modulus of elasticity and tensile strength. Visit the Web site of the MTS Systems Corporation to obtain information on test machines used to test the strength of materials. Write a brief report discussing your findings.
22. *Endoscopy* refers to medical examination of the inside of a human body by means of inserting a lighted optical instrument through a body opening. Fiberscopes operate in the visible wavelengths and consist of two major components. One component consists of a bundle of fibers that illuminates the examined area, and the other component transmits the images of the examined area to the eye of the physician or to some display device. Investigate the design of fiberscopes or the fiber-optic endoscope, and discuss your findings in a brief report.
23. Crystal tableware glass that sparkles is sought after by many people as a sign of affluence. This crystal commonly contains lead monoxide. Investigate the properties of crystal glass in detail, and write a brief report discussing your findings.
24. You all have seen grocery bags that have labels and printed information on them. Investigate how information is printed on plastic bags. For example, a common practice includes using a wet-inking process; another process makes use of lasers and heat transfer decals. Discuss your findings in a brief report.
25. Teflon and Nylon are trade names of plastics that are used in many products. Look up the actual chemical name of these products, and give at least five examples of where they are used.
26. Investigate how the following basic wood products are made: plywood, particle board, veneer, and fiberboard. Discuss your findings in a brief report. Also investigate common methods of wood preservation, and discuss your findings in your report. What is the environmental impact of both the production and use of treated wood products in this question?
27. Investigate the common uses of cotton and its typical properties. Discuss your findings in a brief report.
28. As most of you know, commercial transport planes cruise at an altitude of approximately 10,000 m (~33,000 ft). The power required to maintain level flight depends on air drag, or resistance, at that altitude, which may be estimated by the following relationship:

$$\text{power} = \frac{1}{2}\rho_{\text{air}} C_{\text{D}} A U^3$$

where ρ_{air} is density of air at the given altitude, C_{D} represents the drag coefficient of the plane, A is the planform area, and U represents the cruising speed of the plane. Assume that a plane is moving at constant speed and, C_{D} remaining constant, determine the ratio of power that would be required if the plane is cruising at 8000 m and when the plane may be cruising at 11,000 m.
29. Investigate the average daily water consumption per capita in the United States. Discuss the personal and

public needs in a brief report. Also discuss factors such as geographical location, time of the year, time of the day, and cost of consumption patterns. For example, more water is consumed during the early morning hours. Civil engineers need to consider all these factors when designing water systems for cities. Assuming that the life expectancy of people has increased by five years over the past decades, how much additional water is needed to sustain the lives of 50 million people?

30. When a ring gets stuck on a finger, most people resort to water and soap as a lubricant to get the ring off. In earlier times, animal fat was a common lubricant used in wheel axles. Moving parts in machinery, the piston inside your car's engine, and bearings are examples of mechanical components that require lubrication. A lubricant is a substance that is introduced between the parts that have relative motion to reduce wear and friction. The lubricants must have characteristics that are suitable for a given situation. For instance, for liquid lubricants, viscosity is one of the important properties. The flash and fire and the cloud and pour points are examples of other characteristics that are examined when selecting lubricants. Investigate the use of petroleum-based lubricants in reducing wear and friction in today's mechanical components. Write a brief report discussing the application and characteristics of liquid-petroleum-based and solid lubricants that are commonly used, such as SAE 10W-40 oil and graphite.

CHAPTER

12 Engineering Standards and Codes

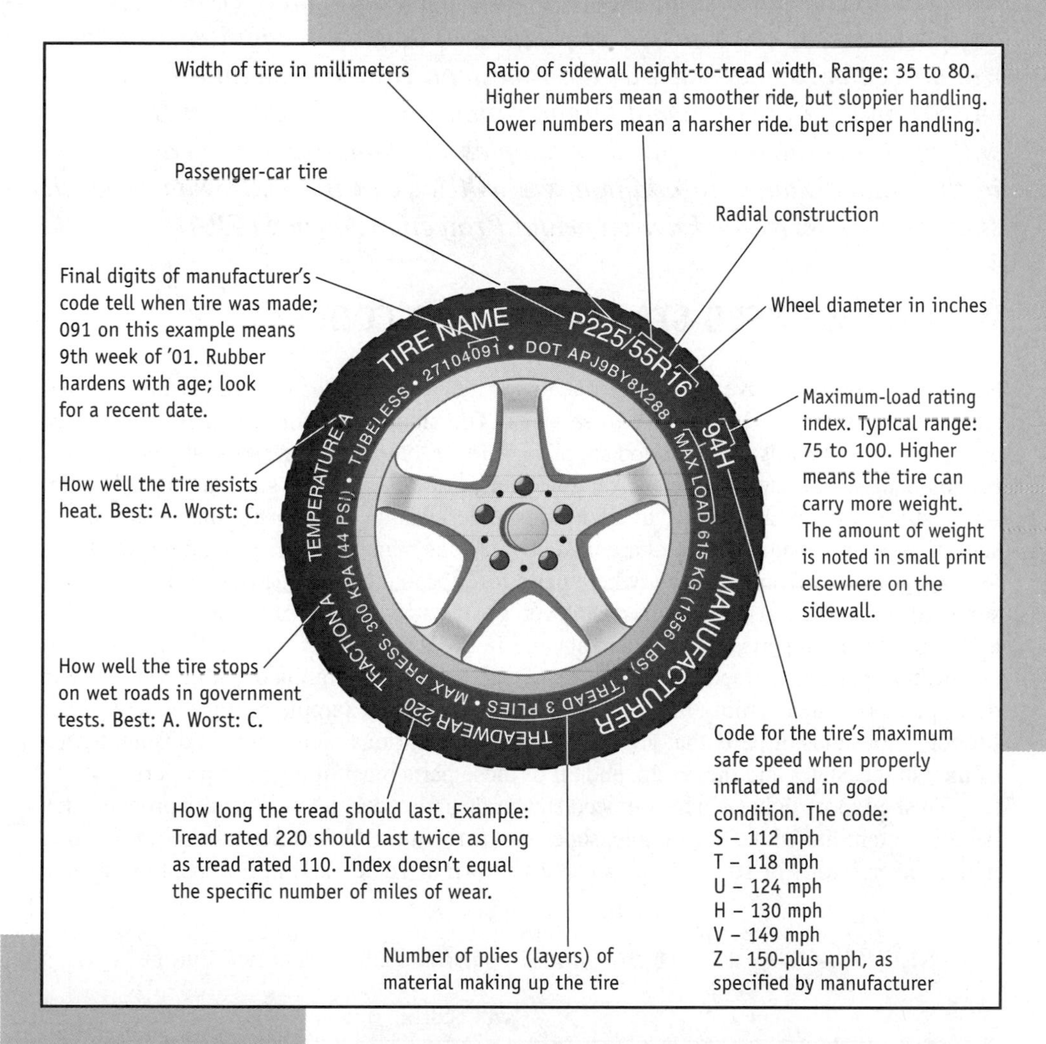

Standards and codes have been developed over the years to ensure that we have safe structures, safe transportation systems, safe electrical systems, safe drinking water, and safe indoor/outdoor air quality. A tire is a good example of an engineered product that adheres to such standards. There are many national and international standardization organizations that set these authoritative standards. Standards and codes also ensure uniformity in the size of parts and components made by various manufacturers around the world.

Source: Rubber Manufacturer's Association.

Today's existing standards and codes ensure that we have safe structures, safe transportation systems, safe drinking water, safe indoor/outdoor air quality, safe products, and reliable services. Standards also encourage uniformity in the size of parts and components that are made by various manufacturers around the world. The objectives of this chapter are to introduce the concepts of standards and codes in engineering and to introduce the organizations that develop standards and safety codes. You will be introduced to the function of the internationally recognized standardization organizations, such as the American National Standards Institute (ANSI) and the International Organization for Standardization (ISO). We will also look at some specific examples of standards and codes in use in the United States. In addition, we will look at the U.S. water and air standards as set by the Environmental Protection Agency (EPA). ■

12.1 WHY DO WE NEED STANDARDS AND CODES?

Standards and codes have been developed over the years by various organizations to ensure product safety and reliability in services. The standardization organizations set the authoritative standards for safe food supplies, safe structures, safe water systems, safe and reliable electrical systems, safe and reliable transportation systems, safe and reliable communication systems, and so on. In addition, standards and codes ensure uniformity in the size of parts and components that are made by various manufacturers around the world. In today's globally driven economy where parts for a product are made in one place and assembled somewhere else, there exists an even greater need than ever before for uniformity and consistency in parts and components and in the way they are made. These standards ensure that parts manufactured in one place can easily be combined with parts made in other places on an assembly line. An automobile is a good example of this concept. It has literally thousands of parts that are manufactured by various companies in different parts of the United States and the world, and all of these parts must fit together properly.

To shed more light on why we need standards and codes, let us consider products that we all are familiar with, for example, shoes or shirts. In the United States, you are familiar with shoe sizes of 9 or 10 or 11 and so on, as shown in Table 12.1. In Europe, the standard

TABLE 12.1 Standard Shoe and Shirt Sizes in the United States and Europe

	Men's Shirts						
Europe	36	37	38	39	41	42	43
U.S.	14	$14\frac{1}{2}$	15	$15\frac{1}{2}$	16	$16\frac{1}{2}$	17

	Men's Shoes							
Europe	38	39	41	42	43	44	45	46
U.S.	5	6	7	8	9	10	11	12

shoe sizes are 43 or 44 or 45 and so on. Similarly, the standard shirt sizes in the United States are 15 or $15\frac{1}{2}$ or 16 and so on, whereas in Europe the standard shirt sizes are 38 or 39 or 41 and so on. If a shirt manufacturer in Europe wants to sell shirts in the United States, it has to label them such that people understand the sizes so that they can choose a shirt of the correct size. Conversely, if a shoe manufacturer from the United States wants to sell shoes in Europe, it has to label them such that the shoe sizes are understood by European customers. Would it not be easier if every shirt or shoe manufacturer in the world used uniform size identifications to eliminate the need for cross referencing? These simple examples demonstrate the need for uniformity in the size and the way products are labeled. Now, think about all possible parts and components that are manufactured every day by thousands of companies around the world: parts and components such as bolts, screws, nuts, cables, tubes, pipes, beams, gears, paints, adhesives, springs, wires, tools, lumber, fasteners, and so on. You see that if every manufacturer built products using its own standards and specifications, this practice could lead to chaos and many misfit parts! Fortunately, there are existing international standards that are followed by many manufacturers around the world.

A good example of a product that uses international standards is your credit card or your bankcard. It works in all the ATM machines or store credit card readers in the world. The size of the card and the format of information on the card conform to the International Organization of Standards (ISO), thus allowing the card to be read by ATM machines everywhere. The 35-mm camera film speed (e.g., 100, 200, 400) is another example of ISO standards being used by film manufacturers. As another example, warning and functional symbols based on ISO standards on the instrument panel of your car have become commonplace. The ISO standards are being implemented by more and more companies around the world every day.

There are many standardization organizations in the world, among them various engineering organizations. Recall from our discussion in Chapter 1 that most national/ international engineering organizations create, maintain, and distribute codes and standards that deal with uniformity in size of parts and correct engineering design practices so that public safety is ensured. In fact, the American Society of Mechanical Engineers (ASME) discussed at its first meeting in 1880 the need for standardized sizes for screws. Here we will focus on some of the larger standardization organizations in the United States, Canada, Europe, and Asia. We will briefly describe the role of these organizations and how they may interact. Among the more well-known and internationally recognized organizations

Examples of products conforming to the ISO

are the American National Standards Institute (ANSI), the American Society for Testing and Materials (ASTM), the Canadian Standards Association (CSA), the British Standards Institute (BSI), the German *Deutsches Institut für Normung* (DIN), the French *Association Française de normalisation* (AFNOR), the Swedish *Standardiserigen I Sverige* (SIS), the China State Bureau of Quality and Technical Supervision (CSBTS), and the International Organization for Standardization (ISO). We will briefly describe these organizations in the following sections.

12.2 EXAMPLES OF STANDARDS AND CODES ORGANIZATIONS IN THE UNITED STATES

The American National Standards Institute

The American National Standards Institute (ANSI) was founded in 1918 by five engineering societies and three government agencies to administer and coordinate standards in the United States. The ANSI is a not-for-profit organization, which is supported by various public and private organizations. The institute itself does not develop the standards, but instead it assists qualified groups, such as various engineering organizations, with the development of the standards and sets the procedures to be followed. Today, the American National Standards Institute represents the interests of well over a thousand companies and other members. According to the American National Standards Institute, there are over 13,000 approved ANSI standards in use today, and more standards are being developed.

The American Society for Testing and Materials (ASTM)

Founded in 1898, the American Society for Testing and Materials (ASTM) is another not-for-profit organization. It publishes standards and test procedures that are considered authoritative technical guidelines for product safety, reliability, and uniformity. The testing is performed by its national and international member laboratories. The ASTM collects and publishes the work of over 100 standards-writing committees dealing with material test methods. For example, ASTM sets the standard procedures for tests and practices to determine elastic properties of materials, impact testing, fatigue testing, shear and torsion properties, residual stress, bend and flexure testing, compression, ductility, and linear thermal expansion. The ASTM also sets the standards for medical devices and equipment, including bone cements, screws, bolts, pins, prostheses, and plates, and specifications for alloys used in surgical implants. Electrical insulation and electronics-related standards are also set by ASTM. Additional examples of ASTM work include the following:

- Test guidelines for evaluating mechanical properties of silicon or other procedures for testing semiconductors, such as germanium dioxide.
- Testing methods for trace metallic impurities in electronic-grade aluminum–copper and aluminum–silicon.
- Standards related to the chemical analysis of paints or detection of lead in paint, along with tests to measure the physical properties of applied paint films, such as film thickness, physical strength, and resistance to environmental or chemical surroundings.

- Standard procedures for evaluating properties of motor, diesel, and aviation fuels, crude petroleum, hydraulic fluids, and electric insulating oils.
- Test procedures for measurements of insulation properties of materials.
- Standard procedures for soil testing, such as density characteristics, soil texture, and moisture content.
- Building-construction-related tests and procedures, such as measuring the structural performance of sheet metal roofs.
- Tests for evaluating the properties of textile fibers, including cotton and wool.
- Standards for steel piping, tubing, and fittings.

You can find specific standards dealing with any of the materials in these examples in the *Annual Book* of ASTM, which includes large volumes of standards and specifications in the following areas:

Iron and steel products
Nonferrous metal products
Metals test methods and analytical procedures
Construction
Petroleum products, lubricants, and fossil fuels
Paints, related coatings, and aromatics
Textiles
Plastics
Rubber
Electric insulation and electronics
Water and environmental technology
Nuclear, solar, and geothermal energy
Medical devices and services
General methods and instrumentation
General products, chemical specialties, and end-use products

ASTM Standards are becoming available on CD-ROM and on-line. ASTM also publishes a number of journals:

Cement, Concrete & Aggregates
Geotechnical Testing Journal
Journal of Composites Technology and Research
Journal of Forensic Sciences
Journal of Testing and Evaluation

National Fire Protection Association (NFPA)

Losses from fires total billions of dollars per year. Fire, formally defined as a process during which rapid oxidization of a material occurs, gives off radiant energy that can not only be felt but also seen. Fires can be caused by malfunctioning electrical systems, hot surfaces, and overheated materials. The National Fire Protection Association (NFPA) is a not-for-profit organization that was established in 1896 to provide codes and standards to reduce the burden of fire. The NFPA publishes the *National Electrical Code*®, the *Life Safety Code*®, the *Fire Prevention Code*™, the *National Fuel Gas Code*®, and the *National Fire Alarm Code*®. It also provides training and education.

Underwriters Laboratories (UL)

The Underwriters Laboratories Inc. (UL) is a nonprofit organization that performs product safety tests and certifications. Founded in 1894, today Underwriters Laboratories has laboratories in the United States, England, Denmark, Hong Kong, India, Italy, Japan, Singapore, and Taiwan. Its certification mark is one of the most recognizable marks on products.

12.3 EXAMPLES OF INTERNATIONAL STANDARDS AND CODES

The International Organization for Standardization (ISO)

As the name implies, the International Organization for Standardization, established in 1947, consists of a federation of national standards from various countries. The International Organization for Standardization promotes and develops standards that can be used by all countries in the world, with the objective of facilitating standards that allow for free, safe exchange of goods, products, and services among countries. It is recognized by its abbreviation, or short form, ISO, which is derived from '*isos*,' a Greek word meaning "equal." As you take more engineering classes, you will see the prefix *iso* in many engineering terms; for example, *iso*bar, meaning equal pressure, or *iso*therm, meaning equal temperature. ISO was adopted instead of IOS (International Organization of Standards) so that there would not be any nonuniformity in the way the abbreviation is presented in other languages.

Other Internationally Recognized Standardization Organizations

The British Standards Institute (BSI) is another internationally known organization that deals with standardization. In fact, BSI, founded in 1901, is one of the oldest standardization bodies in the world. It is a nonprofit organization that organizes and distributes British, European, and International standards. Other internationally recognized standardization organizations include the German *Deutsches Institut für Normung* (DIN), the French *Association Française de normalisation* (AFNOR), the Swedish *Standardiserigen I Sverige* (SIS), and the China State Bureau of Quality and Technical Supervision (CSBTS). Visit the Web sites of these organizations to obtain more information about them. A comprehensive list of various standardization organizations and their Web addresses (if available) that you may find useful when searching for information about organizations dealing with engineering standards and codes follows.

American Boiler Manufacturers Association (ABMA)
www.abma.org

Air Conditioning Contractors of America (ACCA)
www.acca.org

American Gas Association (AGA)
www.aga.org

Association of Home Appliance Manufacturers (AHAM)
www.aham.org

Air Movement and Control Association (AMCA)
www.amca.org

American National Standards Institute (ANSI)
www.ansi.org

Air Conditioning and Refrigeration Institute (ARI)
www.ari.org

American Institute of Aeronautics and Astronautics
www.aiaa.org

American Institute of Chemical Engineers
www.aiche.org

The American Society of Agricultural Engineers
www.asae.org

American Society of Civil Engineers
www.asce.org

American Nuclear Society
www.ans.org

American Society of Heating, Refrigerating and Air-Conditioning Engineers (ASHRAE)
www.ashrae.org

American Society of Mechanical Engineers (ASME International)
www.asme.org

American Society for Testing and Materials (ASTM)
www.astm.org

American Water Works Association (AWWA)
www.awwa.org

Building Officials and Code Administrators International (BOCA)

British Standards Institute (BSI)
www.bsi.org.uk

China State Bureau of Quality and Technical Supervision (CSBTS)
www.csbts.cn.net/english

Council of American Building Officials (CABO) (see also International Code Council)
www.intlcode.org

Compressed Air and Gas Institute (CAGI)

Canadian Gas Association (CGA)
www.cga.ca

Canadian Standards Association (CSA)
www.csa-international.org

Cooling Tower Institute (CTI)

France—Association Française de normalisation (AFNOR)
www.afnor.fr

Germany-Deutsches Institut für Normung (DIN)
www.din.de

Heat Exchange Institute (HEI)

Hydraulic Institute (HI)

Hydronic Institute (HYDI)

Institute of Electrical and Electronics Engineers
www.ieee.org

Institute of Industrial Engineers
www.iienet.org

International Association of Plumbing and Mechanical Officials (IAPMO)
www.iapmo.org

Illuminating Engineering Society of North America (IESNA)
www.iesna.org

International Fire Code Institute (IFCI)
www.ifci.org

International Organization for Standardization (ISO)
www.iso.ch

Mechanical Contractors Association of America (MCAA)
www.mcaa.org

Manufacturers Standardization Society of the Valve and Fittings Industry (MSS)

North American Insulation Manufacturers Association (NAIMA)
www.naima.org

National Association of Plumbing-Heating-Cooling Contractors (NAPHCC)
www.naphcc.org

National Conference of States on Building Codes and Standards (NCSBCS)

National Electrical Manufacturers Association (NEMA)
www.nema.org

National Fire Protection Association (NFPA)
www.nfpa.org

National Research Council of Canada (NRCC)

Occupational Safety and Health Act (OSHA)
www.osha.gov

Sheet Metal and Air-Conditioning Contractors' National Association (SMACNA)

Society of Automotive Engineers
www.sae.org

Society of Manufacturing Engineering
www.sme.org

Sweden-Standardiserigen I Sverige (SIS)
www.sis.se

Underwriters Laboratories (UL)
www.ul.com

12.4 SPECIFIC EXAMPLES OF STANDARDS AND CODES IN USE IN THE UNITED STATES

In this section, we will look at some specific standards and codes that are used in various engineering products and practices in the United States. Table 12.2 shows examples of these standards and codes.

In the next three sections, we will describe the standards for drinking water and the air that we breathe. The Environmental Protection Agency (EPA) of the United States sets the standards as required by the U.S. Congress. The Clean Air Act was signed into law in 1970, and the Safe Drinking Water Act was enacted by Congress in 1974.

12.5 DRINKING WATER STANDARDS IN THE UNITED STATES

The U.S Environmental Protection Agency (EPA) sets the standards for the maximum level of contaminants that can be in our drinking water and still be considered safe to drink. Basically, the EPA sets two standards for the level of water contaminants: (1) the maximum contaminant level goal (MCLG) and (2) the maximum contaminant level (MCL). The MCLG represents the maximum level of a given contaminant in the water that causes no known harmful health effects. On the other hand, the MCL, which may represent slightly higher levels of contaminants in the water, are the levels of contaminants that are legally enforceable. The EPA attempts to set MCL close to MCLG, but this goal may not be attainable because of economic or technical reasons. Examples of drinking water standards are shown in Table 12.3.

See Problem 15 in this chapter for other standards set by the EPA for the so-called Surface Water Treatment Rule (SWTR).

TABLE 12.2 Examples of Codes and Standards

Subject/Title	Publisher	Reference
Air Conditioning		
Room Air Conditioners	CSA	C22.2 No. 117-1970 ANSI/AHAM (RA -C1)
General Requirements for Helicopter Air Conditioning (1970)	SAE	SAE ARP292B
Boilers		
Heating, Water Supply, and Power Boilers-Electric (1991)	UL	ANSI/UL384-1991
Gas-Fired Low Pressure Steam and Hot Water Boilers	AGA	ANSI Z21.13-1987; Z21.13a-1989
Building Codes		
ASTM Standards Used in Building Codes	ASTM	ASTM BOCA
National Building Code, 11th ed. (1990)	BOCA	BOCA
Safety Codes for Elevators and Escalators	ASME	ANSI/ASME A17.1-1990
Chimneys		
Chimneys, Fireplaces, and Vents, and Solid Fuel Burning Appliances	NFPA	ANSI/NFPA 211-1988
Chimneys, Factory-Built, Residential Type and Building Heating Appliances (1989)	UL	ANSI/UL 103-1988
Clean Rooms		
Procedural Standards for Certified Testing of Cleanrooms	NEBB	NEBB-1988
Compressors		
Safety Standards for Air Compressor Systems	ASME	ANSI/ASME B19.1-1990
Computers		
Protection of Electronic Computer/ Data Processing Equipment	NFPA	ANSI/NFPA75-1989
Controls		
Primary Safety Controls for Gas- and Oil-Fired Appliances (1985)	UL	ANSI/UL 372-1985
Solid State Control for Appliances	UL	ANSI/UL 244A-1987
Gas Appliance Thermostats	AGA	ANSI Z21.23-1989; Z21.23-1991
Coolers		
Milk Coolers	CSA	C22.2 No. 132-1973
Refrigerated Vending Machines (1989)	UL	ANSI/UL 541-1988
Drinking Water Coolers (1987)	UL	ANSI/UL 399-1986
Dehumdifiers		
Dehumidifiers (1987)	UL	ANSI/UL 474-1987
Electrical		
Voltage Ratings for Electrical Power Systems and Equipment	ANSI	ANSI C84.1-1989
Enclosures for Electric Equipment	NEMA	NEMA 250-1991
General Requirements for Wiring Devices	NEMA	NEMA WD1-1983 (R 1989)
National Electric Code (1990)	NEMA	ANSI/NFPA 70-1990
Fuseholders (1987)	UL	ANSI/UL 512-1986
Energy		
Air Conditioning and Refrigerating Equipment Nameplate Voltages	ARI	ARI 110-90
Energy Conservation in Existing Buildings-High Rise Residential	ASHRAE	ANSI/ASHRAE/IES 100.2-1991
Energy Management Equipment (1991)	UL	ANSI/UL 916-1987
Fans		
Duct Design for Residential Winter and Summer Air Conditioning	ACCA	Manual D
Fans	ASME	ANSI/ASME PTC 11-1984 (R 1990)
Electric Fans	UL	ANSI/UL 507
Fire Protection		
Standard Method for Fire Tests of Building Construction and Materials	ASTM	ASTM E 119-88
Uniform Fire Code Standards (1991)	IFCI	IFCI
Fire Doors and Windows	NFPA	ANSI/NFPA 80-1990
Fire Protection Handbook	NFPA	NFPA
National Fire Codes (annual)	NFPA	NFPA

Continued

TABLE 12.2 Examples of Codes and Standards (*continued*)

Subject/Title	Publisher	Reference
Freezers		
Household Refrigerators, Combination Refrigerators-Freezers, and Household Freezers	AHAM	ANSI/AHAM;HRF 1
Household Refrigerators and Freezers (1983)	UL	ANSI/UL 250-1991
Commercial Refrigerators and Freezers (1992)	UL	ANSI/UL 471-1992
Humidifiers		
Humidifiers	UL	ANSI/UL 563-1985
Incinerators		
Incinerators, Waste and Linen Handling Systems and Equipment	NFPA	ANSI/NFPA 82-1990
Residential Incinerators (1973)	UL	UL 791
Insulation		
Test Method for Steady-State and Thermal Performance of Building Assemblies	ASTM	ASTM C236-89
National Commercial and Industrial Insulation Standards	MICA	MICA 1988
Measurements		
Measurement of Fluid Flow in Pipes, Using Orifice, Nozzle, and Venturi	ASME	ASME MFC-3M-1989
Measurement of Industrial Sound	ASME	ANSI/ASME PTC 36-1985
Measurement of Rotary Speed	ASME	ANSI/ASME PTC 19.13-1961
Measurement Uncertainty	ASME	ANSI/ASME PTC 19.1-1985(R 1990)
Pressure Measurement	ASME	ANSI/ASME PTC 19.2-1987
Temperature Measurement	ANSI	ANSI/ASME PTC 19.3-1974 (R 1986)
Mobile Homes and Recreational Vehicles		
Mobile Homes	CSA	CAN/CSA-Z240 MH Series-M1992
Recreational Vehicles	CSA	CAN/CSA-Z240 RV Series-M92
Recreational Vehicles	NFPA	NFPA 501C-1990
Low Voltage Lighting Fixtures for Use in Recreational Vehicles	UL	UL 234
Motors and Generators		
Steam Generating Units	ASME	ANSI/ASME PTC 4.1-1964 (R1991)
Motors and Generators	NEMA	NEMA MG 1-1987
Electric Motors (1989)	UL	ANSI/UL 1004-1988
Thermal Protection for Motors (1986)	UL	ANSI/UL 547-1991
Pipes, Tubings, and Fittings		
Power Piping	ASME	ANSI/ASME B31.1-1989
Refrigeration Piping	ASME	ASME/ANSI B31.5-1987
Specification for Polyvinyl Chloride (PVC) Plastic Pipe	ASTM	ASTM B42-89
Specification for Seamless Copper Pipe, Standard Sizes	ASTM	ASTM B42-89
Plastic Piping Components and Related Materials	NSF	NSF-14
Plumbing		
Uniform Plumbing Code (1991)	IAPMO	IAPMO
National Standard Plumbing Code (NSPC)	NAPHCC	NSPC 1993
Standard Plumbing Code	SBCCI	SBCCI (1991) (R 1992)
Pumps		
Centrifugal Pumps	ASME	ASME PTC 8.2-1990
Centrifugal Pump Test Standard (1988)	HI	HI
Electric Swimming Pool Pumps, Filters and Chlorinators (1986)	UL	ANSI/UL 1081-1985
Motor-Operated Water Pumps	UL	ANSI/UL 343-1985
Sound Measurement		
Measurement of Industrial Sound	ASME	ASME/ANSI PTC 36-1985
Procedural Standards for Measuring Sound and Vibration	NEBB	NEBB-1977
Space Heaters		
Electric Air Heaters (1980)	UL	ANSI/UL 1025-1980

Source: American Society of Heating, Refrigerating, and Air-Conditioning Engineers.

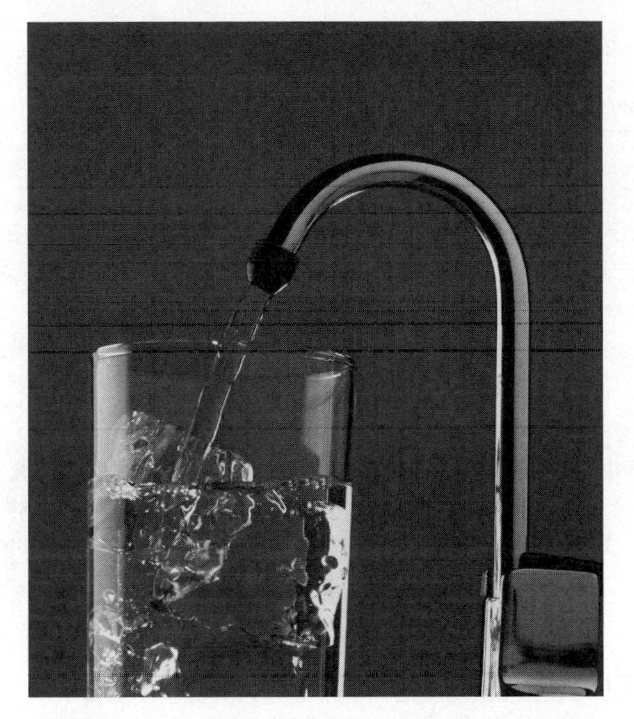

The EPA sets the standard for MCL in our drinking water.

TABLE 12.3 Examples of Drinking Water Standards

Contaminant	MCGL	MCL	Source of Contaminant by Industries
Antimony	6 ppb	6 ppb	copper smelting, refining, porcelain plumbing fixtures, petroleum refining, plastics, resins, storage batteries
Asbestos	7 M.L. (million fibers per liter)	7 M.L.	asbestos products, chlorine, asphalt felts and coating, auto parts, petroleum refining, plastic pipes
Barium	2 ppm	2 ppm	copper smelting, car parts, inorganic pigments, gray ductile iron, steel works, furnaces, paper mills
Beryllium	4 ppb	4 ppb	copper rolling and drawing, nonferrous metal smelting, aluminum foundries, blast furnaces, petroleum refining
Cadmium	5 ppb	5 ppb	zinc and lead smelting, copper smelting, inorganic pigments.
Chromium	0.1 ppm	0.1 ppm	pulp mills, inorganic pigments, copper smelting, steel works.
Copper	1.3 ppm	1.3 ppm	primary copper smelting, plastic materials, poultry slaughtering, prepared feeds

Continued

TABLE 12.3 Examples of Drinking Water Standards (*continued*)

Contaminant	MCGL	MCL	Source of Contaminant by Industries
Cyanide	0.2 ppm	0.2 ppm	metal heat treating, plating, and polishing
Lead	zero	15 ppb	lead smelting, steel works and blast furnaces, storage batteries, china plumbing fixtures
Mercury	2 ppb	2 ppb	electric lamps, paper mills
Nickel	0.1 ppm	0.1 ppm	petroleum refining, gray iron foundries, primary copper, blast furnaces, steel
Nitrate	10 ppm	10 ppm	nitrogenous fertilizer, fertilizing mixing, paper mills, canned foods, phosphate fertilizers
Nitrite	1 ppm	1 ppm	
Selenium	0.05 ppm	0.05 ppm	metal coatings, petroleum refining
Thallium	0.5 ppb	2 ppb	primary copper smelting, petroleum refining, steel works, blast furnaces

12.6 OUTDOOR AIR QUALITY STANDARDS IN THE UNITED STATES

The source of outdoor air pollution may be classified into three broad categories: the *stationary sources, the mobile sources, and the natural sources*. Examples of stationary sources include power plants, factories, and dry cleaners. The mobile sources of air pollution consist of cars, buses, trucks, planes, and trains. As the name implies, the sources of natural air pollution could include windblown dust, volcanic eruptions, and forest fires. The Clean Air Act, which sets the standard for six major air pollutants, was signed into law in 1970. The Environmental Protection Agency is responsible for setting standards for these six major air pollutants: carbon monoxide (CO), lead (Pb), nitrogen dioxide (NO_2), ozone (O_3), sulfur dioxide (SO_2), and particulate matter (PM). The EPA measures the concentration level of these pollutants in many urban areas and collects air quality information by actual measurement of pollutants from thousands of monitoring sites located throughout the country. According to a study performed by the EPA (1997), between 1970 and 1997, the U.S. population increased by 31% and the vehicle miles traveled increased by 127%. During this period, the total emission of air pollutants from stationary and mobile sources decreased by 31% because of improvements made in the efficiency of cars and in industrial practices, along with the enforcement of the Clean Air Act regulations.

However, there are still approximately 107 million people who live in areas with unhealthy air quality. The EPA is continuously working to set standards and monitor the emission of pollutants that cause acid rain and damage to bodies of water and fish (currently there are over 2000 bodies of water in the United States that are under fish consumption advisories), damage to the stratospheric ozone layer, and damage to our buildings and our national parks. The unhealthy air has more pronounced adverse health effects on children and elderly people. The human health problems associated with poor air

Outdoor pollution!

quality include various respiratory illnesses and heart or lung diseases. Congress passed amendments to the Clean Air Act in 1990, which required the EPA to address the effect of many toxic air pollutants by setting new standards. Since 1997, the EPA has issued 27 air standards that are to be fully implemented in the coming years. The EPA is currently working with the individual states in this country to reduce the amount of sulfur in fuels and setting more stringent emission standards for cars, buses, trucks, and power plants.

We all need to understand that air pollution is a global concern that can affect not only our health, but also affect our climate. It may trigger the onset of global warming that could lead to unpleasant natural events. Because we all contribute to this problem, we need to be aware of the consequences of our lifestyles and find ways to reduce pollution. We could carpool or take public transportation when going to work or school. We should not leave our cars running idle for long periods of time, and we can remind others to consume less energy. We should conserve energy around home and school. For example, turn off the light in a room that is not in use. When at home, in winter, set the thermostat at 65°F or slightly lower and wear a sweater to feel warm. During summer, at home set the air-conditioning thermostat at 78°F or slightly higher. By consuming less energy and driving less we can help our environment and reduce air pollution.

12.7 INDOOR AIR QUALITY STANDARDS IN THE UNITED STATES

In the previous section, we discussed outdoor air pollution and the related health effects. Indoor air pollution can also create health risks. According to EPA studies of human exposure to air pollutants, the indoor levels of pollutants may be 2–5 times higher than outdoor levels. Indoor air quality is important in homes, schools, and workplaces. Because most of us spend approximately 90% of our time indoors, the indoor air quality is very important to our short-term and long-term health. Moreover, lack of good indoor air quality can reduce productivity at the workplace or create an unfavorable learning environment at school, causing sickness or discomfort to building occupants. Failure to monitor indoor air quality (IAQ) or failure to prevent indoor air pollution can also have adverse effects on equipment and the physical appearance of buildings. In recent years, liability issues related to people who suffer dizziness or headaches or other illness related to "sick buildings" are becoming a concern for building managers. According to the EPA, some common health symptoms caused by poor indoor air quality are

- headache, fatigue, and shortness of breath;
- sinus congestion, coughing, and sneezing;
- eye, nose, throat, and skin irritation;
- dizziness and nausea.

As you know, some of these symptoms may also be caused by other factors and are not necessarily caused by poor air quality. Stress at school, work, or at home could also create health problems with symptoms similar to the ones mentioned. Moreover, individuals react differently to similar problems in their surroundings.

The factors that influence air quality can be classified into several categories: the heating, ventilation, and air-conditioning (HVAC) system; sources of indoor air pollutants; and occupants. In recent years, we have been exposed to more indoor air pollutants for the following reasons. (1) In order to save energy we are building tight houses that have lower air infiltration or exfiltration compared to older structures. In addition, the ventilation rates have also been reduced to save additional energy. (2) We are using more synthetic building materials in newly built homes that could give off harmful vapors. (3) We are using more chemical pollutants, such as pesticides and household cleaners.

As shown in Table 12.4, indoor pollutants can be created by sources within the building or they can be brought in from the outdoors. It is important to keep in mind that the level of contaminants within a building can vary with time. For example, in order to protect floor surfaces from wear and tear, it is customary to wax them. During the period when waxing is taking place, it is possible, based on the type of chemical used, that anyone near the area to be exposed to harmful vapors. Of course, one simple remedy to this indoor air problem is to wax the floor late on Friday afternoons to avoid exposing too many occupants to harmful vapors. Moreover, this approach will provide some time for the vapor to be exhausted out of the building by the ventilation system over the weekend when the building is not occupied.

The primary purpose of a well-designed heating, ventilation, and air-conditioning (HVAC) system is to provide thermal comfort to its occupants. Based on the building's heating or cooling load, the air that is circulated through the building is conditioned by heating, cooling, humidifying, or dehumidifying. The other important role of a well-designed HVAC system is to filter out the contaminants or to provide adequate ventilation to dilute air-contaminant levels.

The air flow patterns in and around a building will also affect the indoor air quality. The air flow pattern inside the building is normally created by the HVAC system. However, the outside air flow around a building envelope that is dictated by wind patterns could also affect the air flow pattern within the building as well. When looking at air flow

TABLE 12.4 Typical Sources of Indoor Air Pollutants

Outside Sources	Building Equipment	Component/Furnishings	Other Indoor Sources
Polluted Outdoor Air Pollen, dust, fungal spores, industrial emissions and vehicle emissions **Nearby Sources** Loading docks, odors from dumpsters, unsanitary debris or building exhausts near outdoor air intakes **Underground Sources** Radon, pesticides, and leakage from underground storage tanks	**HVAC Equipment** Microbiological growth in drip pans, ductwork, coils, and humidifiers, improper venting of combustion products, and dust or debris in ductwork **Non-HVAC Equipment** Emissions from office equipment, and emissions from shops, labs and cleaning processes	**Components** Microbiological growth on soiled or water-damaged materials, dry traps that allow the passage of sewer gas, materials containing volatile organic compounds, inorganic compounds, damaged asbestos, and materials that produce particles (dust) **Furnishings** Emissions from new furnishings and floorings and microbiological growth on or in soiled or water-damaged furnishings	Science laboratories, copy and print areas, food preparation areas, smoking lounges, cleaning materials, emission from trash, pesticides, odors and volatile organic compounds from paint, chalk, and adhesives, occupants with communicable diseases, dry-erase markers and similar pens, insects and other pests, personal hygiene products

Source: EPA Fact sheets, EPA-402-F-96-004, October 1996.

patterns, the important concept to keep in mind is that air will always move from a high-pressure region to a low-pressure region.

Methods to Manage Contaminants

There are several ways to control the level of contaminants: (1) source elimination or removal, (2) source substitution, (3) proper ventilation, (4) exposure control, and (5) air cleaning.

A good example of source elimination is not allowing people to smoke inside the building or not allowing a car engine to run idle near a building's outdoor air intake. In other words, eliminate the source before it spreads out! It is important for engineers to keep that idea in mind when designing the HVAC systems for a building—avoiding placing the outdoor air intakes near loading docks or dumpsters, for example. A good example of source substitution is to use a gentle cleaning product rather than a product that gives off harmful vapors when cleaning bathrooms and kitchens. Local exhaust control means removing the sources of pollutants before they can be spread through the air distribution system into other areas of a building. Everyday examples include use of an exhaust fan in restrooms to force out harmful contaminants. Fume hoods are another example of local exhaust removal in many laboratories. Clean outdoor air can also be mixed with the inside air to dilute the contaminated air. The American Society of Heating, Ventilating and Air Conditioning Engineers (ASHRAE) has established a set of codes and standards for how much fresh outside air must be introduced for various applications. Air cleaning means removing harmful particulate and gases from the air as it passes through some cleaning system. There are various methods that deal with air contaminant removal, including absorption, catalysis, and use of air filters.

Finally, we you can bring the indoor air quality issues to the attention of friends, classmates, and family. We all need to be aware and try to do our part to create and maintain healthy indoor air quality.

SUMMARY

Now that you have reached this point in the text

- You should know why we need to have standards and codes in engineering.
- You should be familiar with the role and mission of some of the larger standardization organizations in the world, such as ANSI, ASTM, BSI, CSA, ISO, CSBTS, SIS, AFNOR, and DIN.
- You should be familiar with the role of the EPA and the standards it sets for drinking water, outdoor air quality, and indoor air quality.
- You should be able to name some of the sources of indoor and outdoor air pollutants.
- You should be able to name some of the sources of water pollutants.

PROBLEMS

1. Identify and make a list of at least ten products around your home that are certified by the Underwriters Laboratories.
2. Create a table showing hat sizes in the United States and Europe.

Continental	51	52	53	.	.	.
U.S.	$6\frac{1}{4}$	$6\frac{3}{8}$	$6\frac{1}{2}$	.	.	.

3. Create a table showing wrench sizes in metric and U.S units.
4. Obtain information about what the colors on an electrical resistor mean. Create a table showing the electrical resistor codes. Your table should have a column with colors: Black, Brown, Red, Orange, Yellow, Green, Blue, Violet, Gray, White, Gold, and Silver, and a column showing the values. Imagine that you are making this table for others to use; therefore, include at least two examples of how to read the codes on electrical resistors at the bottom of your table.
5. Collect information on U.S. standard steel pipes (1/4 in. to 20 in.). Create a table showing nominal size, schedule number, inside diameter, outside diameter, wall thickness, and cross-sectional area.
6. Collect information on the American Wire Gage (AWG) standards. Create a table for annealed copper wires showing the gage number, diameter in mils, cross-sectional area, and resistance per 1000 ft.
7. Write a brief memo to your instructor explaining the role and function of the U.S. Department of Transportation (DOT).
8. Obtain information on the standard sign typefaces used for highway signs in the United States. The Federal Highway Administration publishes a set of standards called *Standard Alphabets for Highway Signs*. Write a brief memo to your instructor explaining your findings.
9. Obtain information about the Nuclear Regulatory Commission (NRC), which sets the standards for handling and other activities dealing with radioactive materials. Write a brief report to your instructor regarding your findings.
10. Obtain information about the classification of fire extinguishers. Write a brief report explaining what is meant by Class A, Class B, Class C, and Class D fires.
11. Write a brief report detailing the development of safety belts in cars. When was the first safety belt designed? Which was the first car manufacturer to incorporate safety belts as standard items in its cars?

12. Investigate the mission of each of the following standards organizations. For each of the organizations listed, write a one-page memo to your instructor about its mission and role.
    a. the European Committee for Electrotechnical Standardization (CENELEC)
    b. European Telecommunication Standards Institute (ETSI)
    c. Pan American Standards Commission (COPANT)
    d. Bureau of Indian Standards (BID)
    e. Hong Kong Standards and Testing Centre Ltd. (HKSTC)
    f. Korea Academy of Industrial Technology (KAITECH)
    g. Singapore Academy of Industrial Technology (PSB)
    h. Standards New Zealand (SNZ)
13. Write a brief report explaining what is meant by ISO 9000 and ISO 14000 certification.
14. Ask your local city water supplier to give you a list of the chemicals that it tests for in your water. Also ask how your city water is being treated. You may want to contact your state department of health/environment to get additional information. For help in locating state and local agencies, or for information on drinking water in general, you can call the EPA's Safe Drinking Water Hotline: (800) 426-4791. You can also visit the EPA Web site at www.epa.org. You can also obtain information about the uses and releases of chemicals in your state by contacting the Community Right-to-Know Hotline: (800) 535-0202.
15. Collect information on the Surface Water Treatment Rule (SWTR) standards set by the EPA. Write a brief memo to your instructor explaining your findings.
16. Obtain the EPA's consumer fact sheets (they are now available on the Web) on antimony, barium, beryllium, cadmium, cyanide, and mercury. After reading the fact sheets, prepare a brief report explaining what they are, how they are used, and what health effects are associated with them.
17. In 1970 the U.S. Congress passed the Occupational Safety and Health Act. The following is a duplicate of the act, which reads:

> Public Law 91-596
> 91st Congress, S. 219
> December 29, 1970
> As Amended by Public Law 101-552,
> Section 3101, November 5, 1990
> As Amended by Public Law 105-198,
> July 16, 1998
> As Amended by Public Law 105-241
> September 29, 1998
>
> **An Act**
>
> To assure safe and healthful working conditions for working men and women; by authorizing enforcement of the standards developed under the Act; by assisting and encouraging the States in their efforts to assure safe and healthful working conditions; by providing for research, information, education, and training in the field of occupational safety and health; and for other purposes.
>
> Be it enacted by the Senate and House of Representatives of the United States of America in Congress assembled, that this Act may be cited as the "Occupational Safety and Health Administration Compliance Assistance Authorization Act of 1998."

Visit the Department of Labor's Occupational Safety and Health Act (OSHA) home page at www.osha.gov, and write a brief report describing the type of safety and health standards that are covered by OSHA.

13 Engineering Symbols and Abbreviations

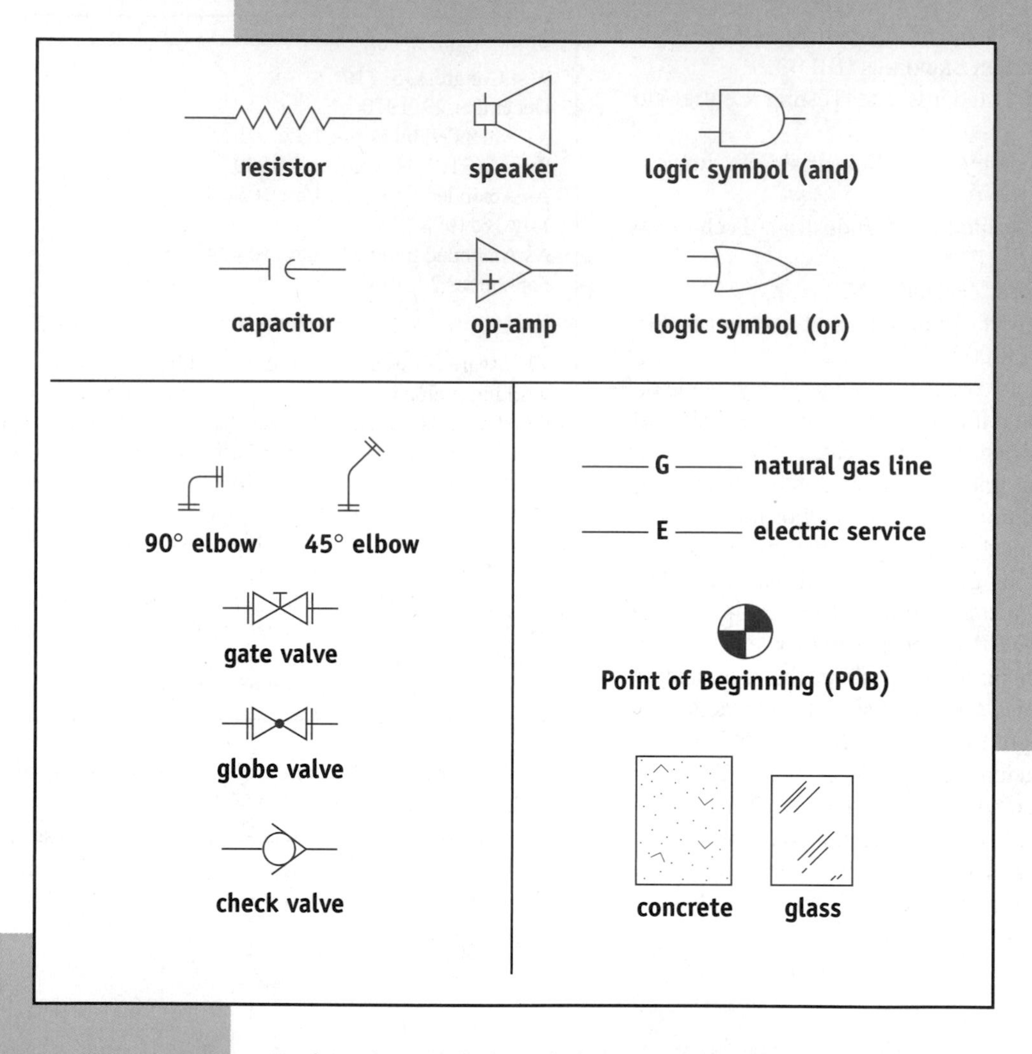

Engineering symbols and signs provide valuable information. These symbols are a "language" used by engineers to convey their ideas, their solutions to problems, or analyses of certain situations. As engineering students you will learn about the graphical ways by which engineers communicate among themselves.

In this chapter, we will discuss the need for conventional engineering symbols as a means to convey information and to effectively communicate to other engineers. We will begin by explaining why there exists a need for engineering signs and symbols. We will then discuss some common symbols used in civil, electrical, and mechanical engineering. Examples of math symbols will also be discussed. This chapter concludes with a discussion of the importance of knowing the Greek alphabet and its use in engineering formulas and engineering drawings. ■

13.1 WHY DO WE NEED ENGINEERING SYMBOLS?

Almost all of you have a driver's license by now; perhaps you even own a car. If this is the case, then you are already familiar with conventional traffic signs that provide valuable information to drivers. Examples of traffic signs are shown in Figure 13.1. These signs are designed and developed, based on some acceptable national and international standards, to convey information not only effectively but also quickly. For example, a stop sign, which has an octagon shape with a red background, tells you to bring your car to a complete stop.

A sign that signals a possibly icy road ahead is another example that warns you to slow down because the road conditions may be such that you may end up in a hazardous situation. That same information could have been conveyed to you in other ways. In place of the sign indicating a slippery road, the highway department could have posted the following words on a board: "Hey you, be careful. The road is slippery, and you could end up in a ditch!" Or they could have installed a loudspeaker warning drivers: "Hey, be careful, slippery road ahead. You could end up in a ditch." Which is the most effective, efficient, and least expensive way of conveying the information? You understand the point of these examples and the question. Road and traffic information can be conveyed inexpensively, quickly, and effectively using signs and symbols. Of course, to understand the meaning of the signs, you had to study them and learn what they mean before you could

■ **Figure 13.1**
Examples of traffic signs

take your driving test. Perhaps in the future the new advancements in technology will reach a stage that road and traffic information can be conveyed directly in a wireless digital format to a computer in your car to allow your car to respond accordingly to the given information! In that case, would you still need to know the meaning of the signs, and would the highway departments need to post them?

13.2 EXAMPLES OF COMMON SYMBOLS IN CIVIL, ELECTRICAL, AND MECHANICAL ENGINEERING

As the examples in the previous section demonstrated, valuable information can be provided in a number of ways: through a long written sentence, orally, graphically or symbolically, or any logical combination of these. But which is the more effective way? As you study various engineering topics, not only will you learn many new concepts but you will also learn about the graphical way that engineers communicate among themselves. You will learn about engineering signs and symbols that provide valuable information and save time, money, and space. These symbols and signs are like a language that engineers use to convey their ideas, their solutions to a problem, or their analyses of certain situations. For example, electrical engineers use various symbols to represent the components that make up an electrical or an electronic system, such as a television set, a cellular phone, or a computer. Examples of engineering symbols are shown in Table 13.1.

Mechanical engineers use diagrams to show the layout of piping networks in buildings or to show the placement of air supply ducts, air return ducts, and fans in a heating or cooling system.

For more detailed information on symbols, see the following documents:

Graphic Electrical Symbols for Air-Conditioning and Refrigeration Equipment by ARI (ARI 130-88). *Graphic Symbols for Electrical and Electronic Diagrams* by IEEE

TABLE 13.1 Examples of Engineering Symbols

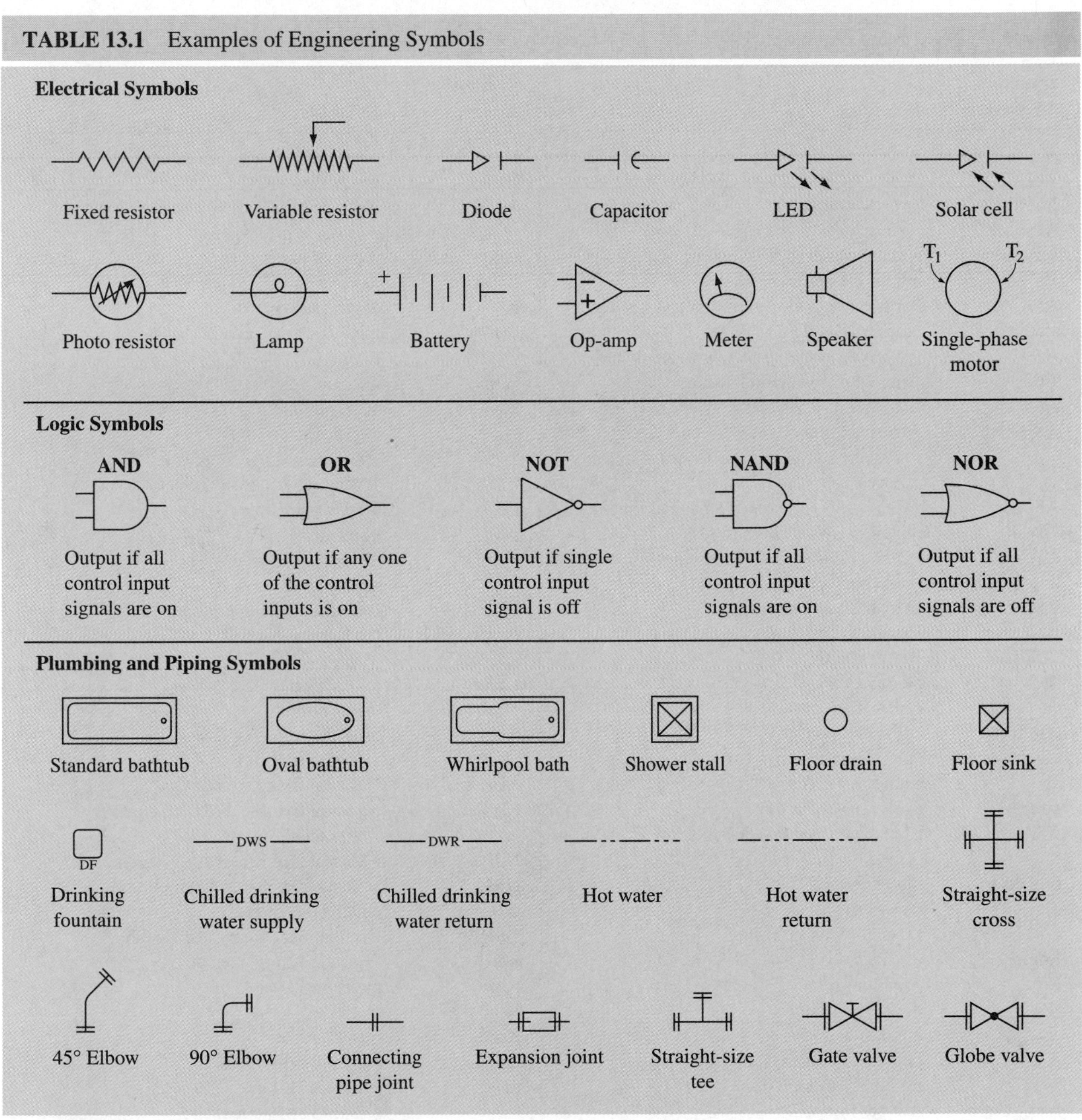

Source: © 1999 American Technical Publishers, Ltd.

(ANSI/IEEE 315-1975). *Graphic Symbols for Pipe Fittings, Valves, and Piping* by ASME (ANSI/ASME ASME Y32.2.2.3-1949 (R 1988)).

Symbols for Mechanical and Acoustical Elements as Used in Schematic Diagrams by ASME (ANSI/ASME Y32.18-1972 (R 1985)). A list of other useful engineering symbols and abbreviations is given in Table 13.2.

TABLE 13.2 Some Engineering Symbols and Abbreviations

Symbol/ Abbreviation	Definition
A°	Angstrom unit = 10^{-10} m
AIChE	American Institute of Chemical Engineers
AIME	American Institute of Mining Engineers
AISI	American Iron and Steel Institute
a.m.	ante meridian (before noon)
am	amplitude modulation
AMCA	Air Moving and Conditioning Association
amu	atomic mass unit
ANSI	American National Standards Institute
API	American Petroleum Institute
ARI	Air Conditioning and Refrigeration Institute
ASHRAE	American Society of Heating, Refrigerating, and Air-Conditioning Engineers
ASME	American Society of Mechanical Engineers
AWG	American Wire Gage
AWWA	American Water Works Association
bhp	brake horsepower
BSI	British Standards Institute
Btu	British thermal units
Btu/h	Btu per hour
°C	degree Celsius
cal	calories
cfh, ft^3/h	cubic feet per hour
cfs, ft^3/s	cubic feet per second
cg	center of gravity
cps	cycles per second
CSA	Canadian Standards Association
db	decibel
d-c, dc	direct current
dia	diameter
DST	Daylight Saving Time
Engr.	engineer
Eq.	equation
°F	degree Fahrenheit
F	farad
FAA	Federal Aviation Adminstration
f-m, fm	frequency modulation
fpm, ft/min	feet per minute
g	acceleration due to gravity
gal	gallons
gpm, gal/min	gallons per minute
H	henry
hp	horsepower
hr, h	hour
Hz	hertz, same as cps
ID	inside diameter
IEEE	Institute of Electrical and Electronics Engineers
in·lb	inch-pounds
J	joule, N·m.
K	degree Kelvin
kcps	kilocycles per second
kg	kilogram
kg·m	kilogram-meters
kip	1000 pound
km	kilometer
kpsi	1000 pounds per square inch
ksi	1000 pounds per square inch
kVA	kilovolt-amperes
kW	kilowatts
kWh	kilowatt-hours
mA	milliamperes
MBh	thousand of Btu per hour
Mcf	thousand cubic feet
mi	miles
mph	miles per hour
MW	megawatts
NASA	National Aeronautics and Space Administration
NEC	National Electric Code
NEMA	National Electrical Manufacturers Association
NFPA	National Fire Protection Association
NRC	Nuclear Regulatory Commission
OD	outside diameter
OSHA	Occupational Safety and Health Administration
p, pp	page, pages
Pa	pascal
p.m.	post merdian
ppb	parts per billion
ppm	parts per million
psi	pounds per square inche
psia	absolute pressure expressed in psi
psig	gauge pressure expressed in psi
PVC	polyvinyl chloride
R	degree Rankine
R&D	research and development
r-f, rf	radio frequency

Continued

TABLE 13.2 Some Engineering Symbols and Abbreviations (*continued*)

Symbol/ Abbreviation	Definition	Symbol/ Abbreviation	Definition
rms	root mean square	USDA	U.S. Department of Agriculture
rpm	revolutions per minute	V	volt
s	seconds	VHF	very high frequency
SAE	Society of Automotive Engineers	vol	volume
SI	International System of Units	W	watt
SME	Society of Manufacturing Engineers	yd	yard
temp	temperature	yr	year
USAF	U.S. Air Force	μ	micro $= 10^{-6}$

13.3 EXAMPLES OF MATH SYMBOLS

As you already know, mathematics is a language that has its own symbols and terminology. In elementary school, you learned about the arithmetic operational symbols such as plus, minus, division, and multiplication. Later you learned about degree symbols, trigonometry symbols, and so on. In the next four years, you will learn additional mathematical symbols and their meanings. Make sure that you understand what they mean and use them properly when communicating with other students or with your instructor. Examples of math symbols are shown in Table 13.3

TABLE 13.3 Some Math Symbols

Symbol	Definition	Symbol	Definition
$+$	plus or positive	$\geq$	equal to or greater than
$-$	minus or negative	$\|x\|$	absolute value of x
$\pm$	plus or minus	$\propto$	proportional to
$\times$ or $\cdot$	multiplication	$\therefore$	therefore
$\div$ or /	division	Σ	summation
:	ratio	$\int$	integral
$<$	less than	!	factorial, for example: $5! = 5 \times 4 \times 3 \times 2 \times 1$
$>$	greater than	Δ	delta, indicating difference
$\ll$	much less than	∂	partial
$\gg$	much greater than	π	pi: 3.1415926 . . .
$=$	equal to	∞	infinity
$\approx$	approximately equal to	$^\circ$	degree
$\neq$	not equal to	( )	parentheses
$\equiv$	identical with	[ ]	brackets
$\leq$	equal to or less than	{ }	braces

13.4 THE GREEK ALPHABET AND ROMAN NUMERALS

As you take more and more mathematics and engineering classes, you will see that the Greek alphabetic characters are quite commonly used to express angles, dimensions, and

TABLE 13.4 The Greek Alphabet

Uppercase Letter	Lowercase Letter	Name	Uppercase Letter	Lowercase Letter	Name
A	α	alpha	N	ν	nu
B	β	beta	Ξ	ξ	xi
Γ	γ	gamma	O	o	omicron
Δ	δ	delta	Π	π	pi
E	ε	epsilon	P	ρ	rho
Z	ζ	zeta	Σ	σ	sigma
H	η	eta	T	τ	tau
Θ	θ	theta	Y	υ	upsilon
I	ι	iota	Φ	ϕ	phi
K	κ	kappa	X	χ	chi or khi
Λ	λ	lambda	Ψ	ψ	psi
M	μ	mu	Ω	ω	omega

physical variables in drawings and in mathematical equations and expressions. Take a few moments to learn and memorize these characters. Knowing these symbols will save you time when communicating with other students or when asking a question of your professor. You don't want to refer to ζ (zeta) as that "curly thing" when speaking to your professor! You may also find Roman numerals in use to some extent in science and engineering. The Greek alphabet and the Roman numerals are shown in Tables 13.4 and 13.5.

TABLE 13.5 Roman Numerals

Roman Numeral	Arabic Numeral	Roman Numeral	Arabic Numeral
I	1	XX	20
II	2	XXX	30
III	3	XL	40
IIII or IV	4	L	50
V	5	LX	60
VI	6	LXX	70
VII	7	LXXX	80
VIII	8	XC	90
IX	9	C	100
X	10	CC	200
XI	11	CCC	300
XII	12	CCCC or CD	400
XIII	13	D	500
XIV	14	DC	600
XV	15	DCC	700
XVI	16	DCCC	800
XVII	17	CM	900
XVIII	18	M	1000
XIX	19	MM	2000

SUMMARY

Now that you have reached this point in the text

- You should know why we need and use engineering symbols to communicate among ourselves.
- You should be familiar with some of the common civil, electrical, and mechanical engineering symbols so that you can understand and interpret a technical drawing.
- You should be familiar with the examples of math symbols given in this chapter and be prepared to learn new math symbols as you take additional math and engineering classes.
- You should understand the role of the Greek alphabet in engineering and their importance in terms of representing angles, dimensions, or a physical variable in a formula. You should also memorize the Greek alphabets so that you can communicate with others effectively.

PROBLEMS

1. Using Table 13.1, identify the engineering symbols shown in the accompanying figure.

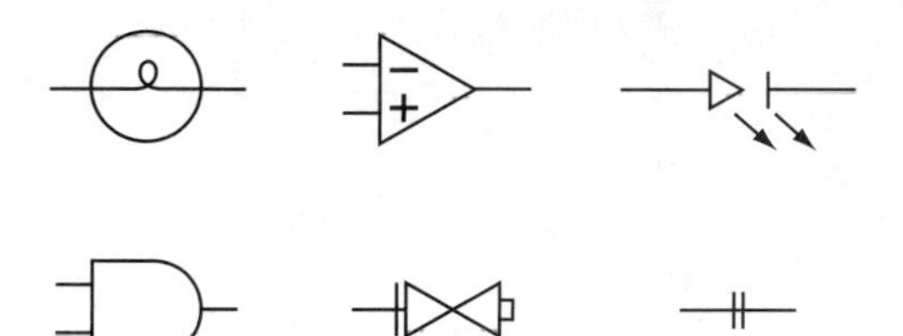

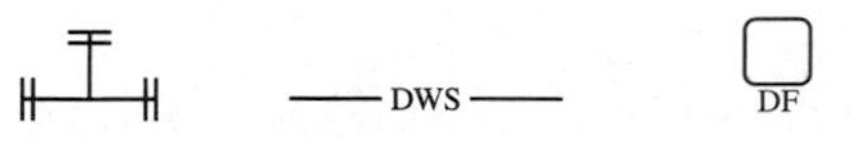

2. Using Table 13.1, identify the components of the logic system shown in the accompanying figure.

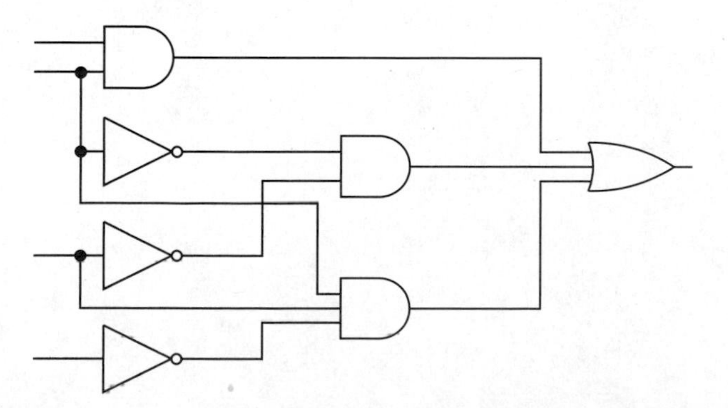

3. What do the following symbols mean?
   - Σ ____________
   - ∂ ____________
   - Δ ____________
   - $\propto$ ____________
   - $\equiv$ ____________
   - $\int$ ____________

4. What do the following engineering symbols and abbreviations mean?
   - ASHRAE ____________
   - ASME ____________
   - ASCE ____________
   - Btu ____________
   - bhp ____________
   - ANSI ____________
   - AIChE ____________
   - IEEE ____________
   - OSHA ____________
   - gpm ____________
   - kpsi ____________
   - ksi ____________
   - SAE ____________
   - ppm ____________
   - pp ____________
   - SME ____________

5. Write the names of the following Greek alphabetic characters and how they are pronounced.

Γ γ ______________________
Δ δ ______________________
Ε ε ______________________
Ζ ζ ______________________
Η η ______________________
Λ λ ______________________
Μ μ ______________________
Ν ν ______________________
Ξ ξ ______________________
Ρ ρ ______________________
Σ σ ______________________
Τ τ ______________________
Φ ϕ ______________________
Χ χ ______________________
Ψ ψ ______________________

6. What numbers do the following Roman numerals represent?

MCMLXXX ______________________
XXIX ______________________
XLVIII ______________________

CHAPTER

14 Electronic Spreadsheets

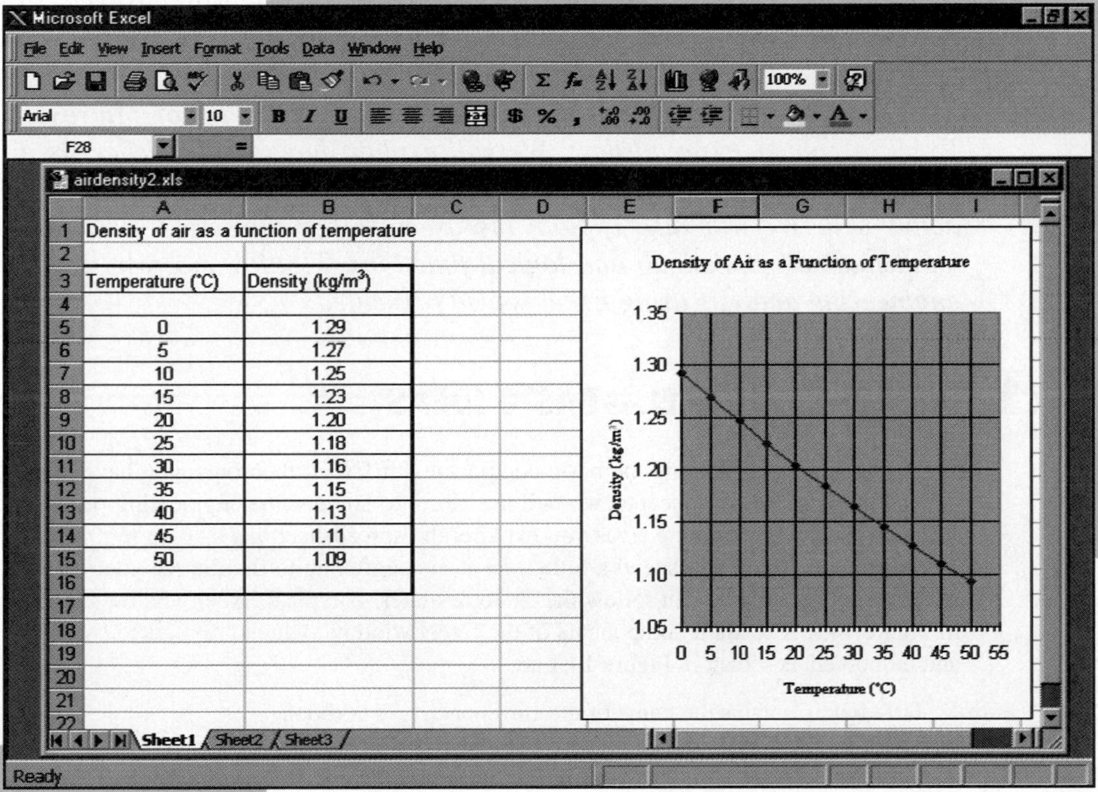

An electronic spreadsheet is a tool that can be used to solve an engineering problem. Spreadsheets are commonly used to record, organize, and analyze data using formulas. Spreadsheets are also used to present the results of an analysis in chart form. Although engineers still write computer programs to solve complex engineering problems, simpler problems can be solved with the help of a spreadsheet.

In this chapter, we will discuss the use of spreadsheets in solving engineering problems. Before the introduction of electronic spreadsheets, engineers wrote their own computer programs. Computer programs were typically written for problems where more than a few hand calculations were required. FORTRAN was a common programming language that was used by many engineers to perform numerical computations. Although engineers still write computer programs to solve complex engineering problems, simpler problems can be solved with the help of a spreadsheet. Compared to writing a computer program and debugging it, spreadsheets are much easier to use, to record, organize, and analyze data using formulas that are input by the user. Spreadsheets are also used to show the results of an analysis in the form of charts. Because of their ease of use, spreadsheets are common in many other disciplines, including business, marketing, and accounting.

This chapter begins by discussing the basic makeup of Microsoft® Excel, a common spreadsheet. We will explain how a spreadsheet is divided into rows and columns, and how to input data or a formula into an active cell. We will also explain the use of other tools such as Excel's mathematical, statistical, and logical functions. Plotting the results of an engineering analysis using Excel is also presented. ■

14.1 MICROSOFT® EXCEL—BASIC IDEAS

We will begin by explaining the basic components of Excel; then once you have a good understanding of these concepts, we will use Excel to solve some engineering problems. As is the case with any new areas you explore, the spreadsheet has its own terminology. Therefore, make sure you spend a little time at the beginning to familiarize yourself with the terminology, so you can follow the examples later. A typical Excel window is shown in Figure 14.1. The main components of the Excel window, which are marked by arrows and numbered as shown in Figure 14.1 are

1. **Title bar**: Contains the name of the current active workbook.
2. **Menu bar**: Contains the commands used by Excel to perform certain tasks.
3. **Toolbar buttons**: Contains push buttons that execute commands used by Excel.
4. **Active cell**: A worksheet is divided into rows and columns. A cell is the box that you see as the result of the intersection of a column and a cell. *Active cell* refers to a specific selected cell.
5. **Formula bar**: Shows the data or the formula used in the active cell.
6. **Name box**: Contains the address of the active cell.
7. **Column header**: A worksheet is divided into rows and columns. The columns are marked by A, B, C, D, and so on.
8. **Row header**: The rows are identified by numbers 1, 2, 3, 4, and so on.
9. **Worksheet tabs**: Allow you to move from one sheet to another sheet. As you will learn later, you can name these worksheets.

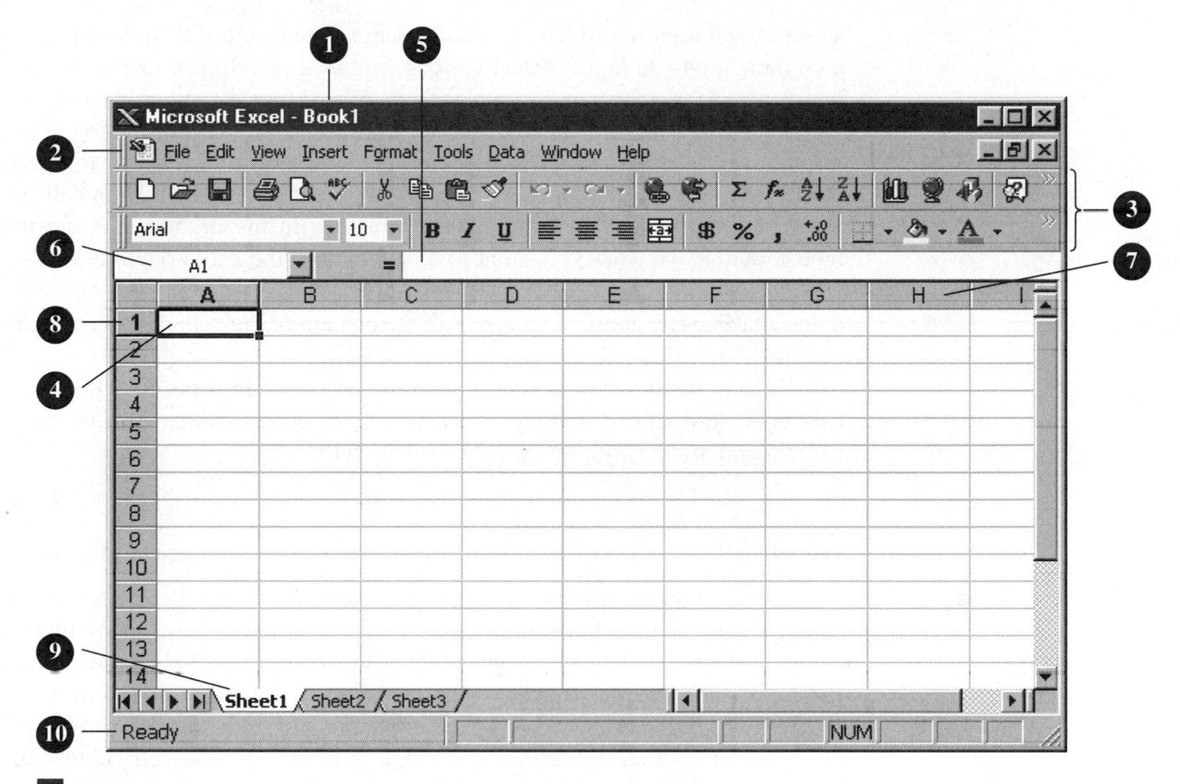

Figure 14.1 The components of the Excel window

10. **Status bar**: Gives information about the command mode. For example, "Ready" indicates the program is ready to accept input for a cell or "Edit" indicates Excel is in an edit mode.

A ***workbook*** is the spreadsheet file that you create and save. A workbook could consist of many worksheets and charts. A ***worksheet*** represents the rows and columns where you input information such as data, formulas, and the result of various calculations. As you will see soon, you may also include charts as a part of a given worksheet as well.

Naming Worksheets

To name a worksheet, double-click the sheet tab to be named, type the desired name, and hit the enter key. You can use the **Edit** or the **Insert** menu to delete or insert a worksheet. You can move (or change the position of) a worksheet in the workbook by selecting the sheet tab and while holding down the button on the mouse move the tab to the desired position among other sheets.

14.2 CELLS AND THEIR ADDRESSES

As shown in Figure 14.1, a worksheet is divided into rows and columns. The columns are marked by A, B, C, D, and so on, while the rows are identified by numbers 1, 2, 3, 4, and so on. A ***cell*** represents the box that one sees as the result of the intersection of a row and a column. You can input (enter) various entities in a cell. For example, you can type in words or enter numbers or a formula. To enter words or a number in a cell, simply choose the cell where you want to enter the information, type the information, and then hit the Enter key on your keyboard. Perhaps the simplest and the easiest way to move around in a worksheet is to use a mouse. For example, if you want to move from cell A5 to cell C8, move the mouse such that the mouse pointer is in the desired cell and then click the mouse button. To edit the content of a cell, choose the cell, double-click the mouse button, and then similar to editing a word processing document, use any combination of delete, backspace, or arrow keys to edit the content of the cell. As an alternative to double clicking, you can use the F2 key to select the edit mode.

Keep in mind that as you become more proficient in using Excel, you will learn that for certain tasks there is more than one way to do something. In this chapter, we will explain one of the ways, which can be easily followed.

A Range

As you will soon see when formatting, analyzing, or plotting data, it is often convenient to select a number of cells simultaneously. The cells that are selected simultaneously are called a ***range***. To define a range, begin with the first cell that you want included in the range and then drag the mouse (while pressing down the left button) to the last cell that should be included in the range. An example of selecting a range is shown in Figure 14.2. Note that in spreadsheet language, a range is defined by the cell address of the top-left

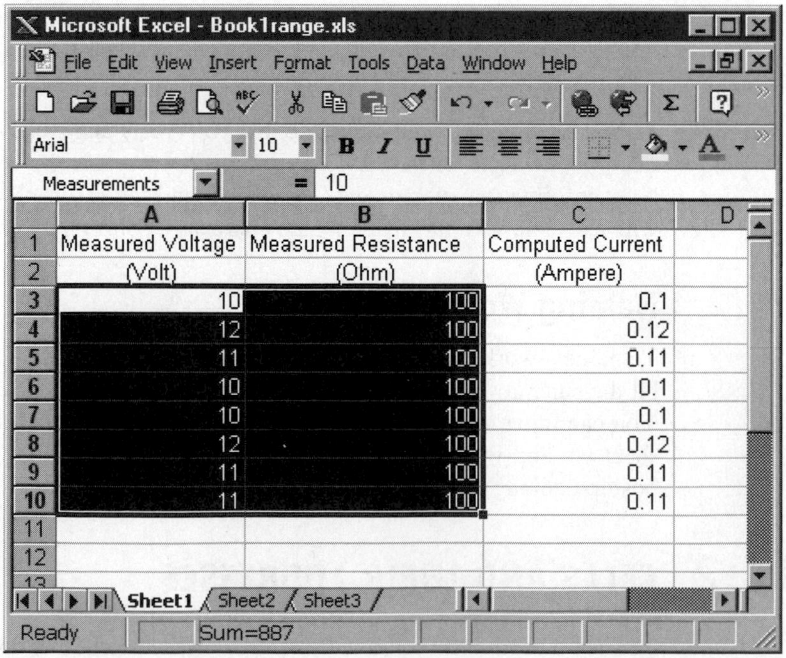

Figure 14.2
An example showing the selection of a range of cells

selected cell in the range followed by a colon, :, and ends with the address of the bottom-right cell in the range. For example, to select cells A3 through B10, we first select A3 and then drag the mouse diagonally to B10. In spreadsheet language, this range is specified in the following manner—A3:B10. There are situations where you may want to select a number of cells that are not side by side. In such cases, you must first select the contiguous cells, and then while holding (pressing) the Ctrl key select the other noncontiguous cells by dragging the mouse button.

Excel allows the user to assign names to a range (selected cells). To name a range, first select the range as just described, and then click on the Name box in the Formula bar and type in the name you want to assign to the range. You can use upper- or lowercase letters along with numbers, but no spaces are allowed between the characters or the numbers. For example, as shown in Figure 14.2, we have grouped the measured voltages and the resistance into one range, which we have called *measurements*.

Inserting Cells, Columns, and Rows

After entering data into a spreadsheet you may realize that you should have entered some additional data in between two cells, columns, or rows that you have just created. In such a case, you can always insert new cells, column(s), or row(s) among already existing data cells, columns, and rows in a worksheet. To insert new cell(s) between other existing cells, you must first select the cell(s) where the new cell(s) are to be inserted. Next, from the **Insert** menu choose the **Cells** option. Indicate whether you want the selected cells to be shifted to the right or down. For example, let's say we want to insert three new cells in the location E8 through E11 (E8:E11) and shift the existing content of E8:E11 down. We first select cells E8:E11; then from the **Insert** menu we choose the **Cells** option, and then choose the Shift cell down option. To insert a column, click on the column indicator button to the right of where you would like to have the new column inserted. From the **Insert** menu, choose **Columns**. The procedure is similar for inserting a new row among already existing rows. For example, if you would like to insert a new column between columns D and E, you must first select column E, and then from the **Insert** menu choose **Columns**; the new column will be inserted to the left of column E. To insert more than one column or row simultaneously, you should select as many column indicator buttons as necessary to the right of where you would like to have the columns inserted. For example, if you would like to insert three new columns between columns D and E, then you must first select columns E, F, G; then from the **Insert** menu choose **Columns**, and three new columns will be inserted to the left of column E.

14.3 CREATING FORMULAS IN EXCEL

By now you know that engineers use formulas that represent physical and chemical laws governing our surroundings to analyze various problems. You can use Excel to input engineering formulas and compute the results. In Excel a formula always begins with an equal sign, =. To enter a formula, select the cell where you want the result of the formula to be displayed. In the Formula bar, then type the equal sign and the formula. Remember when typing your formula to use parentheses to dictate the order of operation. For exam-

TABLE 14.1 The Basic Excel Arithmetic Operations

Operation	Symbol	Example: Cells A5 and A6 contain the values 10 and 2, respectively	Cell A7 contains the result of the formula given in the example
Addition	+	=A5+A6+20	32
Subtraction	−	=A5−A6	8
Multiplication	*	=(A5*A6)+9	29
Division	/	=(A5/2.5)+A6	6
Raised to a power	^	=(A5^A6)^0.5	10

ple, if you were to type =100+5*2, Excel will perform the multiplication first, which results in a value of 10, and then this result is added to 100, which yields an overall value of 110 for the formula. If however, you wanted Excel to add the 100 to 5 first and then multiply the resulting 105 by 2, you should have placed parentheses around the 100 and 5 in the following manner: =(100+5)*2, which results in a value of 210. The basic Excel arithmetic operations are shown in Table 14.1.

EXAMPLE 14.1

As we explained in the previous chapters, thermophysical properties of a substance, including density, viscosity, thermal conductivity, and heat capacity, play a key role in engineering calculations. As discussed, the thermophysical property values represent information such as how compact the material is for a given volume (density), or how easily a fluid flows (viscosity), or how good a material is in conducting heat (thermal conductivity), or how good the material is in storing thermal energy (heat capacity). The values of thermophysical properties are commonly measured in laboratories at given conditions. Moreover, the values of thermophysical properties of a substance generally change with temperature. The following example will show how the density of standard air changes with temperature. The density of standard air is a function of temperature and may be approximated using the ideal gas law according to

$$\rho = \frac{P}{RT}$$

where

P = standard atmospheric pressure (101.3 kPa)

$R \equiv$ gas constant and its value for air is 286.9 $\left(\frac{\text{J}}{\text{kg}\cdot\text{K}}\right)$

$T \equiv$ air temperature in Kelvin

Using Excel, we want to create a table that shows the density of air as a function of temperature in the range of 0°C (273.15 K) to 50°C (323.15 K) in increments of 5°C.

Refer to the Excel sheets shown in the accompanying figures when following the steps.

1. In cell A1 type **Density of air as a function of temperature**.
2. In cells A3 and B3 type **Temperature (C), Density (kg/m^3)**, respectively.
3. In cells A5 and A6 type **0** and **5**, respectively.

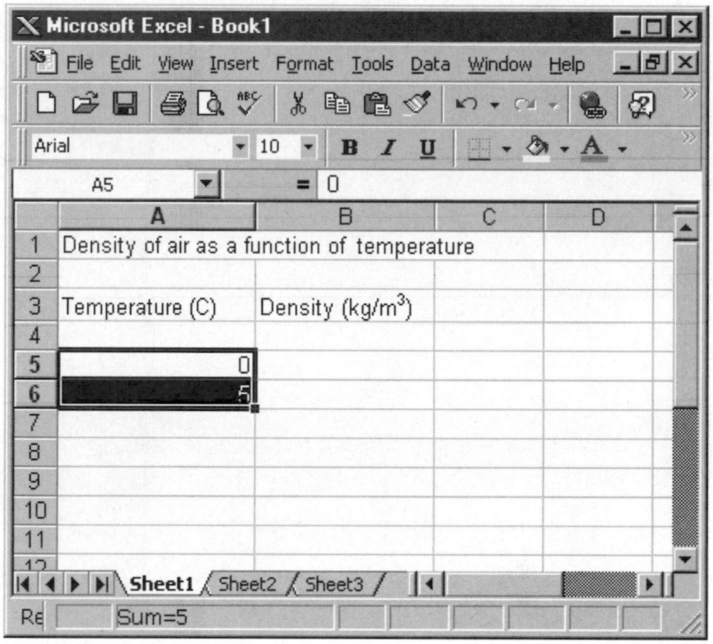

Steps 1 through 3

4. Pick cells A5 and A6 and use the fill command with the + handle to copy the pattern into cells A7 to A15.

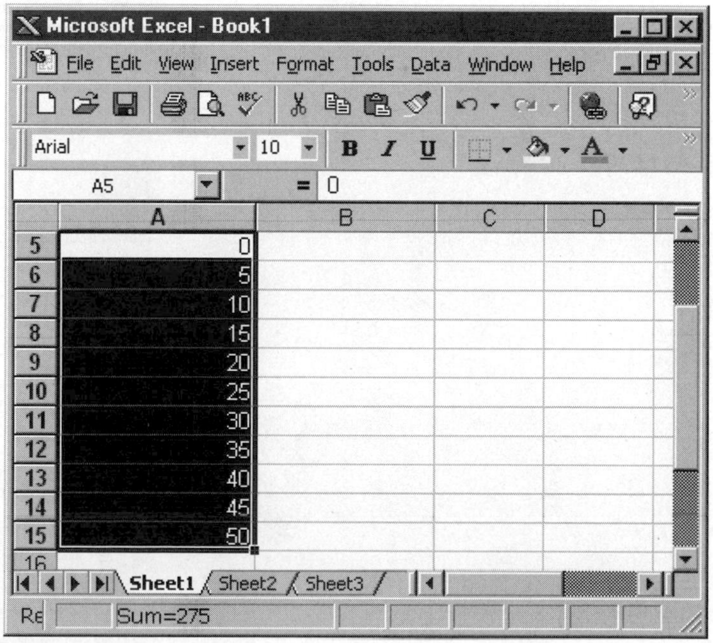

Step 4

5. In cell B5 type the formula =(101300)/((286.9)*(A5+273)), as shown.

Microsoft Excel - Book1

B5 = =(101300)/((286.9)*(A5+273))

	A	B	C	D
1	Density of air as a function of temperature			
2				
3	Temperature (C)	Density (kg/m^3)		
4				
5	0	1.293350544		
6	5			
7	10			
8	15			
9	20			
10	25			
11	30			

Step 5

6. Use the **Edit** menu and the **Fill** command to copy the formula into cells B6 to B15.

Microsoft Excel - Book1

B5 = =(101300)/((286.9)*(A5+273))

	A	B	C	D
1	Density of air as a function of temperature			
2				
3	Temperature (C)	Density (kg/m^3)		
4				
5	0	1.293350544		
6	5	1.270088844		
7	10	1.247649111		
8	15	1.225988536		
9	20	1.20506723		
10	25	1.184847982		
11	30	1.165296035		
12	35	1.146378891		
13	40	1.128066129		
14	45	1.110329241		
15	50	1.093141481		

Sum=13.07020402

Step 6

7. Use the **Format** menu and the **Cells** command to change the number of significant digits, as shown.

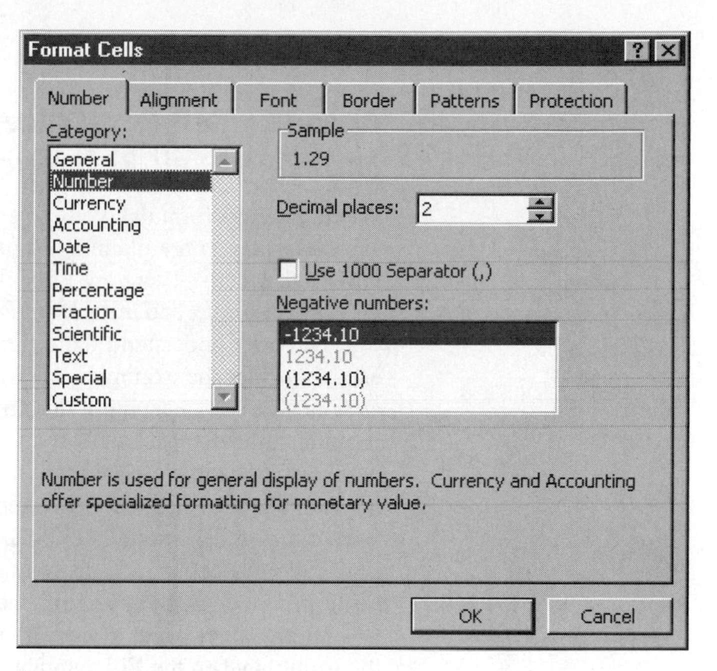

Step 7

The final results for Example 14.1 are shown in Figure 14.3.

Microsoft Excel - airdensity4.xls

	A	B
1	Density of air as a function of temperature	
2		
3	Temperature (C)	Density (kg/m^3)
4		
5	0	1.29
6	5	1.27
7	10	1.25
8	15	1.23
9	20	1.20
10	25	1.18
11	30	1.16
12	35	1.15
13	40	1.13
14	45	1.11
15	50	1.09
16		
17		
18		

Figure 14.3 The final result for Example 14.1

Absolute Cell Reference, Relative Cell Reference, and Mixed Cell Reference

When creating formulas you have to be careful how you refer to the address of a cell, especially if you are planning to use the **Fill** command to copy the pattern of formulas, in the other cells. There are three ways that you can refer to a cell address in a formula: *absolute, relative*, and *mixed reference*.

To better understand the differences among the absolute, relative, and mixed reference, consider the examples shown in Figure 14.4. As the name implies, ***absolute reference*** is absolute, meaning it does not change when the **Fill** command is used to copy the formula into other cells. Absolute reference to a cell is made by $column-letter$row-number. For example, $A$3 will always refer to the content of cell A3, regardless of how the formula is copied. In the example shown, cell A3 contains the value 1000, and if we were to input the formula =0.06*$A$3 in cell B3, the result would be 60. Now if we were to use the **Fill** command and copy the formula down in cells B4 through B11, this would result in a value of 60 appearing in cells B4 through B11, as shown in Figure 14.4(a).

On the other hand, if we were to make a ***relative reference*** to A3, that would change the formula when the **Fill** command is used to copy the formula into other cells. To make a relative reference to a cell, a special character, such as $, is not needed. You simply refer to the cell address. For example, if we were to input the formula =0.06*A3 in cell B3, the result would be 60; and if we use the **Fill** command to copy the formula into cell B4, the A3 in the formula will automatically be substituted by A4, resulting in a value 75. Note that the formula in cell B4 now becomes =0.06*A4. The result of applying the **Fill** command to cells B4 through B11 is shown in Figure 14.4(b).

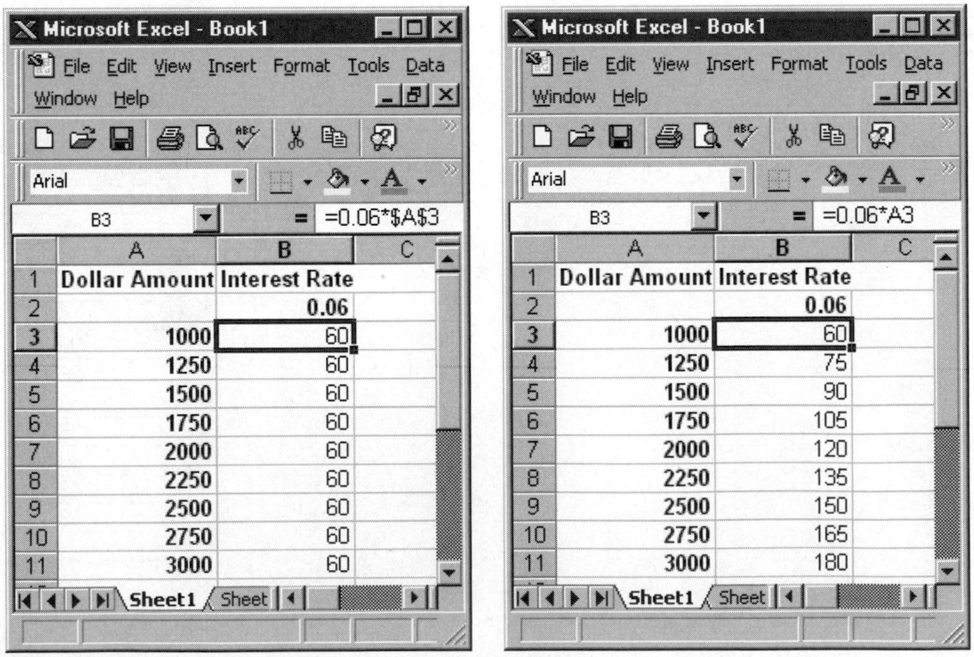

B3 = =0.06*$A$3

	A	B	C
1	Dollar Amount	Interest Rate	
2		0.06	
3	1000	60	
4	1250	60	
5	1500	60	
6	1750	60	
7	2000	60	
8	2250	60	
9	2500	60	
10	2750	60	
11	3000	60	

(a)

B3 = =0.06*A3

	A	B	C
1	Dollar Amount	Interest Rate	
2		0.06	
3	1000	60	
4	1250	75	
5	1500	90	
6	1750	105	
7	2000	120	
8	2250	135	
9	2500	150	
10	2750	165	
11	3000	180	

(b)

Figure 14.4 Examples showing the difference between the results of a formula when absolute and relative cell references are made in the formula

The ***mixed cell reference*** could be done in one of two ways: (1) You can keep the column as absolute (unchanged) and have a relative row, or (2) you can keep the row as absolute and have a relative column. For example, if you were to use $A3 in a formula, it would mean that column A remains absolute and unchanged, but row 3 is a reference row and changes as the formula is copied into other cells. On the other hand, A$3 means row 3 remains absolute while column A changes as the formula is copied into other cells. The use of mixed cell reference is demonstrated in the following example.

EXAMPLE 14.2

Using Excel, create a table that shows the relationship between the interest earned and the amount deposited, as shown in Table 14.2.

TABLE 14.2 The Relationship Between the Interest Earned and the Amount Deposited

	Interest Rate			
Dollar Amount	**0.06**	**0.07**	**0.075**	**0.08**
1000	60	70	75	80
1250	75	87.5	93.75	100
1500	90	105	112.5	120
1750	105	122.5	131.25	140
2000	120	140	150	160
2250	135	157.5	168.75	180
2500	150	175	187.5	200
2750	165	192.5	206.25	220
3000	180	210	225	240

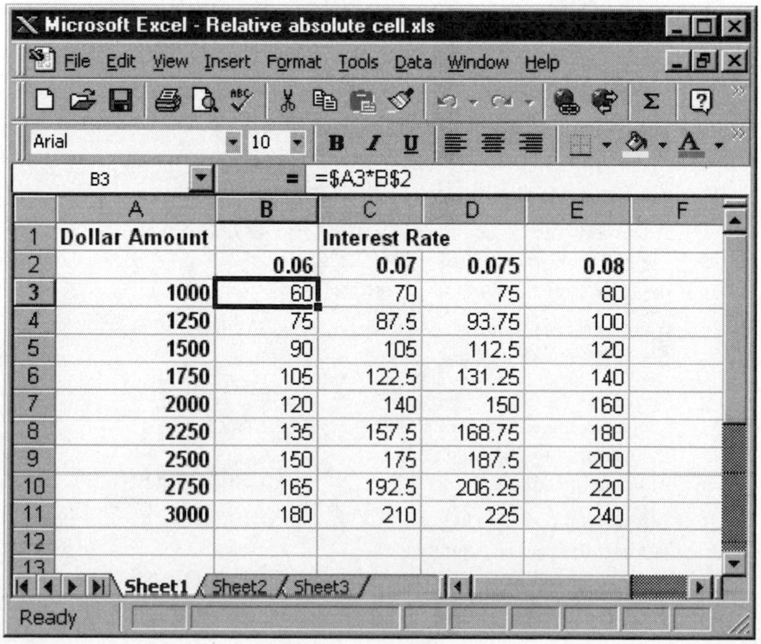

Microsoft Excel - Relative absolute cell.xls

B3 = =$A3*B$2

	A	B	C	D	E	F
1	Dollar Amount		Interest Rate			
2		0.06	0.07	0.075	0.08	
3	1000	60	70	75	80	
4	1250	75	87.5	93.75	100	
5	1500	90	105	112.5	120	
6	1750	105	122.5	131.25	140	
7	2000	120	140	150	160	
8	2250	135	157.5	168.75	180	
9	2500	150	175	187.5	200	
10	2750	165	192.5	206.25	220	
11	3000	180	210	225	240	
12						

Figure 14.5
Excel spreadsheet for Example 14.2

In order to create the table for Example 14.2, using Excel we will first create the dollar amount column and the interest row, as shown in Figure 14.5. Next we will type into cell B3 the formula =**$A3*B$2**. We can now use the **Fill** command to copy the formula in other cells, resulting in the table shown in Figure 14.5. Note that the dollar sign before A3 means column A is to remain unchanged in the calculations when the formula is copied into other cells. Also note that the dollar sign before 2 means that row 2 is to remain unchanged in calculations when the **Fill** command is used.

TABLE 14.3 Some Excel Functions That You May Use in Engineering Analyses

Function	Description of the Function	Example	Result of the Example
SUM(range)	It sums the values in the given range.	=SUM(A1:B10) or =SUM(values)	164
AVERAGE(range)	It calculates the average value of the data in the given range.	=AVERAGE(A1:B10) or =AVERAGE(values)	8.2
COUNT	It counts the number of values in the given range.	=COUNT (A1:B10) or =COUNT(values)	20
MAX	It determines the largest value in the given range.	=MAX(A1:B10) or =MAX(values)	10
MIN	It determines the smallest value in the given range.	=MIN(A1:B10) or =MIN(values)	6
STDEV	It calculates the standard deviation for the values in the given range.	=STDEV(A1:B10) or =STDEV(values)	1.105
PI	It returns the value of π, 3.141519265358979, accurate to 15 digits.	=PI()	3.141519265358979
DEGREES	It converts the value in the cell from radians to degrees.	=DEGREES(PI())	180
RADIANS	It converts the value from degrees to radians.	RADIANS(90) or =RADIANS(D1)	1.57079 3.14159
COS	It returns the cosine value of the argument. The argument must be in radians.	=COS(PI()/2) or =COS(RADIANS(D1))	0 −1
SIN	It returns the sine value of the argument. The argument must be in radians.	=SIN(PI()/2) or =SIN(RADIANS(D1))	1 0

14.4 USING EXCEL FUNCTIONS

Excel offers a large selection of built-in functions that you can use to analyze data. By built-in functions we mean standard functions such as the sine or cosine of an angle as well as formulas that calculate the total value, the average value, or standard deviation of a set of data points. The Excel functions are grouped into various categories, including mathematical and trigonometric, statistical, financial, and logical functions. In this chapter, we will discuss some of the common functions that you may use during your engineering education or later as a practicing engineer. You can enter a function in any cell by simply typing the name of the function if you already know it. If you do not know the name of the function, then you can press the **Paste Function** (f_x) button on the toolbar, and then from the menu select the Function category and the Function name. There is also a Help button, marked by a question mark, **?**, on the lower left corner of the **Paste Function** menu, which once activated and followed leads to information about what the function computes and how the function is to be used.

Some examples of commonly used Excel functions, along with their proper use and descriptions, are shown in Table 14.3. Refer to Example 14.3 and Figure 14.6 when studying Table 14.3.

EXAMPLE 14.3

A set of values is given in the worksheet shown in Figure 14.6. Familiarize yourself with some of Excel built-in functions as described in Table 14.3. When studying Table 14.3,

Microsoft Excel - Book1

E14 = =SIN(RADIANS(D1))

	A	B	C	D	E
1	10	8		180	164
2	8	8			8.2
3	8	7			20
4	10	9			10
5	9	9			6
6	7	6			1.105012503
7	6	8			3.141592654
8	8	8			180
9	8	9			1.570796327
10	9	9			3.141592654
11					6.12574E-17
12					-1
13					1
14					1.22515E-16
15					

Figure 14.6 The Excel worksheet for Example 14.3

note that columns A and B contain the data range, which we have named *values*; cell D1 contains the angle 180. Also note that the functions were typed in the cells E1 through E14; consequently, the results of the executed Excel functions are shown in those cells.

More examples of Excel's functions are shown in Table 14.4.

TABLE 14.4 More Examples of Additional Excel Functions

Function	Description of the Function
SQRT(x)	Returns the square root of value x.
FACT(x)	Returns the value of the factorial of x. For example, FACT (5) will return: (5)(4)(3)(2)(1)=120.
Trigonometric Functions	
TAN(x)	Returns the value for the tangent of x. The argument must be in radians.
DEGREES (x)	Converts the value of x from radians to degrees. It returns the value of x in degrees.
ACOS(x)	This is the inverse cosine function of x. It is used to determine the value of an angle when its cosine value is known. It returns the angle value in radians, when the value of cosine between -1 and 1 is used for argument x.
ASIN(x)	This is the inverse sine function of x. It is used to determine the value of an angle when its sine value is known. It returns the angle value in radians when the value of sine falls between -1 and 1.
ATAN(x)	This is the inverse tangent of the x function. It is used to determine the value of an angle when its tangent value is known.
Exponential and Logarithmic Functions	
EXP(x)	Returns the value of e^x.
LN(x)	Returns the value of the natural logarithm of x. Note that x must be greater than 0.
LOG(x)	Returns the value of the common logarithm of x.

EXAMPLE 14.4

Using Excel, compute the average (arithmetic mean) and the standard deviation of the density of water data given in the accompanying table. Refer to Chapter 6, Section 6.8 to refresh your memory about what the value of standard deviation for a set of data points represents.

Refer to Figure 14.7 when following the steps.

1. In cell B1 type **Group A findings**, and in cell C1 type **Group B findings**.
2. In each of cells B3 and C3 type **Density (kg/m3)**. Highlight the 3 in the kg/m3, and use the following command to make 3 a superscript. Choose **Cells** from the **Format**, Cells menu, and turn on the superscript toggle switch.

Data for Example 14.4

Group A Findings	Group B Findings
ρ (kg/m^3)	ρ (kg/m^3)
1020	950
1015	940
990	890
1060	1080
1030	1120
950	900
975	1040
1020	1150
980	910
960	1020

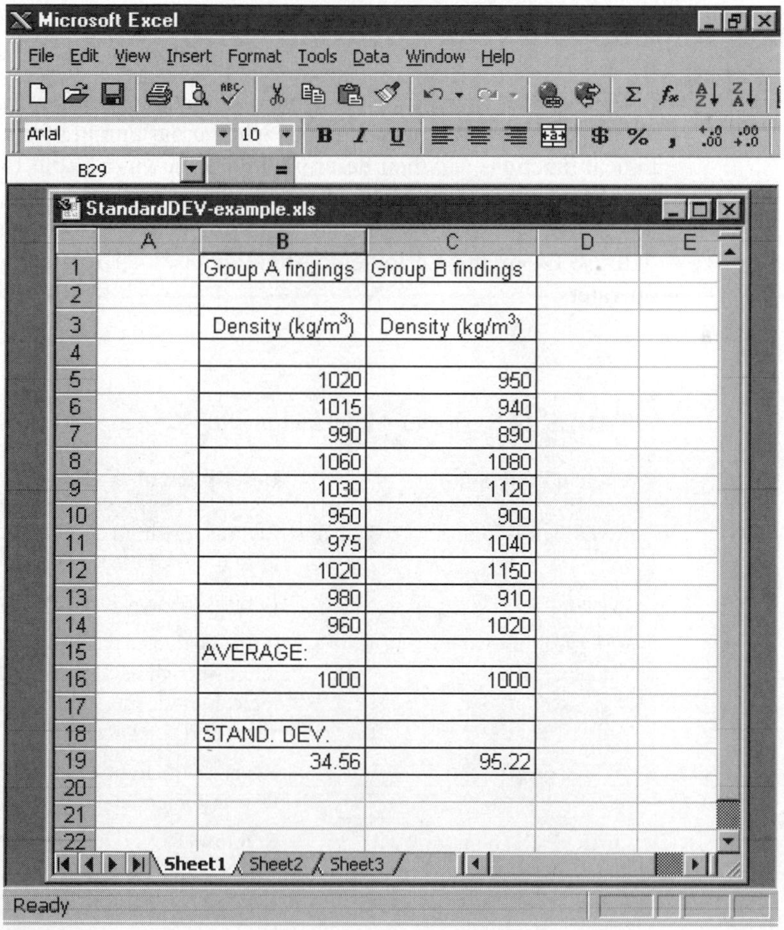

Microsoft Excel — StandardDEV-example.xls

	A	B	C	D	E
1		Group A findings	Group B findings		
2					
3		Density (kg/m^3)	Density (kg/m^3)		
4					
5		1020	950		
6		1015	940		
7		990	890		
8		1060	1080		
9		1030	1120		
10		950	900		
11		975	1040		
12		1020	1150		
13		980	910		
14		960	1020		
15		AVERAGE:			
16		1000	1000		
17					
18		STAND. DEV.			
19		34.56	95.22		
20					
21					
22					

Figure 14.7 The Excel spreadsheet for Example 14.4

3. Next, we want to compute the arithmetic means for the Group A and Group B data, but first we need to create a title for this computation. Because we are calculating the average, we might as well just use the word AVERAGE for the title of our calculations, thus in cell B15 type **AVERAGE:**.
4. In order to have Excel compute the average, we use the AVERAGE function in the following manner. In cell B16, we type **=AVERAGE(B5:B14)**, and similarly in cell C16, we type **=AVERAGE(C5:C14)**.
5. Next, we will make a title for the standard deviation calculation by simply typing in cell B18 **STAND. DEV**.
6. To compute the standard deviation for the Group A findings, in cell B19 type **=STDEV(B5:B14)**, and similarly to calculate the standard deviation for the Group B findings, in cell C19 type **=STDEV(C5:C14)**. Note that we used the function STDEV and the appropriate data range.

The final results for Example 14.4 are shown in Figure 14.7.

14.5 USING EXCEL LOGICAL FUNCTIONS

In this section, we will look at some of Excel's logical functions. These are functions that allow you to test various conditions when programming formulas to analyze data. Excel's logical functions and their descriptions are shown in Table 14.5.

Excel also offers relational or comparison operators that allow for testing of relative magnitude of various arguments. These relational operators are shown n Table 14.6. We will use Example 14.5 to demonstrate the use of Excel's logical functions and relational operators.

TABLE 14.5 Excel's Logical Functions

Logical Functions	Description of the Function
AND(logic1, logic2, logic3, . . .)	Returns true if all arguments are true and returns false if any of the arguments are false.
False( )	Returns the logical value false.
IF(logical test, value_if_true, value_if_false)	It first evaluates the logical test; if true, then it returns the value_if_true; if the evaluation of the logical test deems false, then it returns the value_if_false value.
NOT(logical)	Reverses the logic of its argument; returns true for a false argument and false for the true argument.
OR(logical1, logical2, . . .)	Returns TRUE if any argument is true and returns FALSE if all arguments are false.
TRUE( )	Returns the logical value TRUE.

TABLE 14.6 Excel's Relational Operators and Their Descriptions

Relational Operator	Description
<	less than
<=	less than or equal to
=	equal to
>	greater than
>=	greater than or equal to
<>	not equal to

EXAMPLE 14.5

The pipeline shown in Figure 14.8 is connected to a control (check) valve that opens when the pressure in the line reaches 20 psi. Various readings were taken at different times and recorded. Using Excel's logical functions, create a list that shows the corresponding open and closed position of the check valve (see Figure 14.9).

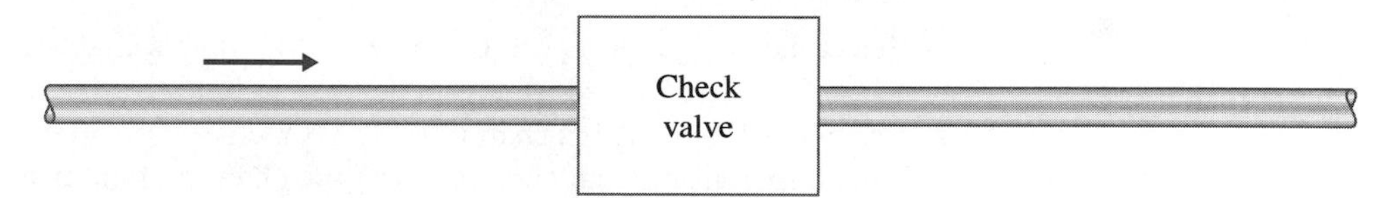

Figure 14.8 A schematic diagram for Example 14.5

Microsoft Excel - Logical example12.xls

B3 = =IF(A3>=20,"OPEN","CLOSED")

	A	B	C	D
1	Pressure Reading			
2	(psi)			
3	20	OPEN		
4	18	CLOSED		
5	22	OPEN		
6	26	OPEN		
7	10	CLOSED		
8	19	CLOSED		
9	21	OPEN		
10	12	CLOSED		
11				
12				

Sheet1 / Sheet2 / Sheet3

Figure 14.9 The solution to Example 14.5

The solution to Example 14.5 is shown in the figure. The pressure readings were entered in column A. In cell B3 we type the formula **=IF(A3 >= 20,"OPEN","CLOSED")** and use the **Fill** command to copy the formula in cells B4 through B10. Note that we made use of the relational operator >= and relative reference in the If function.

14.6 PLOTTING WITH EXCEL

Today's spreadsheets offer many choices when it comes to creating charts. You can create column charts (or histograms), pie charts, line charts, or *xy* charts. As an engineering student, and later as a practicing engineer, most of the charts that you will create will be of *xy*-type charts. Therefore, next we will explain in detail how to create an *xy* chart.

Excel offers Chart Wizard, which is a series of dialogue boxes that walks you through the necessary steps to create a chart. To create a chart using the Excel Chart Wizard, follow the procedure explained here.

- Select the data range as was explained earlier in this chapter.
- Click the Chart Wizard icon from the toolbar buttons.
- Select the **XY (Scatter)** chart type. The XY chart type offers four chart subtype options. (It is important to note here that the **Line** chart is often mistakenly used instead of **XY (Scatter)**).
- From the four chart subtype options, select the **"data points connected by smooth lines"** chart option.
- Click on the **Next** button.
- The selected data range will show in the data range box.
- Click the **Next** button.
- In the chart option dialog boxes enter the Chart title, Category (X) axis (this is the x axis title), Value (Y) axis (this is the y axis title), and click the OK button.

When creating an engineering chart, whether you are using Excel or using freehand methods, you must include proper labels with proper units for each axis. The chart must also contain a figure number with a title explaining what the chart represents. If more than one set of data is plotted on the same chart, the chart must also contain a legend or list showing symbols used for different data sets.

EXAMPLE 14.6

Using the results of Example 14.1, create a graph showing the value of air density as a function of temperature.

1. First we will **select** the data range, as shown.
2. Next **click** the Chart Wizard icon from the toolbar buttons and **select** the XY chart type. Also **select** the chart subtype, as shown. **Click** the Next button.
3. You will now see the Chart Source Data window with the selected data range in the data range box. **Click** the Next button.
4. Now enter the chart title, the x axis title, and the y axis title, as shown. **Click** the Next button.

Microsoft Excel - airdensity4.xls

File Edit View Insert Format Tools Data
Window Help

Arial

A5 = 0

	A	B
1	Density of air as a function of temperature	
2		
3	Temperature (C)	Density (kg/m^3)
4		
5	0	1.29
6	5	1.27
7	10	1.25
8	15	1.23
9	20	1.20
10	25	1.18
11	30	1.16
12	35	1.15
13	40	1.13
14	45	1.11
15	50	1.09
16		

Sheet1 Sheet2

Sum=288.0635911

Step 1

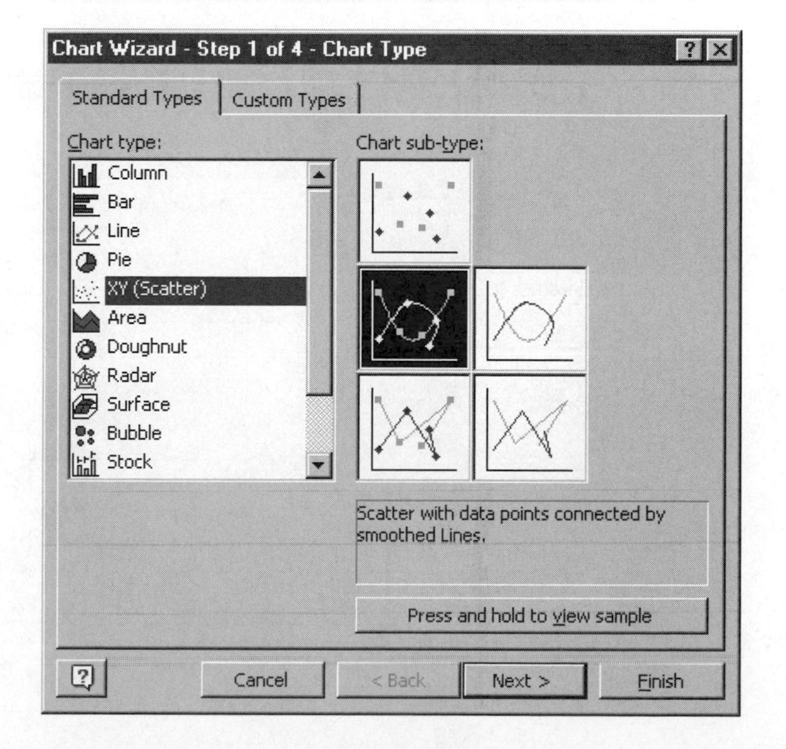

Step 2

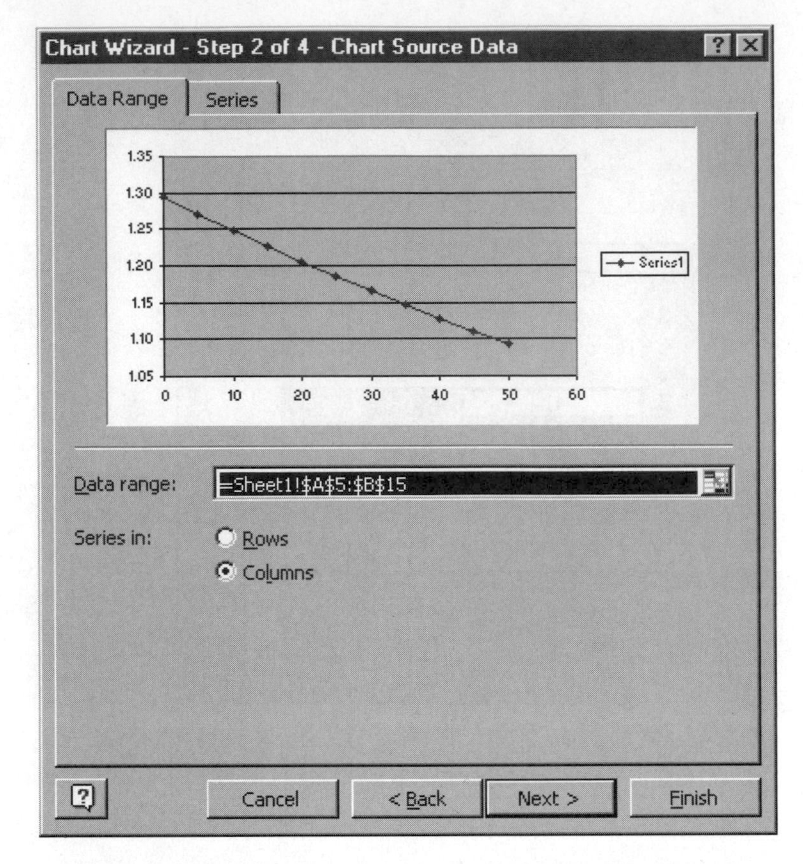
Chart Wizard - Step 2 of 4 - Chart Source Data
Data Range
Series
1.35
1.30
1.25
1.20
1.15
1.10
1.05
0
10
20
30
40
50
60
Series1
Data range:
=Sheet1!$A$5:$B$15
Series in:
Rows
Columns
Cancel
< Back
Next >
Finish

Step 3

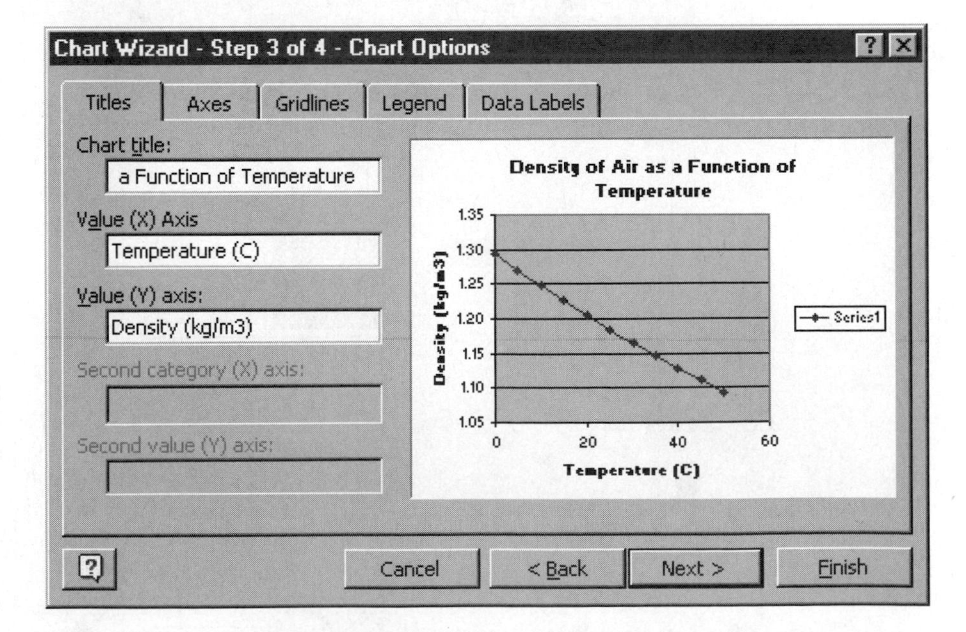
Chart Wizard - Step 3 of 4 - Chart Options
Titles
Axes
Gridlines
Legend
Data Labels
Chart title:
a Function of Temperature
Value (X) Axis
Temperature (C)
Value (Y) axis:
Density (kg/m3)
Second category (X) axis:
Second value (Y) axis:
Density of Air as a Function of Temperature
Density (kg/m3)
1.35
1.30
1.25
1.20
1.15
1.10
1.05
0
20
40
60
Temperature (C)
Series1
Cancel
< Back
Next >
Finish

Step 4

5. You now can save the chart as an object in the worksheet or as a new sheet. **Select** the "As object in" button and **click** Finish.
6. If for any reason you want to modify or edit the content of the chart, **click** the Chart menu and choose the Chart Options. Now you can modify or edit title, axis labels, spacing among the gridlines, legend, and so on. Just choose the pertinent tab and proceed.

Chart Wizard - Step 4 of 4 - Chart Location

Place chart:

As new sheet: Chart1

As object in: Sheet1

Cancel < Back Next > Finish

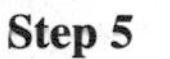

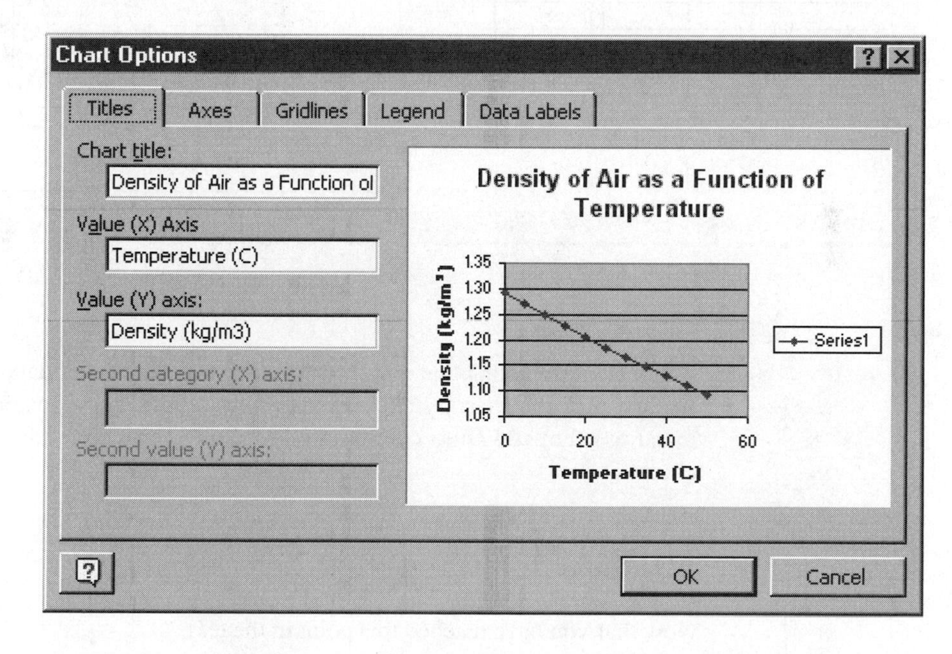

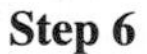

7. Finally, you can place the chart in an appropriate location, as shown.

Density of air as a function of temperature

Temperature (C)	Density (kg/m^3)
0	1.29
5	1.27
10	1.25
15	1.23
20	1.20
25	1.18
30	1.16
35	1.15
40	1.13
45	1.11
50	1.09

Density of Air as a Function of Temperature

Step 7

It is worth noting that you can plot more than one set of data on the same chart. To do so, first pick the chart by clicking anywhere on the chart area, and then from the **Chart** menu use the **Add Data** command and follow the steps to plot the other data set to the chart.

SUMMARY

Now that you have reached this point in the text

- You should know that a spreadsheet is a tool that can be used to solve an engineering problem. Spreadsheets are commonly used to record, organize, and analyze data using formulas. Moreover, you can use a spreadsheet such as Excel to present the results of an analysis in chart form. You can input your own formulas or use the built-in functions provided by the spreadsheet.

- You should know how to move around in a workbook and input data into different cells. You should also know how to edit the content of a cell.
- You should know how to select multiple cells and create a range. You should also realize that you can name a range and use the name in your formulas or in plotting data.
- You should understand how to refer to a cell by its address. You should also know the differences among a cell's relative address, absolute address, and mixed address, and remember to use the proper address when creating formulas.
- You should be familiar with Excel's built-in functions.
- You should know how to insert cells, columns, and rows in an existing worksheet.
- You should know how to create a proper engineering chart using Excel.

PROBLEMS

1. Using the Excel **Help** menu, discuss how the following functions are used. Create a simple example and demonstrate the proper use of the function.
   a. TRUNC(number, num_digits)
   b. ROUND(number, num_digits)
   c. COMBIN(number, number_chosen)
   d. DEGREES(angle)
   e. SLOPE(known_y's, known_x's)
   f. CEILING(number, significance)
2. In Chapter 16, we will cover engineering economics. For now, using the Excel **Help** menu, familiarize yourself with the following functions. Create a simple example and demonstrate the proper use of the function.
   a. FV(rate,nper,pmt,pv,type)
   b. IPMT(rate,per,nper,pv,fv,type)
   c. NPER(rate,pmt,pv,fv,type)
   d. PV(rate,nper,pmt,fv,type)
3. In Chapter 7, we discussed fluid pressure and the role of water towers in small towns. Recall that the function of a water tower is to create a desirable municipal water pressure for household and other usage in a town. To achieve this purpose, water is stored in large quantities in elevated tanks. Also recall that the municipal water pressure may vary from town to town, but it generally falls somewhere between 50 and 80 lb/in^2 (psi). In this assignment, use Excel to create a table that shows the relationship between the height of water above ground in the water tower and the water pressure in a pipeline located at the base of the water tower. The relationship is given by

$$P = \rho g h$$

where

P = the water pressure at the base of the water tower in pounds per square foot

ρ = the density of water in slugs per cubic foot (ρ = 1.94 slugs/ft^3)

g = the acceleration due to gravity (g = 32.2 ft/s^2)

h = the height of water above ground in feet

Create a table that shows the water pressure in lb/in^2 in a pipe located at the base of the water tower as you vary the height of water in increments of 5 ft. Also plot water pressure (lb/in^2) vs. the height of water in feet. What should be the water level in the water tower to create 80 psi water pressure in a pipe at the base of the water tower?

4. As we explained in Chapter 7, viscosity is a measure of how easily a fluid flows. For example, honey has a higher value of viscosity than does water because if you were to pour water and honey side by side on an inclined surface, the water will flow faster. The viscosity of a fluid plays a significant role in the analysis of many fluid dynamics problems. The viscosity of water can be determined from the following correlation:

$$\mu = c_1 10^{c_2/(T-c_3)}$$

where

$\mu \equiv$ viscosity (N/s·m^2)

$T \equiv$ temperature (K)

$c_1 \equiv 2.414 \times 10^{-5}$(N/s·m^2)

$c_2 = 247.8$ K

$c_3 = 140$ K

Using Excel, create a table that shows the viscosity of water as a function of temperature in the range of 0°C (273.15 K) to 100°C (373.15 K) in increments of 5°C. Also create a graph showing the value of viscosity as a function of temperature.

5. Using Excel, create a table that shows the relationship between the units of temperature in degrees Celsius and Fahrenheit in the range of −50° to 150°C. Use increments of 10°C.
6. Using Excel, create a table that shows the relationship among the units of height of people in centimeters, inches, and feet in the range of 150 cm to 2 m. Use increments of 5 cm.
7. Using Excel, create a table that shows the relationship among the units of mass to describe people's mass in kilogram, slugs, and pound mass in the range of 20 kg to 120 kg. Use increments of 5 kg.
8. Using Excel, create a table that shows the relationship among the units of pressure in Pa, psi, and inches of water in the range of 1000 to 10,000 Pa. Use increments of 500 Pa.
9. Using Excel, create a table that shows the relationship between the units of pressure in Pa and psi in the range of 10 kPa to 100 kPa. Use increments of 5 kPa.
10. Using Excel, create a table that shows the relationship between the units of power in watts and horsepower in the range of 100 W to 10,000 W. Use smaller increments of 100 W up to 1000 W, and then use increments of 1000 W all the way up to 10,000 W.
11. As we explained in Chapter 4, the air resistance to the motion of a vehicle is something important that engineers investigate. As you may also know, the

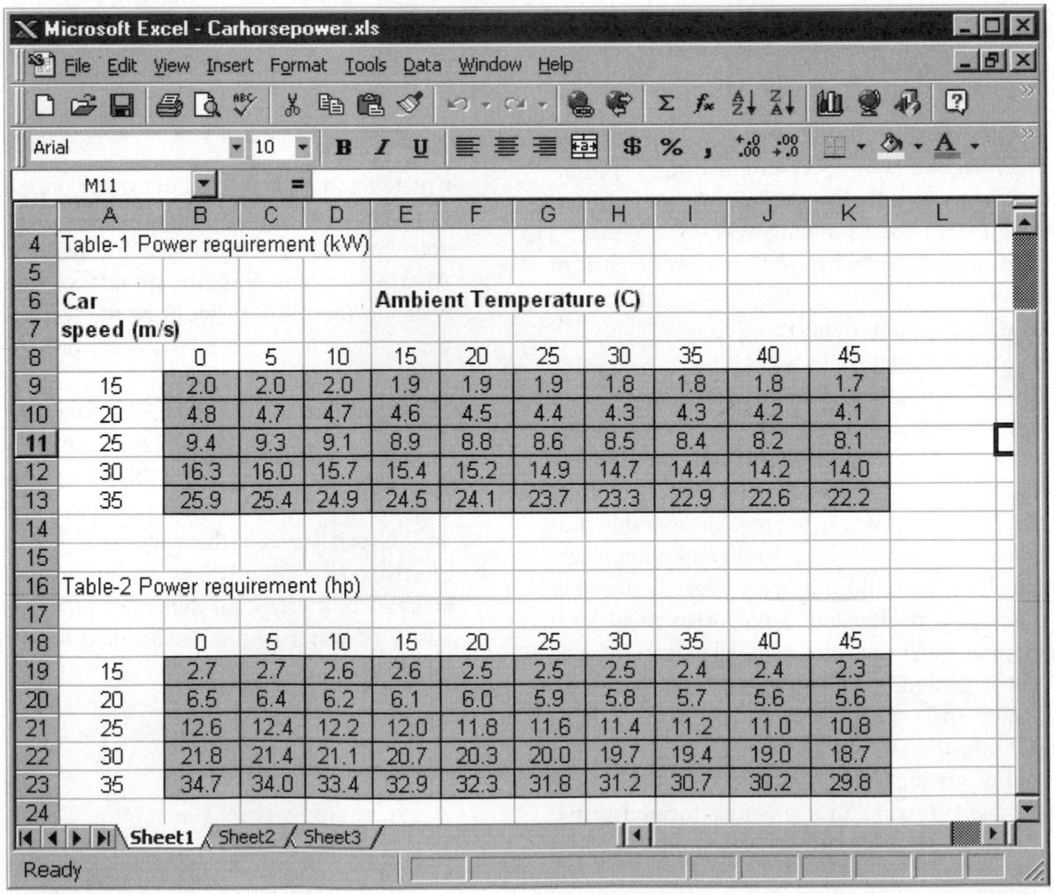

Microsoft Excel - Carhorsepower.xls

File Edit View Insert Format Tools Data Window Help

Arial 10

M11

Table-1 Power requirement (kW)

Car speed (m/s)	Ambient Temperature (C)									
	0	5	10	15	20	25	30	35	40	45
15	2.0	2.0	2.0	1.9	1.9	1.9	1.8	1.8	1.8	1.7
20	4.8	4.7	4.7	4.6	4.5	4.4	4.3	4.3	4.2	4.1
25	9.4	9.3	9.1	8.9	8.8	8.6	8.5	8.4	8.2	8.1
30	16.3	16.0	15.7	15.4	15.2	14.9	14.7	14.4	14.2	14.0
35	25.9	25.4	24.9	24.5	24.1	23.7	23.3	22.9	22.6	22.2

Table-2 Power requirement (hp)

	0	5	10	15	20	25	30	35	40	45
15	2.7	2.7	2.6	2.6	2.5	2.5	2.5	2.4	2.4	2.3
20	6.5	6.4	6.2	6.1	6.0	5.9	5.8	5.7	5.6	5.6
25	12.6	12.4	12.2	12.0	11.8	11.6	11.4	11.2	11.0	10.8
30	21.8	21.4	21.1	20.7	20.3	20.0	19.7	19.4	19.0	18.7
35	34.7	34.0	33.4	32.9	32.3	31.8	31.2	30.7	30.2	29.8

Sheet1 Sheet2 Sheet3

Ready

Problem 11

drag force acting on a car is determined experimentally by placing the car in a wind tunnel. The air speed inside the tunnel is changed, and the drag force acting on the car is measured. For a given car, the experimental data is generally represented by a single coefficient that is called the *drag coefficient*. It is defined by the following relationship:

$$C_d = \frac{F_d}{\frac{1}{2}\rho V^2 A}$$

where

C_d = drag coefficient (unitless)

F_d = measured drag force (N or lb)

ρ = air density (kg/m^3 or slugs/ft^3)

V = air speed inside the wind tunnel (m/s or ft/s)

A = frontal area of the car (m^2 or ft^2)

The frontal area A represents the frontal projection of the car's area and could be approximated simply by multiplying 0.85 times the width and the height of a rectangle that outlines the front of a car. This is the area that you see when you view the car from a direction normal to the front grills. The 0.85 factor is used to adjust for rounded corners, open space below the bumper, and so on. To give you some idea, typical drag coefficient values for sports cars are between 0.27 to 0.38, and for sedans are between 0.34 to 0.5.

The power requirement to overcome air resistance is computed by

$$P = F_d V$$

where

P = power (watts or ft·lb/s)

1 horsepower (hp) = 550 ft·lb/s

and

1 horsepower (hp) = 746 W

The purpose of this exercise is to see how the power requirement changes with the car speed and the air temperature. Determine the power requirement to overcome air resistance for a car that has a listed drag coefficient of 0.4 and width of 74.4 in. and height of 57.4 in. Vary the air speed in the range of 15 m/s $< V <$ 35 m/s, and change the air density range of 1.11 kg/m^3 $< \rho <$ 1.29 kg/m^3. The given air density range corresponds to 0° to 45°C. You may use the ideal gas law to relate the density of the air to its temperature. Present your findings in both kilowatts and horsepower as shown in the accompanying spreadsheet. Discuss your findings in terms of power consumption as a function of speed and air temperature.

12. The cantilevered beam shown in the accompanying figure is used to support a load acting on a balcony. The deflection of the centerline of the beam is given by the following equation:

$$y = \frac{-wx^2}{24EI}(x^2 - 4Lx + 6L^2)$$

where

y = deflection at a given x location (m)

w = distributed load (N/m)

E = modulus of elasticity (N/m^2)

I = second moment of area (m^4)

x = distance from the support as shown (x)

L = length of the beam (m)

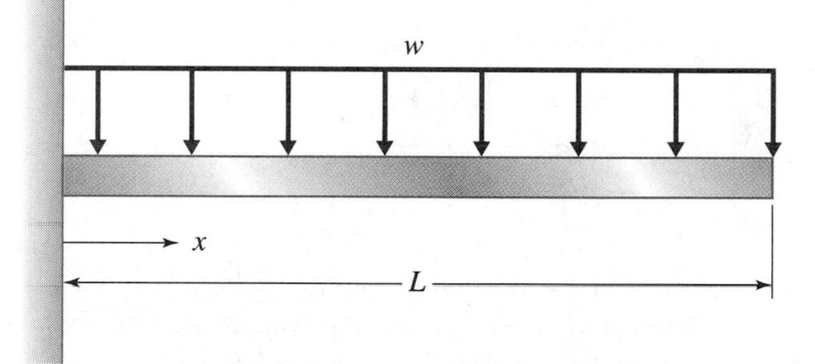

Using Excel, plot the deflection of a beam whose length is 5 m with the modulus of elasticity of E = 200 GPa and $I = 99.1 \times 10^6$ mm^4. The beam is designed to carry a load of 10,000 N/m. What is the maximum deflection of the beam?

13. Fins, or extended surfaces, are commonly used in a variety of engineering applications to enhance cooling. Common examples include a motorcycle or lawn mower engine head, extended surfaces used in electronic equipment, and finned tube heat exchangers in room heating and cooling applications. Consider a rectangular profile of aluminum fins shown in the accompanying figure, which are used to remove heat from a surface whose temperature is 100°C (T_{base} = 100°C). The temperature of the ambient air is 20°C. We are interested in determining how the temperature of the fin varies along its length and plotting this temperature variation. For long fins, the temperature distribution along the fin is given by

$$T - T_{ambient} = (T_{base} - T_{ambient})e^{-mx}$$

where

$$m = \sqrt{\frac{hp}{kA}}$$

h = the heat transfer coefficient (W/m²·K)

p = perimeter of the fin (2*a+b), (m)

A = cross-sectional area of the fin (a*b), (m²)

k = thermal conductivity of the fin material (W/m·K)

Plot the temperature distribution along the fin using the following data: k = 168 W/m·K, h = 12 W/m²·K, a = 0.05 m, b = 0.01 m. Vary x from 0 to 0.1 m in increments of 0.01 m.

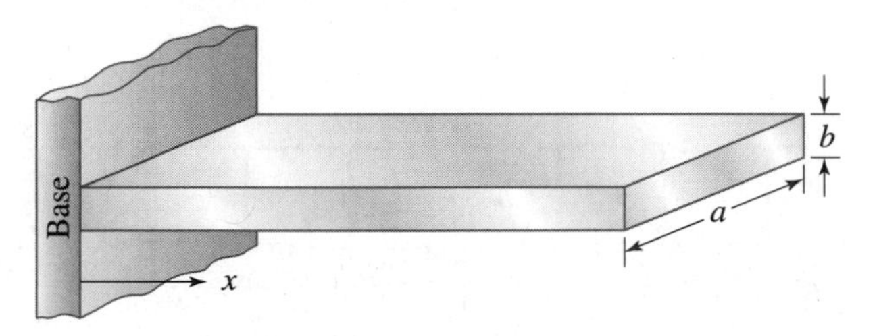

14. A person by the name of Huebscher developed a relationship between the equivalent size of round ducts and rectangular ducts according to

$$D = 1.3\frac{(ab)^{0.625}}{(a+b)^{0.25}}$$

where

D = diameter of equivalent circular duct (mm)

a = dimension of one side of the rectangular duct (mm)

b = the other dimension of the rectangular duct (mm)

Using Excel, create a table that shows the relationship between the circular and the rectangular duct, similar to the one shown in the accompanying table.

	Length of One Side of Rectangular Duct (length a), mm				
Length b	100	125	150	175	200
400	159				
450					
500					
550					
600					

15. A Pitot tube is a device commonly used in a wind tunnel to measure the speed of the air flowing over a model. The air speed is measured from the following equation:

$$V = \sqrt{\frac{2P_d}{\rho}}$$

where

V = air speed (m/s)

P_d = dynamic pressure (Pa)

ρ = density of air (1.23 kg/m³)

Using Excel, create a table that shows the air speed for the range of dynamic pressure of 500 to 800 Pa. Use increments of 50 Pa.

16. Use Excel to solve Example 4.1. Recall we applied the trapezoidal rule to determine the area of the shape given.

17. We will discuss engineering economics in Chapter 16. Using Excel, create a table that can be used to look up monthly payments on a car loan for a period of 5 years. The monthly payments are calculated from

$$A = P\left[\frac{\left(\frac{i}{1200}\right)\left(1+\frac{i}{1200}\right)^{60}}{\left(1+\frac{i}{1200}\right)^{60}-1}\right]$$

	Interest Rate				
Loan	7%	7.5%	8%	8.5%	9%
10,000					
15,000					
20,000					
25,000					

where

A = monthly payments in dollars

P = the loan in dollars

i = interest rate, e.g., 7, 7.5, . . . , 9

18. A person by the name of Sutterland has developed a correlation that can be used to evaluate the viscosity of air as a function of temperature. It is given by

$$\mu = \frac{c_1 T^{0.5}}{1 + \frac{c_2}{T}}$$

where

μ = viscosity (N/s·m^2)

T = temperature (K)

$$c_1 = 1.458 \times 10^{-6} \left(\frac{\text{kg}}{\text{m} \cdot \text{s} \cdot \text{K}^{1/2}} \right)$$

$$c_2 = 110.4 \text{ K}$$

Create a table that shows the viscosity of air as a function of temperature in the range of 0°C (273.15 K) to 100°C (373.15 K) in increments of 5°C. Also create a graph showing the value of viscosity as a function of temperature as shown in the accompanying spreadsheet.

19. In Chapter 8, we explained the concept of windchill factors. We said that the heat transfer rates from your body to the surroundings increase on a cold, windy day. Simply stated, you lose more body heat on the cold, windy day than you do on a calm day. The windchill index accounts for

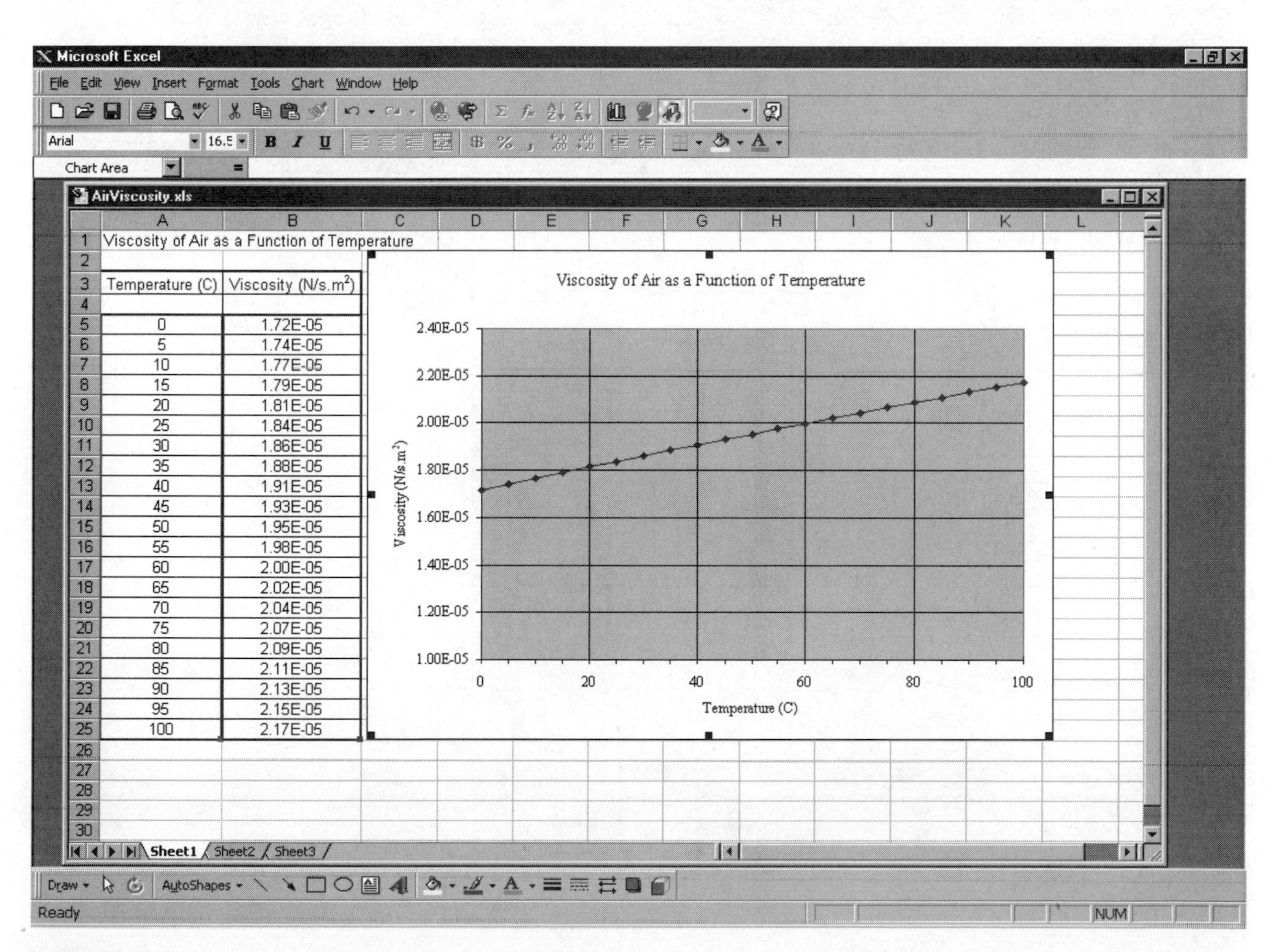

Viscosity of Air as a Function of Temperature

Temperature (C)	Viscosity (N/s.m^2)
0	1.72E-05
5	1.74E-05
10	1.77E-05
15	1.79E-05
20	1.81E-05
25	1.84E-05
30	1.86E-05
35	1.88E-05
40	1.91E-05
45	1.93E-05
50	1.95E-05
55	1.98E-05
60	2.00E-05
65	2.02E-05
70	2.04E-05
75	2.07E-05
80	2.09E-05
85	2.11E-05
90	2.13E-05
95	2.15E-05
100	2.17E-05

Problem 18

the combined effect of wind speed and the air temperature. It accounts for the additional body heat loss that occurs on a cold, windy day. The windchill values are determined empirically, and a common correlation used to determine the windchill index is

$$WCI = (10.45 - V + 10\sqrt{V})(33 - T_a)$$

where

WCI = windchill index (kcal/m²·h)

V = wind speed (m/s)

T_a = ambient air temperature (°C)

and the value 33 is the body surface temperature in degrees Celsius.

The more common equivalent windchill temperature $T_{equivalent}$ (°C) is given by

$$T_{equivalent} = 0.045(5.27V^{0.5} + 10.45 - 0.28V)(T_a - 33) + 33$$

Note that V is expressed in km/h.

Create a table that shows the windchill temperatures for the range of ambient air temperature $-30°C < T_a < 10°C$ and wind speed of 20 km/h < V < 80 km/h as shown in the accompanying spreadsheet.

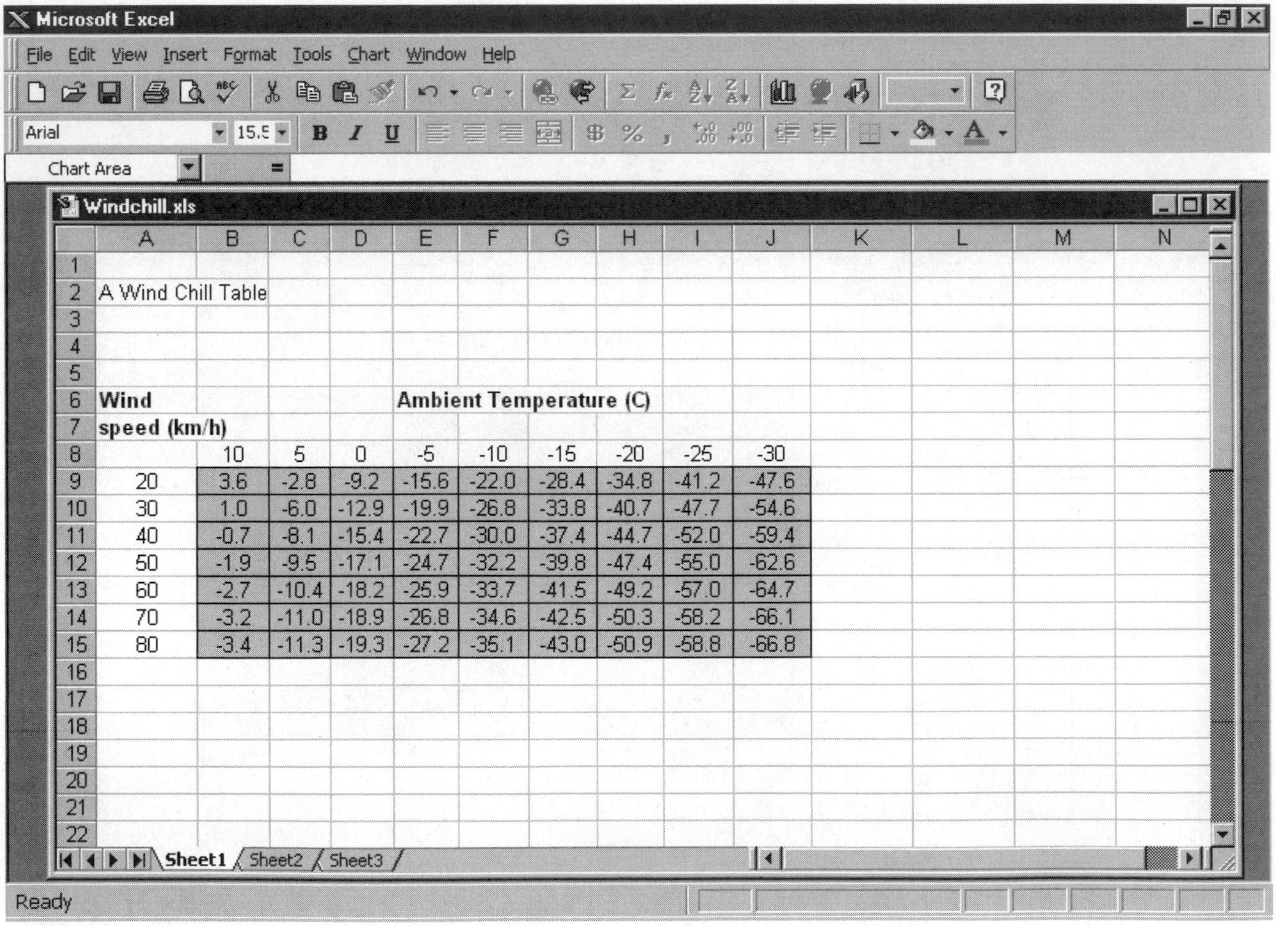

Microsoft Excel — Windchill.xls

A Wind Chill Table

Wind speed (km/h)	Ambient Temperature (C)								
	10	5	0	-5	-10	-15	-20	-25	-30
20	3.6	-2.8	-9.2	-15.6	-22.0	-28.4	-34.8	-41.2	-47.6
30	1.0	-6.0	-12.9	-19.9	-26.8	-33.8	-40.7	-47.7	-54.6
40	-0.7	-8.1	-15.4	-22.7	-30.0	-37.4	-44.7	-52.0	-59.4
50	-1.9	-9.5	-17.1	-24.7	-32.2	-39.8	-47.4	-55.0	-62.6
60	-2.7	-10.4	-18.2	-25.9	-33.7	-41.5	-49.2	-57.0	-64.7
70	-3.2	-11.0	-18.9	-26.8	-34.6	-42.5	-50.3	-58.2	-66.1
80	-3.4	-11.3	-19.3	-27.2	-35.1	-43.0	-50.9	-58.8	-66.8

Sheet1 / Sheet2 / Sheet3

Ready

Problem 19

CHAPTER 15 Engineering Drawings

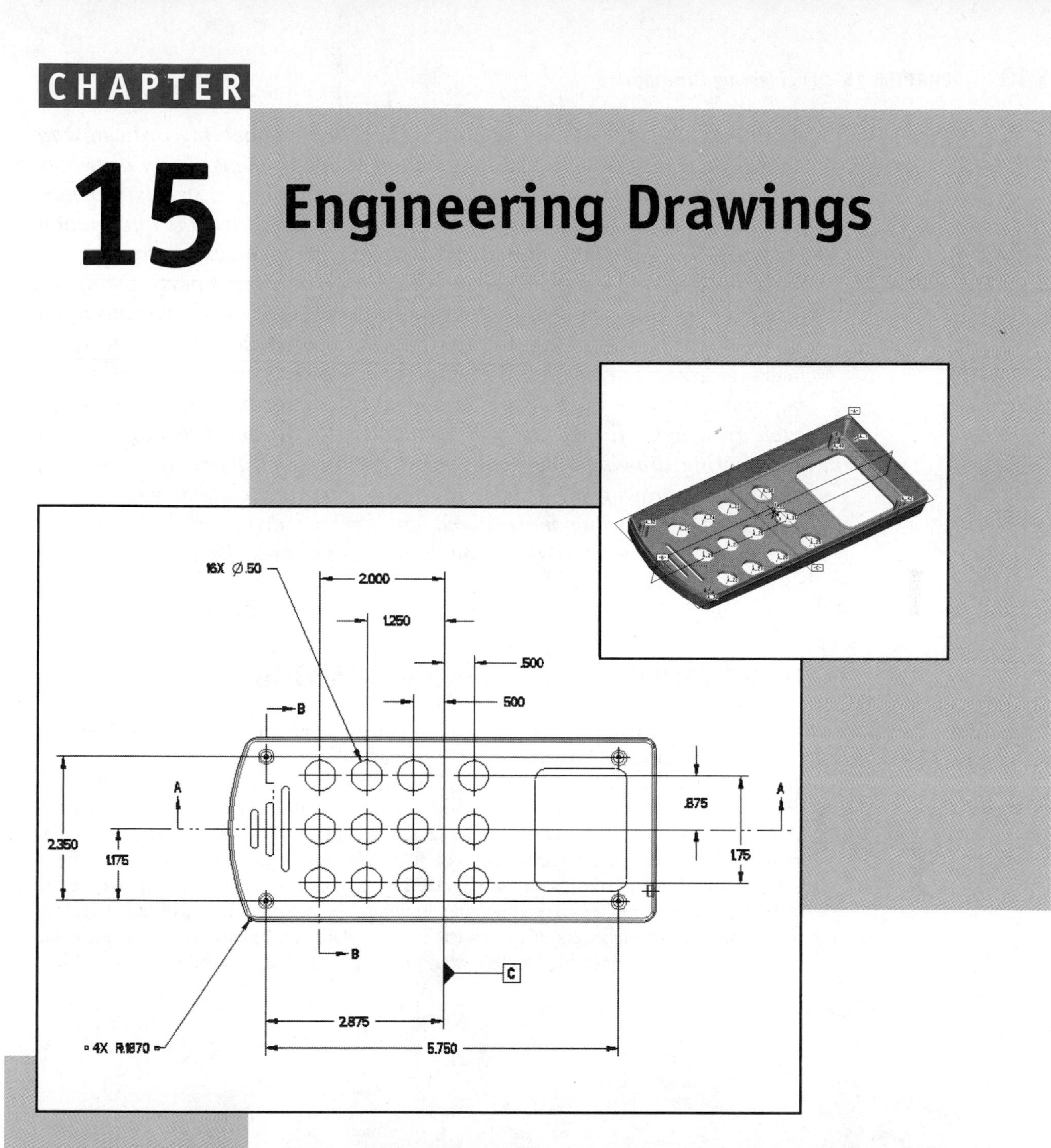

Engineering drawings, such as the cellular phone schematics above, are important in conveying useful information to other engineers or machinists in a standard manner that allows for visualization of the proposed product. Important information, such as the shape of the product, its size, type of material used, and the assembly steps required, are provided by these drawings.

Source: From *Pro/ENGINEER® 2000i*², by Louis Gary Lamit, pp. L10-19 and L10-20, Brooks/Cole, 2001. Reprinted with permission of Brooks/Cole, an imprint of the Wadsworth Group, a division of Thomson learning. Fax: 800-730-2215.

Engineers use special kind of drawings, called engineering drawings, to convey their ideas and design information about products. These drawings portray vital information, such as shape of the product, its size, type of material used, and assembly steps. Moreover, machinists use the information provided by engineers or drafts persons on the engineering drawings to make the parts. For complicated systems made of various parts, the drawings also serve as a how-to-assemble guide, showing how the various parts fit together. The following sections provide a brief introduction to engineering drawing principles. We will discuss why engineering drawings are important, how they are drawn, and what rules must be followed to create such drawings. Most of you will eventually take a semester-long class in engineering drawing where you will learn in much more detail how to create such drawings. For now, if after reading this chapter you are still interested in learning more about engineering drawings, talk to your instructor to obtain a list of good textbooks that discuss engineering drawings in detail. ■

15.1 IMPORTANCE OF ENGINEERING DRAWING

Have you ever had an idea about a new product that could make a certain task easier? How did you get your idea across to other people? What were the first things you did to make your idea clearly known to your audience? Imagine you are having a cup of coffee with a friend, and you decide to share your idea about a product with her. After talking about the idea for a while, to clarify your idea, you will naturally draw a picture or a diagram to show what the product would look like. You have heard the saying "a picture is worth a thousand words"; well, in engineering, a good drawing is worth even more words! Technical drawings or engineering drawings are important in conveying useful information to other engineers or machinists in a standard acceptable manner to allow the readers of these drawings to visualize what the proposed product would look like. More

significantly, information such as the dimensions of the proposed product, or what it would look like when viewed from the top or from the side or the front is provided. The drawings will also specify what type of material is to be used to make this product. In order to draw or read engineering drawings, you must first learn a set of standard rules that are followed by all engineers, draftspersons, and machinists. In the next sections, we will briefly discuss these rules.

15.2 ORTHOGRAPHIC VIEWS

Orthographic views (diagrams) show what an object's projection looks like when seen from the top, the front, or the side. To better understand what we mean by orthographic views, imagine that you have placed the object shown in Figure 15.1 in the center of a glass box. Now if you were to draw perpendicular lines from the corners of the object into the faces of the glass box, you would see the outlines shown in Figure 15.1. The outlines are called the orthographic projection of the object into the *horizontal, vertical*, and the *profile planes*.

Now imagine that you open up or unfold the faces of the glass box that have the projections of the object. The unfolding of the glass faces will result in the layout shown in

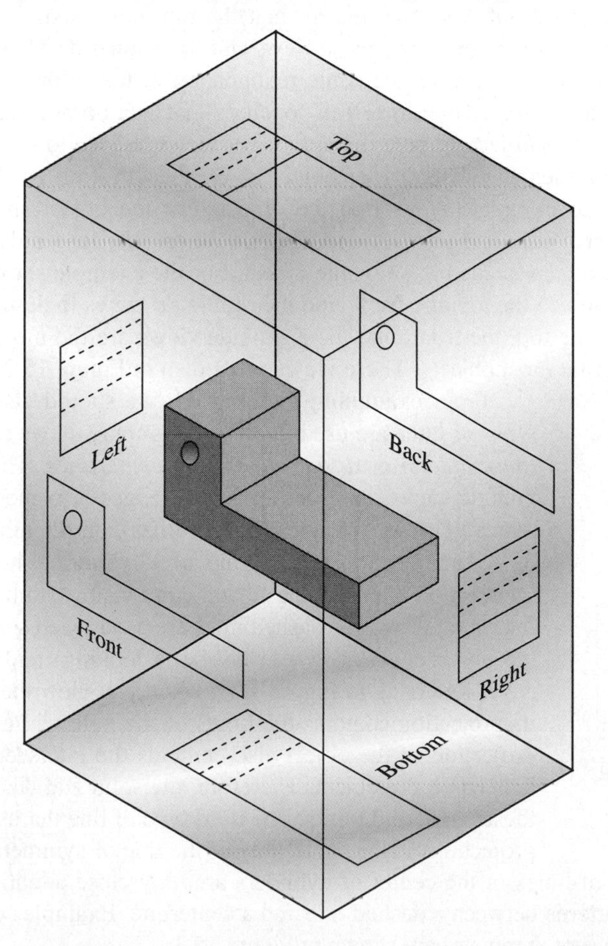

Figure 15.1 The orthographic projection of an object into the horizontal, vertical, and profile planes

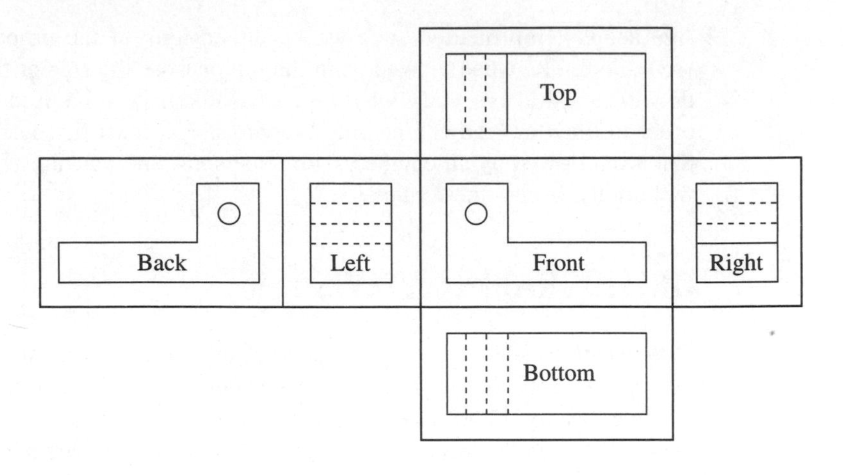

Figure 15.2 The relative locations of the top, bottom, front, back, right-side, and left-side views

Figure 15.2. Note the relative locations of the top view, the bottom view, the front view, the back view, the right-side view, and the left-side view.

At this point, you may realize that the top view is similar to the bottom view, the front view is similar to the back view, and the right-side view is similar to the left-side view. Therefore, you notice some redundancy in the information provided by these six views (diagrams). Therefore, you conclude that you do not need to draw all six views to describe this object. In fact, the number of views needed to describe an object depends on how complex the shape of an object is. So the question is, then, how many views are needed to completely describe the object. For the object shown in Figure 15.1, three views are sufficient to fully describe the object, because only three principle planes of projection are needed to show the object. For the example shown in Figure 15.1, we may decide to use the top, the front, and the right-side views to describe the object completely. In fact, the top, the front, and the right-side views are the most commonly created views to describe most objects. These views are shown in Figure 15.3.

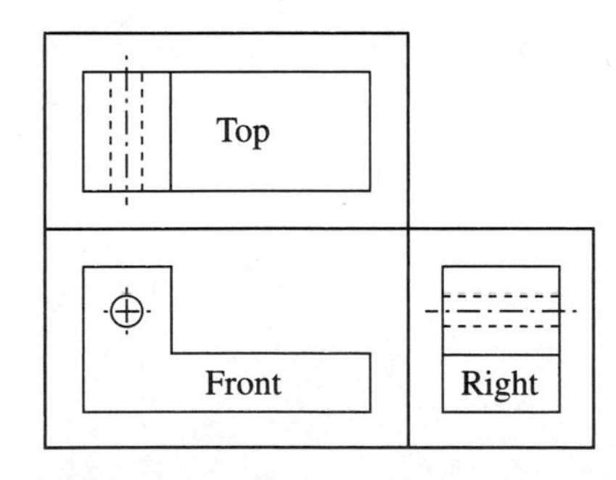

Figure 15.3 The top, front, and the right-side views of an object

From examining Figure 15.3, we should also note that three different types of lines are used in the orthographic views to describe the object: *solid lines, hidden* or *dashed lines*, and *centerlines*. The solid lines on the orthographic views represent the visible edges of planes or the intersection of two planes. The dashed lines (hidden lines), on the other hand, represent an edge of a plane or the extreme limits of a cylindrical hole inside the object, or the intersection of two planes that are not visible from the direction you are looking. In other words, dashed lines are used when some material exists between the observer (from where he/she is looking) and the actual location of the edge. Referring to Figure 15.3, when you view the object from the right side, its projection contains the limits of the hole within the object. The right-side projection of the object also contains the intersection of two planes that are located on the object. Therefore, the solid and dashed lines are used to show these edges and limits. The third type of line that is employed in orthographic projections is the centerline, or the line of symmetry, which shows where the center of holes or the center of cylinders are. Pay close attention to the difference in the line patterns between a dashed line and a centerline. Examples of solid lines, dashed lines, and lines of symmetry are shown in Figure 15.3.

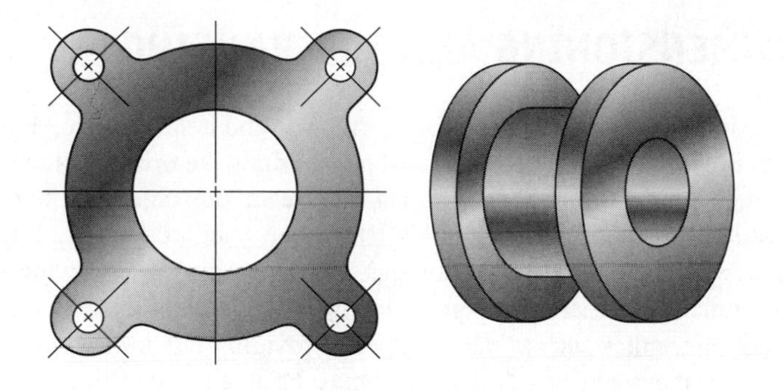

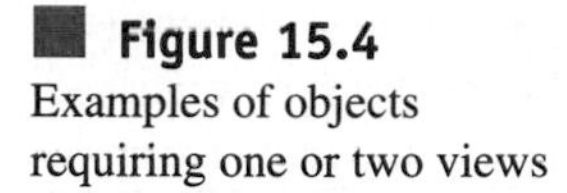

Figure 15.4
Examples of objects requiring one or two views

As we said earlier, the number of views that you should draw to represent an object will depend on how complex the object is. For example, if you want to show a bolt washer or a gasket, you need to draw only a single top view and specify the thickness of the washer or the gasket. For other objects, such as bolts, we may draw only two views. Examples of objects requiring one or two views are shown in Figure 15.4.

EXAMPLE 15.1

Draw the orthographic views of the object shown in Figure 15.5(a).

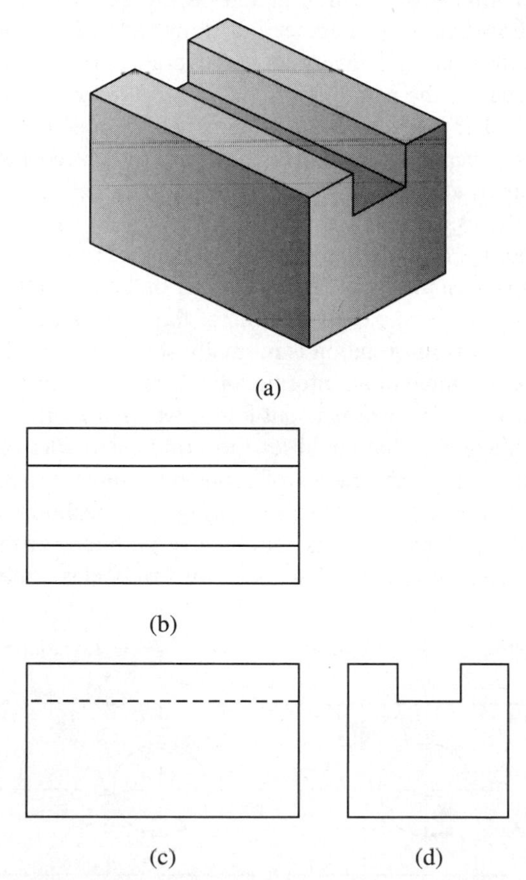

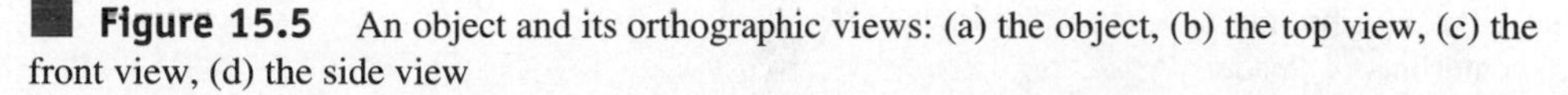

Figure 15.5 An object and its orthographic views: (a) the object, (b) the top view, (c) the front view, (d) the side view

15.3 DIMENSIONING AND TOLERANCING

Engineering drawings provide information about shape, size, and material of a product. In the previous section, we discussed how to draw the orthographic views. We did not say anything about how to show the actual size of the object on the drawings. The American National Standards Institute (ANSI) sets the standards for the dimensioning and tolerancing practices for engineering drawings. Every engineering drawing must include dimensions, tolerances, the materials from which the product will be made, the finished surfaces marked, and other notes such as part numbers. Providing this information on the diagrams is important for many reasons. A machinist must be able to make the part from the detailed drawings without needing to go back to the engineer or the draftsperson who drew the drawings to ask questions regarding the size, or the tolerances, or what type of material the part should be made from. There are basically two concepts that you need to keep in mind when specifying dimensions in an engineering drawing: *size* and *location*. As shown in Figure 15.6, not only do you need to specify how wide or how long an object is but you must also specify the location of the center of a hole or center of a fillet in the part. Moreover, a drawing is dimensioned with the aid of *dimension lines, extension lines, centerlines*, and *leaders*.

Dimension lines provide information on the size of the object; for example, how wide it is and how long it is. You need to show the overall dimensions of the object because the machinist can then determine the overall size of the stock material from which to make the piece. As the name implies, ***extension lines*** are those lines that extend from the points to which the dimension or location is to be specified. Extension lines are drawn parallel to each other, and the dimension lines are placed between them, as shown in Figure 15.6. The ***leaders*** are the arrows that point to a circle or a fillet for the purpose of specifying their sizes. Often the drawings are shown *Not To Scale* (NTS), and therefore a scaling factor for the drawing must also be specified. In addition to dimensions, all engineering drawings must also contain an information box with the following items: name of the person who prepared the drawing, title of the drawing, date, scale, sheet number, and drawing number. This information is normally shown on the upper- or lower-right corner of a drawing. An example of an information box is shown in Figure 15.7.

Let us now say a few words about fillets, which are often overlooked in engineering drawings, a shortcoming that could lead to problems. ***Fillet*** refers to the rounded edges of an object; their sizes, the radius of roundness, must be specified in all drawings. If the size of fillets is not specified in a drawing, the machinist may not round the edges; consequently, the absence of fillets could create problems or failure in parts. As some of you will learn later in your mechanics of materials class, mechanical parts with sharp

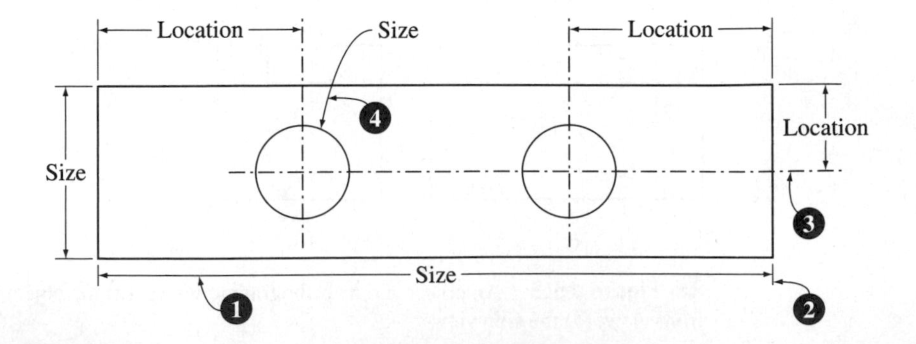

Figure 15.6 The basics of dimensioning practices: (1) dimension line, (2) extension line, (3) centerline, (4) leader

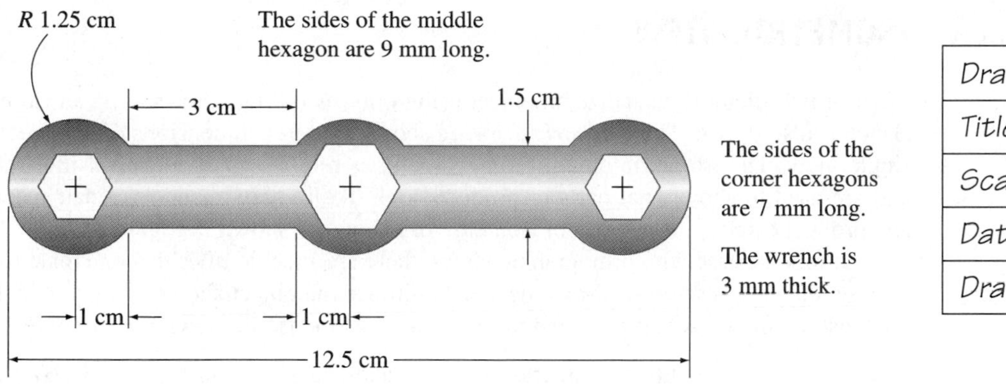

Figure 15.7 An example of engineering drawing with an information box

edges or a sudden reduction in their cross-sectional areas could fail when subjected to loads because of high stress concentrations near the sharp regions. As you will learn later, a simple way of reducing the stress in these regions is by rounding the edges and creating a gradual reduction in cross-sectional areas.

Engineered products generally consist of many parts. In today's globally driven economy, some of the parts made for a product in one place must be easily assembled with parts made elsewhere. When you specify a dimension on a drawing—say, 2.50 centimeter—how close does the actual dimension of the machined part need to be to the specified 2.50 cm for the part to fit properly with other parts in the product? Would everything fit correctly if the actual dimension of the machine part were 2.49 cm or 2.51 cm? If so, then you must specify a tolerance of $\pm$ 0.01 cm on your drawing regarding this dimension. Tolerancing is a broad subject with its own rules and symbols, and as we mentioned earlier, the American National Standard Institute sets the tolerencing standards that must be followed by those creating or reading engineering drawings. Here, we have briefly introduced these ideas; you must consult the standards if you are planning to prepare an actual engineering drawing.

EXAMPLE 15.2

Show the dimensions of the object in Figure 15.8 on its orthographic views.

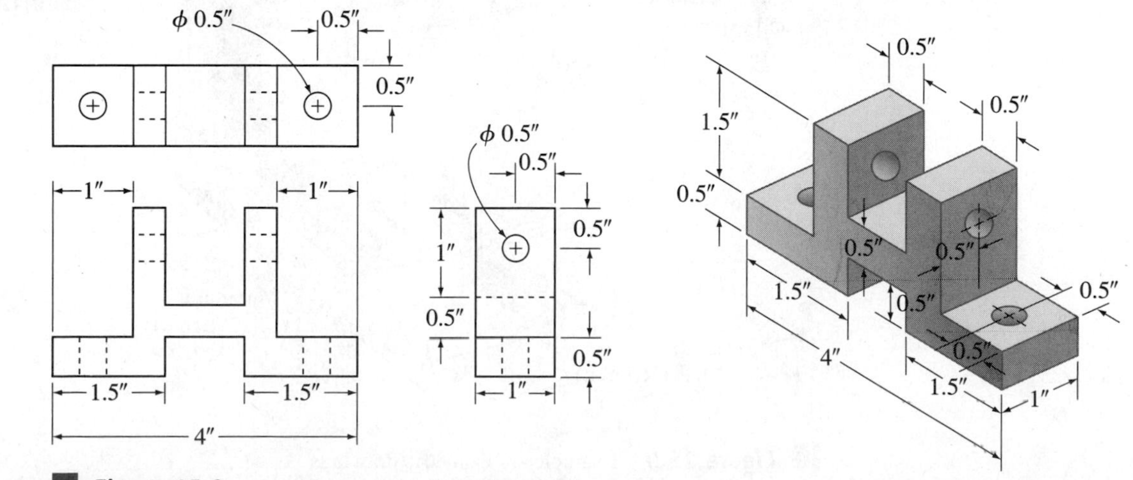

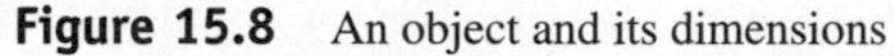
Figure 15.8 An object and its dimensions

15.4 ISOMETRIC VIEW

When it is difficult to visualize an object using only its orthographic views, an isometric sketch is also drawn. The ***isometric drawing*** shows the three dimensions of an object in a single view. The isometric drawings are sometimes referred to as technical illustrations and are used to show what parts or products look like in parts manuals, repair manuals, and product catalogs. Examples of isometric drawings are shown in Figure 15.9.

As was the case with orthographic views, there are specific rules that one must follow to draw the isometric view of an object. We will use the object shown in Figure 15.10 to demonstrate the steps that you need to follow to draw the isometric view of an object.

Step 1: Draw the width, height, and the depth axes, as shown in Figure 15.11(a). Note that the isometric grid consists of the width and the depth axes; they form a 30° angle

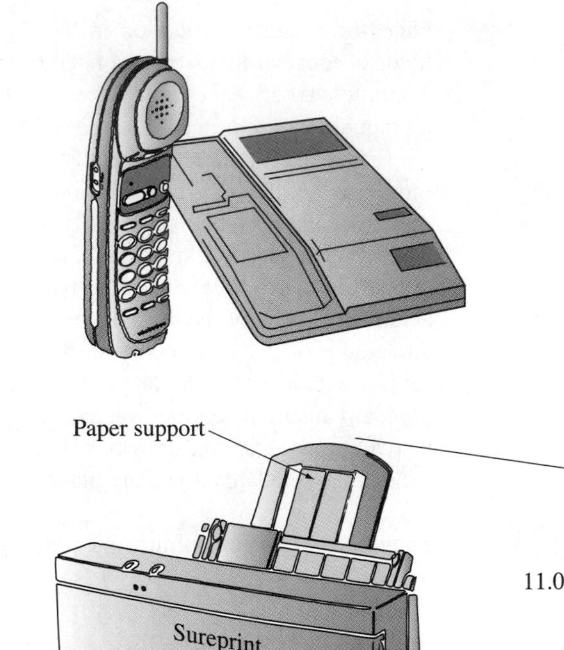

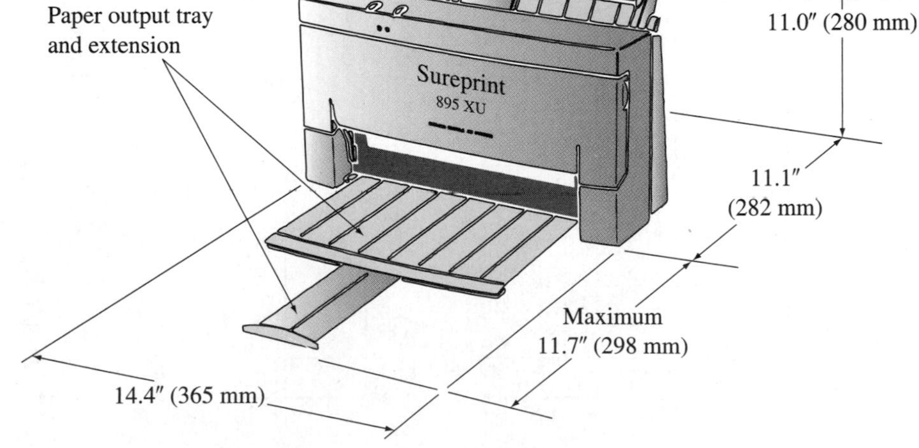

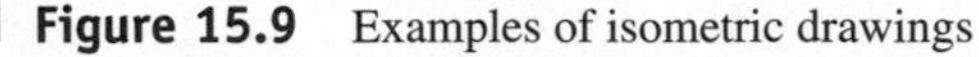

Figure 15.9 Examples of isometric drawings

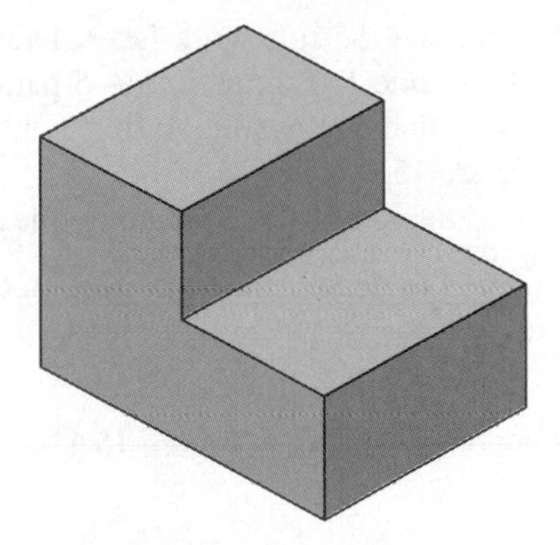

Figure 15.10 The object used in demonstrating the steps in creating an isometric view

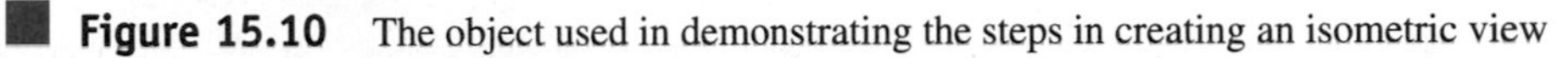

with a horizontal line. Also note that the height axis makes a 90° angle with a horizontal line and a 60° angle with each of the depth and width axes.

Step 2: Measure and draw the total width, height, and depth of the object. Hence, draw lines 1–2, 1–3, and 1–4 as shown in Figure 15.11(b).

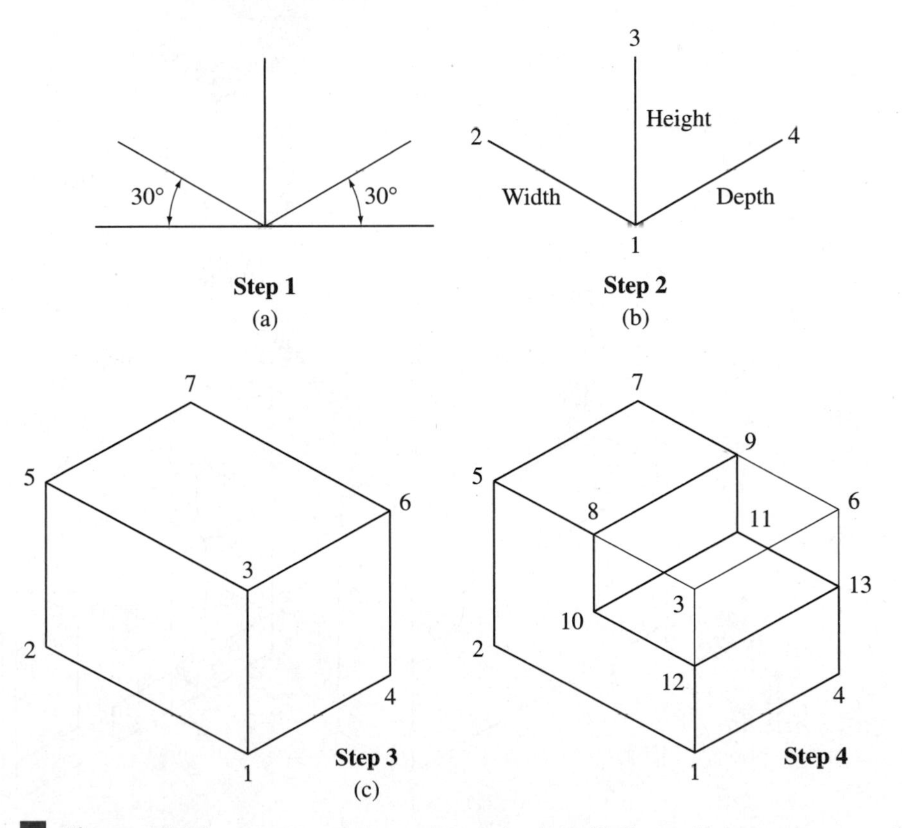

Figure 15.11 Steps in creating an isometric view: (1) Create the isometric axes and grid; (2) mark the height, width, and depth of the object on the isometric grid; (3) create the front, the top, and the side work faces; (4) complete the drawing.

Step 3: Create the front, the top, and the side work faces. Draw line 2–5 parallel to line 1–3; draw line 4–6 parallel to line 1–3; draw line 3–5 parallel to line 1–2; draw line 3–6 parallel to line 1–4; draw line 5–7 parallel to line 3–6; and draw line 6–7 parallel to line 3–5 as shown in Figure 15.11(c).

Step 4: Complete the drawing as marked by the remaining line numbers.

Next, we will use an example to demonstrate the steps in creating an isometric view of an object.

EXAMPLE 15.3

Draw the isometric view of the object shown in Figure 15.12.

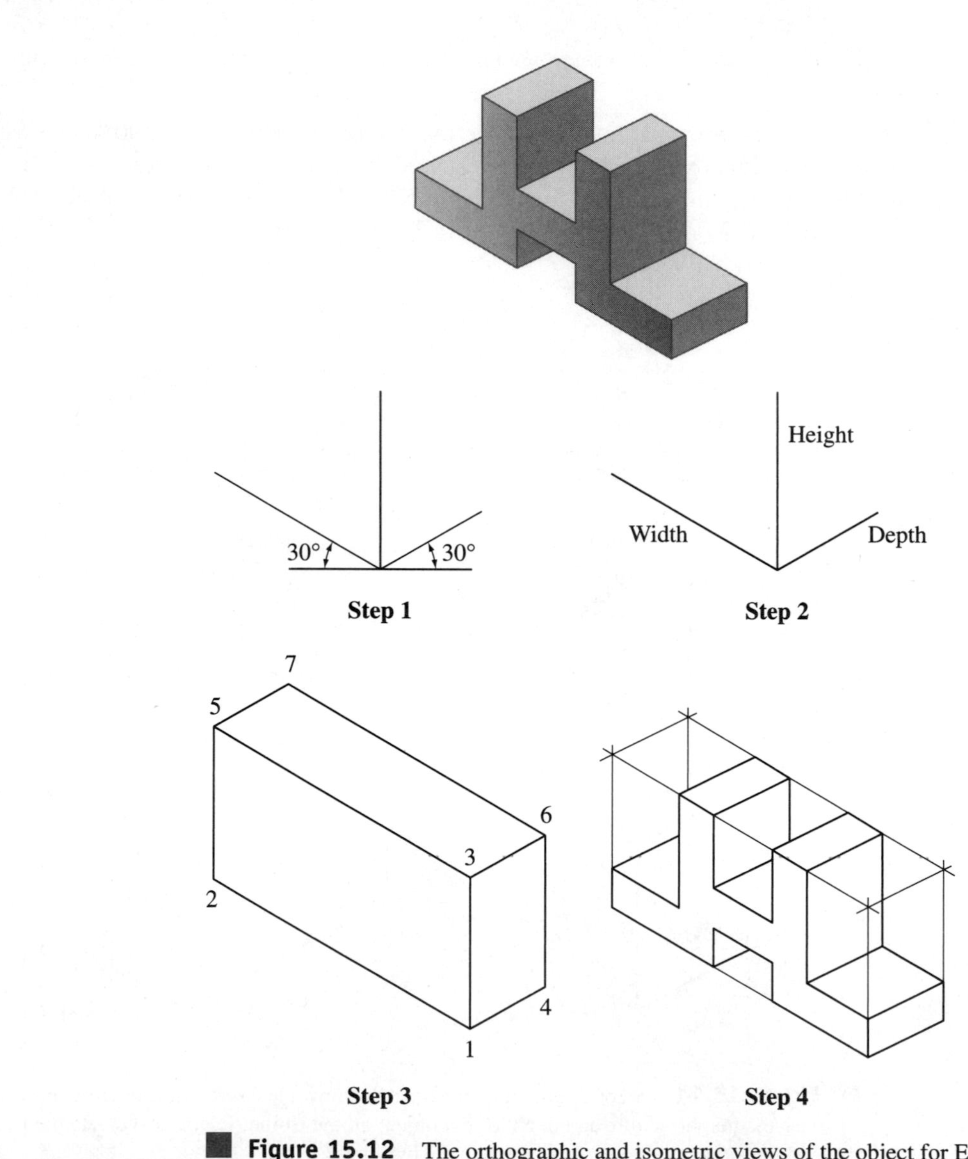

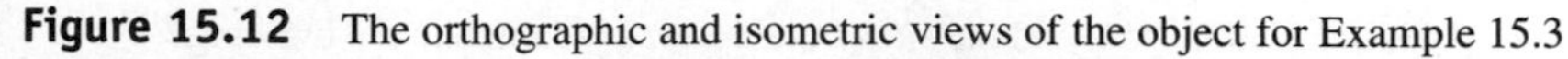

Figure 15.12 The orthographic and isometric views of the object for Example 15.3

15.5 SECTIONAL VIEWS

You recall from Section 15.2 that you use the dashed (hidden) lines to represent the edge of a plane or intersection of two planes or limits of a hole that are not visible from the direction you are looking. Of course, as mentioned earlier, this happens when some material exists between you and the edge. For objects with a complex interior—for example, with many interior holes or edges—the use of dashed lines on orthographic views could result in a confusing drawing. The dashed lines could make the drawing difficult to read and thus difficult for the reader to form a visual image of what the inside of the object looks like. An example of an object with a complex interior is shown in Figure 15.13. For objects with complex interiors, ***sectional views*** are used. Sectional views reveal the inside of the object. A sectional view is created by making an imaginary cut through the object, in a certain direction, to reveal its interior. The sectional views are drawn to show clearly the solid portions and the voids within the object.

Let us now look at the procedure that you need to follow to create a sectional view. The first step in creating a sectional view involves defining the cutting plane and the direction of the sight. The direction of the sight is marked using directional arrows, as shown in Figure 15.14. Moreover, identifying letters are used with the directional arrows to name the section. The next step involves identifying and showing on the sectional view which portion of the object is made of solid material and which portion has the voids. The solid section of the view is then marked by parallel inclined lines. This method of marking the solid portion of the view is called *cross-hatching*. An example of a cutting plane, its directional arrow, its identifying letter, and cross-hatching is shown in Figure 15.14.

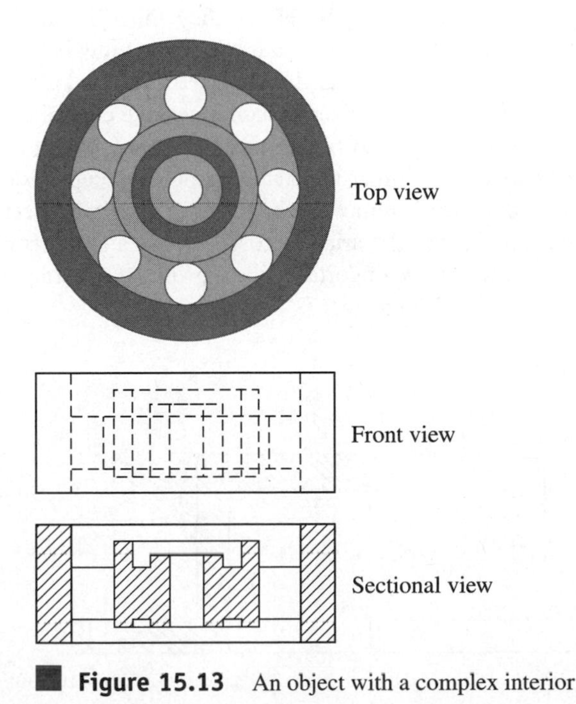

Figure 15.13 An object with a complex interior

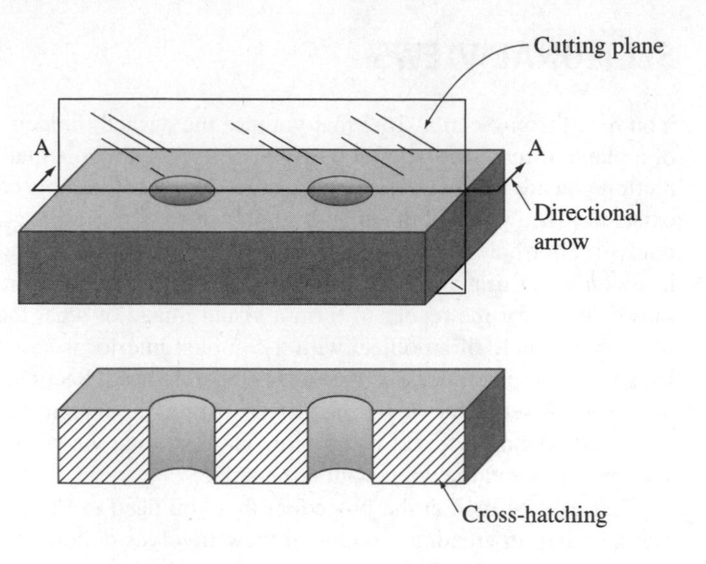

Figure 15.14 A sectional view of an object. Crosshatch patterns also indicate the type of material that parts are made from.

Based on how complex the inside of an object is, different methods are used to show sectional views. Some of the common section types include: *full section, half section, broken-out section, rotated section*, and *removed section.*

- ***Full-section views*** are created when the cutting plane passes through the object completely, as shown in Figure 15.14.
- ***Half-sectional views*** are used for symmetrical objects. For such objects, it is customary to draw half of the object in sectional view and the other half as exterior view. The main advantage of half-sectional views is that they show the interior and exterior of the object using one view. An example of a half-section view is shown in Figure 15.15.
- ***Rotated section view*** may be used when the object has a uniform cross section with a shape that is difficult to visualize. In such cases, the cross section is rotated by 90° and is shown in the plane of view. An example of a rotated section is shown in Figure 16.16.
- ***Removed sections*** are similar to rotated sections, except instead of drawing the rotated view on the view itself, removed sections are shown adjacent to the view. They may be used for objects with a variable cross section, and generally many cuts through the section are shown. It is important to note that the cutting planes must be properly marked, as shown in Figure 15.17.

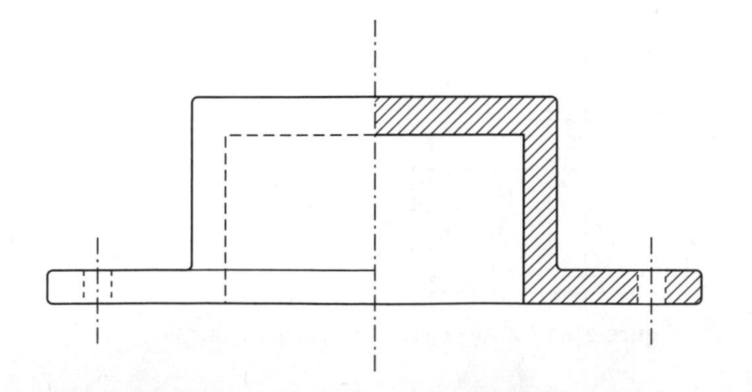

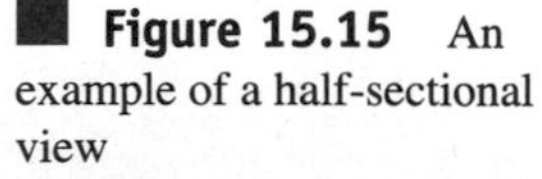

Figure 15.15 An example of a half-sectional view

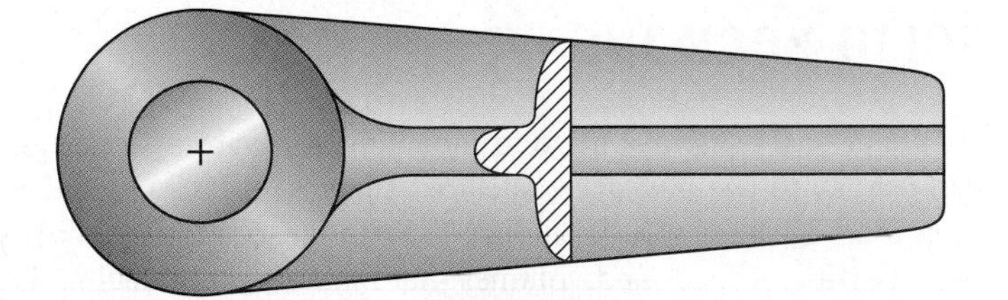

Figure 15.16 An example of a rotated section view

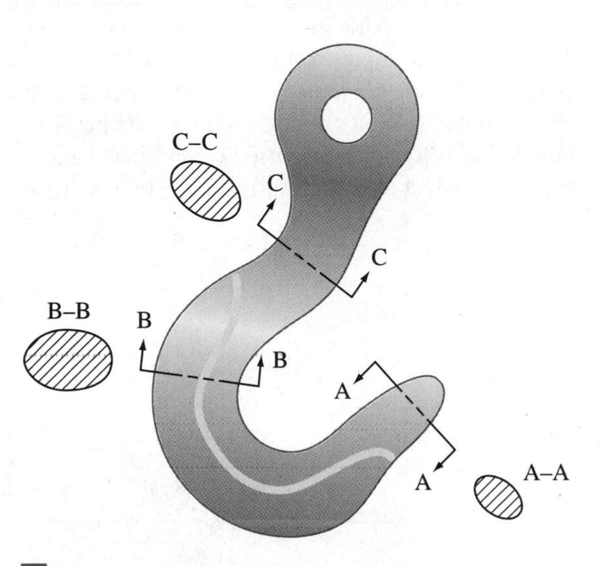

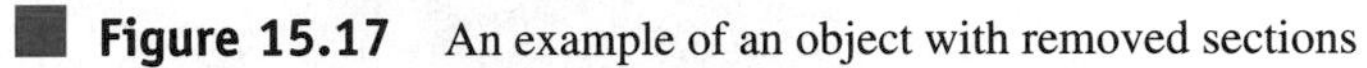

Figure 15.17 An example of an object with removed sections

EXAMPLE 15.4

Draw the sectional view of the object shown in Figure 15.18, as marked by the cutting plane.

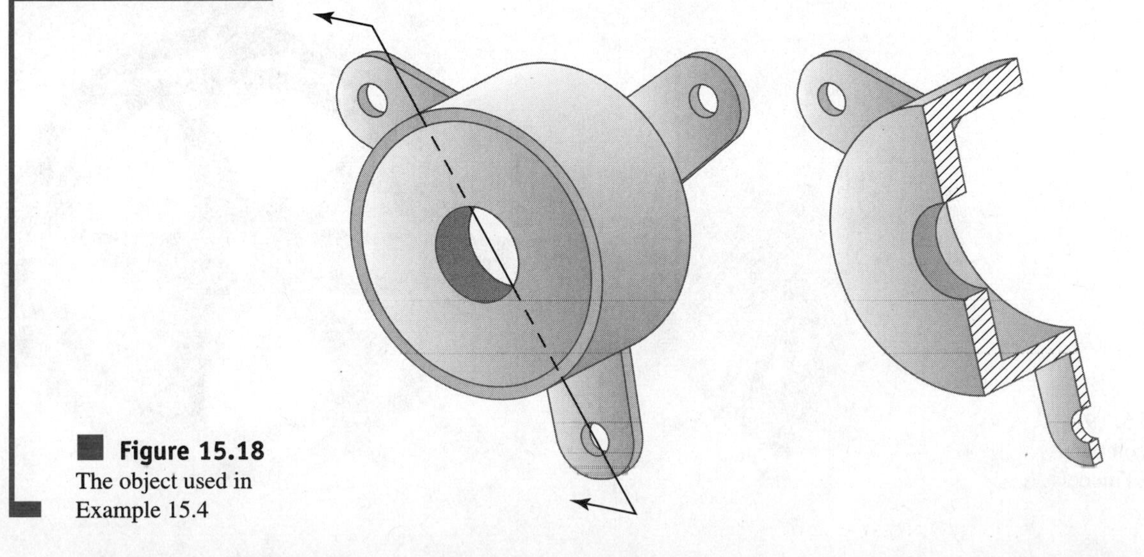

Figure 15.18 The object used in Example 15.4

15.6 SOLID MODELING

In recent years, the use of solid modeling software as a design tool has grown dramatically. Easy-to-use packages, such as AutoCAD, IDEAS, and Pro-E, have become common tools in the hands of engineers. With these software tools you can create models of objects with surfaces and volumes that look almost indistinguishable from the actual objects. These solid models provide great visual aids for what the parts that make up a product look like before they are manufactured. The solid modeling software also allows for experimenting on a computer screen with the assembly of parts to examine any unforeseen problems before the parts are actually made and assembled. Moreover, changes to the shape and the size of a part can be made quickly with such software. Once the final design is agreed upon, the computer-generated drawings can be sent directly to computer numerically controlled (CNC) machines to make the parts.

Solid modeling software is also used by architects and engineers to present concepts. For example, an architect uses such software to show a client a model of what the exterior

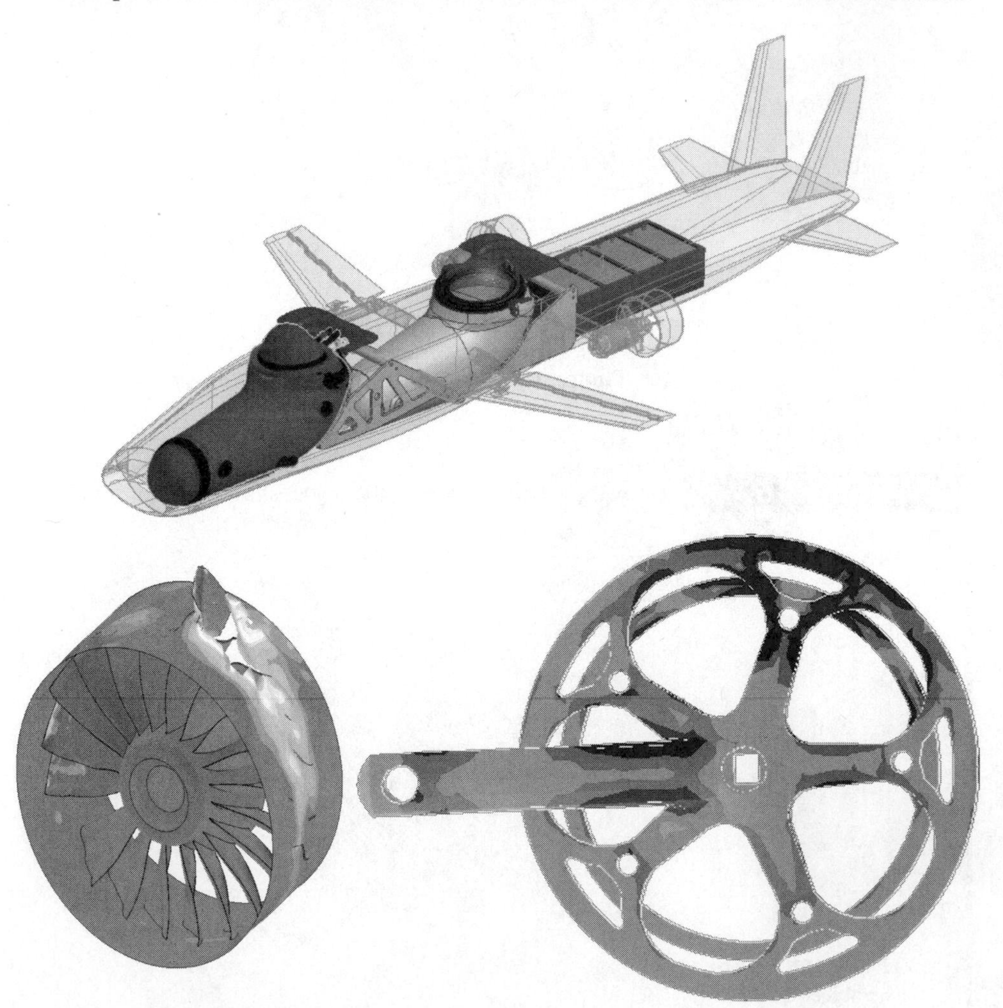

Figure 15.19
Examples of computer-generated solid models

or interior of a proposed building would look like. Design engineers employ the solid modeling software to show concepts for shapes of cars, boats, computers, and so on. The computer-generated models save time and money. Moreover, there is additional software that makes use of these solid models to perform additional engineering analysis, such as stress calculations or temperature distribution calculations for products subjected to loads and/or heat transfer. Examples of solid models generated by commonly used software are shown in Figure 15.19.

Let us now briefly look at how solid modeling software generates solid models. There are two ways to create a solid model of an object: *bottom-up modeling* and *top-down modeling*. With bottom-up modeling you start by defining keypoints first, then lines, areas, and volumes in terms of the defined keypoints. Keypoints are used to define the vertices of an object. Lines, next in the hierarchy of bottom-up modeling, are used to represent the edges of an object. You can then use the created lines to generate a surface. For example, to create a rectangle, you first define the corner points by four keypoints, next you connect the keypoints to define four lines, and then you define the area of the rectangle by the four lines that enclose the area. There are additional ways to create areas: (1) dragging a line along a path, (2) rotating a line about an axis, (3) creating an area fillet, (4) skinning a set of lines, and (5) offsetting areas. With the area-fillet operation, you can create a constant-radius fillet tangent to two other areas. You can generate a smooth surface over a set of lines by using the skinning operation. Using the area offset command, you can generate an area by offsetting an existing area. These operations are all shown in Figure 15.20.

The created areas then may be put together to enclose and create a volume. As with areas, you can also generate volumes by dragging or extruding an area along a line (path) or by rotating an area about a line (axis of rotation). Examples of these volume-generating operations are shown in Figure 15.21.

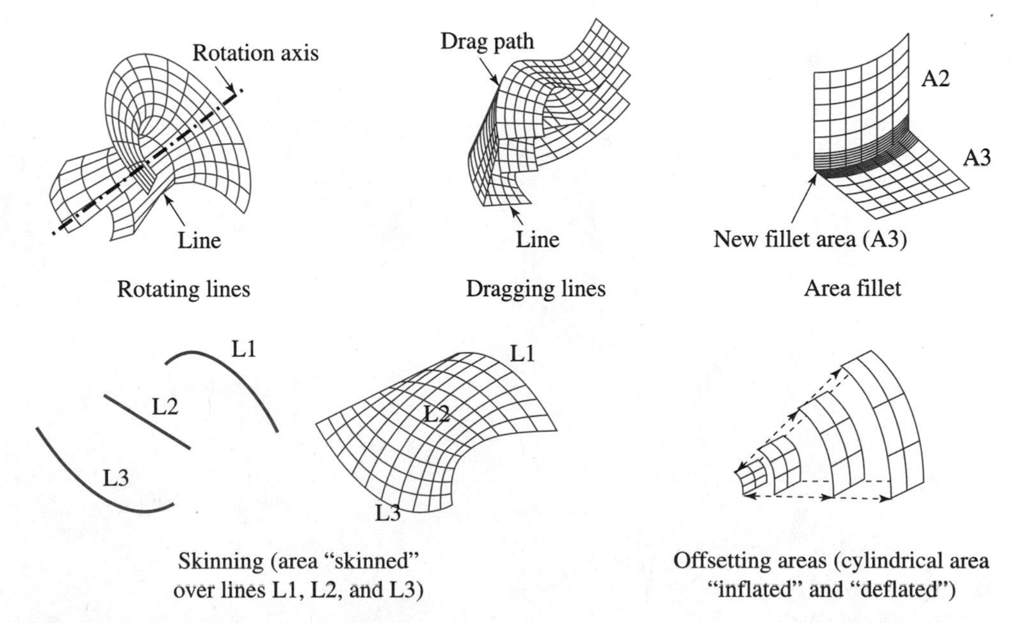

Figure 15.20 Additional area-generation methods

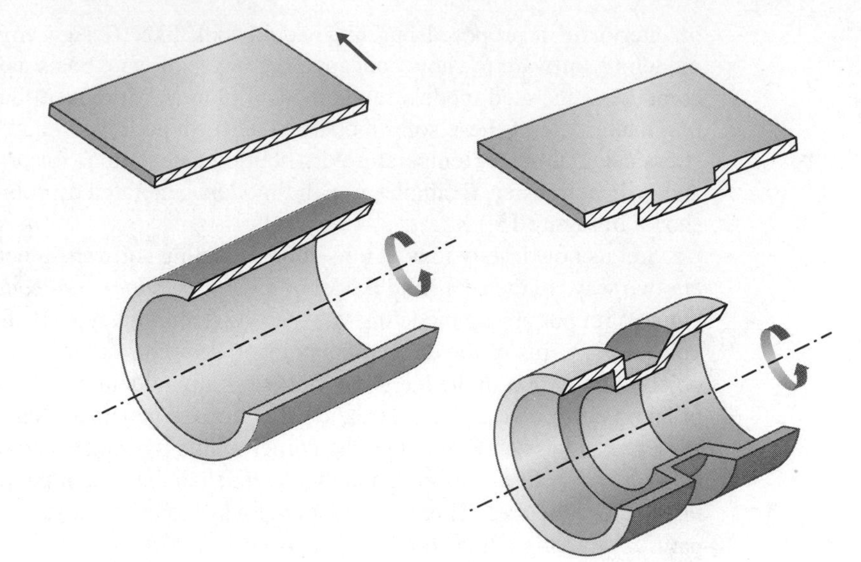

Figure 15.21
Examples of volume-generating operations

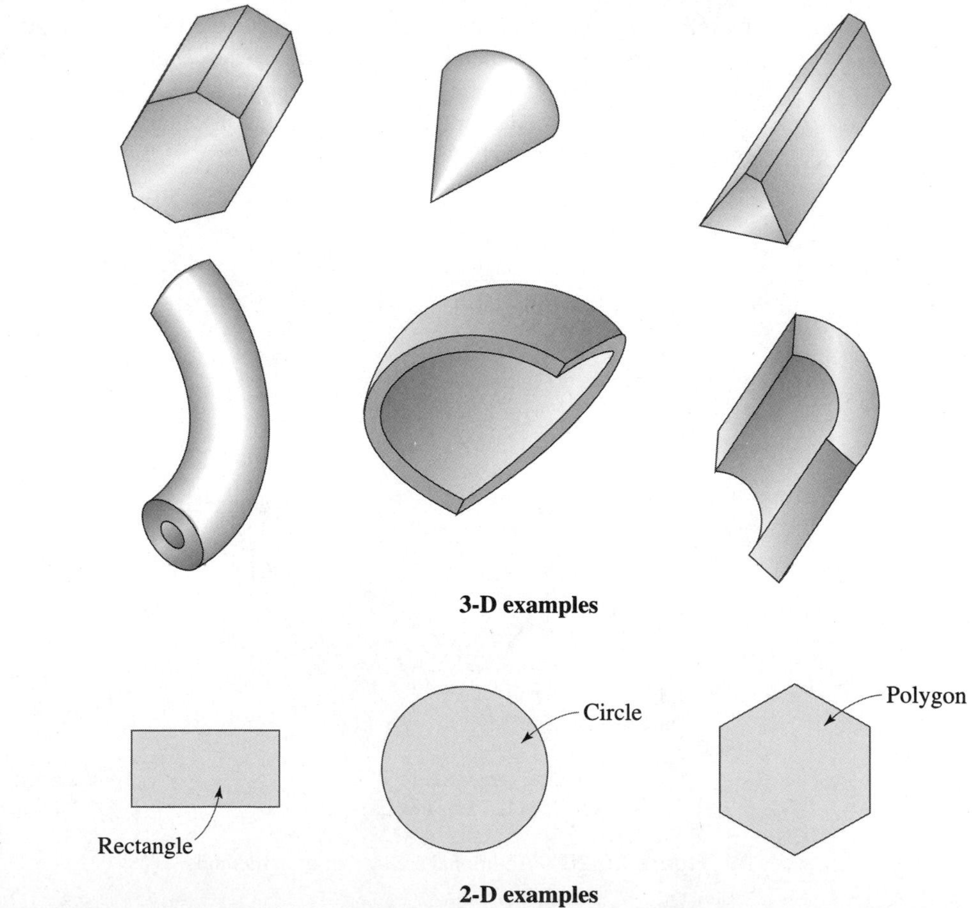

Figure 15.22
Examples of two- and three-dimensional primitives

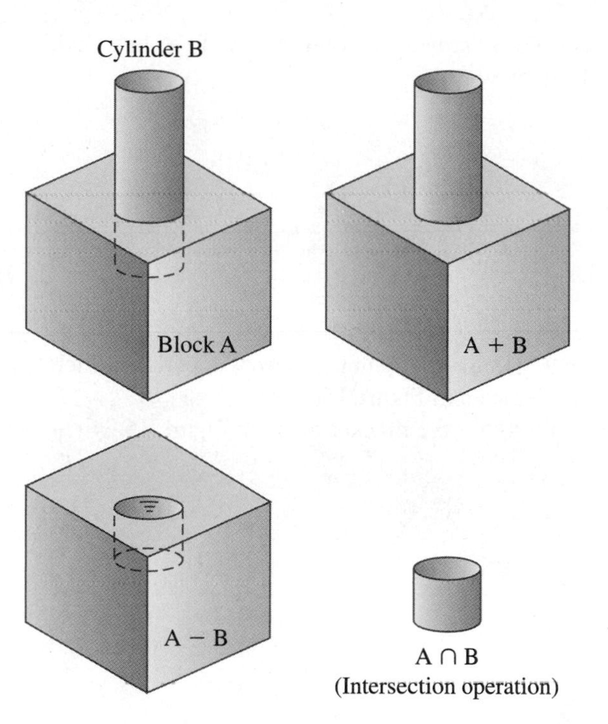

Figure 15.23 Examples of Boolean union (add) and subtract operations

With top-down modeling, you can create surfaces or three-dimensional solid objects using area and volume *primitives*. Primitives are simple geometric shapes. Two-dimensional primitives include rectangles, circles, and polygons, and three-dimensional volume primitives include blocks, prisms, cylinders, cones, and spheres, as shown in Figure 15.22.

Regardless of how you generate areas or volumes, you can use Boolean operations to add (union) or subtract entities to create a solid model. Examples of Boolean operations are shown in Figure 15.23.

EXAMPLE 15.5

Discuss how to create the solid model of the objects shown in Figure 15.24, using the operations discussed in this section.

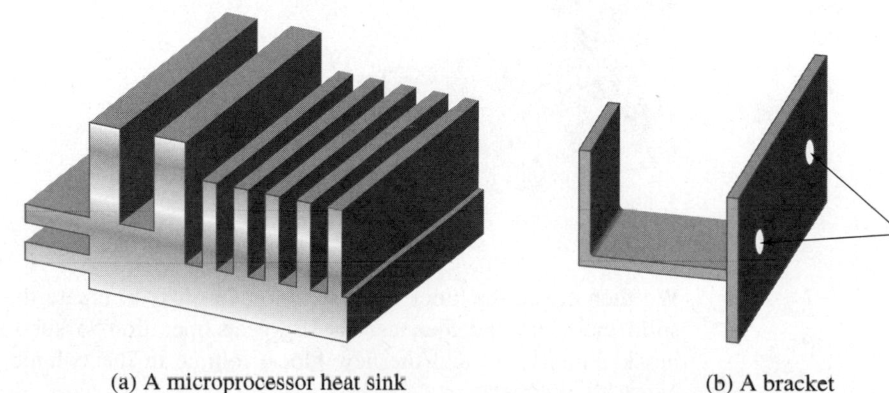

(a) A microprocessor heat sink

(b) A bracket

Figure 15.24 The objects for Example 15.5

(a) In order to create the solid model of the heat sink shown in Figure 15.24(a), we draw its front view first.

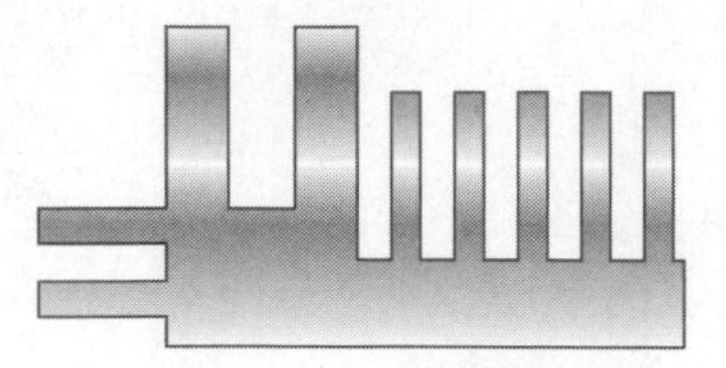

We then extrude this profile in the normal direction, which leads to the solid model of the heat sink as depicted in Figure 15.24(a).

(b) Similarly for the bracket given in Figure 15.24 (b), we begin by creating the profile shown.

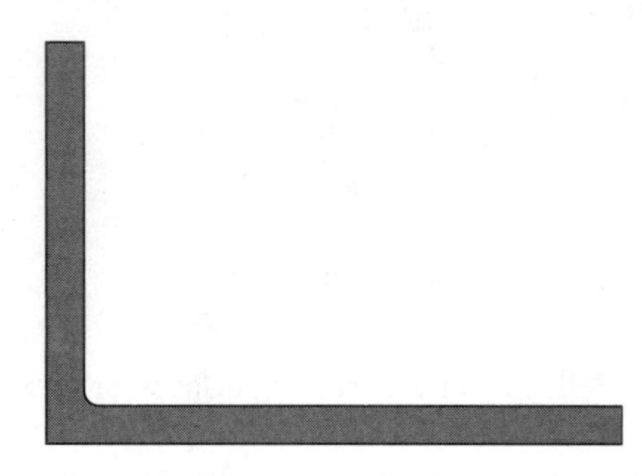

Next, we will extrude the profile in the normal direction.

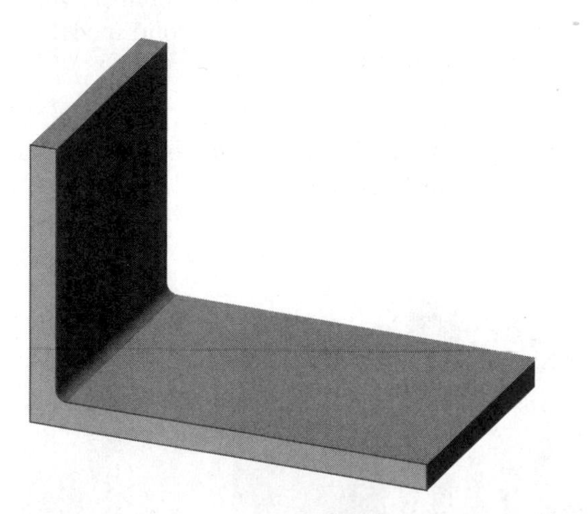

We then create the block and the holes. In order to create the holes, first we create two solid cylinders and then use the Boolean operation to subtract the cylinders from the block. Finally, we add the new block volume to the volume, which we created by the extrusion method.

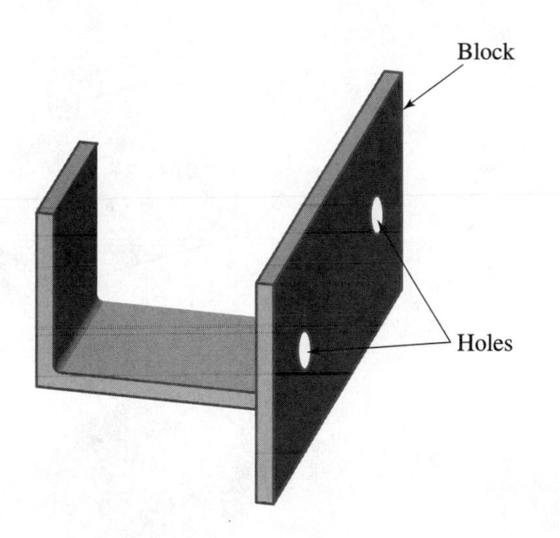

SUMMARY

Now that you have reached this point in the text

- You should have a good understanding of the importance of engineering drawings in conveying information to other engineers, machinists, and assembly personnel.
- You should understand what is meant by orthographic views, isometric drawing, and solid modeling.
- You should understand the basic rules required for an engineering drawing, including showing dimensions, specifying material size, and indicating finished surfaces.
- You should know when to use isometric views and sectional views.
- You should be familiar with the different types of sectional views.
- You should understand the importance of solid modeling in conveying concepts and examining parts for their ability to fit with other parts.

PROBLEMS

For Problems 1 through 10, draw the top, the front, and the right-side orthographic views of the objects shown. Indicate when an object needs only one or two views to be fully described.

1.

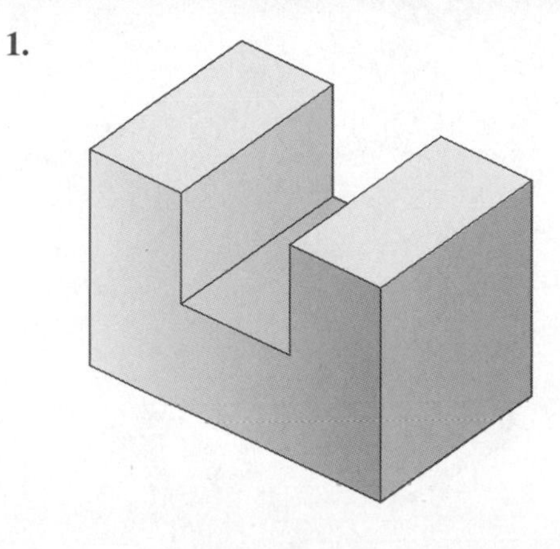

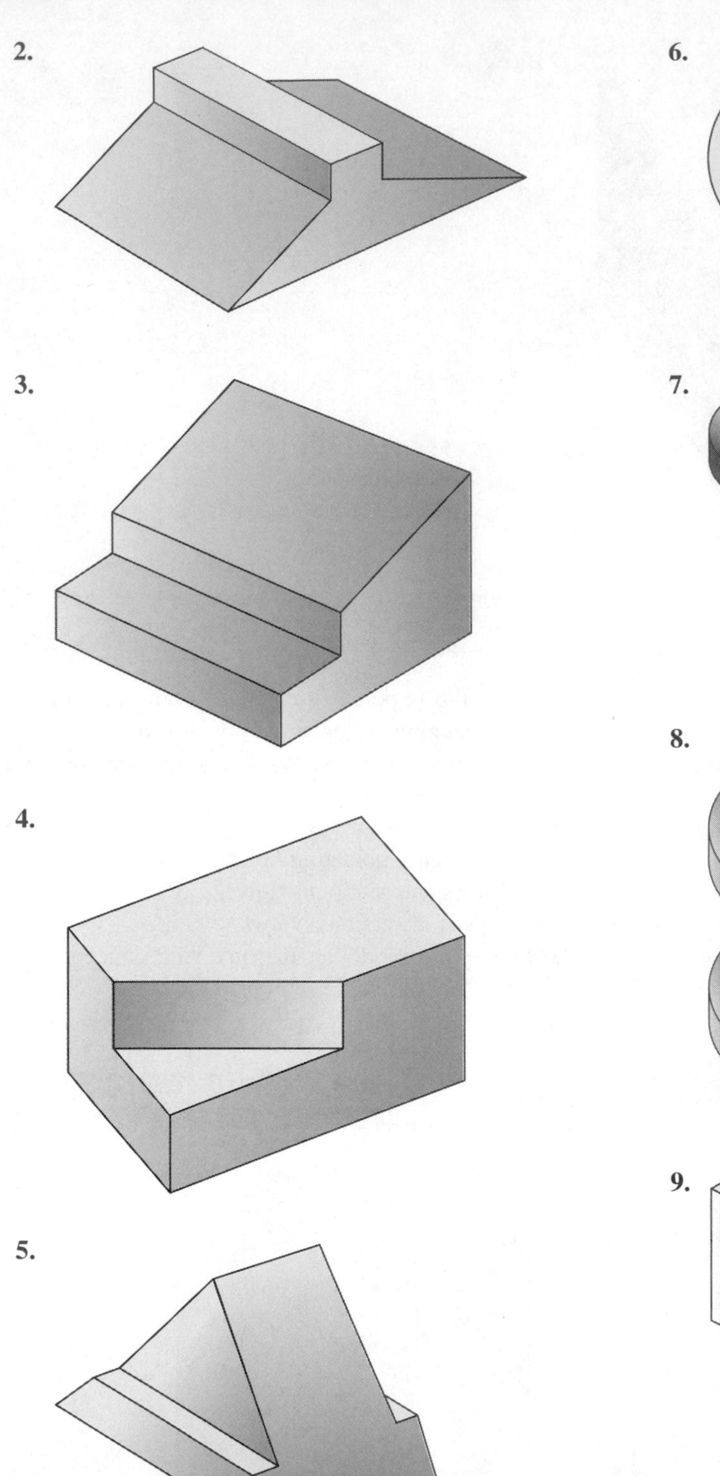

2.
3.
4.
5.

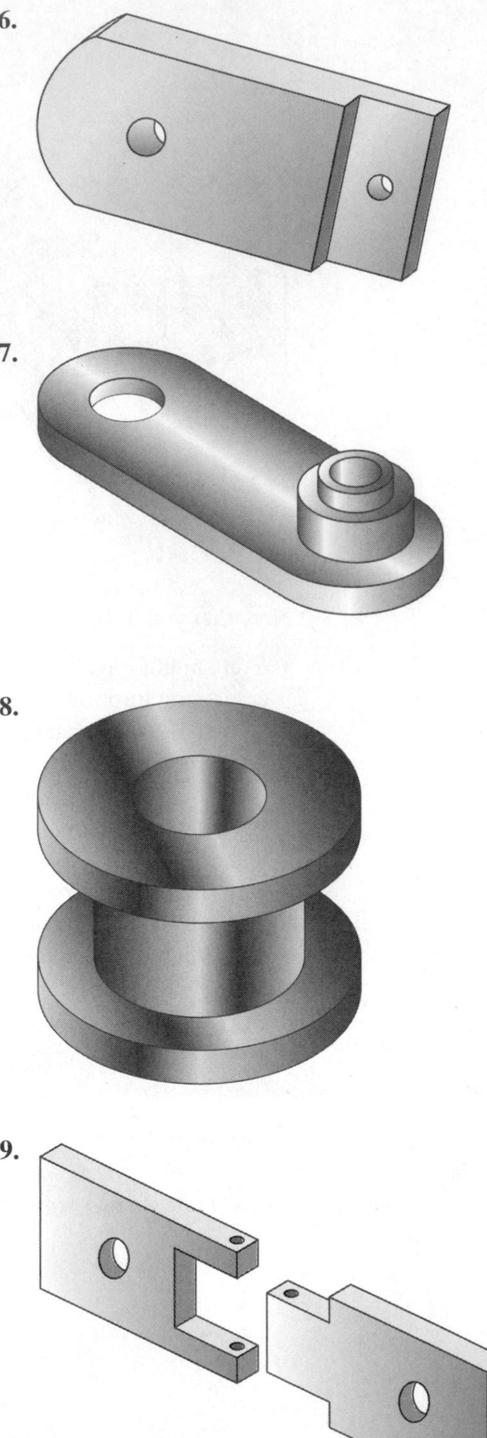

6.
7.
8.
9.

10.

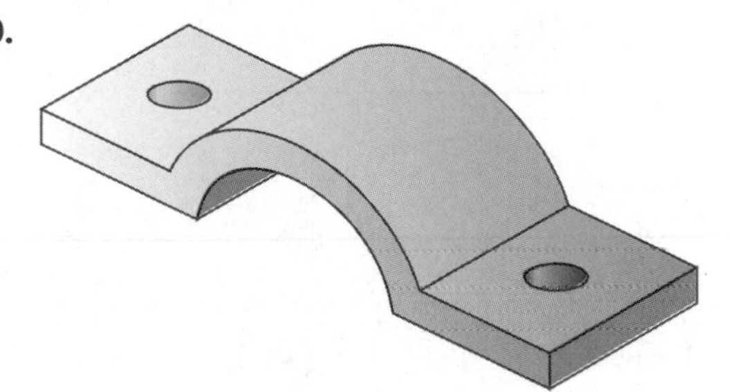

For Problems 11 through 15, use the cutting planes shown to draw the sectional views.

11.

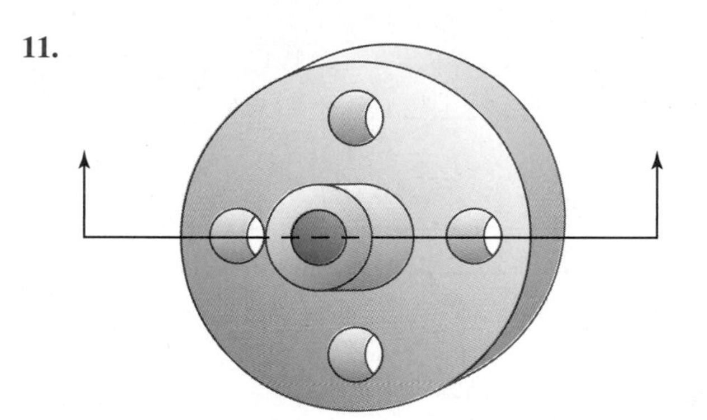

12.

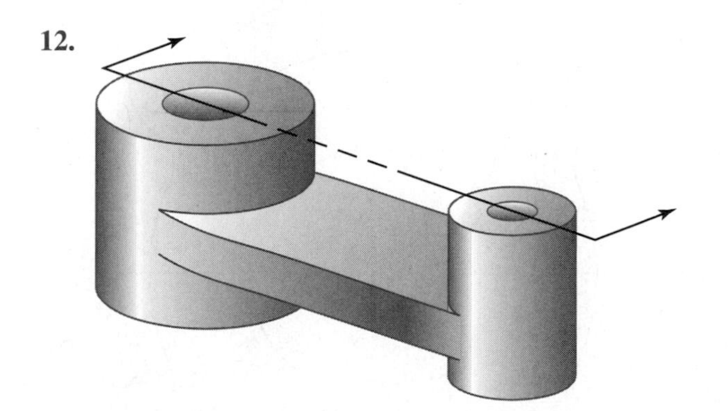

13.

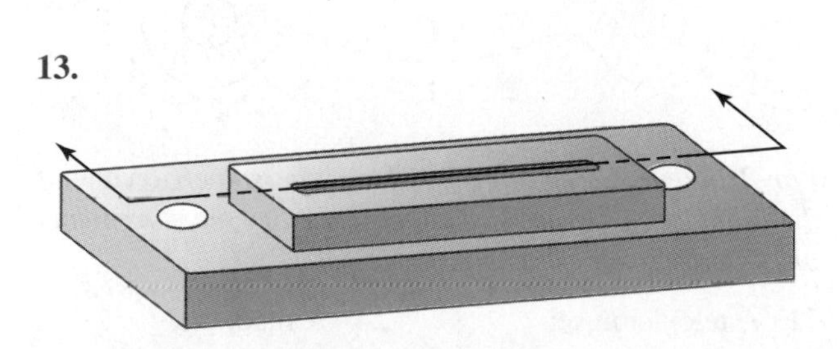

14.

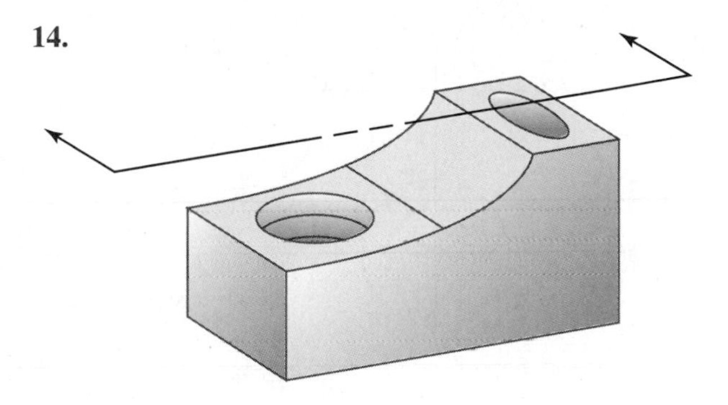

15.

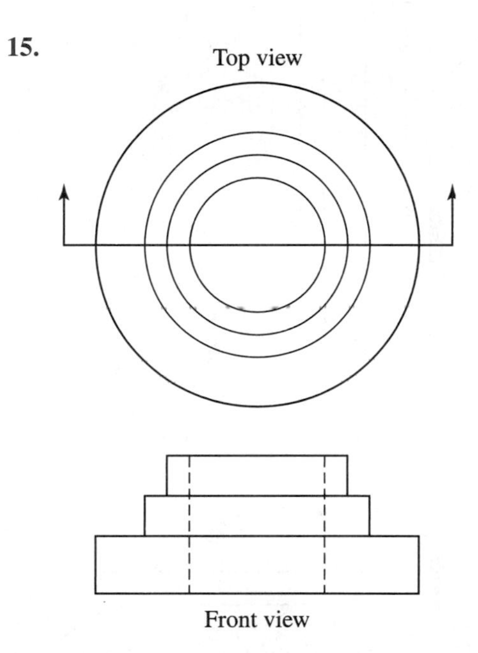

For Problems 16 through 20, using the rules discussed in this chapter, show the dimensions of the views shown.

16.

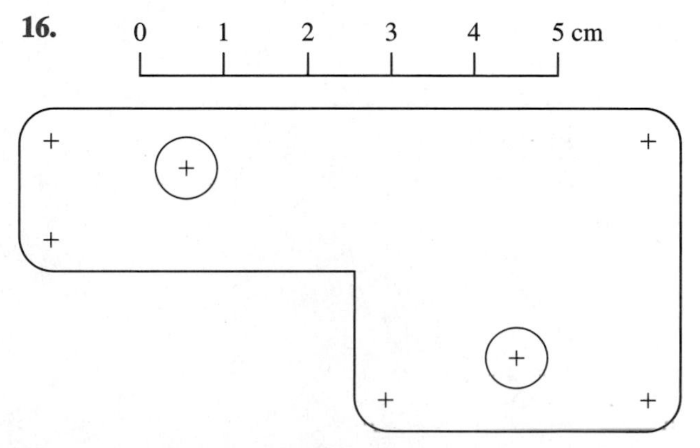

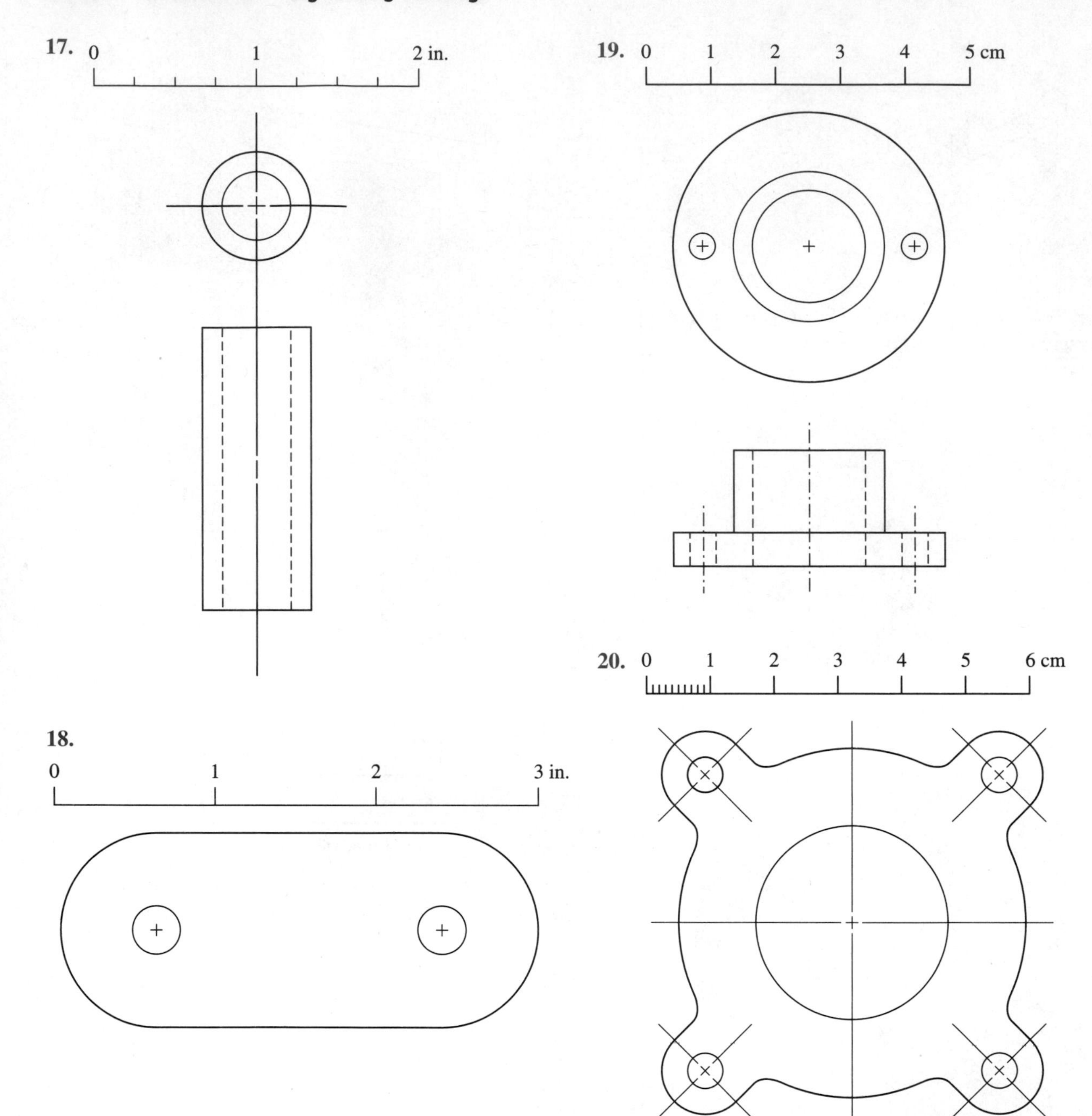

For Problems 21 through 26, draw the isometric view of the following objects. Make the necessary measurements or estimations of dimensions.

21. A television set

22. A computer

23. A telephone

24. A razor

25. A chair

26. A car

For Problems 27 through 31, discuss how you would create the solid model of the given objects. See Example 15.5 to better understand what you are being asked to do.

27. A bracket

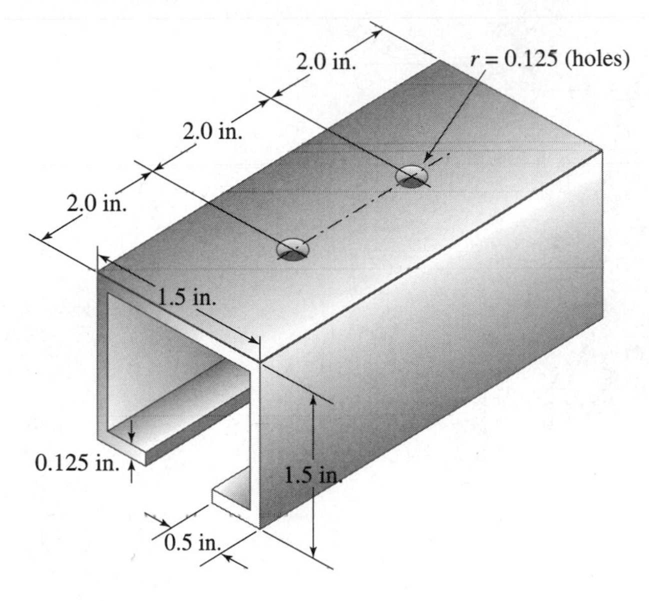

28. A wheel

10.0
9.0
6.75
5.25
2.0
R 0.8
R 0.7
8 holes
1.0 dia.
R 0.8
R 0.7
7.0
6.0
4.25
2.0
1.0
y
x
z

Dimensions are in inches.

29. A pipe

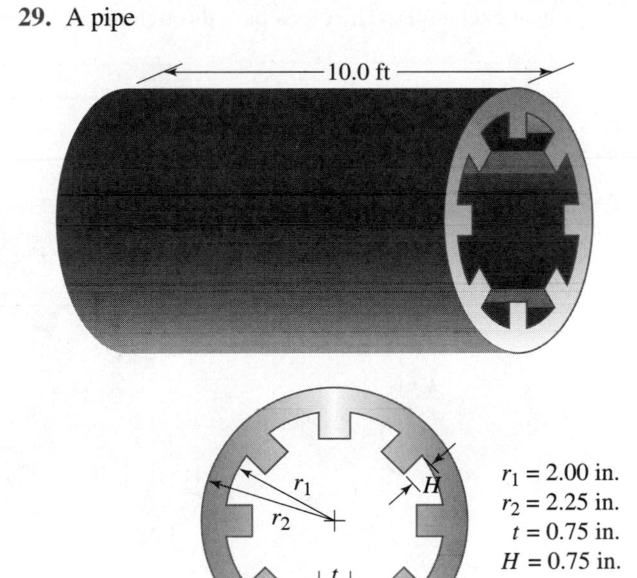

30. A socket

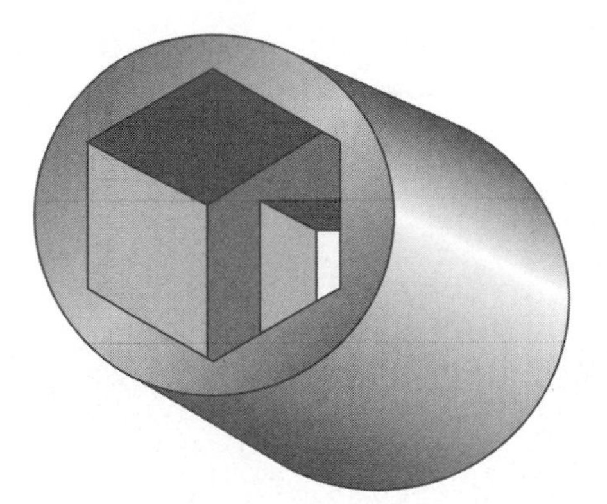

31. A heat exchanger. The pipes pass through all of the fins.

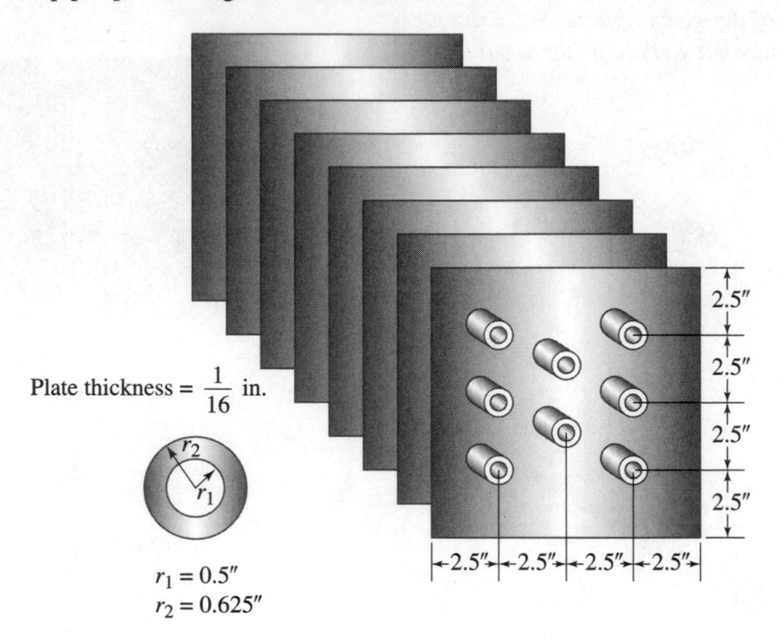

AN ENGINEERING MARVEL

Boeing 777* Commercial Airplane

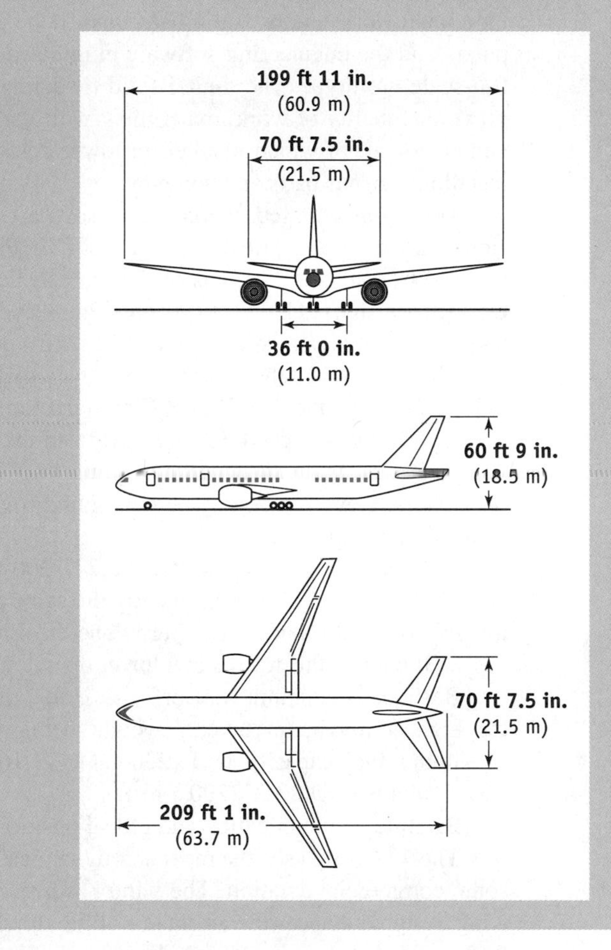

*Materials were adapted with permission from Boeing documents.

OVERVIEW

The Boeing 777 is the first commercial airplane that was fully designed using three-dimensional digital solid modeling technology. The core of the design group consisted of 238 teams that included engineers of various backgrounds. A number of international aerospace companies, from Europe, Canada, and Asia/Pacific, contributed to the design and production of the 777. The Japanese aerospace industry was among the largest of the overseas participants. Representatives from airline customers such as Nippon Airways and British Airways also provided input to the design of 777. Throughout the design process, the different components of the airplane were designed, tested, assembled (to ensure proper fitting), and disassembled on a network of computers. Approximately 1700 individual workstations and 4 IBM mainframe computers were used. The use of the computers and the engineering software eliminated the need for the development of a costly, full-scale prototype. The digital solid modeling technology allowed the engineers to improve the quality of work, experiment with various design concepts, and reduce changes and errors, all of which resulted in lower costs and increased efficiency in building and installing various parts and components.

The engineers used, among other software, CATIA (Computer-Aided Three-Dimensional Interactive Application) and ELFINI (Finite Element Analysis System), both developed by Dassault Systems of France and licensed in the United States through IBM. Designers also used EPIC (Electronic Preassembly Integration on CATIA) and other digital preassembly applications developed by Boeing.

The 777 series, the world's largest twinjet, is available in three models: the initial model, 777-200; the 777-200ER (Extended Range) model; and the larger 777-300 model. The 777-300 is stretched 33 ft (10 m) from the initial 777-200 model to a total of 242 ft 4 in. (73.9 m). In an all-economy layout, the 777-300 can accommodate as many as 550 passengers. However, it may be configured for 368 to 386 passengers in three classes to provide more comfort.

In terms of range capability, the 777-300 can serve routes up to 6450 statute miles (10,370 km). The 777-300 has nearly the same passenger capacity and range capability as the 747-100/-200 models but burns one-third less fuel and has 40% lower maintenance costs. Of course, this results in a lower operating cost.

Baseline maximum takeoff mass for the 777-300 is 580,000 lb (263,080 kg); the highest maximum takeoff weight being offered is 660,000 pounds (299,370 kg). Maximum fuel capacity is 45,220 gal (171,160 L). The 777-300 has a total available cargo volume of 7080 ft^3 (200.5 m^3).

Satellite communication and global positioning systems are basic to the airplane.

The 777 wing uses the most aerodynamically efficient airfoil ever developed for subsonic commercial aviation. The wing has a span of 199 ft 11 in. (60.9 m). The advanced wing design enhances the airplane's ability to climb quickly and cruise at higher altitudes than its predecessor airplanes. The wing design also allows the airplane to carry full passenger payloads out of many high-elevation, high-temperature airfields. Fuel is stored entirely within the wing and its structural center section. The longer-range model and the 777-300 model can carry up to 45,220 gal (171,155 L).

The Boeing Company, upon request from airplane buyers, can install engines from three leading engine manufacturers, namely, Pratt & Whitney, General Electric, and Rolls-Royce. These engines are rated in the 74,000 to 77,000-pound thrust class. For the longer-range model and the 777-300, these engines will be capable of thrust ratings in the 84,000 to 98,000-pound category.

New, lightweight, cost-effective structural materials are used in several 777 applications. For example, an improved aluminum alloy, 7055, is used in the upper wing skin and stringers. This alloy offers greater compression strength than previous alloys, enabling designers to save weight and also improve resistance to corrosion and fatigue. Lightweight composites are found in the vertical and horizontal tails. The floor beams of the passenger cabin also are made of advanced composite materials.

The principal flight, navigation, and engine information is all presented on six large, liquid-crystal, flat-panel displays. In addition to saving space, the new displays weigh less and require less power, and because they generate less heat, less cooling is required compared to the older conventional cathode-ray-tube screens. The flat-panel displays remain clearly visible in all conditions, even direct sunlight.

The Boeing 777 uses an Integrated Airplane Information Management System that provides flight and maintenance crews all pertinent information concerning the overall condition of the airplane, its maintenance requirements, and its key operating functions, including flight, thrust, and communication management.

The flight crew transmits control and maneuvering commands through electrical wires, augmented by computers, directly to hydraulic actuators for the elevators, rudder, ailerons, and other controls surfaces. The three-axis, "fly-by-wire" flight-control system saves weight, simplifies factory assembly compared to conventional mechanical systems relying on steel cables, and requires fewer spares and less maintenance in airline service.

A key part of the 777 system is a Boeing-patented, two-way digital data bus, which has been adopted as a new industry standard: ARINC 629. It permits airplane systems and their computers to communicate with one another through a common wire path (a twisted pair of wires) instead of through separate one-way wire connections. This further simplifies assembly and saves weight, while increasing reliability through a reduction in the amount of wires and connectors. There are 11 of these ARINC 629 pathways in the 777.

The interior of the Boeing 777 is one of the most spacious passenger cabins ever developed; the 777 interior offers configuration flexibility. Flexibility zones have been designed into the cabin areas specified by the airlines, primarily at the airplane's doors. In 1-in. increments, galleys and lavatories can be positioned anywhere within these zones, which are preengineered to accommodate wiring, plumbing, and attachment fixtures. Passenger service units overhead stowage compartments are designed for quick removal without disturbing ceiling panels, air-conditioning ducts, or support structure. A typical 777 configuration change is expected to take as little as 72 hours, while such a change might take two to three weeks on other aircraft. For improved, more efficient, in-flight service, the 777 is equipped with an advanced cabin management system. Linked to a computerized control console, the cabin management system assists cabin crews with many tasks and allows airlines to provide new services of passengers, including a digital sound system comparable to the most state-of-the-art home stereo or compact disc players.

The main landing gear for the 777 is in a standard two-post arrangement but features six-wheel trucks, instead of the conventional four-wheel units. This provides the main landing gear with a total of 12 wheels for better weight distribution on runways and taxi areas and avoids the need for a supplemental two-wheel gear under the center of the

TABLE 1 Boeing 777-200/300 Specifications

Design Variable	777-200	777-300
Seating	305 to 320 passengers in three classes	368 to 386 passengers in three classes
Length	209 ft 1 in. (63.7 m)	242 ft 4 in. (73.9 m)
Wingspan	199 ft 11 in. (60.9 m)	199 ft 11 in. (60.9 m)
Tail height	60 ft 9 in. (18.5 m)	60 ft 8 in. (18.5 m)
Engines	Pratt & Whitney 4000 General Electric GE90 Rolls-Royce Trent 800	Pratt & Whitney 4000 General Electric GE90 Rolls-Royce Trent 800
Maximum takeoff mass	506,000 lb (229,520 kg)	580,000 lb (263,080 kg)
Fuel capacity	31,000 U.S. gallons (117,335 L)	45,200 U.S. gallons (171,160 L)
Altitude capability	39,300 ft (11,975 m)	36,400 ft (11,095 m)
Cruise speed	555 mph (893 km/h) Mach 0.84	555 mph (893 km/h) Mach 0.84
Cargo capacity	5656 ft^3 (160 m^3)	7552 ft^3 (214 m^3)
Maximum range	5150 nautical miles; 5952 statute miles; 9525 km	5600 nautical miles; 6450 statute miles; 10,370 km

fuselage. Another advantage is that the six-wheel trucks allow for a more economical brake design. The 777 landing gear is the largest ever incorporated into a commercial airplane.

The Boeing–United Airlines 1000-cycle flight tests for the Pratt & Whitney engine were completed on May 22, 1995. In addition, engine makers and the many suppliers of parts for the airplane intensified their own development and testing efforts to ensure that their products met airline requirements. This thorough test program demonstrated the design features needed to obtain approval for extended-range twin-engine operations (ETOPS). All 777s are ETOPS-capable, as part of the basic design. To ensure reliability, the 777 with Pratt & Whitney engines was tested and flown under all appropriate conditions to prove it is capable of flying ETOPS missions. A summary of Boeing 777 specifications is shown in Table 1.

PROBLEMS

1. Using the data given for the Boeing 777, estimate the flight time from New York City to London.
2. Estimate the mass of passengers, fuel, and cargo for a full flight.
3. Using the maximum range and fuel capacity data, estimate the fuel consumption of the Boeing 777 on a per hour and per mile basis.

4. Calculate the linear momentum of the Boeing 777 at cruise speed and at two-thirds maximum takeoff mass.
5. Calculate the Mach number of the Boeing 777 at cruise speed using

$$\text{Mach} = \frac{\text{cruise speed}}{\sqrt{kRT}}$$

where

k = specific heat ratio = 1.4

R = air gas constant = 287 J/kg·K

T = air temperature at cruising altitude (K)

Compare your Mach number to the one given in the Boeing data table.

6. As mentioned previously, the flight crew transmits control and maneuvering commands through electrical wires, augmented by computers, directly to hydraulic actuators for the *elevators*, *rudder*, *ailerons*, and other controls surfaces. The elevators, rudder, and ailerons for a small plane are shown in the accompanying figure. In a small plane the ailerons are moved by turning the control wheel in the cockpit. When the wheel is turned left, the left aileron moves up and the right aileron moves down. This is how the pilots starts a turn to the left. In a small plane the rudder is operated by the pilot's feet. When the pilot presses the left rudder pedal, the nose of the plane moves left. When the right rudder pedal is pressed the nose moves right. The elevators make the nose of the plane move up and down. When the pilot pulls back on the control wheel in the cockpit, the nose of the plane moves up. When the control wheel is pushed forward, the nose moves down.

   Investigate the aerodynamics of maneuvering flight in more detail. Explain what happens to air pressure distribution over these surfaces as their orientations are changed. What are the directions of the resulting force due to the pressure distributions over these surfaces? Write a brief report explaining your findings.

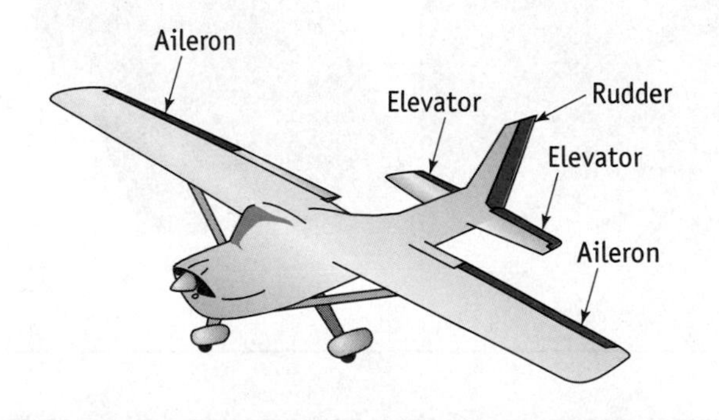

CHAPTER 16 Introduction to Engineering Design

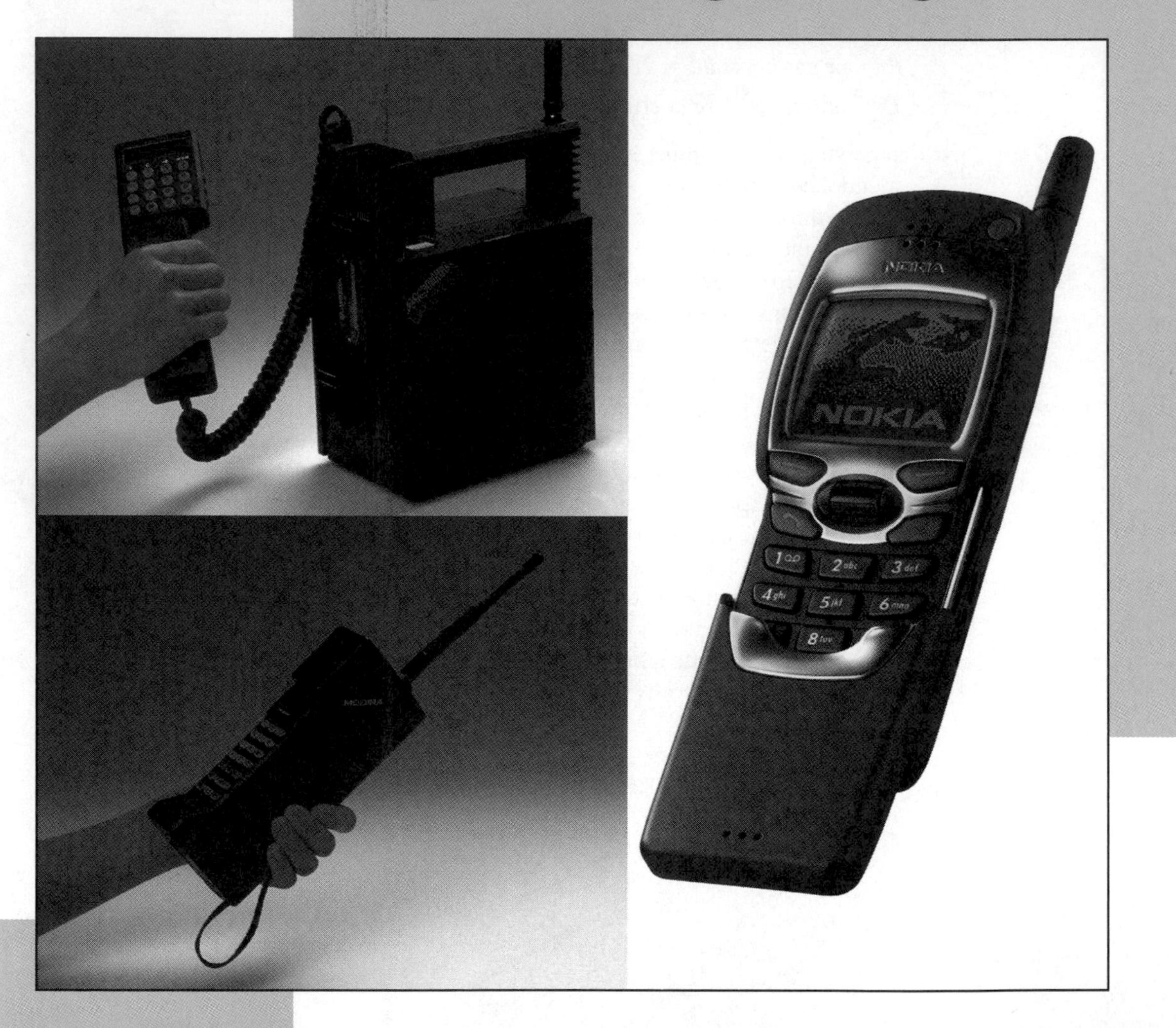

Engineers, regardless of their backgrounds, follow certain steps when designing the products and services we use in our everyday lives. These steps are: (1) recognizing the need for a product or service, (2) defining and understanding the problem (the need) completely, (3) doing the preliminary research and preparation, (4) conceptualizing ideas for possible solutions, (5) synthesizing the results, (6) evaluating good ideas in more detail, (7) optimizing the result to arrive at the best possible solution, and (8) presenting the solution.

Engineers are problem solvers. *In this chapter, we will introduce you to the engineering design process. As we discussed in Chapter 2, engineers apply physical and chemical laws and principles and mathematics to* design *millions of products and services that we use in our everyday lives. Here we will look more closely at what the term* design *means and learn more about how engineers go about designing these products and services. We will discuss the basic steps that most engineers follow when designing something. We will also introduce you to the economic considerations of product and service development. We will then conclude this chapter with a discussion of engineering ethics and why it is so important and an explanation of why engineers are expected to practice engineering using the highest standards of honesty and integrity.* ■

16.1 ENGINEERING DESIGN PROCESS

Let us begin by emphasizing what we said in Chapter 2 about what engineers do. Engineers apply physical laws, chemical laws and principles, and mathematics to *design* millions of products and services that we use in our everyday lives. These products include cars, computers, aircraft, clothing, toys, home appliances, surgical equipment, heating and cooling equipment, health care devices, tools and machines that make various products, and so on. Engineers consider important factors such as cost, efficiency, reliability, and safety when designing the products, and they perform tests to make certain that the products they design withstand various loads and conditions. Engineers are continuously searching for ways to improve already existing products as well. Engineers also *design* and supervise the construction of buildings, dams, highways, and mass transit systems. They also *design* and supervise the construction of power plants that supply power to manufacturing companies, homes, and offices. Engineers play a significant role in the *design* and maintenance of nations' infrastructures, including communication systems, utilities, and transportation. They continuously develop new advanced materials to make products lighter and stronger for different applications. Engineers are also responsible for finding *suitable ways* and *designing* the necessary equipment to extract petroleum, natural gas, and raw materials from the earth.

Let us now look more closely at what constitutes the ***design process***. These are the basic steps that engineers, regardless of their background, follow to arrive at solutions to problems. The steps include: (1) recognizing the need for a product or a service, (2) defining and understanding the problem (the need) completely, (3) doing preliminary research and preparation, (4) conceptualizing ideas for possible solutions, (5) synthesizing the findings, (6) evaluating good ideas in more detail, (7) optimizing solutions to arrive at the best possible solution, (8) and presenting the final solution.

Keep in mind that these steps, which we will discuss soon, are not independent of one another and do not necessarily follow one another in the order in which they are presented here. In fact, engineers often need to return to steps 1 and 2 when clients decide to change design parameters. Quite often, engineers are also required to give oral and written progress reports on a regular time basis. Therefore, be aware of the fact that even though we listed presentation of the design process as step 8, it could well be an integral part of many other design steps. Let us now take a closer look at each step, starting with the need for a product or a service.

Step 1: Recognizing the Need for a Product or a Service

All you have to do is look around to realize the large number of products and services—designed by engineers—that you use every day. Most often we take these products and services for granted until, for some reason, there is an interruption in the services they provide. Some of these existing products are constantly being modified to take advantage of new technologies. For example, cars and home appliances are constantly being redesigned to incorporate new technologies. In addition to the products and the services already in use, new products are being developed every day for the purpose of making our lives more comfortable, more pleasurable, and less laborious. There is also that old saying that "every time someone complains about a situation, or about a task, or curses a product, right there is an opportunity for a product or a service." As you can tell, the need for products and services exists; what one has to do is to identify them. The need may be identified by you, the company that you may eventually work for, or by a third-party client who needs a solution to a problem or a new product to make what it does easier and more efficient.

Step 2: Problem Definition and Understanding

One of the first things you need to do as a design engineer is to fully understand the problem. *This is the most important step in any design process.* If you do not have a good grasp of what the problem is or of what the client wants, you will not come up with a solution that is relevant to the need of the client. The best way to fully understand a problem is by asking many questions. You may ask the client questions such as: How much money are you willing to spend on this project? Are there restrictions on the size or the type of materials that can be used? When do you need the product or the service? How many of these products do you need? Questions often lead to more questions that will better define the problem. Moreover, keep in mind that engineers generally work in a team environment where they consult each other to solve complex problems. They divide up the task into smaller, manageable problems among themselves; consequently, productive engineers must be good team players. Good interpersonal and communication skills are increasingly important now because of the global market. You need to make sure you clearly understand your portion of the problem and how it fits with the other problems. For example, various parts of a product could be made by different companies located in different states or countries. In order to ensure that all components fit and work well together, cooperation and coordination are essential, which demands good teamwork and strong communication skills. Make sure you understand the problem, and make sure that the problem is well defined before you move on to the next step. *This point cannot be emphasized enough*. Good problems solvers are those who first fully understand what the problem is.

Step 3: Research and Preparation

Once you fully understand the problem, as a next step you need to collect useful information. Generally speaking, a good place to start is by searching to determine if a product already exists that closely meets the need of your client. Perhaps a product, or components of a product, already has been developed by your company that you could modify to meet the need. You do not want to "reinvent the wheel"! As mentioned earlier, depending on the scope, some projects require collaboration with other companies, so you need to find out what is available through these other companies as well. Try to collect as much information as you can. This is where you spend lots of time not only with the client but also with other engineers and technicians. Internet search engines are becoming increasingly important tools to gather such information. Once you have collected all pertinent information, you must then review it and organize it in a suitable manner.

Step 4: Conceptualization

During this phase of design, you need to generate some ideas or concepts that could offer reasonable solutions to your problem. In other words, without performing any detailed analysis, you need to come up with some possible ways of solving the problem. You need to be creative and perhaps develop several alternative solutions. At this stage of design, you do not need to rule out any reasonable working concept. If the problem consists of a complex system, you need to identify the components of the system. You do not need to look at details of each possible solution yet, but you need to perform enough analysis to see whether the concepts that you are proposing have merit. Simply stated, you need to ask yourself the following question: Would the concepts be likely to work if they were pursued further? Throughout the design process, you must also learn to budget your time. Good engineers have time-management skills that enable them to work productively and efficiently. You must learn to create a milestone chart detailing your time plan for completing the project. You need to show the time periods and the corresponding tasks that are to be performed during these time periods.

Step 5: Synthesis

Recall from our discussion in Chapter 2 that good engineers have a firm grasp of the fundamental principles of engineering, which they can use to solve many different problems. Good engineers are analytical, detailed oriented, and creative. During this stage of design, you begin to consider details. You need to perform calculations, run computer models, narrow down the type of materials to be used, size the components of the system, and answer questions about how the product is going to be fabricated. You will consult pertinent codes and standards and make sure that your design will be in compliance with these codes and standards. Recall that we discussed engineering codes and standards in Chapter 12.

Step 6: Evaluation

Analyze the problem in more detail. You may have to identify critical design parameters and consider their influence in your final design. At this stage, you need to make sure that all calculations are performed correctly. If there are some uncertainties in your analysis, you must perform experimental investigation. When possible, working models must be created and tested. At this stage of the design procedure, the best solution must be identified from alternatives. Details of how the product is to be fabricated must be fully worked out.

Step 7: Optimization

Optimization means minimization or maximization. There are two broad types of design: a functional design and an optimized design. A functional design is one that meets all of the preestablished design requirements but allows for improvement to be made in certain areas. To better understand the concept of a functional design, we will consider an example. Let us assume that we are to design a 10-foot-tall ladder to support a person who weighs 1335 newtons (300 pounds) with a certain factor of safety. We will come up with a design that consists of a steel ladder that is 10 feet tall and can safely support the load of 1335 N (300 lb) at each step. The ladder would cost a certain amount of money. This design would satisfy all of the requirements, including those of strength and size, and thus constitutes a functional design. Before we can consider improving our design, we need to ask ourselves what criterion we should use to optimize the design. Design optimization is always based on some particular criterion, such as cost, strength, size, weight, reliability, noise, or performance. If we use the weight as an optimization criterion, then the problem becomes one of minimizing the weight of the ladder without jeopardizing its strength. For example, we may consider making the ladder from aluminum. We would also perform stress analysis on the new ladder to see if we could remove material from certain sections of the ladder without compromising the loading and safety requirements.

Another important fact to keep in mind is that optimizing individual components of an engineering system does not necessarily lead to an optimized system. Refer to Chapter 3 to refresh your memory about what we mean by a system and its components. For example, consider a thermal-fluid system such as a refrigerator. Optimizing the individual

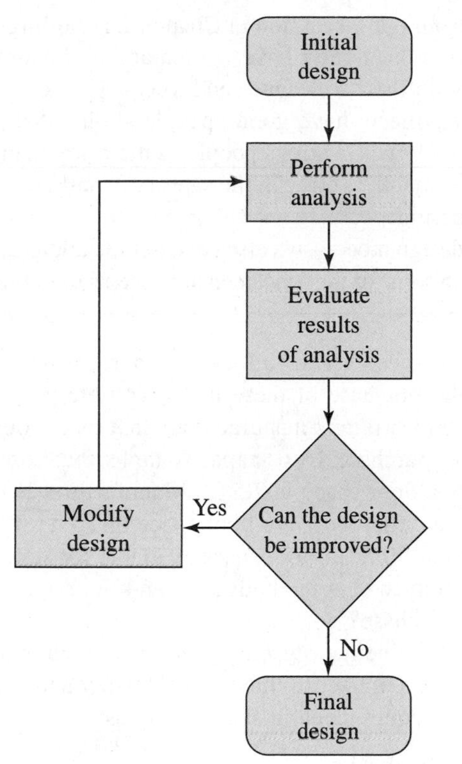

Figure 16.1 An optimization procedure

components independently—such as the compressor, the evaporator, or the condenser—with respect to some criterion does not lead to an optimized overall system.

Traditionally, improvements in a design come from the process of starting with an initial design, performing an analysis, looking at results, and deciding whether or not we can improve the initial design. This procedure is shown in Figure 16.1. In the past few decades, the optimization process has grown into a discipline that ranges from linear to nonlinear programming techniques. As is the case with any discipline, the optimization field has its own terminology. There are advanced classes that you can take to learn more about the design optimization process.

Step 8: Presentation

Now that you have a final solution, you need to communicate your solution to the client, who may be your boss, another group within your company, or an outside customer. You may have to prepare not only an oral presentation but also a written report. As we said in Chapter 2, engineers are required to write reports. Depending on the size of the project, these reports might be lengthy, detailed technical reports containing graphs, charts, and engineering drawings, or they may take the form of a brief memorandum or executive summaries. A reminder again that although we have listed the presentation as step 8 of the design process, quite often engineers are required to give oral and written progress reports on a regular time basis to various groups. Consequently, presentation could well be an integral part of many other design steps.

Finally, recall from our discussion in Chapter 2 regarding the attributes of good engineers, we said that good engineers have written and oral communication skills that equip them to work well with their colleagues and to convey their expertise to a wide range of clients. Moreover, engineers have good "people skills" that allow them to interact and communicate effectively with various people in their organization. For example, they are able to communicate equally well with the sales and marketing experts and with their own colleagues in engineering.

In step 7 of the design process, we discussed optimization. Let us now use the next example to introduce you to some of the fundamental concepts of optimization and its terminology.

EXAMPLE 16.1

Assume that you have been asked to look into purchasing some storage tanks for your company, and for the purchase of these tanks, you are given a budget of $1680. After some research, you find two tank manufacturers that meet your requirements. From Manufacturer A, you can purchase 16-ft^3-capacity tanks that cost $120 each. Moreover, the type of tank requires a floor space of 7.5 ft^2. Manufacturer B makes 24-ft^3-capacity tanks that cost $240 each and that require a floor space of 10 ft^2. The tanks will be placed in a section of a lab that has 90 ft^2 of floor space available for storage. You are looking for the greatest storage capacity within the budgetary and floor-space limitations. How many of each tank must you purchase?

First, we need to define the *objective function*, which is the function that we will attempt to minimize or maximize. In this example, we want to maximize storage capacity. We can represent this requirement mathematically as

$$\text{maximize } Z = 16\,x_1 + 24\,x_2 \qquad (16.1)$$

subject to the following constraints:

$$120\,x_1 + 240\,x_2 \leq 1680 \qquad (16.2)$$

$$7.5\,x_1 + 10\,x_2 \leq 90 \qquad (16.3)$$

$$x_1 \geq 0 \qquad (16.4)$$

$$x_2 \geq 0 \qquad (16.5)$$

In Equation (16.1), Z is the objective function, while the variables x_1 and x_2 are called *design variables*, and represent the number of 16-ft^3-capacity tanks and the number of 24-ft^3-capacity tanks, respectively. The limitations imposed by the inequalities in (16.2)–(16.5) are referred to as a set of *constraints*. Although there are specific techniques that deal with solving linear programming problems (the objective function and constraints are linear), we will solve this problem graphically to illustrate some additional concepts.

Let us first review how you would plot the regions given by the inequalities. For example, to plot the region as given by the linear inequality $120\,x_1 + 240\,x_2 \leq 1680$, we must first plot the line $120\,x_1 + 240\,x_2 = 1680$ and then determine which side of the line represents the region. For example, after plotting the line $120\,x_1 + 240\,x_2 = 1680$, we can test points $x_1 = 0$ and $x_2 = 0$ to see if they fall inside the inequality region; because substitution of these points into the inequality satisfies the inequality, that is, $(120)(0) + (240)(0) \leq 1680$, the shaded region represents the given inequality. Note that if we were to substitute a set of points outside the region, such as $x_1 = 15$

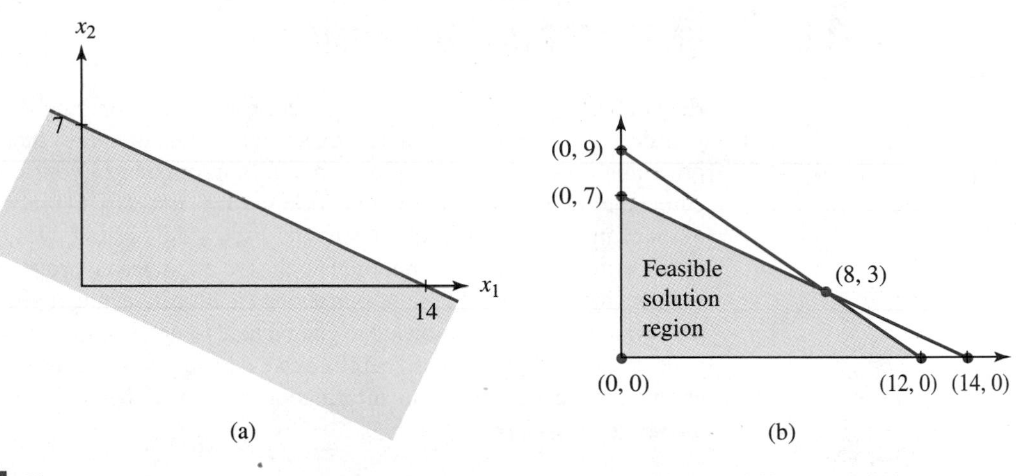

Figure 16.2 (a) The region as given by the linear inequality $120\,x_1 + 240\,x_2 \leq 1680$. (b) The feasible solution for Example 16.1.

and $x_2 = 0$, into the inequality, we would find that the inequality is not satisfied (see Figure 16.2(a)). The inequalities in (16.2)–(16.5) are plotted in Figure 16.2(b).

The shaded region shown in Figure 16.2(b) is called a *feasible solution region*. Every point within this region satisfies the constraints. However, our goal is to maximize the objective function given by Equation (16.1). Therefore, we need to move the objective function over the feasible region and determine where its value is maximized. It can be shown that the maximum value of the objective function will occur at one of the corner points of the feasible region. By evaluating the objective function at the corner points of the feasible region, we see that the maximum value occurs at $x_1 = 8$ and $x_2 = 3$. This evaluation is shown in Table 16.1.

Thus, we should purchase eight of the 16-ft^3-capacity tanks from Manufacturer A and three of the 24-ft^3-capacity tanks from Manufacturer B to maximize the storage capacity within the given constraints.

It is worth noting here that most of you will take specific design classes during the next four years. In fact, most of you will work on a relatively comprehensive design project during your senior year. Therefore, you will learn more in depth about design process and its application specific to your discipline. For now our intent has been to introduce you to the design process, but keep in mind that more design is coming your way.

TABLE 16.1 Values of the Objective Function at the Corner Points of the Feasible Region

Corner Points (x_1, x_2)	Value of $Z = 16\,x_1 + 24\,x_2$
0, 0	0
0, 7	168
12, 0	192
8, 3	200 (max.)

16.2 ENGINEERING ECONOMICS

Economic factors always play important roles in engineering design decision making. If you design a product that is too expensive to manufacture, then it can not be sold at a price that consumers can afford and still be profitable to your company. The fact is that companies design products and provide services not only to make our lives better but also to make money! In this section, we will discuss the basics of engineering economics. The information provided here not only applies to engineering projects but can also be applied to financing a car or a house or borrowing from or investing money in banks. Some of you may want to apply the knowledge gained here to determine your student loan payments or your credit card payments. Therefore, we advise you to develop a good understanding of engineering economics; the information presented here could help you manage your money more wisely.

Cash Flow Diagrams

Cash flow diagrams are visual aids that show the flow of costs and revenues over a period of time. Cash flow diagrams show *when the cash flow occurs, the cash flow magnitude, and whether the cash flow is out of your pocket (cost) or into your pocket (revenue).* It is an important visual tool that shows the timing, the magnitude, and the direction of cash flow. To shed more light on the concept of the cash flow diagram, imagine that you are interested in purchasing a new car. Being a first-year engineering student, you may not have too much money in your savings account at this time; for the sake of this example, let us say that you have \$1200 to your name in a savings account. The car that you are interested in buying costs \$15,500; let us further assume that including the sales tax and other fees, the total cost of the car would be \$16,880. Assuming you can afford to put down \$1000 as a down payment for your new shiny car, you ask your bank for a loan. The bank decides to lend you the remainder, which is \$15,880 at 8% interest. You will sign a contract that requires you to pay \$315.91 every month for the next 5 years. You will soon learn how to calculate these monthly payments, but for now let us focus on how to draw the cash flow diagram. The cash flow diagram for this activity is shown in Figure 16.3. Note in Figure 16.3 the direction of the arrows representing the money given to you by the bank and the payments that you must make to the bank over the next 5 years (60 months).

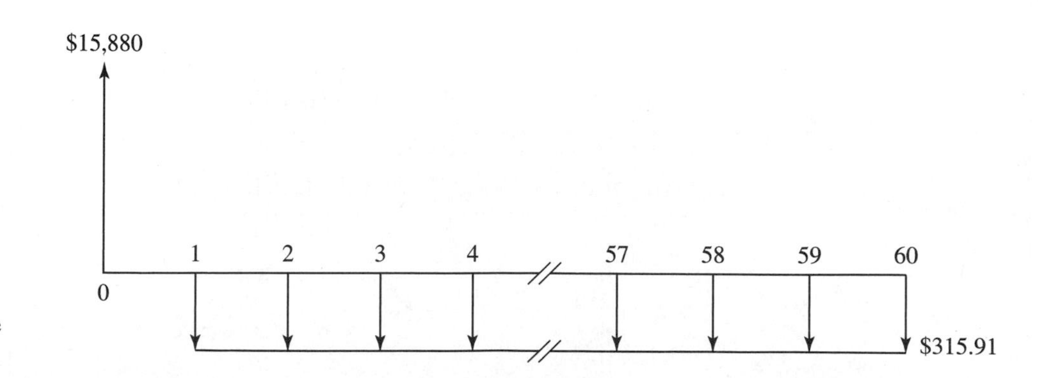

Figure 16.3 A cash flow diagram for borrowed money and the monthly payments

EXAMPLE 16.2

Draw the cash flow diagram for an investment that includes purchasing a machine that costs $50,000 with a maintenance and operating cost of $1000 per year. It is expected that the machine will generate revenues of $15,000 per year for 5 years. The expected salvage value of the machine at the end of 5 years is $8000.

The cash flow diagram for the investment is shown in Figure 16.4. Again, note the directions of arrows in the cash flow diagram. We have represented the initial cost of $50,000 and the maintenance cost by arrows pointing down, while the revenue and the salvage value of the machine are shown by arrows pointing up.

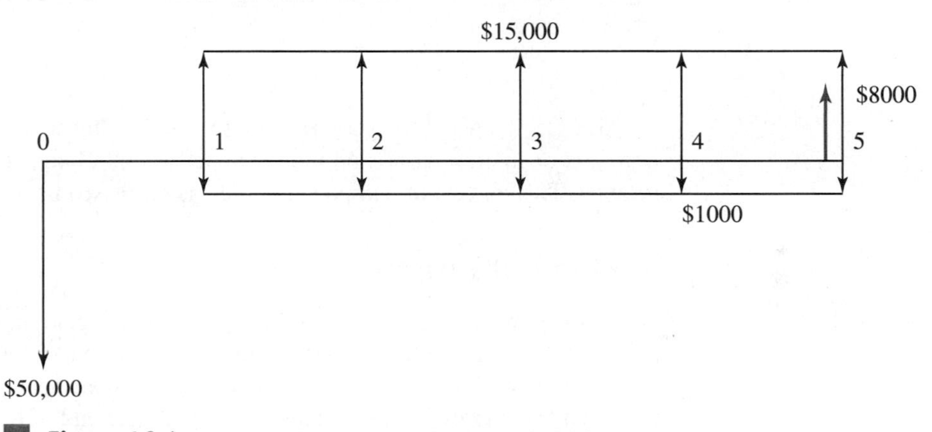

Figure 16.4 The cash flow diagram for Example 16.2.

Simple Interest

Interest is the extra money in addition to the borrowed amount that one must pay for the purpose of having access to the borrowed money. ***Simple interest*** is the interest that would be paid only on the initial borrowed or deposited amount. For simple interest, the interest accumulated on the principle each year will not collect interest itself. Only the initial principal will collect interest. For example, if you deposit $100 in a bank at 6% simple interest, after 6 years you will have $136 in your account. In general, if you deposit the amount P at a rate of i% for a period of n years, then the total future value F of the P at the end of the nth year is given by

$$F = P + (P)(i)(n) = P(1 + ni) \tag{16.6}$$

EXAMPLE 16.3

Compute the future value of a $1500 deposit, after 8 years, in an account that pays a simple interest rate of 7%. How much interest will be paid to this account?

You can determine the future value of the deposited amount using Equation (16.6), which results in

$$F = P(1 + ni) = 1500[1 + 8(0.07)] = \$2340$$

And the total interest to be paid to this account is

$$\textit{interest} = (P)(n)(i) = (1500)(8)(0.07) = \$840$$

TABLE 16.2 The Effect of Compounding Interest

Year	Balance at the Beginning of the Year (dollars and cents)	Interest for the Year at 6% (dollars and cents)	Balance at the End of the Year, Including the Interest (dollars and cents)
1	100.00	6.00	106.00
2	106.00	6.36	112.36
3	112.36	6.74	119.10
4	119.10	7.14	126.24
5	126.24	7.57	133.81
6	133.81	8.02	141.83

Simple interests are very rare these days! Almost all interest charged to borrow accounts or interest earned on money deposited in a bank is computed using ***compound interest***. The concept of compound interest is discussed next.

Compound Interest

Under the compounding interest scheme, the interest paid on the initial principal will also collect interest. To better understand how the compound interest earned or paid on a principal works, consider the following example. Imagine that you put \$100 in a bank that pays you 6% interest compounding annually. At the end of the first year (or the beginning of the second year) you will have \$106.00 in your bank account. You have earned interest in the amount of \$6.00 during the first year. However, the interest earned during the second year is determined by (\$106.00)(0.06) = \$6.36. That is because the \$6 interest of the first year also collects 6% interest, which is 36 cents itself. Thus, the total interest earned during the second year is \$6.36, and the total amount available in your account at the end of the second year is \$112.36. Computing the interest and the total amount for the third, fourth, fifth and the sixth year in a similar fashion will lead to \$141.83 in your account at the end of the sixth year. Refer to Table 16.2 for detailed calculations. Note the difference between \$100 invested at 6% simple interest and 6% interest compounding annually for a duration of 6 years. For the simple interest case, the total interest earned, after 6 years, is \$36.00, whereas the total interest accumulated under the annual compounding case is \$41.83 for the same duration.

Future Worth of a Present Amount

Now we will develop a general formula that you can use to compute the future value F of any present amount (principal) P, after n years collecting i% interest compounding annually. The cash flow diagram for this situation is shown in Figure 16.5. In order to demonstrate, step-by-step, the compounding effect of the interest each year, Table 16.3 has been developed. As shown in Table 16.3, starting with the principal P, at the end of the first year we will have $P + Pi$ or $P(1 + i)$. During the second year, the $P(1 + i)$ collects interest in an amount of $P(1 + i)i$, and by adding the interest to the $P(1 + i)$ amount that we started with in the second year, we will have a total amount of $P(1 + i) + P(1 + i)i$. Factoring out the $P(1 + i)$ term, we will have $P(1 + i)^2$ dollars at the end of the second year. Now

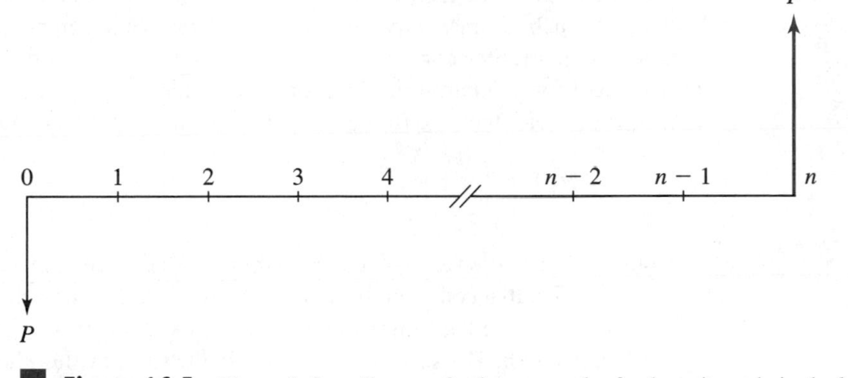

Figure 16.5 The cash flow diagram for future worth of a deposit made in the bank today

by following Table 16.3 you can see how the interest earned and the total amount are computed for the third, fourth, fifth, . . . , and the nth year. Consequently, you can see that the relationship between the present worth P and the future value F of an amount collecting $i\%$ interest compounding annually after n years is given by

$$F = P(1 + i)^n \tag{16.7}$$

TABLE 16.3 The Relationship Between Present Value P and the Future Value F

Year	Balance at the Beginning of the Year	Interest for the Year	Balance at the End of the Year, Including the Interest
1	P	$(P)(i)$	$P + (P)(i) = P(1 + i)$
2	$P(1 + i)$	$P(1 + i)(i)$	$P(1 + i) + P(1 + i)(i) = P(1 + i)^2$
3	$P(1 + i)^2$	$P(1 + i)^2(i)$	$P(1 + i)^2 + P(1 + i)^2(i) = P(1 + i)^3$
4	$P(1 + i)^3$	$P(1 + i)^3(i)$	$P(1 + i)^3 + P(1 + i)^3(i) = P(1 + i)^4$
5	$P(1 + i)^4$	$P(1 + i)^4(i)$	$P(1 + i)^4 + P(1 + i)^4(i) = P(1 + i)^5$
.	....	....	....
n	$P(1 + i)^{n-1}$	$P(1 + i)^{n-1}(i)$	$P(1 + i)^{n-1} + P(1 + i)^{n-1}(i) = P(1 + i)^n$

EXAMPLE 16.4

Compute the future value of a \$1500 deposit made today, after 8 years, in an account that pays an interest rate of 7% that compounds annually. How much interest will be paid to this account?

The future value of the \$1500 deposit is computed by substituting in Equation (16.7) for P, i, and n, which results in the amount that follows:

$$F = P(1 + i)^n = 1500(1 + .07)^8 = \$2577.27$$

The total interest earned during the 8-year life of this account is determined by calculating the difference between the future value and the present deposit value.

$$\textit{interest} = \$2577.27 - \$1500 = \$1077.27$$

Many financial institutions pay interest that compounds more than once a year. For example, a bank may pay you an interest rate that compounds semiannually (twice a year), or quarterly (four times a year), or monthly (12 periods a year). If the principal P is deposited for a duration of n years and the interest given is compounded m periods (or m times) per year, then the future value F of the principal P is determined from

$$F = P\left(1 + \frac{i}{m}\right)^{nm} \tag{16.8}$$

EXAMPLE 16.5

Compute the future value of a \$1500 deposit, after 8 years, in an account that pays an interest rate of 7% that compounds monthly. How much interest will be paid to this account?

To determine the future value of the \$1500 deposit, we substitute in Equation (16.8) for P, i, m and n. The substitution results in the future value shown next.

$$F = 1500\left(1 + \frac{0.07}{12}\right)^{(8)(12)} = 1500\left(1 + \frac{0.07}{12}\right)^{96} = \$2621.73$$

And the total interest is

$$interest = \$2621.73 - \$1500 = \$1121.73$$

The results of Examples 16.3, 16.4, and 16.5 are compared and summarized in Table 16.4. Note the effects of simple interest, interest compounding annually, and interest compounding monthly on the total future value of the \$1500 deposit.

Effective Interest Rate

If you deposit \$100 in a savings account, at 6% compounding monthly, then, using Equation (16.8), at the end of 1 year you will have \$106.16 in your account. The \$6.16 earned during the first year is higher than the stated 6% interest, which could be understood as \$6 for a \$100 deposit over a period of 1 year. In order to avoid confusion, the stated or the quoted interest rate is called the ***nominal interest rate***, and the actual earned interest rate is called the ***effective interest rate***. The relationship between the nominal rate, i, and the effective rate, i_{eff}, is given by

$$i_{eff} = \left(1 + \frac{i}{m}\right)^{m} - 1 \tag{16.9}$$

where m represents the number of compounding periods per year. To better understand the compounding effect of interest, let us see what happens if we deposit \$100 in an

TABLE 16.4 Comparison of Results for Examples 16.3, 16.4, and 16.5

Example Number	Principal (dollars)	Interest Rate	Duration (years)	Future Value (dollars and cents)	Interest Earned (dollars and cents)
Example 16.3	1500	7% simple	8	2340.00	840.00
Example 16.4	1500	7% compounding annually	8	2577.27	1077.27
Example 16.5	1500	7% compounding monthly	8	2621.73	1121.73

TABLE 16.5 The Effect of the Frequency of Interest Compounding Periods

Compounding Period	Total Number of Compounding Periods	Total Amount after 1 Year (dollars and cents)	Interest (dollars and cents)	Effective Interest Rate
Annually	1	$100(1 + 0.06) = 106.00$	6.00	6%
Semiannually	2	$100\left(1 + \frac{0.06}{2}\right)^2 = 106.09$	6.09	6.09%
Quarterly	4	$100\left(1 + \frac{0.06}{4}\right)^4 = 106.13$	6.13	6.13%
Monthly	12	$100\left(1 + \frac{0.06}{12}\right)^{12} = 106.16$	6.16	6.16%
Daily	365	$100\left(1 + \frac{0.06}{365}\right)^{365} = 106.18$	6.18	6.18%

account for a year based on one of the following quoted interests: 6% compounding annually, 6% semiannually, 6% quarterly, 6% monthly, and 6% daily. Table 16.5 shows the difference among these compounding periods, the total amount of money at the end of 1 year, the interest earned, and the effective interest rates for each case.

When comparing the five different interest compounding frequencies, the difference in the interests earned on a $100 investment, over a period of a year, may not seem much to you, but as the principal and the time of deposit are increased this value becomes significant. To better demonstrate the effect of principal and time of deposit, consider the following example.

EXAMPLE 16.6

Determine the interest earned on $5000 deposited in a savings account, for 10 years, based on one of the following quoted interest rates: 6% compounding annually, semiannually, quarterly, monthly, and daily. The solution to this problem is presented in Table 16.6.

TABLE 16.6 The Solution of Example 16.6

Compounding Period	Total Number of Compounding Periods	Total Future Amount Using Eq. (16.8) (dollars and cents)	Interest (dollars and cents)
Annually	10	$5000(1 + 0.06)^{10} = 8954.23$	3954.23
Semiannually	20	$5000\left(1 + \frac{0.06}{2}\right)^{20} = 9030.55$	4030.55
Quarterly	40	$5000\left(1 + \frac{0.06}{4}\right)^{40} = 9070.09$	4070.09
Monthly	120	$5000\left(1 + \frac{0.06}{12}\right)^{120} = 9096.98$	4096.98
Daily	3650	$5000\left(1 + \frac{0.06}{365}\right)^{3650} = 9110.14$	4110.14

EXAMPLE 16.7

Determine the effective interest rates corresponding to the nominal rates: (a) 7% compounding monthly, (b) 16.5% compounding monthly, (c) 6% compounding semiannually, (d) 9% compounding quarterly.

We can compute the i_{eff} for each case by substituting for i and m in Equation (16.9).

$$\text{(a)}\ i_{eff} = \left(1 + \frac{i}{m}\right)^m - 1 = \left(1 + \frac{0.07}{12}\right)^{12} - 1 = 0.0722 \text{ or } 7.22\%$$

$$\text{(b)}\ i_{eff} = \left(1 + \frac{0.165}{12}\right)^{12} - 1 = 0.1780 \text{ or } 17.80\%$$

$$\text{(c)}\ i_{eff} = \left(1 + \frac{0.06}{2}\right)^{2} - 1 = 0.0609 \text{ or } 6.09\%$$

$$\text{(d)}\ i_{eff} = \left(1 + \frac{0.09}{4}\right)^{4} - 1 = 0.0930 \text{ or } 9.30\%$$

Present Worth of a Future Amount

Let us now consider the following situation. You would like to have \$2000 available to you for a down payment on a car when you graduate from college in, say, 5 years. How much money do you need to put in a certificate of deposit (CD) with an interest rate of 6.5% (compounding annually) today? The relationship between the future and present value was developed earlier and is given by Equation (16.7). Rearranging Equation (16.7), we have

$$P = \frac{F}{(1 + i)^n} \tag{16.10}$$

and substituting in Equation (16.10) for the future value F, the interest rate i, and the period n, we have

$$P = \frac{2000}{(1 + 0.065)^5} = \$1459.76$$

This may be a relatively large sum to put aside all at once, especially for a first-year engineering student. A more realistic option would be to put aside some money each year. Then the question becomes, how much money do you need to put aside every year for the next 5 years at the given interest rate to have that \$2000 available to you at the end of the fifth year? To answer this question, we need to develop the formula that deals with a series of payments or series of deposits. This situation is discussed next.

Present Worth of Series Payment or Annuity

In this section, we will first formulate the relationship between a present lump sum, P, and future uniform series payments, A, and then from that relationship we will develop the formula that relates the uniform series payments A to a future lump sum F. This approach is much easier to follow as you will see. To derive these relationships, let us first consider

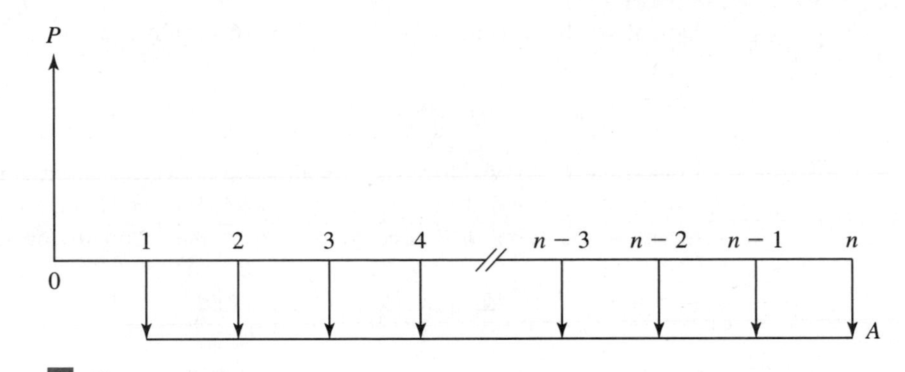

Figure 16.6 The cash flow diagram for a borrowed sum of money and its equivalent series payments

a situation where we have borrowed some money, denoted by P, at an annual interest rate i from a bank, and we are planning to pay the loan yearly, in equal amounts A, in n years, as shown in Figure 16.6.

To obtain the relationship between P and A, we will treat each future payment separately and relate each payment to its present equivalent value using Equation (16.10); we then add all the resulting terms together. This approach leads to the following relationship:

$$P = \frac{A}{(1+i)} + \frac{A}{(1+i)^2} + \frac{A}{(1+i)^3} + \cdots + \frac{A}{(1+i)^{n-1}} + \frac{A}{(1+i)^n} \tag{16.11}$$

As you can see, Equation (16.11) is not very user-friendly, so we need to simplify it somehow. What if we were to multiply both sides of Equation (16.11) by the term $(1 + i)$? This operation results in the following relationship:

$$P(1+i) = A + \frac{A}{(1+i)} + \frac{A}{(1+i)^2} + \frac{A}{(1+i)^3} + \cdots + \frac{A}{(1+i)^{n-2}} + \frac{A}{(1+i)^{n-1}} \tag{16.12}$$

Now if we subtract Equation (16.11) from Equation (16.12), we have

$$P(1+i) - P = A + \frac{A}{(1+i)} + \frac{A}{(1+i)^2} + \frac{A}{(1+i)^3} + \cdots + \frac{A}{(1+i)^{n-2}} + \frac{A}{(1+i)^{n-1}}$$

$$- \left[\frac{A}{(1+i)} + \frac{A}{(1+i)^2} + \frac{A}{(1+i)^3} + \cdots + \frac{A}{(1+i)^{n-1}} + \frac{A}{(1+i)^n}\right] \tag{16.13}$$

Simplifying the right-hand side of Equation (16.13) leads to the following relationship:

$$P(1+i) - P = A - \frac{A}{(1+i)^n} \tag{16.13b}$$

And after simplifying the left-hand side of Equation (16.13), we have

$$P(i) = \frac{A((1+i)^n - 1)}{(1+i)^n} \tag{16.13c}$$

Now if we divide both sides of Equation (16.13c) by i, we have

$$P = A\left[\frac{(1+i)^n - 1}{i(1+i)^n}\right] \tag{16.14}$$

Equation (16.14) establishes the relationship between the present value of a lump sum P and its equivalent uniform series payments A. We can also rearrange Equation (16.14), to represent A in terms of P directly, as given by the following formula:

$$A = \frac{P(i)(1+i)^n}{(1+i)^n - 1} = P\left[\frac{(i)(1+i)^n}{(1+i)^n - 1}\right] \tag{16.15}$$

To develop a formula for computing the future worth of a series of uniform payments, we begin with the relationship between the present worth and the future worth, Equation (16.7), and then we substitute for P in Equation (16.7) in terms of A, using Equation (16.14). This procedure is demonstrated, step-by-step, next. The relation between a present value and a future value is given by Equation (16.7):

$$F = P(1+i)^n \tag{16.7}$$

And the relationship between the present worth and a uniform series is given by Equation (16.14):

$$P = A\left[\frac{(1+i)^n - 1}{i(1+i)^n}\right] \tag{16.14}$$

Substituting into Equation (16.7) for P in terms of A using Equation (16.14), we have

$$F = P(1+i)^n = \overbrace{A\left[\frac{(1+i)^n - 1}{i(1+i)^n}\right]}^{P}(1+i)^n \tag{16.16}$$

Simplifying Equation (16.16) results in the direct relationship between the future worth F and the uniform payments or deposits A, which follows:

$$F = A\left[\frac{(1+i)^n - 1}{i}\right] \tag{16.17}$$

And by rearranging Equation (16.17), we can obtain a formula for A in terms of future worth F:

$$A = F\left[\frac{i}{(1+i)^n - 1}\right] \tag{16.18}$$

Now that we have all the necessary tools, we turn our attention to the question we asked earlier about how much money you need to put aside every year for the next 5 years to have \$2000 for the down payment of your car when you graduate. Recall that the interest rate is 6.5% compounding annually. The annual deposits are calculated from Equation (16.18), which leads to the following amount:

$$A = 2000\left[\frac{0.065}{(1+0.065)^5 - 1}\right] = \$351.26$$

Putting aside \$351.26 in a bank every year for the next 5 years may be more manageable than depositing a lump sum of \$1459.76 today, especially if you don't currently have access to that large a sum!

It is important to note that Equations (16.14), (16.15), (16.17), and (16.18) apply to a situation wherein the uniform series of payments or revenues *occur annually*. Well, the next question is, how do we handle situations where the payments are made monthly? For example, a car or a house loan payments occur monthly. Let us now modify our findings by considering the relationship between present value P and uniform series payments or revenue A that occur more than once a year at the same frequency as the frequency of compounding interest per year. For this situation, Equation (16.14) is modified to incorporate the frequency of compounding interest per year, m, in the following manner:

$$P = A\left[\frac{\left(1+\frac{i}{m}\right)^{nm} - 1}{\frac{i}{m}\left(1+\frac{i}{m}\right)^{nm}}\right] \tag{16.19}$$

Note that in order to obtain Equation (16.19), we simply substituted in Equation (16.14) for i, i/m, and for n, nm. Equation (16.19) can be rearranged to solve for A in terms of P according to

$$A = P\left[\frac{\left(\frac{i}{m}\right)\left(1+\frac{i}{m}\right)^{nm}}{\left(1+\frac{i}{m}\right)^{nm} - 1}\right] \tag{16.20}$$

Similarly, Equations (16.17) and (16.18) can be modified for situations where A occurs more than once a year, at the same frequency as the compounding interest, leading to the following relationship:

$$F = A\left[\frac{\left(1+\frac{i}{m}\right)^{(m)(n)} - 1}{\frac{i}{m}}\right] \tag{16.21}$$

$$A = F\left[\frac{\frac{i}{m}}{\left(1+\frac{i}{m}\right)^{(m)(n)} - 1}\right] \tag{16.22}$$

Finally, when the frequency of uniform series is different from the frequency of compounding interest, i_{eff} must first be calculated to match the frequency of the uniform series.

EXAMPLE 16.8

Let us return to the question we asked earlier about how much money you need to put aside for the next 5 years to have \$2000 for the down payment on your car when you graduate. Now consider the situation where you make your deposits every month, and the interest rate is 6.5% compounding monthly.

The deposits are calculated from Equation (16.22), which leads to the following:

$$A = F\left[\frac{\frac{i}{m}}{\left(1+\frac{i}{m}\right)^{mn}-1}\right] = 2000\left[\frac{\frac{0.065}{12}}{\left(1+\frac{0.065}{12}\right)^{(12)(5)}-1}\right] = \$28.29$$

Putting aside $28.29 in the bank every month for the next 5 years is even more manageable than depositing $351.26 in a bank every year for the next 5 years, and it is certainly more manageable than depositing a lump sum of $1459.76 in the bank today!

EXAMPLE 16.9

Determine the monthly payments for a 5-year, $10,000 loan at an interest rate of 8% compounding monthly.

To calculate the monthly payments, we use Equation (16.15).

$$A = P\left[\frac{\left(\frac{i}{m}\right)\left(1+\frac{i}{m}\right)^{nm}}{\left(1+\frac{i}{m}\right)^{nm}-1}\right] = 10000\left[\frac{\left(\frac{0.08}{12}\right)\left(1+\frac{0.08}{12}\right)^{60}}{\left(1+\frac{0.08}{12}\right)^{60}-1}\right] = \$202.76$$

Summary of Engineering Economics Analysis

The engineering economics formulas that we have developed so far are summarized in Tables 16.7 and 16.8. The definitions of the terms in the formulas are given here:

P = present worth, or present cost—lump sum ($)
F = future worth, or future cost—lump sum ($)
A = uniform series payment, or uniform series revenue ($)

TABLE 16.7 A Summary of Formulas for Situations when i Compounds Annually and the Uniform Series A Occurs Annually

To Find	Given	Use This Formula	Interest–Time Factor
F	P	$F = P(1+i)^n$	$(F/P, i, n) = (1+i)^n$
P	F	$P = \frac{F}{(1+i)^n}$	$(P/F, i, n) = \frac{1}{(1+i)^n}$
P	A	$P = A\left[\frac{(1+i)^n-1}{i(1+i)^n}\right]$	$(P/A, i, n) = \left[\frac{(1+i)^n-1}{i(1+i)^n}\right]$
A	P	$A = P\left[\frac{(i)(1+i)^n}{(1+i)^n-1}\right]$	$(A/P, i, n) = \left[\frac{(i)(1+i)^n}{(1+i)^n-1}\right]$
F	A	$F = A\left[\frac{(1+i)^n-1}{i}\right]$	$(F/A, i, n) = \left[\frac{(1+i)^n-1}{i}\right]$
A	F	$A = F\left[\frac{(i)}{(1+i)^n-1}\right]$	$(A/F, i, n) = \left[\frac{(i)}{(1+i)^n-1}\right]$

TABLE 16.8 A Summary of Formulas for Situations when i Compounds m Times per Year and the Uniform Series A Occurs at the Same Frequency

To Find	Given	Use This Formula
i_{eff}	i	$i_{eff} = \left(1 + \frac{i}{m}\right)^m - 1$
F	P	$F = P\left(1 + \frac{i}{m}\right)^{nm}$
P	F	$P = \frac{F}{\left(1 + \frac{i}{m}\right)^{nm}}$
P	A	$P = A\left[\frac{\left(1 + \frac{i}{m}\right)^{nm} - 1}{\frac{i}{m}\left(1 + \frac{i}{m}\right)^{nm}}\right]$
A	P	$A = P\left[\frac{\left(\frac{i}{m}\right)\left(1 + \frac{i}{m}\right)^{nm}}{\left(1 + \frac{i}{m}\right)^{nm} - 1}\right]$
F	A	$F = A\left[\frac{\left(1 + \frac{i}{m}\right)^{(m)(n)} - 1}{\frac{i}{m}}\right]$
A	F	$A = F\left[\frac{\frac{i}{m}}{\left(1 + \frac{i}{m}\right)^{(m)(n)} - 1}\right]$

i = nominal interest rate
i_{eff} = effective interest rate
n = number of years
m = number of interest compounding periods per year

The interest–time factors shown in the fourth column of Table 16.7 are used as shortcuts to avoid writing long formulas when evaluating equivalent values of various cash flow occurrences. For example, when evaluating the series payment equivalence of a present principal, instead of writing,

$$A = P\left[\frac{(i)(1 + i)^n}{(1 + i)^n - 1}\right]$$

we write $A = P(A/P, i, n)$, where, of course,

$$(A/P, i, n) = \left[\frac{(i)(1 + i)^n}{(1 + i)^n - 1}\right]$$

In this example, the $(A/P,i,n)$ term is called the *interest–time factor*, and it reads A given P at $i\%$ interest rate, for a duration of n years. It is used to find A, when the present principal value P is given, by multiplying P by the value of the interest–time factor $(A/P,i,n)$. As an example, the numerical values of interest–time factors for $i = 8\%$ are calculated and shown in Table 16.9.

TABLE 16.9 The Interest–Time Factors for $i = 8\%$

n	*(F/P, i, n)*	*(P/F, i, n)*	*(P/A, i, n)*	*(A/P, i, n)*	*(F/A, i, n)*	*(A/F, i, n)*
1	1.08000000	0.92592593	0.92592593	1.08000000	1.00000000	1.00000000
2	1.16640000	0.85733882	1.78326475	0.56076923	2.08000000	0.48076923
3	1.25971200	0.79383224	2.57709699	0.38803351	3.24640000	0.30803351
4	1.36048896	0.73502985	3.31212684	0.30192080	4.50611200	0.22192080
5	1.46932808	0.68058320	3.99271004	0.25045645	5.86660096	0.17045645
6	1.58687432	0.63016963	4.62287966	0.21631539	7.33592904	0.13631539
7	1.71382427	0.58349040	5.20637006	0.19207240	8.92280336	0.11207240
8	1.85093021	0.54026888	5.74663894	0.17401476	10.63662763	0.09401476
9	1.99900463	0.50024897	6.24688791	0.16007971	12.48755784	0.08007971
10	2.15892500	0.46319349	6.71008140	0.14902949	14.48656247	0.06902949
11	2.33163900	0.42888286	7.13896426	0.14007634	16.64548746	0.06007634
12	2.51817012	0.39711376	7.53607802	0.13269502	18.97712646	0.05269502
13	2.71962373	0.36769792	7.90377594	0.12652181	21.49529658	0.04652181
14	2.93719362	0.34046104	8.24423698	0.12129685	24.21492030	0.04129685
15	3.17216911	0.31524170	8.55947869	0.11682954	27.15211393	0.03682954
16	3.42594264	0.29189047	8.85136916	0.11297687	30.32428304	0.03297687
17	3.70001805	0.27026895	9.12163811	0.10962943	33.75022569	0.02962943
18	3.99601950	0.25024903	9.37188714	0.10670210	37.45024374	0.02670210
19	4.31570106	0.23171206	9.60359920	0.10412763	41.44626324	0.02412763
20	4.66095714	0.21454821	9.81814741	0.10185221	45.76196430	0.02185221
21	5.03383372	0.19865575	10.01680316	0.09983225	50.42292144	0.01983225
22	5.43654041	0.18394051	10.20074366	0.09803207	55.45675516	0.01803207
23	5.87146365	0.17031528	10.37105895	0.09642217	60.89329557	0.01642217
24	6.34118074	0.15769934	10.52875828	0.09497796	66.76475922	0.01497796
25	6.84847520	0.14601790	10.67477619	0.09367878	73.10593995	0.01367878
26	7.39635321	0.13520176	10.80997795	0.09250713	79.95441515	0.01250713
27	7.98806147	0.12518682	10.93516477	0.09144810	87.35076836	0.01144810
28	8.62710639	0.11591372	11.05107849	0.09048891	95.33882983	0.01048891
29	9.31727490	0.10732752	11.15840601	0.08961854	103.96593622	0.00961854
30	10.06265689	0.09937733	11.25778334	0.08882743	113.28321111	0.00882743
31	10.86766944	0.09201605	11.34979939	0.08810728	123.34586800	0.00810728
32	11.73708300	0.08520005	11.43499944	0.08745081	134.21353744	0.00745081
33	12.67604964	0.07888893	11.51388837	0.08685163	145.95062044	0.00685163
34	13.69013361	0.07304531	11.58693367	0.08630411	158.62667007	0.00630411
35	14.78534429	0.06763454	11.65456822	0.08580326	172.31680368	0.00580326
36	15.96817184	0.06262458	11.71719279	0.08534467	187.10214797	0.00534467
37	17.24562558	0.05798572	11.77517851	0.08492440	203.07031981	0.00492440
38	18.62527563	0.05369048	11.82886899	0.08453894	220.31594540	0.00453894

Continued

TABLE 16.9 The Interest–Time Factors for $i = 8\%$ (*continued*)

n	$(F/P, i, n)$	$(P/F, i, n)$	$(P/A, i, n)$	$(A/P, i, n)$	$(F/A, i, n)$	$(A/F, i, n)$
39	20.11529768	0.04971341	11.87858240	0.08418513	238.94122103	0.00418513
40	21.72452150	0.04603093	11.92461333	0.08386016	259.05651871	0.00386016
41	23.46248322	0.04262123	11.96723457	0.08356149	280.78104021	0.00356149
42	25.33948187	0.03946411	12.00669867	0.08328684	304.24352342	0.00328684
43	27.36664042	0.03654084	12.04323951	0.08303414	329.58300530	0.00303414
44	29.55597166	0.03383411	12.07707362	0.08280152	356.94964572	0.00280152
45	31.92044939	0.03132788	12.10840150	0.08258728	386.50561738	0.00258728
46	34.47408534	0.02900730	12.13740880	0.08238991	418.42606677	0.00238991
47	37.23201217	0.02685861	12.16426741	0.08220799	452.90015211	0.00220799
48	40.21057314	0.02486908	12.18913649	0.08204027	490.13216428	0.00204027
49	43.42741899	0.02302693	12.21216341	0.08188557	530.34273742	0.00188557
50	46.90161251	0.02132123	12.23348464	0.08174286	573.77015642	0.00174286

Additional values of interest-time factors for other interest rates can be created using Excel. Keep in mind that you can use those tables or other similar tables found in the back of most engineering economy text books to determine interest–time factors for interest rates that compound more frequently than once a year. To do so, however, you must first divide the quoted nominal interest rate i by the number of compounding frequency m and use the resulting number to pick the appropriate interest table to use. You must then multiply the number of years n by the number of compounding frequency m and use the outcome of n times m as the period when looking up interest–time factors. For example, if a problem states an interest rate of 18% compounding monthly for 4 years, you use the 1.5% interest table ($18/12 = 1.5$), and for the number of periods, you will use 48 ($4 \times 12 = 48$).

EXAMPLE 16.10

What is the equivalent present worth of the cash flow given in Figure 16.7? Put another way, how much money do you need to deposit in the bank today in order to be able to make the withdrawals shown? The interest rate is 8% compounding annually.

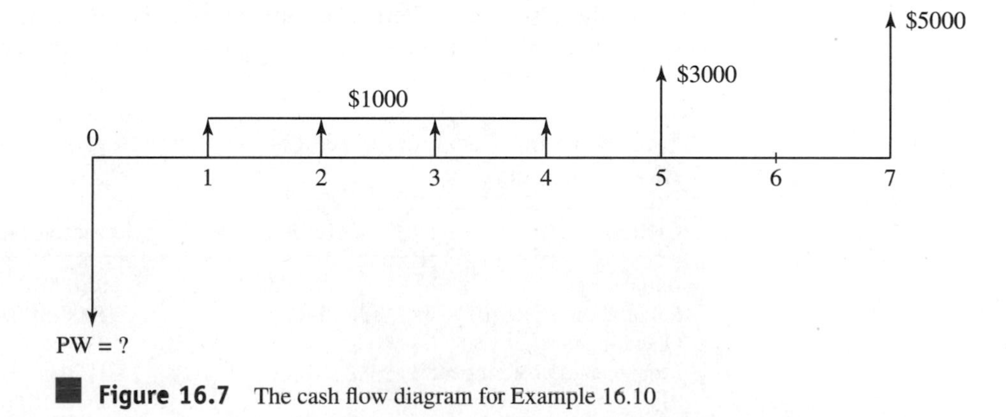

Figure 16.7 The cash flow diagram for Example 16.10

The present worth of the given cash flow is determined from

$$PW = 1000(P/A, 8\%, 4) + 3000(P/F, 8\%, 5) + 5000(P/F, 8\%, 7)$$

We can use Table 16.9 to look up the interest–time factor values, which leads to

$$(P/A, 8\%, 4) = 3.31212684$$

$$(P/F, 8\%, 5) = 0.68058320$$

$$(P/F, 8\%, 7) = 0.58349040$$

$$PW = (1000)(3.31212684) + (3000)(0.68058320) + (5000)(0.58349040)$$

$$PW = \$8271.32$$

Therefore, if today, you put aside $8271.32 in an account that pays 8% interest, you can withdraw $1000 in the next 4 years, and $3000 in 5 years, and $5000 in 7 years.

Choosing the Best Alternatives—Decision Making

Up to this point, we have been discussing general relationships that deal with money, time, and interest rates. Let us now consider the application of these relationships in an engineering setting. Imagine that you are assigned the task of choosing which air-conditioning unit to purchase for your company. After an exhaustive search, you have narrowed your selection to two alternatives, both of which have an anticipated 10 years of working life. Assuming an 8% interest rate, find the best alternative. Additional information is given in Table 16.10. The cash flow diagrams for each alternative are shown in Figure 16.8.

Here we will discuss three different methods that you can use to choose the best economical alternative from many options. The three methods are commonly referred to as (1) present worth or cost analysis, (2) annual worth or cost analysis, and (3) future worth or cost analysis. When these methods are applied to a problem, they all lead to the same conclusion. So in practice, you need only apply one of these methods to evaluate options; however, in order to show you the details of these procedures, we will apply all of these methods to the preceding problem.

Present Worth or Cost Analysis With this approach you compute the total present worth or the cost of each alternative and then pick the alternative with the lowest present cost or choose the alternative with the highest present worth or profit. To employ this

TABLE 16.10 Data to Be Used in Selection of an Air-Conditioning Unit

Criteria	Alternative A	Alternative B
Initial cost	$100,000	$85,000
Salvage value after 10 years	$10,000	$5000
Operating cost per year	$2500	$3400
Maintenance cost per year	$1000	$1200

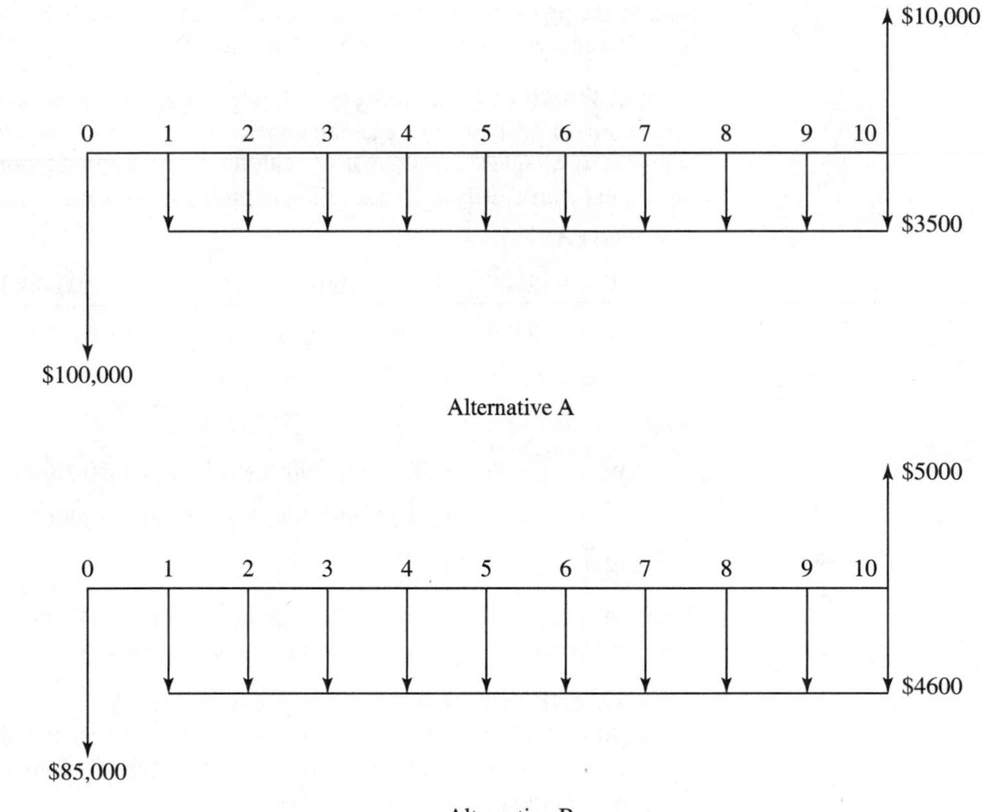

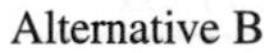

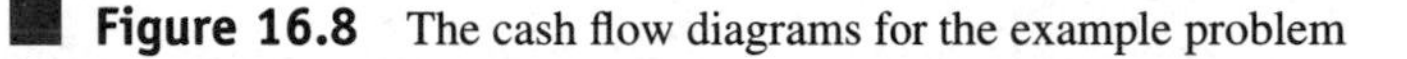
Figure 16.8 The cash flow diagrams for the example problem

method, you begin by calculating the equivalent present value of all cash flow. For the example problem mentioned, the application of the present worth analysis leads to:

Alternative A:

$$PW = -100{,}000 - (2500 + 1000)(P/A, 8\%, 10) + 10{,}000(P/F, 8\%, 10)$$

The time–interest factors for $i = 8\%$ are given in Table 16.9.

$$PW = -100{,}000 - (2500 + 1000)(6.71008140) + (10{,}000)(0.46319349)$$

$$PW = -118{,}853.35$$

Alternative B:

$$PW = -85{,}000 - (3400 + 1200)(P/A, 8\%, 10) + 5000(P/F, 8\%, 10)$$

$$PW = -85{,}000 - (3400 + 1200)(6.71008140) + 5000(0.46319349)$$

$$PW = -113{,}550.40$$

Note we have determined the equivalent present worth of all future cash flow, including the yearly maintenance and operating costs, and the salvage value of the air-conditioning unit. In the preceding analysis, the negative sign indicates cost, and because alternative B has a lower present cost, we choose alternative B.

Annual Worth or Cost Analysis Using this approach, we compute the equivalent annual worth or cost value of each alternative and then pick the alternative with the lowest annual cost or select the alternative with the highest annual worth or revenue. Applying the annual worth analysis to our example problem, we have

Alternative A:

$$AW = -(2500 + 1000) - 100{,}000(A/P, 8\%, 10) + 10{,}000(A/F, 8\%, 10)$$

$$AW = -(2500 + 1000) - (100{,}000)(0.14902949) + (10{,}000)(0.06902949)$$

$$AW = -17{,}712.65$$

Alternative B:

$$AW = -(3400 + 1200) - 85{,}000(A/P, 8\%, 10) + 5000(A/F, 8\%, 10)$$

$$AW = -(3400 + 1200) - 85{,}000(0.14902949) + 5000(0.06902949)$$

$$AW = -16{,}922.35$$

Note that using this method, we have determined the equivalent annual worth of all cash flow, and because alternative B has a lower annual cost, we choose alternative B.

Future Worth or Cost Analysis This approach is based on evaluating the future worth or cost of each alternative. Of course, you will then choose the alternative with the lowest future cost or pick the alternative with the highest future worth of profit. The future worth analysis of our example problem follows.

Alternative A:

$$FW = +10{,}000 - 100{,}000(F/P, 8\%, 10) - (2500 + 1000)(F/A, 8\%, 10)$$

$$FW = +10{,}000 - (100{,}000)(2.15892500) - (2500 + 1000)(14.48656247)$$

$$FW = -256{,}595.46$$

Alternative B:

$$FW = +5000 - 85{,}000(F/P, 8\%, 10) - (3400 + 1200)\,(F/A, 8\%, 10)$$

$$FW = +5000 - (85{,}000)(2.15892500) - (3400 + 1200)\,(14.48656247)$$

$$FW = -245{,}146.81$$

Because alternative B has a lower future cost, again we choose alternative B. Note that regardless of which method we decide to use, alternative B is economically the better option. Moreover, for each alternative, all of the approaches discussed here are related to one another through the interest–time relationships (factors). For example,

Alternative A:

$$PW = AW(P/A, 8\%, 10) = (-17{,}712.65)(6.71008140) = -118{,}853.32$$

or

$$PW = FW(P/F, 8\%, 10) = (-256{,}595.46)(0.46319349) = -118{,}853.34$$

Alternative B:

$$PW = AW(P/A, 8\%, 10) = (-16{,}922.35)(6.71008140) = -113{,}550.40$$

or

$$PW = FW(P/F, 8\%, 10) = (-245{,}146.81)(0.46319349) = -113{,}550.40$$

Finally, it is worth noting that you can take semester-long classes in engineering economics. Some of you will eventually do so. You will learn more in depth about the principles of money–time relationships, including rate-of-return analysis, benefit–cost ratio analysis, general price inflation, depreciation methods, evaluation of alternatives on an after-tax basis, and risk and uncertainty in engineering economics. For now, our intent has been to introduce you to engineering economics, but keep in mind that we have just scratched the surface!

16.3 ENGINEERING ETHICS

As we discussed in Chapter 2, and also at the beginning of this chapter, engineers design many products, including cars, computers, aircraft, clothing, toys, home appliances, surgical equipment, heating and cooling equipment, health care devices, tools and machines that make various products, and so on. Engineers also design and supervise the construction of buildings, dams, highways, and mass transit systems. They also design and supervise the construction of power plants that supply power to manufacturing companies, homes, and offices. Engineers play a significant role in the design and maintenance of nations' infrastructures, including communication systems, utilities, and transportation. Engineers are involved in coming up with ways of increasing crop, fruit, and vegetable yields along with improving the safety of our food products.

As you can see, people rely quite heavily on engineers to provide them with safe and reliable goods and services. There is no room for mistakes or dishonesty in engineering! Mistakes made by engineers could cost not only money but also more importantly lives. Think about the following: An incompetent and unethical surgeon could cause at most the death of one person at one time (when a pregnant woman dies on the operating table, two deaths may result), whereas an incompetent and unethical engineer could cause the deaths of hundreds of people at one time. If an unethical engineer in order to save money designs a bridge or a part for an airplane that does not meet the safety requirements, hundreds of peoples' lives are at risk!

You realize that there are jobs where a person's mistake could be tolerated. For example, if a waiter brings you Coke instead of the Pepsi that you ordered, or instead of french fries brings you onion rings, you can live with that mistake. These are mistakes that usually can be corrected without any harm to anyone. But if an incompetent or unethical engineer incorrectly designs a bridge, or a building, or a plane, he or she could be responsible for killing hundreds of people. Therefore, you must realize why it is so important that as future practicing engineers you are expected to hold to the highest standards of honesty and integrity.

In the section that follows, we will look at an example of a code of ethics, namely, the National Society of Professional Engineers code. The American Society of Mechanical Engineers, the American Society of Civil Engineers, and the Institute of Electrical and Electronics Engineers also have codes of ethics. They are typically posted at their Web sites.

THE CODE OF ETHICS OF THE NATIONAL SOCIETY OF PROFESSIONAL ENGINEERS

The National Society of Professional Engineers (NSPE) ethics code is very detailed. The NSPE ethical code of conduct is used in making judgments about engineering ethic-related cases that are brought before the NSPE's Board of Ethics Review. The NSPE ethical code of conduct follows.

CODE OF ETHICS FOR ENGINEERS*

Preamble

Engineering is an important and learned profession. As members of this profession, engineers are expected to exhibit the highest standards of honesty and integrity. Engineering has a direct and vital impact on the quality of life for all people. Accordingly, the services provided by engineers require honesty, impartiality, fairness and equity, and must be dedicated to the protection of the public health, safety and welfare. Engineers must perform under a standard of professional behavior which requires adherence to the highest principles of ethical conduct.

I. Fundamental Canons

Engineers, in the fulfillment of their professional duties, shall:

1. Hold paramount the safety, health and welfare of the public.
2. Perform services only in areas of their competence.
3. Issue public statements only in an objective and truthful manner.
4. Act for each employer or client as faithful agents or trustees.
5. Avoid deceptive acts.
6. Conduct themselves honorably, responsibly, ethically, and lawfully so as to enhance the honor, reputation and usefulness of the profession.

II. Rules of Practice

1. Engineers shall hold paramount the safety, health, and welfare of the public.

   a. If engineers' judgment is overruled under circumstances that endanger life or property, they shall notify their employer or client and such other authority as may be appropriate.
   b. Engineers shall approve only those engineering documents which are in conformity with applicable standards.
   c. Engineers shall not reveal facts, data, or information without the prior consent of the client or employer except as authorized or required by law or this Code.
   d. Engineers shall not permit the use of their name or associate in business ventures with any person or firm which they believe is engaged in fraudulent or dishonest enterprise.

* *Source*: The National Society of Professional Engineers (NSPE).

e. Engineers having knowledge of any alleged violation of this Code shall report thereon to appropriate professional bodies and, when relevant, also to public authorities, and cooperate with the proper authorities in furnishing such information or assistance as may be required.

2. Engineers shall perform services only in the areas of their competence.

   a. Engineers shall undertake assignments only when qualified by education or experience in the specific technical fields involved.
   b. Engineers shall not affix their signatures to any plans or documents dealing with subject matter in which they lack competence, nor to any plan or document not prepared under their direction and control.
   c. Engineers may accept assignments and assume responsibility for coordination of an entire project and sign and seal the engineering documents for the entire project, provided that each technical segment is signed and sealed only by the qualified engineers who prepared the segment.

3. Engineers shall issue public statements only in an objective and truthful manner.

   a. Engineers shall be objective and truthful in professional reports, statements, or testimony. They shall include all relevant and pertinent information in such reports, statements, or testimony, which should bear the date indicating when it was current.
   b. Engineers may express publicly technical opinions that are founded upon knowledge of the facts and competence in the subject matter.
   c. Engineers shall issue no statements, criticisms, or arguments on technical matters which are inspired or paid for by interested parties, unless they have prefaced their comments by explicitly identifying the interested parties on whose behalf they are speaking, and by revealing the existence of any interest the engineers may have in the matters.

4. Engineers shall act for each employer or client as faithful agents or trustees.

   a. Engineers shall disclose all known or potential conflicts of interest which could influence or appear to influence their judgment or the quality of their services.
   b. Engineers shall not accept compensation, financial or otherwise, from more than one party for services on the same project, or for services pertaining to the same project, unless the circumstances are fully disclosed and agreed to by all interested parties.
   c. Engineers shall not solicit or accept financial or other valuable consideration, directly or indirectly, from outside agents in connection with the work for which they are responsible.
   d. Engineers in public service as members, advisors, or employees of a governmental or quasi-governmental body or department shall not participate in decisions with respect to services solicited or provided by them or their organizations in private or public engineering practice.
   e. Engineers shall not solicit or accept a contract from a governmental body on which a principal or officer of their organization serves as a member.

5. Engineers shall avoid deceptive acts.

   a. Engineers shall not falsify their qualifications or permit misrepresentation of their or their associates' qualifications. They shall not misrepresent or exaggerate their responsibility in or for the subject matter of prior assignments. Brochures or other presentations incident to the solicitation of employment shall not misrepresent pertinent facts concerning employers, employees, associates, joint venturers or past accomplishments.

b. Engineers shall not offer, give, solicit or receive, either directly or indirectly, any contribution to influence the award of a contract by public authority, or which may be reasonably construed by the public as having the effect of intent to influence the awarding of a contract. They shall not offer any gift or other valuable consideration in order to secure work. They shall not pay a commission, percentage, or brokerage fee in order to secure work, except to a bona fide employee or bona fide established commercial or marketing agencies retained by them.

III. Professional Obligations

1. Engineers shall be guided in all their relations by the highest standards of honesty and integrity.

   a. Engineers shall acknowledge their errors and shall not distort or alter the facts.
   b. Engineers shall advise their clients or employers when they believe a project will not be successful.
   c. Engineers shall not accept outside employment to the detriment of their regular work or interest. Before accepting any outside engineering employment they will notify their employers.
   d. Engineers shall not attempt to attract an engineer from another employer by false or misleading pretenses.
   e. Engineers shall not promote their own interest at the expense of the dignity and integrity of the profession.

2. Engineers shall at all times strive to serve the public interest.

   a. Engineers shall seek opportunities to participate in civic affairs; career guidance for youths; and work for the advancement of the safety, health, and well-being of their community.
   b. Engineers shall not complete, sign, or seal plans and/or specifications that are not in conformity with applicable engineering standards. If the client or employer insists on such unprofessional conduct, they shall notify the proper authorities and withdraw from further service on the project.
   c. Engineers shall endeavor to extend public knowledge and appreciation of engineering and its achievements.

3. Engineers shall avoid all conduct or practice that deceives the public.

   a. Engineers shall avoid the use of statements containing a material misrepresentation of fact or omitting a material fact.
   b. Consistent with the foregoing, engineers may advertise for recruitment of personnel.
   c. Consistent with the foregoing, engineers may prepare articles for the lay or technical press, but such articles shall not imply credit to the author for work performed by others.

4. Engineers shall not disclose, without consent, confidential information concerning the business affairs or technical processes of any present or former client or employer, or public body on which they serve.

   a. Engineers shall not, without the consent of all interested parties, promote or arrange for new employment or practice in connection with a specific project for which the engineer has gained particular and specialized knowledge.
   b. Engineers shall not, without the consent of all interested parties, participate in or represent an adversary interest in connection with a specific project or proceeding in which the engineer has gained particular specialized knowledge on behalf of a former client or employer.

5. Engineers shall not be influenced in their professional duties by conflicting interests.

   a. Engineers shall not accept financial or other considerations, including free engineering designs, from material or equipment suppliers for specifying their product.
   b. Engineers shall not accept commissions or allowances, directly or indirectly, from contractors or other parties dealing with clients or employers of the engineer in connection with work for which the engineer is responsible.

6. Engineers shall not attempt to obtain employment or advancement or professional engagements by untruthfully criticizing other engineers, or by other improper or questionable methods.

   a. Engineers shall not request, propose, or accept a commission on a contingent basis under circumstances in which their judgment may be compromised.
   b. Engineers in salaried positions shall accept part-time engineering work only to the extent consistent with policies of the employer and in accordance with ethical considerations.
   c. Engineers shall not, without consent, use equipment, supplies, laboratory, or office facilities of an employer to carry on outside private practice.

7. Engineers shall not attempt to injure, maliciously or falsely, directly or indirectly, the professional reputation, prospects, practice, or employment of other engineers. Engineers who believe others are guilty of unethical or illegal practice shall present such information to the proper authority for action.

   a. Engineers in private practice shall not review the work of another engineer for the same client, except with the knowledge of such engineer, or unless the connection of such engineer with the work has been terminated.
   b. Engineers in governmental, industrial, or educational employ are entitled to review and evaluate the work of other engineers when so required by their employment duties.
   c. Engineers in sales or industrial employ are entitled to make engineering comparisons of represented products with products of other suppliers.

8. Engineers shall accept personal responsibility for their professional activities, provided, however, that engineers may seek indemnification for services arising out of their practice for other than gross negligence, where the engineer's interests cannot otherwise be protected.

   a. Engineers shall conform with state registration laws in the practice of engineering.

b. Engineers shall not use association with a nonengineer, a corporation, or partnership as a "cloak" for unethical acts.

9. Engineers shall give credit for engineering work to those to whom credit is due, and will recognize the proprietary interests of others.

a. Engineers shall, whenever possible, name the person or persons who may be individually responsible for designs, inventions, writings, or other accomplishments.
b. Engineers using designs supplied by a client recognize that the designs remain the property of the client and may not be duplicated by the engineer for others without express permission.
c. Engineers, before undertaking work for others in connection with which the engineer may make improvements, plans, designs, inventions, or other records that may justify copyrights or patents, should enter into a positive agreement regarding ownership.
d. Engineers' designs, data, records, and notes referring exclusively to an employer's work are the employer's property. Employer should indemnify the engineer for use of the information for any purpose other than the original purpose.

As Revised February 2001

"By order of the United States District Court for the District of Columbia, former Section 11(c) of the NSPE Code of Ethics prohibiting competitive bidding, and all policy statements, opinions, rulings or other guidelines interpreting its scope, have been rescinded as unlawfully interfering with the legal right of engineers, protected under the antitrust laws, to provide price information to prospective clients; accordingly, nothing contained in the NSPE Code of Ethics, policy statements, opinions, rulings or other guidelines prohibits the submission of price quotations or competitive bids for engineering services at any time or in any amount."

Statement by NSPE Executive Committee

In order to correct misunderstandings which have been indicated in some instances since the issuance of the Supreme Court decision and the entry of the Final Judgment, it is noted that in its decision of April 25, 1978, the Supreme Court of the United States declared: "The Sherman Act does not require competitive bidding."

It is further noted that as made clear in the Supreme Court decision:

1. Engineers and firms may individually refuse to bid for engineering services.
2. Clients are not required to seek bids for engineering services.
3. Federal, state, and local laws governing procedures to procure engineering services are not affected, and remain in full force and effect.
4. State societies and local chapters are free to actively and aggressively seek legislation for professional selection and negotiation procedures by public agencies.
5. State registration board rules of professional conduct, including rules prohibiting competitive bidding for engineering services, are not affected and remain in full force and effect. State registration boards with authority to adopt rules of professional conduct may adopt rules governing procedures to obtain engineering services.
6. As noted by the Supreme Court, "nothing in the judgment prevents NSPE and its members from attempting to influence governmental action ...".

Note: In regard to the question of application of the Code to corporations vis-a-vis real persons, business form or type should not negate nor influence conformance of individuals to the Code. The Code deals with professional services, which services must be performed by real persons. Real persons in turn establish and implement policies within business structures. The Code is clearly written to apply to the engineer and items incumbent on members of NSPE to endeavor to live up to its provisions. This applies to all pertinent sections of the Code.

SUMMARY

Now that you have reached this point in the text

- You should know the basic design steps that all engineers follow, regardless of their background, to design products and services. These steps are: (1) recognizing the need for a product or a service, (2) defining and understanding the problem (the need) completely, (3) doing the preliminary research and preparation, (4) conceptualizing ideas for possible solutions, (5) synthesizing the results, (6) evaluating good ideas in more detail, (7) optimizing the solution to arrive at the best possible solution, (8) and presenting the solution.
- You should realize that economics plays an important role in engineering decision making. Moreover, a good understanding of the fundamentals of engineering economics could also benefit you in better managing your lifelong financial activities.
- You should know the relationship among money, time, and interest rate. You should be familiar with how these relationships were derived. Moreover, you should also know how to use the engineering economics formulas summarized in Tables 16.7 and 16.8 to solve problems.
- You should understand the importance of engineering ethics and why you should live by these codes of ethics.

PROBLEMS

1. List at least ten products that already exist, which you use, that are constantly being modified to take advantage of new technologies.
2. List five products that are not currently on the market, which could be useful to us, and will most likely be designed by engineers and others.
3. List five sports-related products that you think should be designed to make playing sports more fun.
4. List five internet-based services that are not currently available, but that you think will eventually become realities.
5. You have seen bottle and container caps in use all around you. Investigate the design of bottle caps used in the following products: Pepsi or Coke bottles, aspirin bottles, shampoo bottles, mouthwash bottles, liquid cleaning containers, hand lotion containers, aftershave bottles, ketchup or mustard bottles. Discuss what you think are important design parameters. Discuss the advantages and disadvantages associated with each design.
6. Mechanical clips are used to close bags and keep things together. Investigate the design of various paper clips, hair clips, and potato chip bag clips. Discuss what you think are important design parameters. Discuss the advantages and disadvantages associated with each of three designs.

7. Discuss in detail at least two concepts (for example, activities, processes, or methods) that can be employed during busy hours to better serve customers at grocery stores.
8. Discuss in detail at least two methods or procedures that can be employed by airline companies to pick up your luggage at your home and drop it off at your final destination.
9. In the near future, NASA is planning to send a spaceship with humans to Mars. Discuss important concerns and issues that must be planned for on this trip. Investigate and discuss issues such as how long it would take to go to Mars and when the spaceship should be launched, considering that the distance between the earth and Mars changes based on where they are in their respective orbits around the sun. What type and how much food reserves are needed for this trip? What type of exercise equipment should be on board so muscles won't atrophy on this long trip? What should be done with the waste? What is the energy requirement for such a trip? What do you think are the important design parameters for such a trip? Write a report discussing your findings.
10. You have been using pens and mechanical pencils for many years now. Investigate the design of at least five different pens and mechanical pencils. Discuss what you think are important design parameters. Discuss the advantage and disadvantage associated with each design. Write a brief report explaining your findings.
11. This is an in-class design project. Given a 12 in. by 12 in. aluminum sheet, design a boat that can hold as many pennies as possible. What are some of the important design parameters for this problem? Discuss them among yourselves. You may want to choose a day in advance on which to hold a competition to determine the good designs.
12. This is an in-class design project. Given a bundle of drinking straws and paper clips, design a bridge between two chairs that are 18 in. apart. The bridge should be designed to hold at least 3 lb. You may want to choose a day in advance on which to hold a competition to determine the design that carries maximum load. Discuss some of the important design parameters for this problem.
13. Compute the future value of the following deposits made today:
    a. $10,000 at 6.75% compounding annually for 10 years
    b. $10,000 at 6.75% compounding quarterly for 10 years
    c. $10,000 at 6.75% compounding monthly for 10 years
14. Compute the interest earned on the deposits made in Problem 13.
15. How much money do you need to deposit in a bank today if you are planning to have $5000 in 4 years by the time you get out of college? The bank offers a 6.75% interest rate that compounds monthly.
16. How much money do you need to deposit in a bank each month if you are planning to have $5000 in 4 years by the time you get out of college? The bank offers a 6.75% interest rate that compounds monthly.
17. Determine the effective rate corresponding to the following nominal rates:
    a. 6.25% compounding monthly
    b. 9.25% compounding monthly
    c. 16.9% compounding monthly
18. Using Excel or a spreadsheet of your choice, create interest–time factor tables, similar to Table 16.9, for $i = 6.5\%$ and $i = 6.75\%$.
19. Using Excel or a spreadsheet of your choice, create interest–time factor tables, similar to Table 16.9, for $i = 7.5\%$ and $i = 7.75\%$.
20. Using Excel or a spreadsheet of your choice, create interest–time factor tables, similar to Table 16.9, for $i = 8.5\%$ and $i = 9.5\%$.
21. Using Excel or a spreadsheet of your choice, create interest–time factor tables, similar to Table 16.9, that can be used for $i = 8.5\%$ compounding monthly.
22. Most of you have credit cards, so you already know that if you do not pay the balance on time, the credit card issuer will charge you a certain interest rate each month. Assuming that you are charged 1.25% interest each month on your unpaid balance, what are the nominal and effective interest rates? Also, determine the effective interest rate that your own credit card issuer charges you.
23. You have accepted a loan in the amount of $15,000 for your new car. You have agreed to pay the loan back in 4 years. What is your monthly payment if you agree to pay an interest rate of 9% compounding monthly?

24. How much money will you have available to you after 5 years if you put aside $100 a month in an account that gives you 6.75% interest compounding monthly?
25. How long does it take to double a deposit of $1000
    a. at a compound annual interest rate of 6%
    b. at a compound annual interest rate of 7%
    c. at a compound annual interest rate of 8%
    d. If instead of $1000 you deposit $5000, would the time to double your money be different in parts (a)–(d)? In other words, is the initial sum of money a factor in determining how long it takes to double your money?

    Now use your answers to verify a rule of thumb that is commonly used by bankers to determine how long it takes to double a sum of money. The rule of thumb commonly used by bankers is given by

$$\text{time period to double a sum of money} \approx \frac{72}{\text{interest rate}}$$

26. Imagine that as an engineering intern you have been assigned the task of selecting a motor for a pump. After reviewing motor catalogs, you narrow your choice to two motors that are rated at 1.5 kW. Additional information collected is shown in an accompanying table. The pump is expected to run 4200 hours every year. After checking with your electric utility company, you determine the average cost of electricity is about 11 cents per kWh. Based on the information given here, which one of the motors will you recommend to be purchased?

Criteria	Motor X	Motor Y
Expected useful life	5 years	5 years
Initial cost	$300	$400
Efficiency at the operating point	0.75	0.85
Estimated maintenance cost	$12 per year	$10 per year

27. Plot the following inequalities:
    a. $x \leq -5$
    b. $x \geq 5$
    c. $y \geq 2$
    d. $x + y \geq 4$
    e. $2x + 3y \leq 6$
28. A manufacturer makes a number of products, two of which are discussed here. Product A sells for a $200 profit, while product B sells for a profit of $300. Product A requires 1.5 hours of assembly, while product B takes 2.5 hours to assemble. The assembly line is available for 1950 hours for these products. How many of each product should the manufacturer make to maximize profit?
29. Investigate the construction of the Euro Tunnel, which connects France and Great Britain. Write a brief report discussing the following: How many tunnels make up the Euro Tunnel? How long is each tunnel? How far below the seabed is each channel located? What is the diameter of each tunnel? How far apart are the tunnels that are connected to each other? How many meters per day were excavated? Also discuss the tunnel ventilation, drainage, fire fighting, and cooling systems.
30. How do the design case studies discussed at the end of this chapter relate to the material presented in Chapters 2 through 13?
31. The following is a series of questions pertaining to the NSPE Code of Ethics. Please indicate whether the statements are true or false. These questions are provided by the NSPE.

    Note: This ethics test is intended solely to test individual knowledge of the specific language contained in the NSPE Code of Ethics and is not intended to measure individual knowledge of engineering ethics or the ethics of individual engineers or engineering students.

    1. Engineers in the fulfillment of their professional duties must carefully consider the safety, health and welfare of the public.
    2. Engineers may perform services outside of their areas of competence as long as they inform their employer or client of this fact.
    3. Engineers may issue subjective and partial statements if such statements are in writing and consistent with the best interests of their employer, client or the public.
    4. Engineers shall act for each employer or client as faithful agents or trustees.
    5. Engineers shall not be required to engage in truthful acts when required to protect the public health, safety, and welfare.
    6. Engineers may not be required to follow the provisions of state or federal law when such actions could endanger or compromise their employer or their client's interests.

7. If engineers' judgment is overruled under circumstances that endanger life or property, they shall notify their employer or client and such other authority as may be appropriate.
8. Engineers may review but shall not approve those engineering documents that are in conformity with applicable standards.
9. Engineers shall not reveal facts, data or information without the prior consent of the client or employer except as authorized or required by law or this Code.
10. Engineers shall not permit the use of their name or their associate's name in business ventures with any person or firm that they believe is engaged in fraudulent or dishonest enterprise, unless such enterprise or activity is deemed consistent with applicable state or federal law.
11. Engineers having knowledge of any alleged violation of this Code, following a period of thirty days during which the violation is not corrected, shall report thereon to appropriate professional bodies and, when relevant, also to public authorities, and cooperate with the proper authorities in furnishing such information or assistance as may be required.
12. Engineers shall undertake assignments only when qualified by education or experience in the specific technical fields involved.
13. Engineers shall not affix their signatures to plans or documents dealing with subject matter in which they lack competence, but may affix their signatures to plans or documents not prepared under their direction and control where the engineer has a good faith belief that such plans or documents were competently prepared by another designated party.
14. Engineers may accept assignments and assume responsibility for coordination of an entire project and shall sign and seal the engineering documents for the entire project, including each technical segment of the plans and documents.
15. Engineers shall strive to be objective and truthful in professional reports, statements or testimony, with primary consideration for the best interests of the engineer's client or employer. The engineer's reports shall include all relevant and pertinent information in such reports, statements or testimony, which shall bear the date on which the engineer was retained by the client to prepare the reports.
16. Engineers may express publicly technical opinions that are founded upon knowledge of the facts and competence in the subject matter.
17. Engineers shall issue no statements, criticisms, or arguments on technical matters that are inspired or paid for by interested parties, unless they have prefaced their comments by explicitly identifying the interested parties on whose behalf they are speaking, and by revealing the existence of any interest the engineers may have in the matters.
18. Engineers may not participate in any matter involving a conflict of interest if it could influence or appear to influence their judgment or the quality of their services.
19. Engineers shall not accept compensation, financial or otherwise, from more than one party for services on the same project, or for services pertaining to the same project, unless the circumstances are fully disclosed and agreed to by all interested parties.
20. Engineers shall not solicit but may accept financial or other valuable consideration, directly or indirectly, from outside agents in connection with the work for which they are responsible, if such compensation is fully disclosed.
21. Engineers in public service as members, advisors or employees of a governmental or quasi-governmental body or department may participate in decisions with respect to services solicited or provided by them or their organizations in private or public engineering practice as long as such decisions do not involve technical engineering matters for which they do not posses professional competence.
22. Engineers shall not solicit or accept a contract from a governmental body on which a principal or officer of their organization serves as a member.
23. Engineers shall not intentionally falsify their qualifications or actively permit written misrepresentation of their or their associate's qualifications. Engineers may accept credit for previous work performed where the work was performed during the period the engineer was employed by the previous employer. Brochures or other presentations incident to the solicitation of employment shall specifically indicate the work performed and the dates the engineer was employed by the firm.
24. Engineers shall not offer, give, solicit, or receive, either directly or indirectly, any contribution to influence the award of a contract by a public authority, or which may be reasonably construed by the public as having the effect or intent of influencing the award of a contract unless such contribution is made in accordance with applicable federal or state election campaign finance laws and regulations.
25. Engineers shall acknowledge their errors after consulting with their employer or client.

ENGINEERING ETHICS

A Case Study From NSPE*

The following is an ethics-related case that was brought before NSPE's Board of ethics review.

FACTS

Engineer A is a licensed professional engineer and a principal in a large-sized engineering firm. Engineer B is a graduate engineer who works in industry and has also worked as a student in Engineer A's firm during a summer. Although Engineer B was employed in Engineer A's firm, Engineer A did not have direct knowledge of Engineer B's work. Engineer B is applying for licensure as a professional engineer and requests that Engineer A provide him with a letter of reference testifying as to Engineer B's engineering experience and that the engineer (Engineer A) was in direct charge of Engineer B. Engineer B was under the assumption that Engineer A had personal knowledge of Engineer B's work. Engineer A inquired about Engineer B's experience from someone who had direct knowledge of Engineer B's experience. Based on the inquiry, Engineer A provides the letter of reference explaining the professional relationship between Engineer A and Engineer B.

Question

Was it ethical for Engineer A to provide the letter of reference for Engineer B attesting as to Engineer B's engineering experience even though Engineer A did not have direct control of Engineer B's engineering work?

References

Section II.3.—Code of Ethics: Engineers shall issue public statements only in an objective and truthful manner.

Section II.3.a.—Code of Ethics: Engineers shall be objective and truthful in professional reports, statements or testimony. They shall include all relevant and pertinent information in such reports, statements or testimony, which should bear the date indicating when it was current.

*Materials were adapted with permission from the National Society of Professional Engineers (NSPE).

Section II.5.a.—Code of Ethics: Engineers shall not falsify their qualifications or permit misrepresentation of their, or their associates' qualifications. They shall not misrepresent or exaggerate their responsibility in or for the subject matter of prior assignments. Brochures or other presentations incident to the solicitation of employment shall not misrepresent pertinent facts concerning employers, employees, associates, joint venturers or past accomplishments.

Section III.1.—Code of Ethics: Engineers shall be guided in all their relations by the highest standards of honesty and integrity.

Section III.8.a.—Code of Ethics: Engineers shall conform with state registration laws in the practice of engineering.

DISCUSSION

The Board has, on prior occasions, considered cases involving the misstatement of credentials of an engineer employed in a firm. In BER Case No. 92-1, Engineer A was an EIT who was employed by a medium-sized consulting engineering firm in a small city. Engineer A had a degree in mechanical engineering and had performed services almost exclusively in the field of mechanical engineering. Engineer A learned that the firm had begun a marketing campaign and in its literature listed Engineer A as an electrical engineer. There were other electrical engineers in the firm. Engineer A alerted the marketing director, also an engineer, to the error in the promotional literature and the marketing director indicated that the error would be corrected. However, after a period of six months, the error is not corrected. In ruling that the firm should take actions to correct the error, the Board noted that the firm's marketing director has been informed by the engineer in question that the firm's marketing brochure contains inaccurate information that could mislead and deceive a client or potential client. Under earlier BER Case No. 90-4, the marketing director had an ethical obligation to take expeditious action to correct the error. The Board noted that the marketing director, a professional engineer, had an ethical obligation both to the clients and potential clients, as well as to Engineer A, to expeditiously correct the misimpression which may have been created.

The Board of Ethical Review can certainly understand in the present case the desire of Engineer A to assist another engineer (Engineer B) in enhancing career opportunities and becoming licensed as a professional engineer. Obviously such assistance should not come under misleading or deceptive circumstances. Engineers have an ethical obligation to be honest and objective in their professional reports, and such reports include written assessments of the qualifications and abilities of engineers and others under their direct supervision. Engineers that are not in a position to offer an evaluation of the qualifications and abilities of other individuals should not provide such evaluations or prepare reports that imply that they are providing such evaluations. Claiming to be in responsible charge of another engineer without actually having direct control or personal supervision over that engineer is inconsistent with the letter and the spirit of the NSPE Code.

By providing the report in the manner described, the Board believes Engineer A is sending the right message to Engineer B about what will be expected of Engineer B and his colleagues as professional engineers. Clearly, Engineer B desired the letter of reference from Engineer A, a principal in a consulting firm, in order to improve his chances to become licensed as a professional engineer, and Engineer B is taking conscientious action to address the request. Professional engineers must always be mindful that their conduct and actions as professional engineers set an example for other engineers, particularly those that are beginning their professional careers and who are looking for models and mentors upon which to build their professional identities. A professional engineer providing such a letter of reference should demonstrate that the author has obtained sufficient information about the candidate to write a letter of substance and detail the individual's technical abilities as well as the individual's character. A letter of recommendation for engineering licensure generally requires the recommending professional engineer to state in detail that the candidate possesses legitimate and progressive engineering work experience.

The Board is of the view that an alternative approach could have been for Engineer A to refer Engineer B back to the engineer in the firm that was in responsible charge of engineering for the letter of recommendation. However, the letter provided by Engineer A was just as adequate and ethical.

Conclusion

It was ethical for Engineer A to provide the letter of reference for Engineer B testifying as to Engineer B's engineering experience.

Board of Ethical Review

Lorry T. Bannes, P.E., NSPE
E. Dave Dorchester, P.E., NSPE
John W. Gregorits, P.E., NSPE
Paul E. Pritzker, P.E., NSPE
Richard Simberg, P.E., NSPE
Harold E. Williamson, P.E., NSPE
C. Allen Wortley, P.E., NSPE, Chair

Appendix

A Summary of Formulas Discussed in the Book

Traffic flow: $q = \dfrac{3600n}{T}$

average speed $= \dfrac{\text{distance traveled}}{\text{time}}$

average acceleration $= \dfrac{\text{change in velocity}}{\text{time}}$

volume flow rate $= \dfrac{\text{volume}}{\text{time}}$

angular speed: $\omega = \dfrac{\Delta\theta}{\Delta t}$

the relationship between linear and angular speed:

$V = r\omega$

angular acceleration $= \dfrac{\text{change in angular speed}}{\text{time}}$

density $= \dfrac{\text{mass}}{\text{volume}}$

specific volume $= \dfrac{\text{volume}}{\text{mass}}$

specific gravity $= \dfrac{\text{density of a material}}{\text{density of water@4°C}}$

specific weight $= \dfrac{\text{weight}}{\text{volume}}$

mass flow rate $= \dfrac{\text{mass}}{\text{time}}$

mass flow rate = (density)(volume flow rate)

linear momentum: $\vec{L} = m\vec{V}$

spring force (Hooke's law): $F = kx$

Newton's second law: $\sum F = ma$

Newton's law of gravitational attraction: $F = \dfrac{Gm_1m_2}{r^2}$

weight: $W = mg$

hydrostatic pressure: $P = \rho gh$

buoyancy: $F_B = \rho Vg$

stress−strain relation (Hooke's law): $\sigma = E\varepsilon$

Temperature conversion:

$T(°C) = \frac{5}{9}(T(°F) - 32)$ $\quad T(°F) = \frac{9}{5}T(°C) + 32$

$T(K) = T(°C) + 273.15$ $\quad T(°R) = T(°F) + 459.67$

Fourier's law: $q = kA\dfrac{T_1 - T_2}{L}$

Newton's law of cooling: $q = hA(T_s - T_f)$

radiation: $q = \varepsilon\sigma AT_s^4$

coefficient of thermal linear expansion: $\alpha_L = \dfrac{\Delta L}{L\Delta T}$

coefficient of thermal volumetric expansion:

$\alpha_v = \dfrac{\Delta V}{V\Delta T}$

Coulomb's law: $F_{12} = \dfrac{kq_1q_2}{r^2}$

Ohm's law: $V = RI$

electrical power: $P = VI$

kinetic energy $= \frac{1}{2}mV^2$

change in potential energy $= \Delta PE = mg\Delta h$

elastic energy $= \frac{1}{2}kx^2$

conservation of mechanical energy:

$\Delta KE + \Delta PE + \Delta EE = 0$

conservation of energy—first law of thermodynamics:

$Q - W = \Delta E$

$$\text{power} = \frac{\text{work}}{\text{time}} = \frac{(\text{force})(\text{distance})}{\text{time}} \text{ or power} = \frac{\text{energy}}{\text{time}}$$

$$\text{efficiency} = \frac{\text{actual output}}{\text{required input}}$$

$$\text{standard deviation: } s = \sqrt{\frac{\sum_{i=1}^{n}(x_i - \bar{x})^2}{n-1}}$$

The Greek Alphabet

A	α	alpha
B	β	beta
Γ	γ	gamma
Δ	δ	delta
E	ε	epsilon
Z	ζ	zeta
H	η	eta
Θ	θ	theta
I	ι	iota
K	κ	kappa
Λ	λ	lambda
M	μ	mu
N	ν	nu
Ξ	ξ	xi
O	o	omicron
Π	π	pi
P	ρ	rho
Σ	σ	sigma
T	τ	tau
Y	υ	upsilon
Φ	ϕ	phi
X	χ	chi or khi
Ψ	ψ	psi
Ω	ω	omega

Some Useful Trigonometric Relationships

Pythagorean relation:

$a^2 + b^2 = c^2$

$$\sin\alpha = \frac{\text{opposite}}{\text{hypotenuse}} = \frac{a}{c}$$

$$\sin\beta = \frac{\text{opposite}}{\text{hypotenuse}} = \frac{b}{c}$$

$$\cos\alpha = \frac{\text{adjacent}}{\text{hypotenuse}} = \frac{b}{c}$$

$$\cos\beta = \frac{\text{adjacent}}{\text{hypotenuse}} = \frac{a}{c}$$

$$\tan\alpha = \frac{\sin\alpha}{\cos\alpha} = \frac{\text{opposite}}{\text{adjacent}} = \frac{a}{b}$$

$$\tan\beta = \frac{\sin\beta}{\cos\beta} = \frac{\text{opposite}}{\text{adjacent}} = \frac{b}{a}$$

The sine rule:

$$\frac{a}{\sin\alpha} = \frac{b}{\sin\beta} = \frac{c}{\sin\theta}$$

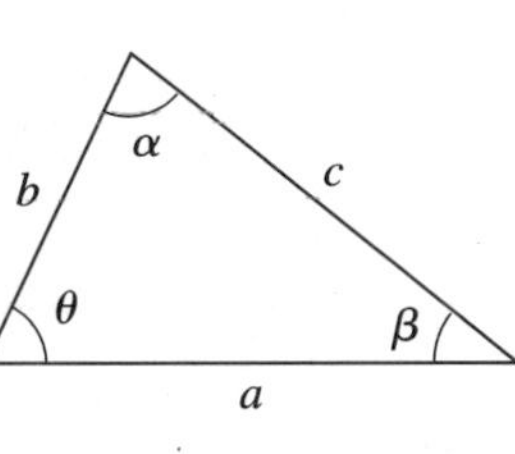

The cosine rule:

$a^2 = b^2 + c^2 - 2bc(\cos\alpha)$

$b^2 = a^2 + c^2 - 2ac(\cos\beta)$

$c^2 = a^2 + b^2 - 2bc(\cos\theta)$

Some other useful trignometry identities:

$\sin^2\alpha + \cos^2\alpha = 1$

$\sin 2\alpha = 2\sin\alpha\cos\alpha$

$\cos 2\alpha = \cos^2\alpha - \sin^2\alpha = 2\cos^2\alpha - 1 = 1 - 2\sin^2\alpha$

$\sin(-\alpha) = -\sin\alpha$

$\cos(-\alpha) = \cos\alpha$

$\sin(\alpha + \beta) = \sin\alpha\cos\beta + \sin\beta\cos\alpha$

$\sin(\alpha - \beta) = \sin\alpha\cos\beta - \sin\beta\cos\alpha$

$\cos(\alpha + \beta) = \cos\alpha\cos\beta - \sin\alpha\sin\beta$

$\cos(\alpha - \beta) = \cos\alpha\cos\beta + \sin\alpha\sin\beta$

$$\theta = \frac{S_1}{R_1} = \frac{S_2}{R_2}$$

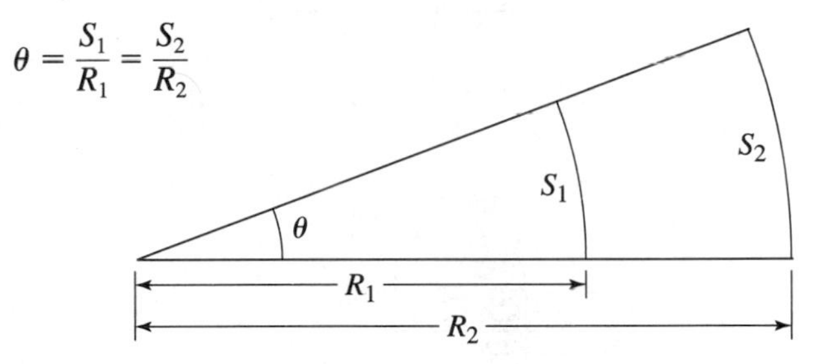

Some Useful Mathematical Relationships

$\pi = 3.14159\ldots$

$2\pi = 360$ degrees

$$1 \text{ radian} = \frac{180}{\pi} = 57.2958°$$

$$1 \text{ degree} = \frac{\pi}{180} = 0.0174533 \text{ rad}$$

$x^n x^m = x^{n+m}$ $\quad (xy)^n = x^n y^n$

$(x^n)^m = x^{nm}$ $\quad x^0 = 1 \quad (x \neq 0)$

$$x^{-n} = \frac{1}{x^n} \qquad \frac{x^n}{x^m} = x^{n-m} \qquad \left(\frac{x}{y}\right)^n = \frac{x^n}{y^n}$$

log = logarithm to the base 10 (common logarithm)

$10^x = y \qquad \log y = x \qquad \log 1 = 0$

$\log 10 = 1 \qquad \log 100 = 2 \qquad \log 1000 = 3$

$\log xy = \log x + \log y$

$$\log\frac{x}{y} = \log x - \log y$$

$\log x^n = n\log x$

$e = 2.71828\ldots$

ln = logarithm to the base e (natural logarithm)

$e^x = y \qquad \ln y = x$

$\ln x = (\ln 10)(\log x) = 2.302585 \log x$

Some Useful Area Formulas

Triangle $A = \frac{1}{2}bh$

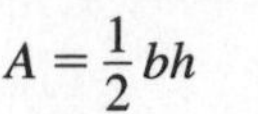

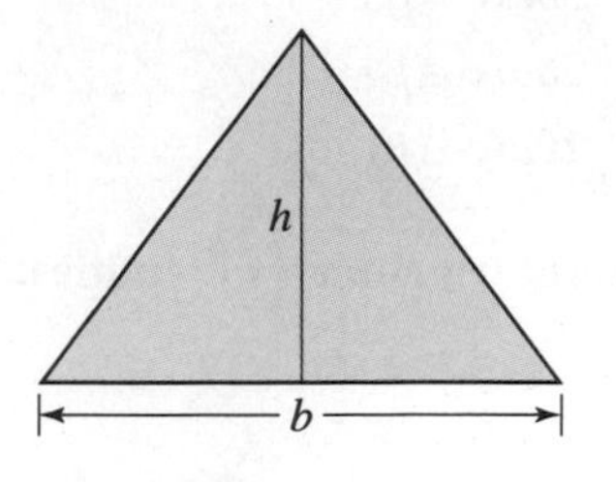

Rectangle $A = bh$

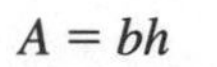

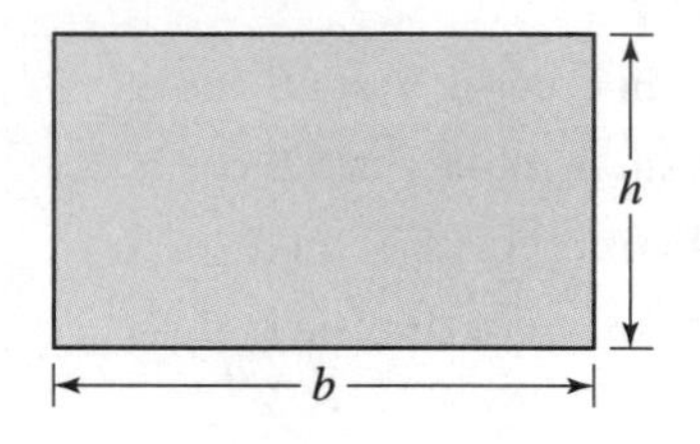

Parallelogram $A = bh$

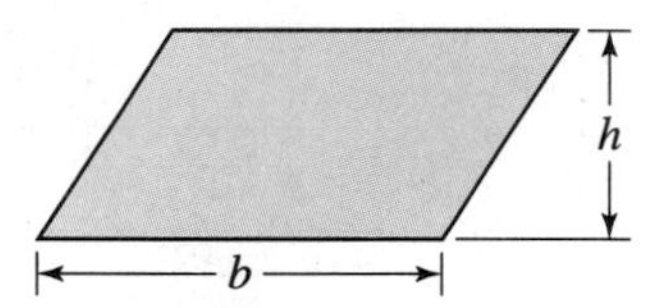

Trapezoid $A = \frac{1}{2}(a + b)h$

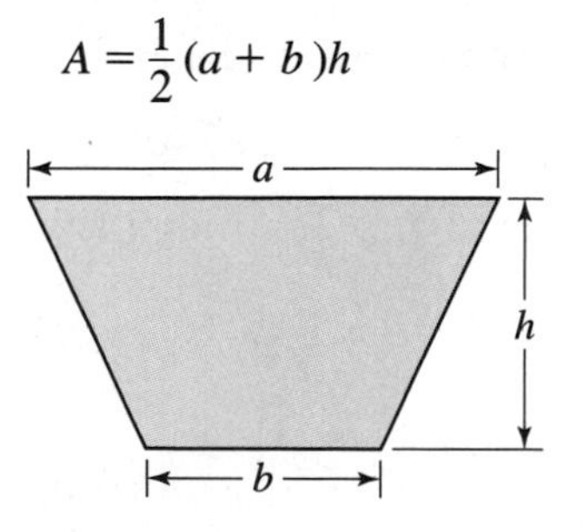

n-sided polygon $A = \left(\frac{n}{4}\right)b^2 \cot\left(\frac{180°}{n}\right)$

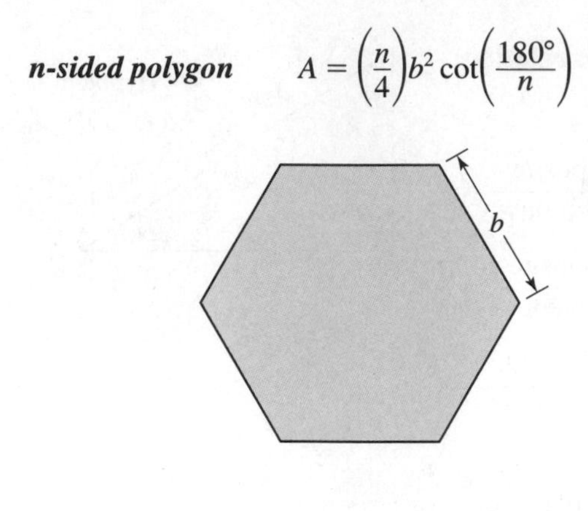

Circle $A = \pi R^2$

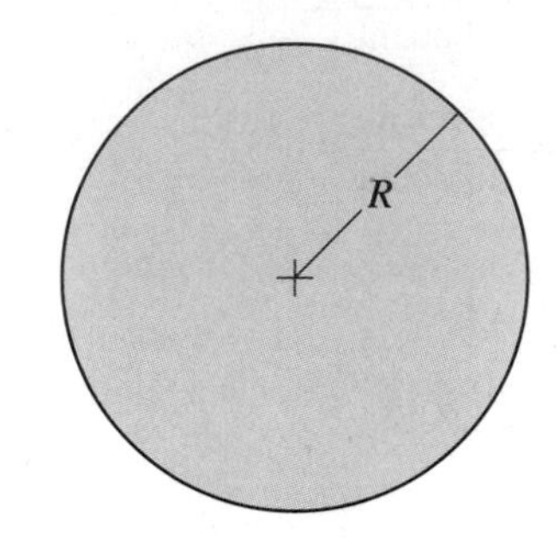

Ellipse $A = \pi ab$

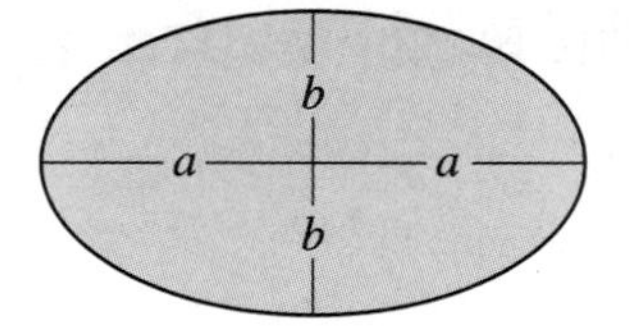

Cylinder $A = 2\pi Rh$

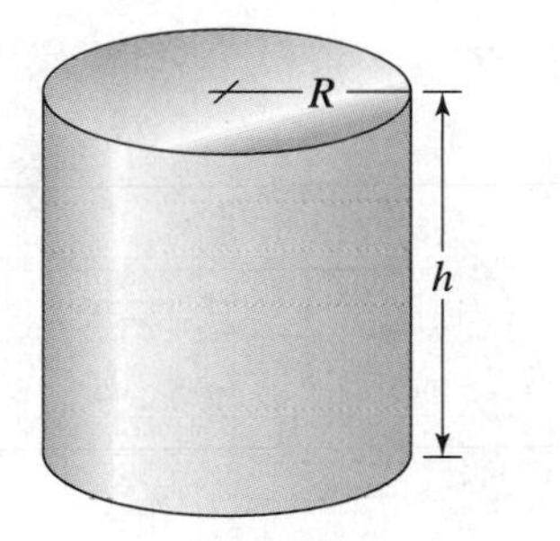

Sphere $A = 4\pi R^2$

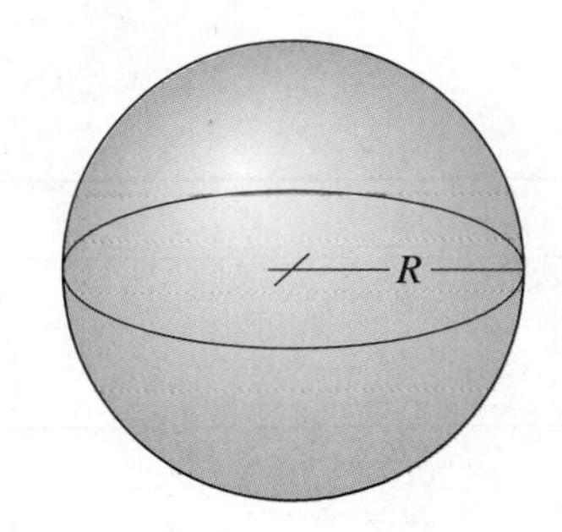

Right circular cone $A = \pi Rs = \pi R\sqrt{R^2 + h^2}$

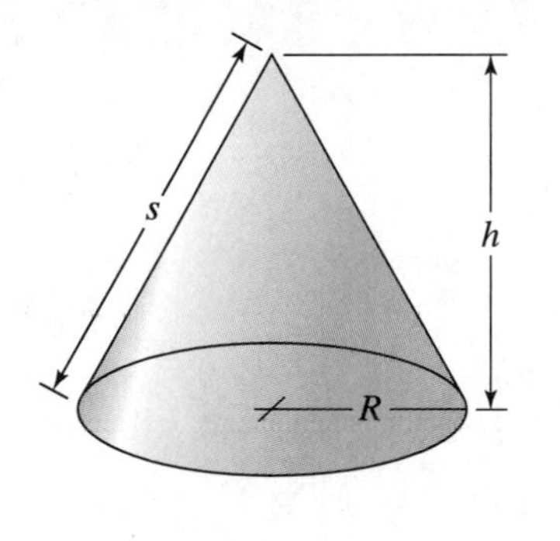

Trapezoidal rule:

$$A \approx h\left(\frac{1}{2}y_0 + y_1 + y_2 + \cdots + y_{n-2} + y_{n-1} + \frac{1}{2}y_n\right)$$

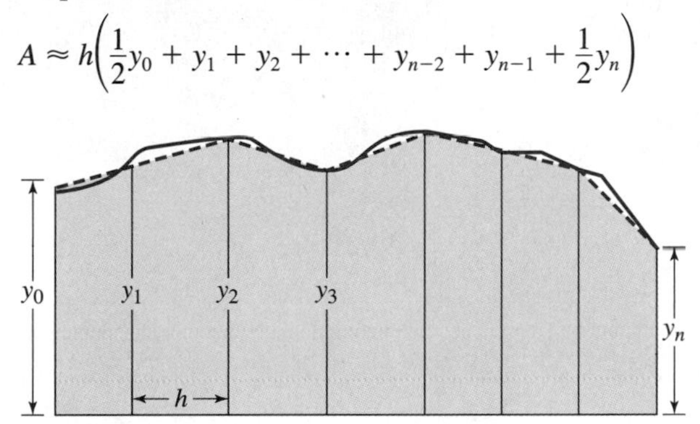

Some Useful Volume Formulas

Cylinder $V = \pi R^2 h$

Right circular cone $V = \frac{1}{3}\pi R^2 h$

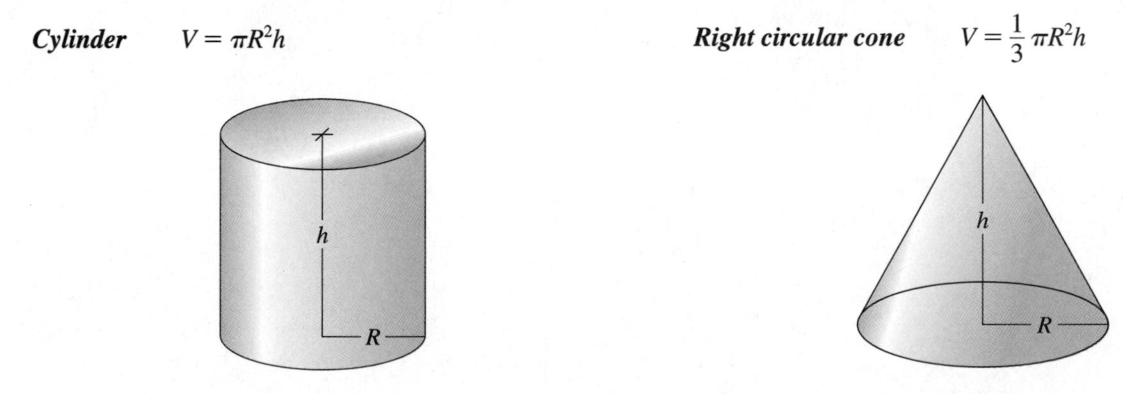

Section of a cone $V = \frac{1}{3}\pi h(R_1^2 + R_2^2 + R_1R_2)$

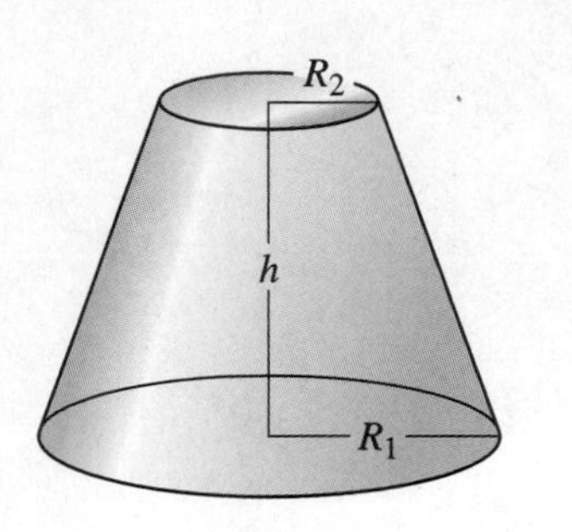

Sphere $V = \frac{4}{3}\pi R^3$

Section of a sphere $V = \frac{1}{6}\pi h(3a^2 + 3b^2 + h^2)$

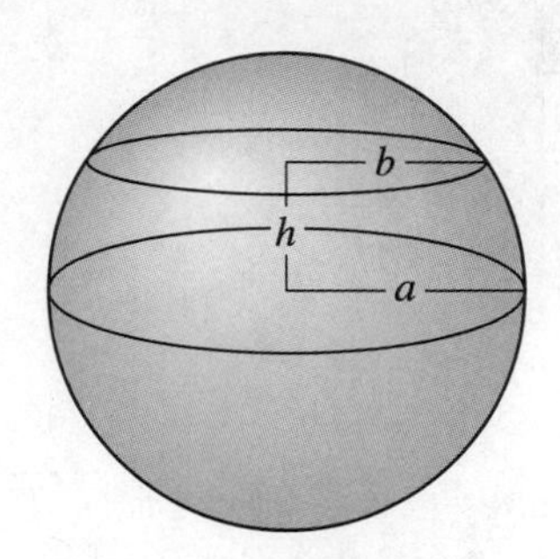

Index

CREDITS

This page constitutes an extension of the copyright page. We have made every effort to trace the ownership of all the copyrighted material and to secure permission from copyright holders. In the event of any question arising as to the use of any material, we will be pleased to make the necessary corrections in future printings. Thanks are due to the following authors, publishers, and agents for permission to use the material indicated.

Text

Chapter 4 **75** Table 4.5: Copyright © 1991 American Society of Heating, Refrigerating, and Air-Conditioning Engineers, Inc. www.ashrae.org. Reprinted by permission from ASHRAE 1991 Handbook—*Fundamentals.*

Chapter 5 **102** Figure 5.3: Map outline © Mountain High Maps. Data supplied by HM Nautical Almanac Office @ Copyright Council for the Central Laboratory of the Research Councils. http://aa.usno.navy.mil/faq/docs/world_tzones.html.

Chapter 7 **157** Figure 7.16: From *Mechanics of Materials,* by R. C. Hibbler, p. 84. Copyright © 1991 Macmillan. Reprinted by permission. **160** Table 7.4: Adapted from *Mechanics of Materials,* Fifth Edition, by J. M. Gere, Table H-2, p. 899, Brooks/Cole, 2001. Reprinted with permission of Brooks/Cole, an imprint for the Wadsworth Group, a division of Thomson Learning. Fax: 800-730-2215.

Chapter 8 **185** Table 8.1: From *Mark's Handbook,* Eighth Edition, by T. Baumeister et al., Table 9, pp. 4–6. Copyright © 1978 The McGraw-Hill Companies. Reprinted with permission. **204** Table 8.7: Copyright © 1989 American Society of Heating, Refrigerating, and Air-Conditioning Engineers, Inc. www.ashrae.org. Reprinted by permission from ASHRAE 1989 Handbook—*Fundamentals.* **205** Table 8.8: Copyright © 1997 American Society of Heating, Refrigerating, and Air-Conditioning Engineers, Inc. www.ashrae.org. Reprinted by permission from ASHRAE 1997 Handbook—*Fundamentals.* **207** Table 8.9: From *Mark's Handbook,* Eighth Edition, by T. Baumeister et al., Table 12, pp. 4–7. Copyright © 1978 The McGraw-Hill Companies. Reprinted with permission. **209** Table 8.11: Copyright © 1985 American Society of Heating, Refrigerating, and Air-Conditioning Engineers, Inc. www.ashrae.org. Reprinted by permission from ASHRAE 1985 Handbook—*Fundamentals.* **210** Tables 8.12 and 8.13: From *Steam: Its Generation and Use,* 38th Edition. Copyright © 1972 Babcock and Wilcox. Reprinted with permission.

Chapter 9 **221** Figure 9.7: From *Electrical Wiring,* Second Edition, 1981, p. 18. Copyright © 1981 AAVIM. Reprinted by permission. **222** Figure 9.8: From *Electrical Wiring,* Second Edition, 1981, p. 10. Copyright © 1981 AAVIM. Reprinted by permission. **222** Figure 9.9: From *Electrical Wiring,* Second Edition, 1981, p. 28. Copyright © 1981 AAVIM. Reprinted by permission. **230** Table 9.3: From *Electric Motors,* Fifth Edition, 1982. Copyright © 1982 AAVIM. Reprinted by permission.

Chapter 11 **266** Tables 11.1 and 11.2: Adapted from *Mechanics of Materials,* Fifth Edition, by J. M. Gere, Table H-2, p. 899, Brooks/Cole, 2001. Reprinted with permission of Brooks/Cole, an imprint for the Wadsworth Group, a division of Thomson Learning. Fax: 800-730-2215. **267** Table 11.3: Adapted from *Mechanics of Materials,* Fifth Edition, by J. M. Gere, Table H-2, p. 899, Brooks/Cole, 2001. Reprinted with permission of Brooks/Cole, an imprint for the Wadsworth Group, a division of Thomson Learning. Fax: 800-730-2215. **268** Table 11.4: Adapted from *Mechanics of Materials,* Fifth Edition, by J. M. Gere, Table H-2, p. 899, Brooks/Cole, 2001. Reprinted with permission of Brooks/Cole, an imprint for the Wadsworth Group, a division of Thomson Learning. Fax: 800-730-2215.

Chapter 12 **294** Table 12.2: Copyright © 1993 American Society of Heating, Refrigerating, and Air-Conditioning Engineers, Inc. www.ashrae.org. Reprinted by permission from ASHRAE 1993 Handbook—*Fundamentals.*

Chapter 15 **352** Figure 15.19: Courtesy of ANSYS, Inc.

Photos

1 Tom Rosenthal/SuperStock, Inc.; **6** © David Weintraub/Stock, Boston/PictureQuest; **7** © SuperStock, Inc.; **9** © Ed Kashi/CORBIS; **13** (upper left) © EyeWire, Inc., (lower left) © Stone/Michael Rosenfield, (right) NASA Ames Home Page; **21** © Michael Newman/Photo Edit; **29** © James A. Sugar/CORBIS; **60** Saeed Moaveni; **61** Courtesy Pentax Corporation; **77** NASA (Marshall Space Flight Center); **92–95** Courtesy of New York City Department of Environmental Protection; **96** © Kristin Finnegan Photography; **104** © 2000 EyeWire, Inc.; **117** © Tempo Sport/CORBIS; **136** U.S. Navy photo courtesy of Electric Boat Div., General Dynamics Corp.; **152** www.comstock.com; **158** Courtesy of MTS Systems Corporation; **170** Kahana's Stunt School; **176** Courtesy of Caterpillar; **180** © DigitalVision/PictureQuest; **198** Photo by Steven Engler; **215** © Steve Allen/ Brand X Pictures/PictureQuest; **235** © DigitalVision/PictureQuest; **252** Courtesy of Rheem Corporation; **257** Photo by Andy Pernick, Bureau of Reclamation; **260** Boeing/courtesy photo; **269** (left) Courtesy of Carroll Shelby Enterprises, (right) © Steve Allen/Brand X Pictures/PictureQuest; **270** (left) Courtesy of Litespeed, (middle) Courtesy Maui Jim, Inc., (right) Courtesy of Seiko Corporation of America; **271** (all) © 2001 PhotoDisc, Inc.; **272** (upper left) © Image Farm Inc./PictureQuest, (lower left) © Image Source/ElektraVision/PictureQuest, (right) © 2001 PhotoDisc, Inc.; **273** (upper left) © Tony Freeman/Photo Edit, (lower left) Image Ideas/Index Stock, (right) © Corbis Images/PictureQuest; **275** (left) © Stockbyte, (upper and lower right) © 2001 PhotoDisc, Inc.; **276** © 2001 PhotoDisc, Inc.; **278** Corning Cable Systems; **287** (left) © DigitalVision/PictureQuest, (right) © Corbis Images/PictureQuest; **295** © 2001 PhotoDisc, Inc.; **297** (upper and lower left) © 2001 PhotoDisc, Inc., (right) © Corbis Images/PictureQuest; **368** (all) Copyright © Nokia, 2001

NOTES

NOTES

NOTES

NOTES

NOTES

NOTES

NOTES

NOTES